NanoScience and Technology

The series NanoScience and Technology is focused on the fascinating nano-world, mesoscopic physics, analysis with atomic resolution, nano and quantum-effect devices, nanomechanics and atomic-scale processes. All the basic aspects and technology-oriented developments in this emerging discipline are covered by comprehensive and timely books. The series constitutes a survey of the relevant special topics, which are presented by leading experts in the field. These books will appeal to researchers, engineers, and advanced students.

Enrico Gnecco · Ernst Meyer
Editors

Fundamentals of Friction and Wear on the Nanoscale

Third Edition

Springer

Editors
Enrico Gnecco
Institute of Physics
Jagiellonian University
Krakow, Poland

Ernst Meyer
Department of Physics
University of Basel
Basel, Switzerland

ISSN 1434-4904 ISSN 2197-7127 (electronic)
NanoScience and Technology
ISBN 978-3-031-63064-4 ISBN 978-3-031-63065-1 (eBook)
https://doi.org/10.1007/978-3-031-63065-1

This Springer imprint is published by the registered company Springer Nature Switzerland AG
The registered company address is: Gewerbestrasse 11, 6330 Cham, Switzerland

Preface

Born as a niche topic at the interface between solid-state physics, surface chemistry and mechanical engineering more than three decades ago, nanotribology, i.e. the study of friction, wear and related phenomena at the nanoscale, has now reached its full maturity as an independent field of research. Because of the multidisciplinary nature of the topic, which now extends far beyond the selected problems addressed in the early studies, it has become almost impossible, even for those who have been active in the field for a long time, to follow all the strands currently being explored in the world. With this in mind, the third edition of Fundamentals of Friction focuses on selected topics that have evolved rapidly enough to warrant a new stand-alone book. These topics include structural lubricity, manipulation of single molecular chains, friction in liquids, colloidal probe force microscopy, friction on strained layers, and bionanotribology. Together with updated versions of key chapters introduced in the previous editions, we hope that the proposed selection will stimulate readers not only to learn interesting facts about sliding motion at the nanoscale, but also to find applications and develop new ideas in this rapidly evolving field.

Krakow, Poland — Enrico Gnecco
Basel, Switzerland — Ernst Meyer

Contents

Contributors

Pulickel Madhava Ajayan Department of Materials Science and Nanoengineering, Rice University, Houston, USA

Rong An Herbert Gleiter Institute of Nanoscience, School of Materials Science and Engineering, Nanjing University of Science and Technology, Nanjing, China; Shandong Laboratory of Advanced Materials and Green Manufacturing, Yantai Zhongke Research Institute of Advanced Materials and Green Chemical Engineering, Yantai, China

Walter Arnold Department of Materials Science and Engineering, Saarland University, Saarbrücken, Germany

Mehmet Z. Baykara Department of Mechanical Engineering, University of California Merced, Merced, CA, USA

Roland Bennewitz INM—Leibniz Institute for New Materials, Saarbrücken, Germany

Ronen Berkovich Department of Chemical Engineering, Ben-Gurion University of the Negev, Beer Sheva, Israel;
The Ilse Katz Institute for Nanoscale Science and Technology, Ben-Gurion University of the Negev, Beer Sheva, Israel;
Marian Smoluchowski Institute of Physics, Jagiellonian University, Krakow, Poland

Diana Berman Department of Materials Science and Engineering, University of North Texas, Denton, USA

Cristina Bernini CNR-SPIN Institute for Superconductivity, Innovative Materials and Devices, National Research Council CNR, Genova, Italy

Renato Buzio CNR-SPIN Institute for Superconductivity, Innovative Materials and Devices, National Research Council CNR, Genova, Italy

Alan B. Dalton Department of Physics and Astronomy, University of Sussex, Brighton, UK

Leonid Dorogin Department of Molecules and Materials, MESA+ Institute for Nanotechnology, University of Twente, Enschede, The Netherlands

Carlos Drummond Centre de Recherche Paul Pascal. CNRS-Université Bordeaux 1, Pessac, France

Mauro Ferrario Department of Physics and Astronomy, University of Bologna, Bologna, Italy

Venkataramana Gadhamshetty Department Civil and Environmental Engineering, South Dakota School of Mines and Technology, Rapid City, SD, USA

Andrea Gerbi CNR-SPIN Institute for Superconductivity, Innovative Materials and Devices, National Research Council CNR, Genova, Italy

Enrico Gnecco Marian Smoluchowski Institute of Physics, Jagiellonian University, Krakow, Poland

Elena V. Gorb Department of Functional Morphology and Biomechanics, Kiel University, Kiel, Germany

Stanislav N. Gorb Department of Functional Morphology and Biomechanics, Kiel University, Kiel, Germany

Nitya Nand Gosvami Department of Materials Science and Engineering, Indian Institute of Technology Delhi, New Delhi, India

Oscar Gutiérrez-Varela Instituto de Ciencia de Materiales de Madrid, Consejo Superior de Investigaciones Científicas, Madrid, Spain;
Departamento de Física Teórica de la Materia Condensada, Universidad Autónoma de Madrid, Madrid, Spain

Urs Gysin Institute of Physics, University of Basel, Basel, Switzerland

Md. Hasan-ur Rahman Department Civil and Environmental Engineering, South Dakota School of Mines and Technology, Rapid City, SD, USA

Kenichi Hiratsuka Faculty of Engineering, Chiba Institute of Technology, Chiba, Japan

Oded Hod Department of Physical Chemistry, School of Chemistry, The Raymond and Beverly Sackler Faculty of Exact Sciences and The Sackler Center for Computational Molecular and Materials Science, Tel Aviv University, Tel Aviv, Israel

Sathvik Ajay Iyengar Department of Materials Science and Nanoengineering, Rice University, Houston, USA

Hesam Khaksar Marian Smoluchowski Institute of Physics, Jagiellonian University, Krakow, Poland

Marcin Kisiel Institute of Physics, University of Basel, Basel, Switzerland

Franciszek Krok Marian Smoluchowski Institute of Physics, Jagiellonian University, Krakow, Poland

Markus Langer Institute of Physics, University of Basel, Basel, Switzerland

Dong-Weon Lee MEMS and Nanotechnology Laboratory, Chonnam National University, Gwangju, South Korea

Mathieu Lizée Laboratoire de Physique de l'Ecole Normale Supérieure, Paris, France

Rynno Lõhmus Tartu Ülikooli Füüsika Instituut, Tartu, Tartumaa, Estonia

Chengfu Ma State Key Laboratory of Solid Lubrication, Lanzhou Institute of Chemical Physics, Chinese Academy of Sciences, Lanzhou, China

Ming Ma Center for Nano and Micro Mechanics, Tsinghua University, Beijing, China;
Department of Mechanical Engineering, and State Key Lab of Tribology in Advanced Equipment (SKLT), Tsinghua University, Beijing, China;
Institute of Superlubricity Technology, Research Institute of Tsinghua University in Shenzhen, Shenzhen, China

Sabine Maier Department of Physics, University of Erlangen-Nuremberg, Erlangen, Germany

Juan J. Mazo Departamento de Estructura de la Materia, Física Térmica y Electrónica and Grupo Interdisciplinar de Sistemas Complejos (GISC), Universidad Complutense de Madrid, Madrid, Spain

Ernst Meyer Department of Physics, University of Basel, Basel, Switzerland

Roman Nevshupa Eduardo Torroja Institute of Construction Sciences, CSIC, Madrid, Spain

Wai H. Oo Department of Mechanical Engineering, University of California Merced, Merced, CA, USA

Sven Oras Tartu Ülikooli Tehnoloogiainstituut, Tartu, Tartumaa, Estonia

Alper Özoğul Institute for Materials Science and Max Bergmann Center for Biomaterials, TU Dresden, Dresden, Germany

Rémy Pawlak Department of Physics, University of Basel, Basel, Switzerland

Boris Polyakov Institute of Solid State Physics, University of Latvia, Riga, Latvia

Giacomo Prampolini Istituto Di Chimica dei Composti Organo Metallici, Consiglio Nazionale delle Ricerche (ICCOM-CNR), Pisa, Italy

Himanshu Rai Department of Materials Science and Engineering, Indian Institute of Technology Delhi, New Delhi, India;

Department of Control Engineering, Faculty of Electrical Engineering, Czech Technical University in Prague, Technicka 2, 16627, Prague 6, Czech Republic

Simon Rast Institute of Physics, University of Basel, Basel, Switzerland

Luca Repetto Physics Department, University of Genova, Genova, Italy

Maria Clelia Righi Department of Physics and Astronomy, University of Bologna, Bologna, Italy

Marina Ruths Department of Chemistry, University of Massachusetts Lowell, Lowell, MA, USA

Gulnaz Sharipov Institute of Petrochemistry and Catalysis, Ufa Federal Research Center of the Russian Academy of Sciences, Ufa, Russia

Andrea Silva CNR-IOM Materials Foundry Institute, National Research Council CNR, Trieste, Italy;
International School for Advanced Studies (SISSA), Trieste, Italy

Alessandro Siria Laboratoire de Physique de l'Ecole Normale Supérieure, Paris, France

Yiming Song Institute of Applied Physics, Justus Liebig University Giessen, Giessen, Germany

Jitendra Soni Department of Materials Science and Engineering, Indian Institute of Technology Delhi, New Delhi, India

Manoj Tripathi Department of Physics and Astronomy, University of Sussex, Brighton, UK

Alina Tukhbatullin Institute of Petrochemistry and Catalysis, Ufa Federal Research Center of the Russian Academy of Sciences, Ufa, Russia

Michael Urbakh Department of Physical Chemistry, School of Chemistry, The Raymond and Beverly Sackler Faculty of Exact Sciences and The Sackler Center for Computational Molecular and Materials Science, Tel Aviv University, Tel Aviv, Israel

Andrea Vanossi CNR-IOM Materials Foundry Institute, National Research Council CNR, Trieste, Italy;
International School for Advanced Studies (SISSA), Trieste, Italy

Jose Guilherme Vilhena Instituto de Ciencia de Materiales de Madrid, Consejo Superior de Investigaciones Científicas, Madrid, Spain

Sergei Vlassov Tartu Ülikooli Füüsika Instituut, Tartu, Tartumaa, Estonia

Dilek Yildiz Institute of Physics, University of Basel, Basel, Switzerland

Quanshui Zheng Center for Nano and Micro Mechanics, Tsinghua University, Beijing, China;

Institute of Superlubricity Technology, Research Institute of Tsinghua University in Shenzhen, Shenzhen, China;
Department of Engineering Mechanics, School of Aerospace Engineering, Tsinghua University, Beijing, China;
Tsinghua Shenzhen International Graduate School, Shenzhen, China

Friction Force Microscopy

Roland Bennewitz

Abstract Friction force microscopy is a key experimental method in nanotribology. The tip of an atomic force microscope is moved in contact over a surfaces and friction forces are detected as deflection of a micro-mechanical force sensor. While the method appears simple, special care must be taken to calibrate the force sensor and to understand the challenges in bridging the gap between molecular forces and macroscopic experiment. We discuss experimental procedures such as measurements of friction as function of load or of temperature, and on inhomogeneous materials.The chapter ends with an overview of dynamic measurements of friction, where the tip is oscillated laterally in contact or above the surface to probe dissipative interactions with the highest sensitivity.

1 Introduction

Friction Force Microscopy (FFM) is a sub-field of scanning force microscopy addressing the measurement of lateral forces in nanometer-scale sliding contacts. In line with all scanning probe methods, the basic idea is to exploit the local interactions with a very sharp probe for obtaining microscopic information on surfaces in lateral resolution. In FFM, the apex of a sharp tip is brought into contact with a sample surface, and the lateral forces are recorded while tip and sample slide relative to each other. There are several areas of motivation to study FFM. First, The understanding of friction between sliding surfaces in general is a very complex problem. The complexity arises from multiple points of contact between surfaces and from the importance of lubricants and third bodies in the sliding process. By reducing one surface to a single asperity, by preparing a well-defined structure of the sample surface, and by controlling the normal load on the contact the complexity of friction studies is greatly reduced and fundamental insights into the relevant processes can be obtained. Second, with the decrease of the size of mechanical devices (MEMS) the

R. Bennewitz (✉)
INM—Leibniz Institute for New Materials, Saarbrücken, Germany
e-mail: roland.bennewitz@leibniz-inm.de

E. Gnecco and E. Meyer (eds.), *Fundamentals of Friction and Wear on the Nanoscale*, NanoScience and Technology, https://doi.org/10.1007/978-3-031-63065-1_1

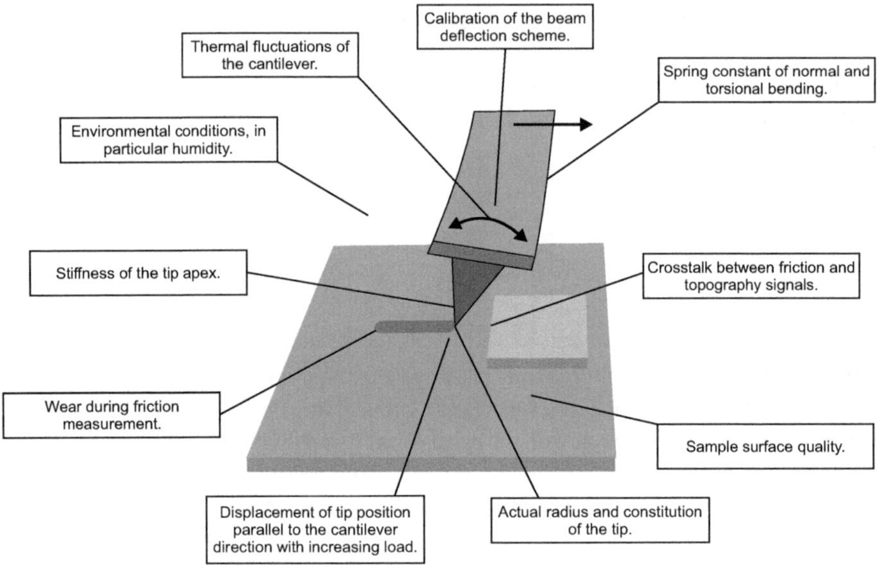

Fig. 1 Critical issues in experimental friction force microscopy which are discussed in this chapter

friction and adhesion of small contacts becomes a technological issue. Finally, the lateral resolution of FFM allows to reveal tribological contrasts caused by material differences on heterogeneous surfaces.

The experimental field of FFM has been pioneered by Mate, McClelland, Erlandsson, and Chiang [1]. The group built a scanning force microscope where the lateral deflection of a tungsten wire could be measured through an optical interferometer. When the etched tip of the tungsten wire slid over a graphite surface, lateral forces exhibited a modulation with the atomic periodicity of the graphite lattice. Furthermore, an essentially linear load dependence of the lateral force could be established.

In this chapter we will describe aspects of instrumentation and measurement procedures. In the course of this description, a series of critical issues in FFM will bee discussed which are summarized in Fig. 1.

2 Instrumentation

2.1 Force Sensors

The force sensor in the original presentation of FFM by Mate et al. was a tungsten wire [1]. Its deflection was detected by an interferometer scheme where the wire constituted one mirror of the optical cavity. A similar concept was later implemented by Hirano et al., who optically detected the deflection of the tungsten wire in a Scanning

Tunneling Microscope when scanning the tip in close proximity to the surface [2]. Mate and Hirano report lateral spring constants from 1.5 to 2500 N/m, depending on the wire thickness and length. Etching the wire to form a tip at its end, mounting the wire, aligning of the light beam, and determination of the spring constant comprise some experimental difficulties. These difficulties are greatly reduced by the use of dedicated micro-fabricated force sensors. A very sophisticated instrumental approach to the solution of those problems has been realized by Dienwiebel et al. [3]. The group has attached a stiff tungsten wire to a micro-fabricated force sensor made of silicon. The central part of the sensor is a pyramid holding the tip. The position of the pyramid is detected in all three dimensions by means of four optical interferometers directed towards the faces of the pyramid. It is suspended in four symmetric high-aspect ratio legs which serve as springs with isotropic spring constant in both lateral directions and a higher spring constant in normal direction. The symmetric design of the instrument allows for determination of normal and lateral forces acting on the tip with minimal cross talk between them. An overview over different experimental realizations of FFM is given in Fig. 2.

The most widely used form of micro-fabricated force sensors for FFM is the micro-fabricated cantilever with integrated tip. The cantilever can be either a rectangular beam or a triangular design based on two beams. The lateral force acting on the tip is detected as torsional deflection of the cantilever. This scheme has been implemented in 1990 by Meyer et al. [4] and Marti et al. [5]. It is interesting to note that the triangular design is more susceptible to deflection by lateral forces than the rectangular beam, contrary to common belief and intuition [6]. However, triangular cantilevers are less prone to the highly unwanted in-plane bending [7]. Modifications of the cantilever's cross section have been suggested to increase the sensitivity to lateral forces acting on the tip [8].

The deflection of cantilever-type force sensors is usually detected by means of a light beam reflected from the back side of the cantilever at the position of the tip. The reflected light beam is directed towards a position-sensitive photodiode which detects normal and torsional bending of the cantilever as a shift in the position of the light beam in orthogonal directions. Realistically, there is always some cross-talk between the signals for normal and torsional bending. It can be detected by exciting the cantilever to oscillate at the fundamental normal and torsional resonance and measuring the oscillation amplitude in the orthogonal channels. The cross-talk can be minimized by rotation of the position-sensitive photodiode or accounted for in the detection electronics or software. Cross-talk can transfer topographic features into the lateral force signal and create topographic artifacts from friction contrast, the latter even amplified by the feedback circuit acting on the sample height.

Calibration of the beam-deflection scheme is not a simple task, however very important in order to compare FFM results from different sources. Many publications in the past have reported on relative changes in frictional properties, without providing any calibration at all. While such relative changes certainly represent important physical findings, it is nevertheless of utmost importance to provide all experimental information available, often allowing for a rough quantitative estimate of the lateral forces. Lateral forces in FFM can easily range from piconewton to micronewton,

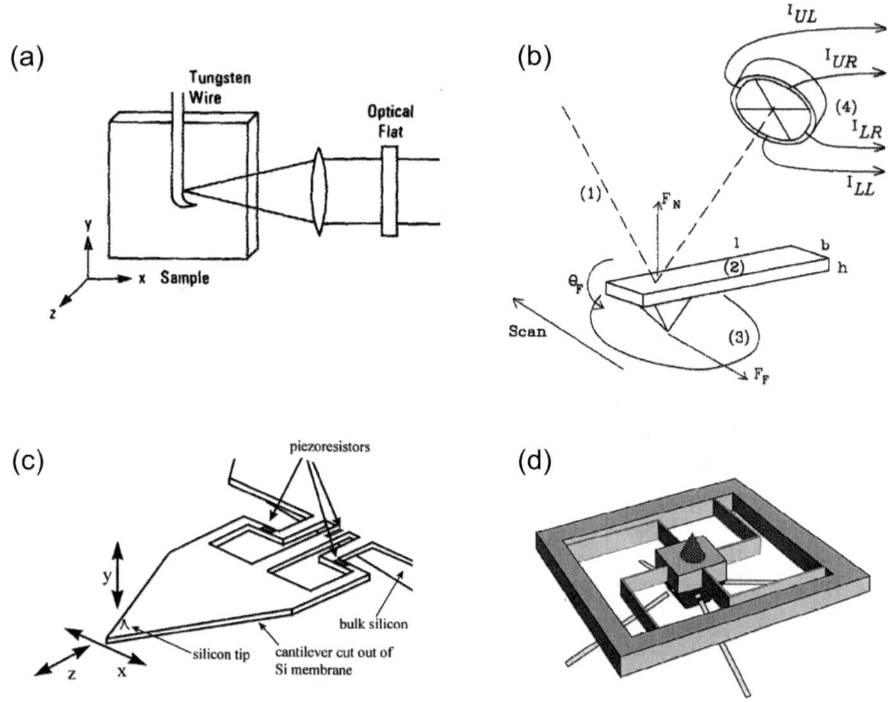

Fig. 2 Four design options for Friction Force Microscopy. **a** Concept of the original instrument used by Mate et al. for their pioneering experiments [1] The deflection of a tungsten wire is detected by an optical interferometer. The bent end of the wire is etched into a sharp tip. **b** Beam-deflection scheme as devised by Marti et al. [5]. Normal force F_N and friction force F_F cause bending and twisting of the cantilever. The deflection of a reflected light beam is recorded by comparing currents from four sections of a photodiode. **c** Cantilever device for the measurement of lateral forces with piezoresistive detection [9]. Lateral forces acting on the tip cause a difference in stress across the piezoresistors. **d** Micro-fabricated force detector for isotropic measurements of friction forces. The block in the center holds a tungsten tip, pointing upwards in this figure. The position of the block in all three dimensions is recorded by four interferometric distance sensors which are indicated by the four light beams below the devices [10]

spanning a range of very different situations in contact mechanics. Knowing at least the order of magnitude of forces helps to sort the results qualitatively into different regimes.

The calibration comprises two steps. First, the spring constant has to be determined for the force sensor. Note that the beam-deflection scheme actually determines the angular deflection of the cantilever. Nevertheless it has become custom to quantify the force constant in N/m, where the length scale refers to the lateral displacement of the tip apex relative to the unbent cantilever. Second, a relation between the deflection of the cantilever and the voltage readout of the instrument has to be established.

For the determination of the spring constant, several methods have been suggested. The easiest is to calculate it with beam mechanics formulae from the dimensions of the cantilever. While width and thickness are easily determined by optical or electron microscopy, thickness is better deduced from the cantilever's resonance frequency. Alternatively, the spring constant can be determined from changes in the resonances caused by the addition of masses to the free end of the cantilever. Also, the analysis of a cantilever's resonance structure in air can provide the required quantities. The latter two methods have recently be described and compared by Green et al. [11].

The relation between tip displacement and voltage readout can be established by trapping the tip in a surface structure and displacing the sample laterally by small distances. For a rough estimate one can also assume that the sensitivity of the position-sensitive photodiode is the same for normal and torsional deflection. Taking into account the geometry of the beam-deflection scheme, the torsional deflection sensitivity can be deduced from the normal deflection sensitivity (See Ref. [12] and page 352 of Ref. [13]). Since the quantification of the thermal-noise driven torsional resonance can be difficult, a combination of thermal noise and beam geometry methods can be useful for the calibration of FFM [14].

A method which provides a direct calibration of the lateral force with respect to the readout voltage is the comparison with a calibrated spring standard. Recent implementations of this approach suggest as calibrated standards optical fibers [15] or micro-fabricated spring-suspended stages with spring constants that can be traced to international standards [16, 17]. Similarly, the lateral stiffness of a magnetically levitated graphite sheet can be used as reference [18]. A particularly elegant method to calibrate FFM experiments is the analysis of friction loops, i.e. lateral force curves from forward and backward scans, recorded across surfaces with well-defined topographic wedges [12, 19, 20]. For reliable results, this wedge method requires the application of normal forces which exceed the adhesion forces [21]. An extension of the calibration by friction loops to lower normal forces has been proposed based on the topography-induced lateral forces recorded on graphene/SiO_x [22].

The torsional deflection of a cantilever can in principle be detected also by an optical interferometer, provided that the beam diameter is smaller than the cantilever width and that the point of reflection is shifted off the torsional axis [23]. However, FFM results including normal and lateral force measurements require the differential reading of multiple interferometers [3, 24].

An alternative to the detection of the cantilever bending via the beam-deflection scheme is the implementation of piezoresistive strain sensors into the cantilever. In order to measure both lateral and normal forces acting on the tip in FFM, two such strain sensors need to be realized on one sensor. Chui et al. have created a piezoresistive sensor which decouples the two degrees of freedom by attaching a normal triangular cantilever to a series of vertical ribs sensing lateral forces [25]. Gotszalk et al. have constructed a U-shaped cantilever with one piezoresistive sensor in each arm, allowing for the the detection of lateral forces at the tip [26]. While the publications presenting these instrumental approaches contain experimental proofs of concept, no further use of piezoresistive sensors in FFM experiments has been reported. This lack of application is probably due to a lack of commercial availability. Furthermore, the signal-to-noise ratio in static force measurements using piezoresistive cantilevers seems not to reach that of optical detection schemes.

2.2 Control over the Contact

Knowledge of the atomic configuration in the contact between tip apex and surface is prerequisite for a complete understanding of the results in Friction Force Microscopy. It is the most severe drawback in FFM that this knowledge is not available in most cases. While sample surfaces can often be prepared with atomic precision and cleanliness, the atomic constitution of the tip apex is usually less controlled. Friction signals vary with tip shape, as has been investigated for steps on graphite [27].

Furthermore, in the course of sliding atoms may be transferred from the tip to the surface or vice versa. Such transfer processes occur even for very gentle contact formation, as shown in experiments combining Scanning Probe Microscopy with a mass spectrometer analysis of the tip apex [28–30]. The transfer of atoms may quite often not only quantitatively but also qualitatively change the lateral forces encountered. Chemical reactions between surface and tip have been found to significantly increase friction between a Pt(111) surface for silicon but not for diamond tips [31]. The occurrence of atomic stick-slip motion can depend on the establishment of a certain degree of structural commensurability between tip and surface in the course of scanning [32, 33]. For atomic stick-slip measurements on graphite surfaces, the role of small graphite flakes attached to the tip has long been discussed and recently confirmed experimentally [1, 34].

The best control over the atomic structure of the tip apex has been achieved for metal tips in vacuum environments. By applying the established procedures of Field Ion Microscopy (FIM), the tip structure can not only be imaged but also conditioned on the atomic scale. Cross et al. have characterized the adhesion between a tungsten tip and a gold surface and proved the conservation of the atomic tip structure by means of FIM [35]. Even with instruments of lower resolution, FIM can at least be used for cleaning procedures and for a determination of the crystalline orientation of the apex cluster [2].

The integrated tips at the end of micro-fabricated silicon cantilevers have a well-defined crystalline orientation, usually pointing with the (100) direction along the tip. However, the tip surface and with it the whole tip apex are at least oxidized and possibly contaminated through packaging, transport, and handling. Furthermore, many tips are sharpened in a oxidation process which introduces large stresses at the apex. While etching in hydrofluoric acid can remove the oxide and for some time passivate silicon surface bonds by hydrogen, a stable formation and reproducible characterization comparable with FIM of metal tips has not yet been reported. Tips integrated into silicon nitride cantilevers are amorphous due to the chemical vapor deposition process and may exhibit an even more complex structure and chemistry at the tip apex.

One way of overcoming the uncertainty of the tip constitution is to use methods of surface chemistry to functionalize the tip [36]. Specific interactions between molecules attached to the tip and molecules on the surface can be sensed by means of FFM [37]. At the same time, very strong adhesion has been reduced by covering the tip with a passivating layer to allow for lateral force imaging for example on silicon [38]. Numerous studies using this method have been published, mainly concentrating on organic monolayers on tip and surface. A review of the field has been given by Leggett et al. [39]. The results of FFM may depend on the attachment flexibility in the functionalization [40]. While most tip functionalization relies on thiol bonding to gold-coated tips, carbon bonding to nanocrystalline diamond tips has also been realized [41]. Schwarz et al. have prepared well-defined tips for FFM by deposition of carbon from residual gas molecules in a Transmission Electron Microscope, keeping control of the tip radius for a quantitative analysis of a contact mechanics study [42]. Force measurements explicitly aiming at interactions between colloidal particles and a surface have been performed by gluing micrometer-sized spheres of the desired size to the cantilever [43, 44]. Atomic control of the contact in FFM can also achieved by using the AFM tip to slide a well-defined nanoscale particle over a crystalline surface [45]. As a final note, one should always be aware of the possible occurrence of major tip wear which has been observed to happen in a concerted action of mechanical and chemical polishing [46, 47].

3 Measurement Procedures

The standard measurement in FFM is the so-called friction loop: The lateral force acting on the tip is recorded for a certain distance of scanning in the direction perpendicular to the long cantilever axis and for the reverse direction. The area in the force-distance loop represents the dissipated energy, and the area divided by twice the distance is the mean friction force. It is always very instructive to record the topography signal of forward and backward scan at the same time, as differences will reveal cross-talk between normal and torsional bending of the cantilever.

Whenever lateral forces are measured as a function of some experimental parameter, the influence of that parameter on adhesion should be studied simultaneously. In order to interpret the experimental results in terms of contact sizes versus dissipation channels the knowledge of adhesion is essential. An excellent example is the jump in lateral forces observed on a C_{60} crystal when cooling across the orientational order-disorder phase transition, which was fully explained by a change in adhesion [48]. For experiments carried out in ambient environment, the dominant contribution to adhesion are usually capillary forces which dependent greatly on the humidity and on the hydrophobicity of the surface [49]. The humidity dependence of FFM results itself can depend again on the temperature [50–52]. Consequently, an enclosure of FFM experiments for humidity control greatly enhances the reproducibility of results.

3.1 Friction as a Function of Load

One of the central experiments in tribology is the quantification of friction, i.e. the change of lateral force with increasing normal load on the sliding contact. One of the questions to be addressed is whether the relation between lateral and normal force is linear for FFM experiments, i.e. whether Amontons' law extends to the nanometer scale [53]. The number of FFM studies reporting lateral force as a function of load is very large, and the overall physical picture is multifaceted, to express it in a positive way. A collection of results is shown in Fig. 3. From a procedural point of view it is extremely important to measure the lateral forces for the full range of small normal forces until the tip jumps out of contact, usually at a negative normal force. In this way the adhesion in the system can be categorized and even maps of adhesion can be produced from friction vs. load experiments [54]. Furthermore, possible nonlinear characteristics at minimal loads are not overlooked. A useful way of analyzing load dependence data from FFM experiments is the representation in lateral force histograms, where for example friction on terraces and friction at steps could automatically be distinguished [55].

When the normal load on the tip is varied the position of the contact may be displaced along the long axis of the cantilever. This effect is caused by the tilt of the cantilever with respect to the surface. On heterogeneous surfaces such displacement may distort the friction measurement and, therefore, has to be compensated [58]. Another effect that can seriously disturb friction experiments is the onset of wear and the concomitant increase of lateral forces. Wear thresholds in FFM can be as low as a few nanonewton normal load, and wear at a constant low load may suddenly start after repeatedly scanning the same area [59].

3.2 Friction as a Function of Material

On inhomogeneous surfaces Friction Force Microscopy can image contrasts between different materials with high lateral resolution. Such contrast has been found to arise from a difference in chemical interactions between different molecular patches at the surface and the tip [60]. FFM can thus serve to identify partial coverage of a surface, for example by graphene patches [61]. As mentioned above, it is crucial to complement lateral friction contrast with local measurements of adhesion in order to elucidate whether adhesion and contact size or different channels of dissipation are dominating the contrast. Care has to be taken regarding topographical artifacts, as different materials on heterogeneous surfaces are often found at different topographic heights. Interestingly, friction contrast is also found between domains of identical molecular layers with anisotropic lateral orientation [62–64]. Friction anisotropy on a given surface has to be clearly distinguished from friction anisotropy for different azimuthal orientations between the tip and the surface. In order to measure the latter, the sample has to be rotated with respect to the tip [34, 65].

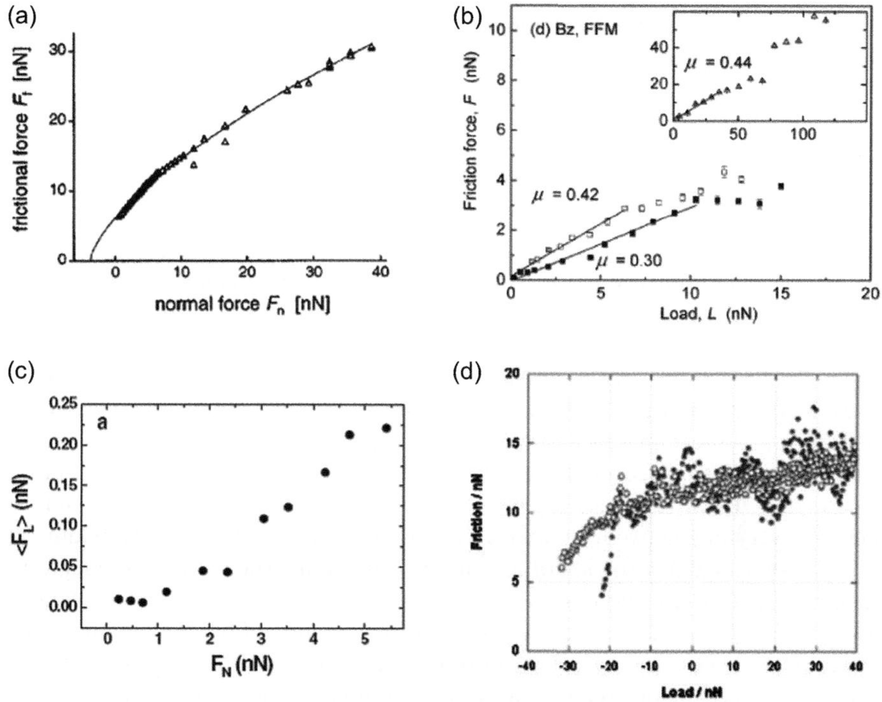

Fig. 3 Examples for the diversity of friction vs. load curves measured by FFM. **a** Amorphous carbon measured in an argon atmosphere [42]. The sub-linear characteristic resembles the results of contact mechanics models. **b** Phenyltrichlorosilane monolayer studied in ethanol [56]. A linear dependence is found until the monolayer collapses under the tip pressure. **c** Atomic friction on NaCl(100) recorded in ultra-high vacuum [57]. A regime of vanishing friction is found for low loads. **d** Friction measurement on a hydrogen-terminated diamond surface with nanometer-scale roughness [58]. The closed circles represent the erratic load dependence of FFM results when the lateral displacement of the tip for increasing load is not compensated. The open circles show the expected sub-linear characteristic after activating the compensation

3.3 Friction Effects in Normal Force Measurements

When the sample is approached towards the tip, the normal force can be determined as a function of distance by measuring the normal bending of the cantilever. In all beam-deflection type FFM the cantilever is tilted with respect to the sample surface to make sure that the tip is the foremost protrusion of the force sensor. Once the tip is in contact, the tilt causes a lateral displacement of the tip position upon further approach. The friction forces arising from this lateral displacement influence the normal force measurement [44]. A detailed analysis of the process proves that one can actually perform a calibrated friction experiment through normal force vs. distance curves, in particular when using extended tips like colloid probes [66]. Even when probing the surface in a dynamic intermittent contact mode these frictional contributions can be detected as a phase shift between excitation and cantilever oscillation [63].

3.4 Fluctuations in Friction Force Microscopy

Friction Force Microscopy is naturally subject to thermal fluctuations. Such thermal fluctuations can influence the frictional behavior of sliding contacts, as evident in the logarithmic dependence of friction on velocity at low scanning velocities [67, 68] which has been linked to thermal fluctuations via its temperature dependence [69]. Cantilever-type force sensors have a distinct resonance structure which dominates the thermal noise spectrum. A full treatment of thermal noise and mechanical vibrations and their influence on FFM have been provided in Ref. [70]. Typically, oscillations at resonances with frequencies of several kHz are averaged out in FFM experiments. However, these resonances influence the experimental result and it is therefore very instructive to study the lateral force signal with high bandwidth [71, 72]. The statistical distribution of lateral forces in atomic stick-slip experiments can be analyzed to reveal the role of thermal fluctuations [73]. The limited scanning velocity of FFM normally separates the frequency regimes of fast fluctuations and of slower occurrence of topographic or even atomic features. The velocity limitations of FFM have been addressed by new designs combining the force sensor of an FFM with a dedicated sample stage [74, 75].

3.5 Friction as a Function of Temperature

The study of friction as a temperature is an obvious field of great interest [48, 50, 69, 76–82]. Thermal drift is a severe problem in the design of Friction Force Microscopes working at variable temperature, since the optical lever of the beam-deflection scheme needs to have a certain length for sensitivity. Variable-temperature instruments with thermal-expansion compensated design comparable to dedicated Scanning Tunneling Microscopes [83] have not been reported so far. One interesting approach to circumvent drift problems is the local heating of the very tip apex [52, 84].

3.6 Dynamic Lateral Force Measurements

3.6.1 Slide-Hold-Slide Experiments

A variable which is often overlooked in FFM experiments is the time for which the contact has been established before sliding sets in. This contact ageing may lead to increased interaction by chemical or structural relaxation. Slide-hold-slide experiments directly address this phenomenon for single asperities in FFM. Examples are the frictional ageing by interfacial bond formation in SiO_x [85] or the universal ageing mechanisms for metal nanoparticles on graphite [86], where the ageing analysis has been extended to stick-slip pattern of friction.

3.6.2 Dynamic Friction Force Microscopy

When the sample is periodically displaced in lateral direction, the lateral force acting on the tip and detected by the cantilever will be modulated with the same periodicity. An early application of such a lateral modulation by Maivald et al. was the enhancement of contrast at step edges [87]. Dynamic Friction Force Microscopy detects the periodic lateral force signal by means of a lock-in amplifier. This idea was implemented by Göddenhenrich et al., who applied the periodic sample displacement along the long axis of the cantilever and detected the lateral force as periodic buckling of the cantilever [88]. Simultaneously, their fiber-interferometer setup could statically measure the deflection of the cantilever caused by normal forces. The same technique was implemented by Colchero et al. for a beam-deflection instrument. The authors provided a detailed analysis for the evaluation of the lateral forces when the sample is displaced in a sinusoidal movement [89]. They also pointed to the fact that using their method of Dynamic Friction Force Microscopy one will obtain quantitative results when taking data, while static experiments need subtraction of forward and backward scan before numbers can be obtained. Carpick et al. have used a similar technique with very small sample displacement amplitudes to avoid any slip of the tip over the surface [90], in some experiments with direct magentic actuation of the torsional oscillation [91]. In such experiments, the amplitude of the lateral force provides a measure for the contact stiffness. Dynamic friction force microscopy has been combined with sophisticated versions of the pulsed-force mode for a simultaneous measurement of all relevant properties of mechanical contacts [92]. Finally, Haugstad has analyzed the non-linear response of the lateral force to the sinusoidal sample displacement in a Fourier analysis [93]. Using this technique he was able to gain new insights into the transition from static to kinetic sliding on a polymer blend.

Dynamic Friction Force Microscopy can gain sensitivity by tuning the periodic excitation to resonances of the cantilever [94, 95]. However, the coupling between the mechanical properties of the contact and the flexural modes of the cantilever require a complex analysis, as provided in a recent review which also references previous work in the field of ultra-sonic force microscopy [96].

3.6.3 Dynamic Non-contact Lateral Force Experiments

The success of dynamic non-contact force microscopy in atomic resolution imaging of insulating surfaces and its prospect of measuring dissipation phenomena with the same resolution [97] has initiated projects which aim at a dynamic non-contact microscopy using lateral oscillation of the tip. Jarvis et al. have constructed a novel force sensor which allows to excite and detect oscillations of the tip in normal as well as in lateral direction [98]. The independent oscillations were achieved by suspending the tip holder in hinges at the end of two normally oscillating cantilevers. The group has controlled the tip-sample distance by changes in the normal oscillation frequency, and simultaneously recorded changes in the amplitude of the lateral oscillation pointing to frictional tip-sample interactions.

A standard rectangular cantilever has been employed by Pfeiffer et al. for the dynamic detection of interactions between a laterally oscillating tip and a surface close to but not in contact [99]. In this study, the cantilever was excited to oscillate at its first torsional resonance, making the tip oscillate laterally. The distance between tip and a copper surface was controlled using the tunneling current as feedback quantity. The lateral interaction between tip and monatomic steps or single impurities could be detected as frequency shift in the torsional oscillation. Giessibl et al. attached a tungsten tip to a quartz tuning fork such that it would oscillate laterally over the surface. Again using tunneling as feedback, they were able to study dissipation in the lateral movement with atomic resolution on a Si(111)7$\times$7 surface, thereby tracing friction to a single atom [100]. The damping of the lateral oscillation has been explained in terms of a fast stick-slip process involving one adatom. The same surface has recently been studied in dynamic lateral force microscopy using a standard rectangular cantilever by Kawai et al. [101]. In this study a small frequency shift in the torsional resonance frequency upon approach was used to control the tip-sample distance. The torsional resonance was detected using a heterodyne interferometer scheme, where the focus of the light beam was positioned on one side of the cantilever in order to be sensitive to the torsional bending. This is actually a very informative method to study the resonance structure of cantilevers which can show significant deviations from ideal modeling due to extra masses and asymmetries [23].

The dynamic non-contact experiments introduced in this section are very interesting tools to study conservative and dissipative interactions in lateral motion even before a repulsive contact is established. Their full strength has been demonstrated by determination of the lateral force needed to move an atom on a surface [102], by relating atomic structure to the anisotropy of lateral forces [103], or by exploring in-plane interactions beside an adsorbed molecule [104]. Implementation in liquids leads to a nanometer-scale rheology which can be applied to hydrogels [105] or confined liquids [106].

4 Outlook

Friction Force Microscopy is now a widely distributed experimental method. The experimental procedures and the calibration have been established and allow for reproducible studies of frictional properties in single-asperity contacts. The biggest drawback within the method is the lack of methods for a reproducible preparation and characterization of tips on atomic scale, as compared to the surface preparation by means of methods of Surface Science. Such control over the atomic constitution of the contact area would greatly advance our understanding of tribological processes on the nanometer scale. Other instrumental challenges in the field include the further improvement of FFM experiments at variable temperatures and in liquid environments, where atomic friction phenomena have been observed with a resolution similar to vacuum experiments [107, 108].

References

1. C. Mate, G. McClelland, R. Erlandsson, S. Chiang, Phys. Rev. Lett. **59**(17), 1942 (1987)
2. M. Hirano, K. Shinjo, R. Kaneko, Y. Murata, Phys. Rev. Lett. **78**(8), 1448 (1997)
3. M. Dienwiebel, E. de Kuyper, L. Crama, J. Frenken, J. Heimberg, D.J. Spaanderman, D. van Loon, T. Zijlstra, E. van der Drift, Rev. Sci. Instr. **76**(4), 43704 (2005)
4. G. Meyer, N. Amer, Appl. Phys. Lett. **57**(20), 2089 (1990)
5. O. Marti, J. Colchero, J. Mlynek, Nanotecnology **1**(2), 141 (1990)
6. J. Sader, R. Sader, Appl. Phys. Lett. **83**(15), 3195 (2003)
7. J. Sader, C. Green, Rev. Sci. Instr. **75**(4), 878 (2004)
8. Y.S. Chen, L. Jiang, L.M. Qian, Friction (2024). https://doi.org/10.1007/s40544-023-0821-z. (<Go to ISI>://WOS:001158577100004)
9. T. Gotszalk, P. Grabiec, I. Rangelow, Ultramicroscopy **82**, 39 (2000)
10. T. Zijlstra, J. Heimberg, E. van der Drift, D. van Loon, M. Dienwiebel, L. de Groot, J. Frenken, Sens. Actuators A-Phys. **84**(1–2), 18 (2000)
11. C. Green, H. Lioe, J. Cleveland, R. Proksch, P. Mulvaney, J. Sader, Rev. Sci. Instr. **75**(6), 1988 (2004)
12. D. Ogletree, R. Carpick, M. Salmeron, Rev. Sci. Instr. **67**(9), 3298 (1996)
13. E. Meyer, R. Overney, K. Dransfeld, T. Gyalog, *Nanoscience: Friction and Rheology on the Nanometer Scale* (World Scientific, Singapore, 1998)
14. R. Alvarez-Asencio, E. Thormann, M.W. Rutland, Rev. Sci. Instr. **84**(9), 096102 (2013)
15. N. Morel, M. Ramonda, P. Tordjeman, Appl. Phys. Lett. **86**(16), 163103 (2005)
16. P. Cumpson, J. Hedley, C. Clifford, J. Vac. Sci. Technol. B (Microelectron. Nanometer Struct.) **23**(5), 1992 (2005)
17. A. Bhattacharjee, N.T. Garabedian, C.L. Evans, D.L. Burris, Tribol. Lett. **68**(4), 111 (2020). https://doi.org/10.1007/s11249-020-01349-y https://doi.org/10.1007/s11249-020-01349-y
18. Q. Li, K. Kim, A. Rydberg, Rev. Sci. Instr. **77**(6), 065105 (2006)
19. M. Varenberg, I. Etsion, G. Halperin, Rev. Sci. Instr. **74**(7), 3362 (2003)
20. H. Wang, M.L. Gee, Ultramicroscopy **136**, 193 (2014)
21. R.D. Ortuso, K. Sugihara, J. Phys. Chem. C **122**(21), 11464 (2018). https://doi.org/10.1021/acs.jpcc.8b03583 https://doi.org/10.1021/acs.jpcc.8b03583
22. M.J. Boland, J.L. Hempel, A. Ansary, M. Nasseri, D.R. Strachan, Rev. Sci. Instr. **89**(11) (2018). https://doi.org/10.1063/1.5044727. <Go to ISI>://WOS:000451735700046
23. M. Reinstaedtler, U. Rabe, V. Scherer, J.A. Turner, W. Arnold, Surf. Sci. **532–535**, 1152 (2003)
24. G. Germann, S. Cohen, G. Neubauer, G. McClelland, H. Seki, J. Appl. Phys. **73**, 163 (1993)
25. B. Chui, T. Kenny, H. Mamin, B. Terris, D. Rugar, Appl. Phys. Lett. **72**(11), 1388 (1998)
26. T. Gotszalk, P. Grabiec, I. Rangelow, Sens. Actuators, A: Phys. **123–124**, 370 (2005)
27. Y. Dong, X.Z. Liu, P. Egberts, Z. Ye, R.W. Carpick, A. Martini, Tribolo. Lett. **50**(1), 49 (2013)
28. U. Weierstall, J. Spence, Surf. Sci. **398**, 267 (1998)
29. T. Shimizu, J.T. Kim, H. Tokumoto, Appl. Phys. A **66**, S771 (1998)
30. A. Wetzel, A. Socoliuc, E. Meyer, R. Bennewitz, E. Gnecco, C. Gerber, Rev. Sci. Instr. **76**(10), 103701 (2005)

31. A. Caron, D.V. Louzguine-Luzguin, R. Bennewitz, ACS Appl. Mater. Interfaces **5**(21), 11341 (2013)
32. A. Livshits, A. Shluger, Phys. Rev. B **56**, 12482 (1997)
33. R. Bennewitz, M. Bammerlin, M. Guggisberg, C. Loppacher, A. Baratoff, E. Meyer, H.J. Güntherodt, Surf. Interface Anal. **27**, 462 (1999)
34. M. Dienwiebel, G. Verhoeven, N. Pradeep, J. Frenken, J. Heimberg, H. Zandbergen, Phys. Rev. Lett. **92**(12), 126101 (2004)
35. G. Cross, A. Schirmeisen, A. Stalder, P. Grütter, M. Tschedy, U. Dürig, Phys. Rev. Lett. **80**, 4685 (1998)
36. T. Nakagawa, K. Ogawa, T. Kurumizawa, J. Vac. Sci. Technol. B (Microelectron. Nanometer Struct.) **12**(3), 2215 (1994)
37. C. Frisbie, L. Rozsnyai, A. Noy, M. Wrighton, C. Lieber, Science **265**(5181), 2071 (1994)
38. L. Howald, R. Lüthi, E. Meyer, P. Güthner, H.J. Güntherodt, Z. Phys. B **93**, 267 (1994)
39. G. Leggett, N.B. NJ, K. Chonga, Phys. Chem. Chem. Phys. **7**, 1107 (2005)
40. J. Blass, M. Albrecht, G. Wenz, R. Guerra, M. Urbakh, R. Bennewitz, J. Phys. Chem. C **121**(29), 15888 (2017). https://doi.org/10.1021/acs.jpcc.7b05412. (<Go to ISI>://WOS:000406726200043)
41. M.E. Drew, A.R. Konicek, P. Jaroenapibal, R.W. Carpick, Y. Yamakoshi, J. Mater. Chem. **22**(25), 12682 (2012)
42. U. Schwarz, O. Zwörner, P. Köster, R. Wiesendanger, Phys. Rev. B **56**, 6987 (1997)
43. W. Ducker, T. Senden, R. Pashley, Nature **353**, 239 (1991)
44. J. Hoh, A. Engel, Langmuir **9**, 3310 (1993)
45. D. Dietzel, M. Feldmann, U.D. Schwarz, H. Fuchs, A. Schirmeisen, Phys. Rev. Lett. **111**(23), 235502 (2013). https://doi.org/10.1103/PhysRevLett.111.235502. (<Go to ISI>://WOS:000328615000006)
46. W. Maw, F. Stevens, S. Langford, J. Dickinson, J. Appl. Phys. **92**(9), 5103 (2002)
47. C. Petzold, M. Koch, R. Bennewitz, Beilstein J. Nanotechnol. **9**, 1647 (2018). https://doi.org/10.3762/bjnano.9.157. (<Go to ISI>://WOS:000434441100002)
48. Q. Liang, O. Tsui, Y. Xu, H. Li, X. Xiao, Phys. Rev. Lett. **90**(14), 146102 (2003)
49. E. Riedo, F. Levy, H. Brune, Phys. Rev. Lett. **88**, 185505 (2002)
50. F. Tian, X. Xiao, M. Loy, C. Wang, C. Bai, Langmuir **15**(1), 244 (1999)
51. R. Szoszkiewicz, E. Riedo, Phys. Rev. Lett. **95**(13), 135502 (2005)
52. C. Greiner, J.R. Felts, Z. Dai, W.P. King, R.W. Carpick, ACS Nano **6**(5), 4305 (2012)
53. J. Gao, W. Luedtke, D. Gourdon, M. Ruths, J. Israelachvili, U. Landman, J. Phys. Chem. B **108**(11), 3410 (2004)
54. R. Alvarez-Asencio, J. Pan, E. Thormann, M.W. Rutland, Tribolo. Lett. **50**(3), 387 (2013)
55. E. Meyer, R. Lüthi, L. Howald, M. Bammerlin, M. Guggisberg, H.J. Güntherodt, J. Vac. Sci. Technol. B **14**, 1285 (1996)
56. M. Ruths, N. Alcantar, J. Israelachvili, J. Phys. Chem. B **107**(40), 11149 (2003)
57. A. Socoliuc, R. Bennewitz, E. Gnecco, E. Meyer, Phys. Rev. Lett. **92**(13), 134301 (2004)
58. R. Cannara, M. Brukman, R. Carpick, Rev. Sci. Instr. **76**, 53706 (2005)
59. A. Socoliuc, E. Gnecco, R. Bennewitz, E. Meyer, Phys. Rev. B (Conden. Matter Mater. Phys.) **68**(11), 115416 (2003)
60. R. Overney, E. Meyer, J. Frommer, D. Brodbeck, R. Luethi, L. Howald, H.J. Guentherodt, M. Fujihira, H. Takano, Y. Gotoh, Nature **359**(6391), 133 (1992)
61. A.J. Marsden, M. Phillips, N.R. Wilson, Nanaotechnology **24**(25), 255704 (2013)
62. M. Liley, D. Gourdon, D. Stamou, U. Meseth, T. Fischer, C. Lautz, H. Stahlberg, H. Vogel, N. Burnham, C. Duschl, Science **280**(5361), 273 (1998)
63. M. Marcus, R. Carpick, D. Sasaki, M. Eriksson, Phys. Rev. Lett. **88**(22), 226103 (2002)
64. M. Kwak, H. Shindo, Phys. Chem. Chem. Phys. **6**(1), 129 (2004)
65. H.S. Liao, B.J. Juang, W.C. Chang, W.C. Lai, K.Y. Huang, C.S. Chang, Rev. Sci. Instr. **82**(11), 113710 (2011)
66. J. Stiernstedt, M. Rutland, P. Attard, Rev. Sci. Instr. **76**(8), 83710 (2005)
67. T. Bouhacina, J. Aime, S. Gauthier, D. Michel, Phys. Rev. B **56**, 7694 (1997)

68. E. Gnecco, R. Bennewitz, T. Gyalog, C. Loppacher, M. Bammerlin, E. Meyer, H. Güntherodt, Phys. Rev. Lett. **84**, 1172 (2000)
69. S. Sills, R. Overney, Phys. Rev. Lett. **91**, 095501 (2003)
70. A. Labuda, M. Lysy, W. Paul, Y. Miyahara, P. Gruetter, R. Bennewitz, M. Sutton, Phys. Rev. E **86**(3), 031104 (2012)
71. T. Kawagishi, A. Kato, Y. Hoshi, H. Kawakatsu, Ultramicroscopy **91**, 37 (2002)
72. S. Maier, Y. Sang, T. Filleter, M. Grant, R. Bennewitz, E. Gnecco, E. Meyer, Phys. Rev. B **72**, 245418 (2005)
73. A. Schirmeisen, L. Jansen, H. Fuchs, Phys. Rev. B **71**, 245403 (2005)
74. N. Tambe, B. Bhushan, Nanotechnology **16**(10), 2309 (2005)
75. E. Tocha, T. Stefanski, H. Schonherr, G. Vancso, Rev. Sci. Instr. **76**(8), 83704 (2005)
76. X. Yang, S.S. Perry, Langmuir **19**(15), 6135 (2003)
77. R.H. Schmidt, G. Haugstad, W.L. Gladfelter, Langmuir **19**(24), 10390 (2003)
78. L. Jansen, A. Schirmeisen, J.L. Hedrick, M.A. Lantz, A. Knoll, R. Cannara, B. Gotsmann, Phys. Rev. Lett. **102**(23), 236101 (2009)
79. L. Jansen, H. Hoelscher, H. Fuchs, A. Schirmeisen, Phys. Rev. Lett. **104**(25), 256101 (2010)
80. X. Zhao, S.R. Phillpot, W.G. Sawyer, S.B. Sinnott, S.S. Perry, Phys. Rev. Lett. **102**(18), 186102 (2009)
81. M. Vorholzer, J.G. Vilhena, R. Perez, E. Gnecco, D. Dietzel, A. Schirmeisen, Phys. Rev. X **9**(4) (2019). https://doi.org/10.1103/PhysRevX.9.041045
82. W. Wang, D. Dietzel, A. Schirmeisen, Sci. Adv. **6**(12) (2020). https://doi.org/10.1126/sciadv.aay0165
83. M. Hoogeman, D. van Loon, R. Loos, H. Ficke, E. de Haas, J. van der Linden, H. Zeijlemaker, L. Kuipers, M. Chang, M. Klik, J. Frenken, Rev. Sci. Instr. **69**(5), 2072 (1998)
84. B. Gotsmann, U. Durig, Langmuir **20**(4), 1495 (2004)
85. Q.Y. Li, T.E. Tullis, D. Goldsby, R.W. Carpick, Nature **480**(7376), 233 (2011). https://doi.org/10.1038/nature10589. (<Go to ISI>://WOS:000297744200049)
86. M. Feldmann, D. Dietzel, A. Tekiel, J. Topple, P. Grutter, A. Schirmeisen, Phys. Rev. Lett. **117**(2), 025502 (2016). https://doi.org/10.1103/PhysRevLett.117.025502. (<Go to ISI>://WOS:000379729000007)
87. P. Maivald, H. Butt, S. Gould, C. Prater, B. Drake, J. Gurley, P. Hansma, Nanotecnology **2**, 103 (1991)
88. T. Göddenhenrich, S. Müller, C. Heiden, Rev. Sci. Instr. **65**, 2870 (1994)
89. J. Colchero, M. Luna, A. Baro, Appl. Phys. Lett. **68**, 2896 (1996)
90. R. Carpick, D. Ogletree, M. Salmeron, Appl. Phys. Lett. **70**, 1548 (1997)
91. N.N. Gosvami, P.C. Nalam, A.L. Exarhos, Q. Tam, J.M. Kikkawa, R.W. Carpick, Appl. Phys. Lett. **105**(9) (2014). https://doi.org/10.1063/1.4894737. <Go to ISI>://WOS:000342749800045
92. H.U. Krotil, T. Stifter, O. Marti, Appl. Phys. Lett. **77**(23), 3857 (2000)
93. G. Haugstad, Tribol. Lett. **19**(1), 49 (2005)
94. M. Reinstadtler, U. Rabe, V. Scherer, U. Hartmann, A. Goldade, B. Bhushan, W. Arnold, Appl. Phys. Lett. **82**(16), 2604 (2003)
95. L. Huang, C. Su, Ultramicroscopy **100**(3–4), 277 (2004)
96. M. Reinstadtler, T. Kasai, U. Rabe, B. Bhushan, W. Arnold, J. Phys. D: Appl. Phys. **38**(18), 269 (2005)
97. S. Morita, R. Wiesendanger, E. Meyer, *Noncontact Atomic Force Microscopy (NanoScience and Technology)* (Springer, Berlin, Germany, 2002)
98. S. Jarvis, H. Yamada, K. Kobayashi, A. Toda, H. Tokumoto, Appl. Surf. Sci. **157**, 314 (2000)
99. O. Pfeiffer, R. Bennewitz, A. Baratoff, E. Meyer, P. Gruetter, Phys. Rev. B **65**, 161403 (2002)
100. F. Giessibl, M. Herz, J. Mannhart, Proc. Natl. Acad. Sci. USA **99**(19), 12006 (2002)
101. S. Kawai, S.I. Kitamura, D. Kobayashi, H. Kawakatsu, Appl. Phys. Lett. **87**(19), 173105 (2005)
102. M. Ternes, C.P. Lutz, C.F. Hirjibehedin, F.J. Giessibl, A.J. Heinrich, Science **319**(5866), 1066 (2008)

103. A.J. Weymouth, D. Meuer, P. Mutombo, T. Wutscher, M. Ondracek, P. Jelinek, F.J. Giessibl, Phys. Rev. Lett. **111**(12), 126103 (2013)
104. S. Nam, E. Riegel, L. Hörmann, O.T. Hofmann, O. Gretz, A.J. Weymouth, F.J. Giessibl, Proc. Nat. Acad. Sci. USA **121**(2) (2024). https://doi.org/10.1073/pnas.2311059120. <Go to ISI>://WOS:001147157500001
105. P.C. Nalam, N.N. Gosvami, M.A. Caporizzo, R.J. Composto, R.W. Carpick, Soft Matter **11**(41), 8165 (2015). https://doi.org/10.1039/c5sm01143d. (<Go to ISI>://WOS:000363204000015)
106. M.D. Krass, N.N. Gosvami, R.W. Carpick, M.H. Müser, R. Bennewitz, J. Phys. Condens. Matter **28**(13), 134004 (2016). http://stacks.iop.org/0953-8984/28/i=13/a=134004
107. A. Labuda, W. Paul, B. Pietrobon, R.B. Lennox, P.H. Gruetter, R. Bennewitz, Rev. Sci. Instr. **81**(8), 083701 (2010)
108. C. Pimentel, E. Gnecco, C.M. Pina, J. Phys. Chem. C **122**(39), 22487 (2018). https://doi.org/10.1021/acs.jpcc.8b05963 https://doi.org/10.1021/acs.jpcc.8b05963

“Surface Forces Apparatus in Nanotribology”

Carlos Drummond and Marina Ruths

Abstract The Surface Forces Apparatus (SFA) has proven to be an excellent tool for studies in nanotribology. The normal load, contact area, and sliding velocity between the surfaces can be controlled and unambiguously measured with higher accuracy than in any conventional tribometer. Furthermore, an image of the surfaces in contact can be obtained as the surfaces are slid, allowing the monitoring of the real size and shape of the contact area and the distance or film thickness profile between the surfaces when atomically smooth surfaces are used. It is relatively simple to perform a comprehensive exploration of the full parameter space to determine the important variables in the frictional behavior of the system. In this chapter, the principles of operation and some experimental details of the Surface Forces Apparatus nanotribometer are described.

1 Introduction

Measurements of the force between solids date back to the 1920s when Tomlinson investigated the interaction between crossed filaments of different metals [1]. Later, research groups in the Netherlands and Russia led by Overbeek and Derjaguin developed different techniques for measuring the force between surfaces of quartz or glass as a function of their separation [2, 3]. The example of these seminal pieces of work was promptly followed by many other groups. Particularly in Cambridge, a remarkable body of work was accomplished, leading to the development of the Surface Forces Apparatus, SFA, by Tabor, Winterton, and Israelachvili [4, 5].

The study of lateral forces between surfaces has a longer history. The problem of friction between surfaces attracted great thinkers, such as Da Vinci, Coulomb,

C. Drummond (✉)
Centre de Recherche Paul Pascal. CNRS-Université Bordeaux 1, Avenue Albert Schweitzer, 33600 Pessac, France
e-mail: carlos.drummond@crpp.cnrs.fr

M. Ruths
Department of Chemistry, University of Massachusetts Lowell, 1 University Avenue, Lowell, MA 01854, USA

E. Gnecco and E. Meyer (eds.), *Fundamentals of Friction and Wear on the Nanoscale*, NanoScience and Technology, https://doi.org/10.1007/978-3-031-63065-1_2

Euler, Amontons, and many others. A fascinating historical account of the history of tribology was compiled by Dowson [6]. Friction phenomena have also been investigated with SFAs modified for that purpose; the first friction measurements using this technique date back to the 1970s [7]. A decade later, Briscoe and Evans reported extensive results on the study of friction of adsorbed monolayers in air [8]. Nevertheless, it was not until the late 1980s that nanotribology studies with the SFA became a very active field of research. Since then, various modifications to the technique have been introduced. In the present chapter, we describe the principles of operations and some experimental details of the SFA-nanotribometer.

2 Surface Forces Apparatus Technique: Generalities

In a typical SFA-nanotribometry experiment, molecularly smooth mica surfaces are glued to cylindrically curved silica lenses and used to confine thin films. The use of mica as a substrate for surface force experiments was originally advanced by Bailey and Courtney-Pratt [9]. The cylindrically shaped silica disks are placed with their axes perpendicular to each other, a configuration that presents several advantages. First, it circumvents the difficult—if not impossible—task of accurately aligning two parallel plates. Unwanted edge effects are easily avoided by this approach. Second, it allows the investigation of different contact spots on the same pair of surfaces by laterally displacing the crossed cylinders. If wear or contamination of the surfaces appears during the experiment, a fresh contact zone can be readily found. Finally, this geometry is convenient for comparing the results of the measurements with theoretical descriptions, typically sketched for flat surfaces. If the separation between the curved surfaces is much smaller than their radius of curvature, R, the SFA cross-cylinder configuration is equivalent to a sphere-on-plate contact. The force between two such surfaces, F, can be related to the energy of interaction between flat surfaces per unit area, E, by using the so-called Derjaguin approximation [10], $E = F/2\pi R$. This provides a normalization method to quantitatively compare data from different experiments. The question of normalization of the measured interaction forces is more involved in friction experiments, as will be discussed below.

One of the major strengths of the SFA technique is the possibility of imaging the area of contact to determine the absolute distance between the surfaces, the refractive index of the film confined between the surfaces, and the geometry of the contact region. The SFA is one of few techniques in the field of tribology that allows imaging in situ and in real time of the geometry of the contact area and is probably the only technique with subnanometric film thickness resolution. Multiple Beam Interferometry (MBI) is used for this purpose [11]. A highly reflective layer is deposited on the back side of the mica surfaces, and white light is passed through this built-in Fabry–Perot interferometer. The intensity of the light transmitted through the stratified media between the two mirrors depends on the optical thickness in a nontrivial way: only wavelengths that interfere constructively after the multiple reflections in the cavity traverse the multilayer system. The emerging beam of light can then be focused on a spectrometer. The resulting constructive interference fringes (Fringes of Equal Chromatic Order, *FECO*) carry information about the thickness and the refractive index of the different layers in the path of the light. In particular, the thickness and refractive index of the film confined between the mica surfaces can be determined with accuracies of 0.1 nm and 0.01, respectively. Israelachvili developed simple explicit expressions to calculate these quantities from the wavelength of the *FECO* for a homogeneous film confined between symmetric mica surfaces [12]. Later, the analysis was extended to asymmetric, adsorbing, anisotropic, or more complicated multilayer systems [13–15]. The potential of extending the analysis to obtain information about the roughness of the surfaces has also been demonstrated [16].

Many different experimental setups for the measurement of surface forces have been reported. One of the oldest versions, the Mk I, was designed by Israelachvili and Adams for the measurement of forces between liquids and vapors [17]. It was based on the earlier designs of Tabor, Winterton, and Israelachvili [4, 5], and was later followed by greatly improved and modified versions, the Mk II and III (SFA3) [18, 19]. Parker and coworkers later developed a circular steel apparatus (Mk IV) that is simpler to clean and assemble than Mk I or II [20]. The stability and reliability of the apparatus, as well as the simplicity of handling, have been progressively improved in each design. Recent improved versions of this setup are the SFA 2000 [21] and the miniature SFA [22], based on the same general principles of operation. The reader is referred to the original publications for the particular details of each apparatus.

To measure the interaction force normal to the surfaces in a typical SFA experiment, one of the surfaces is displaced using a combination of motors and piezoelectric elements. One of the surfaces is coupled to a calibrated spring with a fixed end. Double cantilever springs are typically used to minimize the tilting and/or sliding between the surfaces when the spring is deflected. The interaction force between the surfaces is measured by progressively displacing the fixed end of the double cantilever spring and allowing the separation between the surfaces to reach equilibrium, where the surface forces are balanced by the elastic force of the spring. The difference between the displacement imposed on the surface and the actual change in the separation between the surfaces, Δx (measured by *MBI*), will correspond to the deflection of the spring. It will be used to calculate the interaction force by simply using Hooke's law to calculate the restoring force, $F = k\Delta x$, where k is the elastic constant of the double cantilever spring. Although conceptually simple, measuring the forces by this procedure is limited by spring instabilities. Quickly decaying forces, with a force-distance gradient larger than the spring constant, are inaccessible because of the mechanical instability of the system under such conditions. Derjaguin and coworkers proposed the use of a force feedback technique to overcome this problem [3]. The idea is to control the force applied to the surfaces independently of the displacement. An external force is applied to the surfaces to maintain the spring undeflected. Effectively, this translates into a continuously changing spring constant, which eliminates the mechanical instability mentioned above. Several implementations of this idea have been reported [23, 24]. Steward and Parker modified a Mark IV by incorporating a magnetic force transducer and a bimorph displacement sensor. Tonck et al.

introduced a feedback apparatus with capacitive displacement transducers [25]. An interesting description of the different techniques used for the measurement of the normal force between surfaces was presented by Lodge [26].

3 Surface Forces Apparatus Nanotribometer

In a nanotribology experiment with the SFA, the mica surfaces are brought to a certain separation. By using motors or electromechanical transducers, a lateral displacement between the surfaces is imposed, and the force induced by this displacement is measured. Usually, a certain normal load is applied, L. If the load is high enough, the glue layer under the surfaces undergoes elastic deformation, and a thin film is confined to a flat circular region of uniform thickness and area of contact A, as illustrated in Fig. 1. By using *MBI,* an image of the surfaces in contact can be obtained as the surfaces slide, allowing monitoring of the size and profile of the contact area and the distance between the surfaces by observing the flat region on the *FECO*. Shear-induced elastohydrodynamic deformation can also be distinguished. In addition, damage of the surfaces can be easily detected as soon as it occurs, allowing one to discriminate between undamaged sliding and friction with wear and to independently study the two scenarios.

A subject of major importance in the analysis of an SFA-nanotribology experiment is to identify the area over which the frictional force takes action. Often, the friction force between sliding surfaces will be dominated by the flat contact area. In that case,

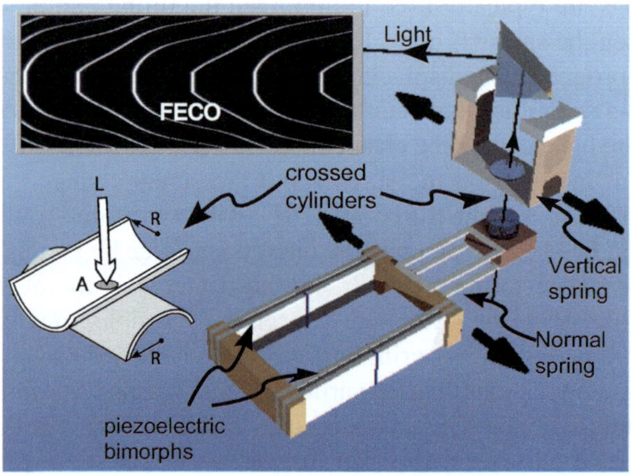

Fig. 1 Schematic illustration of the *SFA* designed by Israelachvili and coworkers, configured for friction experiments. The mica sheets are mounted in crossed-cylinder geometry, and their back surfaces are coated with reflective silver layers to allow for multiple beam interference. The upper and lower surfaces are mounted on cylindrically curved silica discs that are attached to the friction-sensing device and the piezoelectric bimorph slider, respectively

the sharp edge of the *FECO* allows recognition of the "area of contact", which is used to normalize the measured force and to calculate the shear stress. This operation is necessary to quantitatively compare the results of different experiments. From this point of view, the customary friction coefficient is a less fundamental parameter than the shear stress. There is, however, an important caveat to this operation: very often, the measured shear stress depends on the applied pressure. Given that curved surfaces are used in an SFA experiment, the normal pressure is not constant over the flattened area. Its value is given by a nonlinear function of the position in the contact area, a problem that has been extensively treated by the contact mechanics community [27]. It is clear then that the shear stress calculated in an SFA experiment is an average quantity to be treated with caution.

An even more complicated scenario is found when there is a significant contribution to the friction force by regions of the surfaces outside the flattened area. This situation can be envisaged, for example, if there is a contribution to the frictional force coming from breaking and reforming bonds of long molecules that can bridge the two surfaces together. In that case, there is no obvious way to identify the effective contact area. One possibility is to adopt a cutoff length and assume that the contribution to the frictional force is negligible at larger separations. However, at least two problems persist: the choice of the characteristic length is somewhat arbitrary, and the contribution of a given region to the total force will most likely be a function of the local surface separation. It is important to emphasize at this point that the experimental difficulties just outlined are shared by most—if not all—the experimental techniques in nanotribology. In addition, the SFA nanotribometer in its interferometric version is one of few techniques in nanotribology that allows the observation of the contact geometry while rubbing the surfaces.

3.1 Experimental Setup

Several SFA experimental setups have been proposed, each with its capabilities and limitations. In the following, we will briefly describe a few systems that are broadly used in the field. The reader interested in more complete information is referred to the original publications.

Several experimental designs were proposed by the group of Israelachvili [28, 29]; they are all in principle based on the original method proposed by Israelachvili and Tabor [5]. The main features of the most recent versions are illustrated in Fig. 1. The lower surface is mounted on a bimorph-driven slider [29], which moves laterally in a linear fashion when a constant slope voltage ramp is applied between the two electrodes of sectored piezoelectric bimorphs (electromechanical transducers). Alternatively, a constant-frequency sinusoidal input can be imposed on the slider to perform nanorheological experiments. The upper surface is attached to a vertical double cantilever spring, whose deflection is monitored using strain gauges connected to form the arms of a Wheatstone bridge. If the displacement of the lower surface induces a viscous or friction force on the upper surface, the vertical spring will deflect. From the deflection of this spring of known spring constant K, the friction force between the surfaces F can be calculated simply by using Hooke's law [28]. The mechanical properties of the measurement system (e.g., compliance and inertial mass) will influence the results; these factors must be taken into account to obtain meaningful information from the measured signal. This can be done straightforwardly in the SFA because of its mechanical simplicity and easy-to-characterize mechanical properties.

The maximal distance that can be slid with this setup depends on the characteristics of the bimorph strips and is typically several tens of micrometers. Larger displacements can be achieved by mechanically driving the upper surface using a reversible, variable-speed motor-driver micrometer shaft that displaces the translation stage holding the vertical double cantilever spring. The detection limit for the friction force of this setup is typically on the order of several μN. By changing the frequency and amplitude of the input signal to the bimorph slider, the driving speed can typically vary between several Å/s and 0.1 mm/s. This device has been used to study a large number of systems. Some examples included confined simple liquids [30–32], polymer melt and solutions [28, 29, 33–35], self-assembled surfactants, and polymer layers [36–38].

An alternative design conceived for the study of smaller deformations was introduced by Granick and coworkers [39, 40]. The goal of these low-amplitude studies was to focus the investigation on the linear response of the confined films. By applying small deformations, the flow of fresh liquid in the contact zone is avoided. This allows the study of long-term relaxation processes that may occur in the contact region. A schematic of this device is shown in Fig. 2. In this design, the bottom surface remains stationary, while the upper surface is mounted on a holder attached to a double cantilever. Compared with the design of Israelachvili, the vertical metallic cantilevers have been replaced with two piezoelectric bimorph strips. One of the bimorphs is used as an actuator, and the other is used as a sensor. In the experiment, a voltage difference is applied to one of the bimorphs to bend it. Typically, a constant-frequency sinusoidal signal is used, inducing an oscillating force on the bimorph. Simultaneously, the deformation-induced voltage of the second bimorph is measured. These data are used to determine the actual displacement of the surface. The influence of the confined film on the moving surface can be extracted by comparing this response with the one observed when no interaction between the two surfaces is present. The electromechanical characteristics of the system are modeled as a series of effective masses, springs, and dashpots representing the different components of the apparatus. The friction appears as a force acting on the holder of the lower surface, from which an effective viscosity can be extracted [40]. Although mainly conceived for the study of small deformations (of the order of the film thickness), typical displacements range from a few nm to a few μm. The reported sensitivity to the friction force is approximately 5 μN. Many different systems have been explored with this device, including simple liquids [39, 41–43] polymer melts [44–46] and solutions [47, 48].

A third experimental setup widely used in the literature has been developed by Klein and coworkers [49, 50]. A schematic of this device is presented in Fig. 3. In

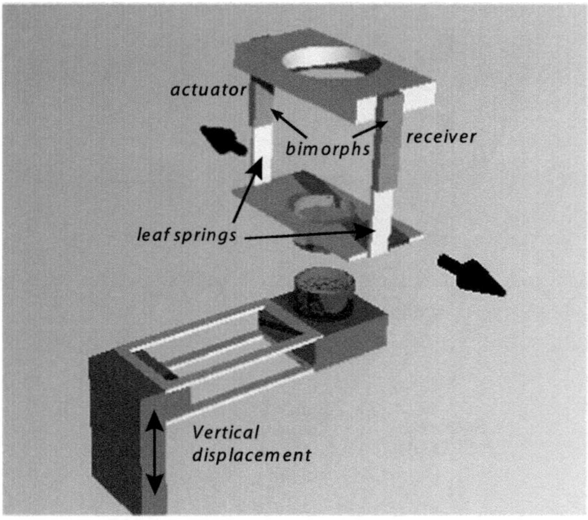

Fig. 2 Schematic illustration of the *SFA* friction setup designed by Granick and coworkers. The shear force is generated by one bimorph (actuator) and the response of the device induces a voltage across the other bimorph (receiver)

this device, the sensitivity to the measured friction forces is greatly improved with respect to the previous designs. A sectored piezoelectric tube is used to produce a normal or lateral displacement of the upper surface. An air-gap capacitor is used to measure the lateral displacement of the lower surface, which is coupled to a calibrated double cantilever spring with a fixed end. The shear-induced frictional force is then directly extracted from this displacement (e.g., the deflection of the spring) by using Hooke's law. The reported sensitivity of the friction force is 50 nN, and the maximal displacement of the upper surface is a few tens of μm. The improved sensitivity of this device comes from the detection method used. This has proven to be very valuable for the study of polymer melts and solutions [51–54] where small forces are typically observed. Research on water and other simple liquids has also been conducted with this experimental setup [55, 56].

Further improvements in different aspects of the experimental technique have been proposed during the last three decades. First, substantial efforts have been devoted to automating the procedure of measurement to improve the accuracy, versatility, and simplicity of the technique. The automatic detection of *FECO* has posed some challenges in the past, but increasingly accurate and affordable charge-coupled devices in the market are currently used in several laboratories for this purpose. Different strategies for the automation of the measurement process have been extensively described by Quon et al. [57] Grünewald and Helm [58], and more recently by Heuberger and coworkers [59], and other teams [60]. Additionally, some laboratories have extended the capabilities of the apparatus to include the movement of the surfaces and the measurement of the friction force between them in two orthogonal directions in the plane of contact [61]. This development should prove to be useful in investigating shear-induced effects (e.g., shear alignment) on confined thin films. More recently, other experimental designs incorporating shear in two orthogonal directions have been described [21, 62]. Finally, actuation methods developed to extend the range of

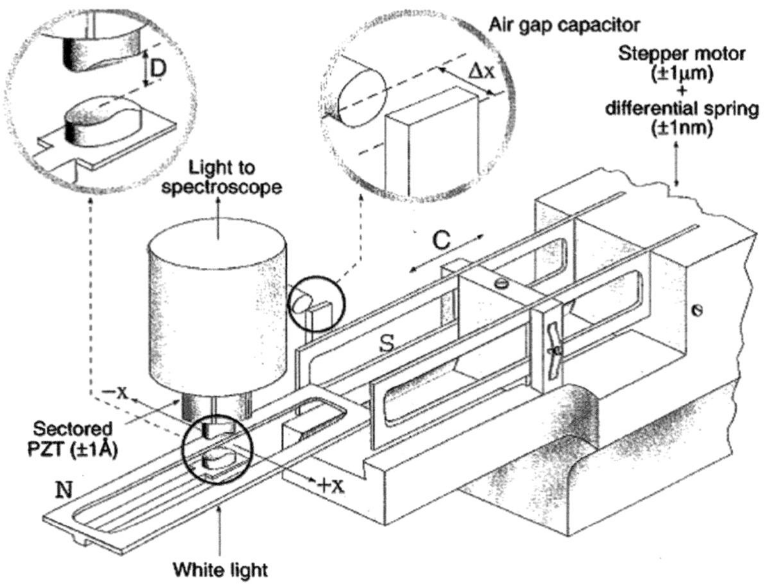

Fig. 3 Schematic illustration of the surface force balance (*SFB*) designed by Klein and coworkers. The separation between the surfaces is controlled via a three-stage mechanism with a sectored piezoelectric tube on which the top surface is mounted. The piezoelectric element produces both normal and lateral displacement. The bending of the shear force spring is detected by an air-gap capacitor. Reprinted with permission from The *Journal of Physical Chemistry B* Volume **105(34)**, 8125–8134 (2001). Uri Raviv, Rafael Tadmor and Jacob Klein

accessible velocities to explore significant conditions in biological and engineering systems have been reported [63]. At high-speed conditions, circular (instead of reciprocating) motion between the surfaces was implemented to avoid wear-prone high accelerations [64].

As mentioned before, efforts have also been devoted to determining the surface displacement by noninterferometric techniques. Several groups have proposed using piezoelectric bimorphs for this purpose [39, 65, 66]. This solution is inadequate for long or quasistatic measurements due to the intrinsic drift and leakage of bimorph sensors (electrical drift and decay). A partial solution to these problems was presented by Parker [67] who suggested the use of an ultrahigh impedance amplifier to lengthen the decay time of the bimorph sensor. A different method was presented by Frantz et al. [68]. They proposed monitoring the capacitance between the silver layers deposited on the back surface of the mica sheets for fast detection of the surface separation and described the use of this setup for the study of contact mechanics. Later, Stewart proposed using capacitance dilatometry to measure the separation between the surfaces: one plate of a parallel-plate capacitor was attached to the moving surface, and the other was attached to the chamber of the apparatus [69]. Improved versions of noninterferometric SFAs based on capacitance distance measurements have been described [70]. More recently, Charlaix and coworkers developed an SFA for static and dynamic measurements incorporating high-resolution displacement sensors based on Nomarski interferometers, allowing the study of the rheology of thin films with unprecedented accuracy [71]. These experimental setups allow a fast and accurate measurement of the displacement of the surface and eliminate the constraint of having to use transparent surfaces. Nevertheless, if opaque surfaces are used, it is impossible to obtain an image of the contact region while shearing, which is one of the major strengths of the interferometric-SFA.

3.2 *Local Structural Information: Combination of the SFA with Other Techniques*

The information gathered in a conventional SFA experiment is limited to the average response of the confined film under shear and compression. To achieve a better understanding of the behavior of confined films under shear, it is desirable to obtain structural information at the molecular level. Obtaining this information implies a significant experimental challenge. On the one hand, the number of molecules involved in a thin film is relatively small, particularly when a localized area is explored, which inevitably reduces the intensity of any measured signal. On the other hand, the investigated thin film is surrounded by layers of different materials that are susceptible to interaction with the used probe (e.g., light, X-ray, or neutrons), increasing the level of noise of the measured signal. Despite these difficulties, several experimental groups have reported encouraging results of experiments combining the capabilities of SFAs with other techniques in situ.

Some efforts in this direction aim to combine SFA with X-ray diffraction (XSFA) [72, 73] The second generation of XSFA combines the force measurement capabilities of the SFA with in situ small-angle X-ray scattering experiments [74]. Films of several liquid crystals were studied with this setup, and the effect of shear on the molecular alignment in confined films was directly evidenced. These earlier studies were limited to films thicker than 500 nm, mainly because of the poor signal-to-noise ratio obtained otherwise. Research in nanotribology calls for much thinner films. Moreover, the results reported with this method were limited to an average investigation of the contact area because of the size of the X-ray beam used. Later studies showed that X-ray reflectivity can be used to obtain structural information of ultraconfined molecular films only a few layers thick [75], and a few teams attempted to incorporate X-ray scattering and reflectivity in modified SFA setups [76]. The first report of successful coupling of X-ray scattering with a fully operational SFA setup under extreme confinements was recently reported by Kékicheff et al. [77]. By using a collimated X-ray beam of a small cross-section and carefully aligning it with the SFA contact area, it was shown that it is possible to obtain lateral structural information of highly confined films. It is reasonable to expect that these techniques, which reveal the local molecular ordering of confined films, will improve our understanding of the friction phenomena in the years to come.

Helm and coworkers [78] showed that *MBI* can be used to obtain structural information of confined thin films without any modification to the original SFA technique. Information about orientation and intermolecular interactions can be extracted from the *FECO* if optically active molecules are investigated. It is possible to study ultrathin films, given that the light absorption by the confined molecules is enhanced by the multiple reflections in the optical cavity. Local information in the contact area can also be obtained. Nevertheless, this technique limits the systems that can be studied to large dye molecules. In addition, for the best determination of the adsorption spectra of the confined dyes, relatively thick mica must be used, reducing the accuracy of the film thickness determination.

Salmeron and coworkers suggested coupling second harmonic and sum-frequency generation to SFAs to study the alignment and relaxation of confined ultrathin films and showed the potential of the application by investigating self-assembled and Langmuir–Blodgett monolayers of several surfactants [79]. The combination of SFA with other optical techniques has been limited by the reflective silver layer used to determine the surface separation by *MBI*. This layer strongly reduces the intensity of the illumination of the confined films, seriously limiting the in situ performance of other optical methods. Granick and coworkers have overcome this limitation by replacing the reflective silver layer with multilayer dielectric coatings, which are transparent in different regions of the optical spectrum. In this manner, they have been able to apply different spectroscopic tools to obtain in situ structural information of ultrathin films under shear [80–82].

By combining the SFA nanotribometer with fluorescence correlation spectroscopy they measured the molecular diffusion coefficient in thin films within spots of submicron size, obtaining spatially resolved measurements [80]. By using confocal geometry, avoiding the problem of the contribution of the bulk to the scattered signal, it was possible to integrate Raman spectroscopy with a conventional SFA setup [81]. They reported spatially resolved Raman scattering before and after shear, evidencing the influence of shear on the orientation of the molecules inside the confined film. A third technique developed in Granick's group is the combination of photoluminescence and absorption dichroism with the SFA [82]. The shear-induced alignment of preadsorbed polymer molecules on mica was quantified by both photoluminescence and absorption spectra. In recent years, several other teams have reported improved designs of interferometric SFA setups, which allow the coupling of optical techniques such as fluorescence microscopy, surface plasmon resonance, and Raman spectroscopy [22, 83].

A valuable line of research has developed based on the incorporation of electrochemical techniques into SFAs [84–86]. Using back-silver layers or metallic thin films as electrodes or developing flat metallic surfaces to replace one or both mica surfaces allows the use of well-established electrochemical techniques to study ultrathin films. In this way, precise control of the surface potential and its influence on surface forces [87] or the study of electroactive lubricants [88] have been investigated.

3.3 Beyond Mica: Alternative Substrates

As mentioned previously, mica surfaces are the most popular substrates for SFA experiments. It gathers a set of properties seldom observed in other materials. It is transparent and can be prepared in the form of thin sheets of molecularly smooth surfaces over large areas by successive cleaving. The combination of these properties is at the heart of the SFA technique: transparent surfaces are needed to determine the geometry of the contact by *MBI*. In addition, mica is a fairly incompressible material, so the forces measured at modest pressures are not flawed by the deformation of the surface. In addition, it is quite inert to chemical reactions, so it is hardly modified during experiments. As a drawback, the process of producing mica surfaces thin enough to fulfill the requirements of the SFA technique calls for a skillful experimentalist. This constraint is greatly relaxed by the careful implementation of automatic thickness measurement. On the other hand, thanks to the smoothness of the surfaces, the geometry of the contact between the two surfaces can be completely described, simplifying the description and interpretation of the results.

The investigation of substrates other than mica is of interest for obvious reasons. The substrate plays a major role in most of the phenomena investigated by SFA, particularly in tribology. It acts not only as a geometrical barrier but also as a major player: the interaction between the surfaces and the confined films determines the general frictional behavior. For these reasons, considerable effort has been devoted to investigating the possibility of modifying or replacing mica surfaces with other substrates to expand the range of applications of the technique. Mica surfaces can be modified by deposition or adsorption of different materials. By properly controlling the modification process, the smoothness of the surfaces can be preserved. In addition, by the deposition of a smooth layer on mica surfaces, it is possible to modify the surface energy of the substrate, preparing surfaces that may be more prone to chemical modifications.

Several groups have investigated the behavior of mica surfaces modified by self-assembly [37, 38, 89, 90] or deposition of Langmuir–Blodgett films [37, 91, 92] of different substances. Mica acquires a negative surface charge when immersed in water, so positively charged species (e.g., cationic surfactants) spontaneously adsorbed on it; the structure of the adsorbed layer and its relationship to the molecular structure of the adsorbed material have been widely investigated [93]. The frictional behavior of the modified surfaces depends strongly on the characteristics of the adsorbed layers: surface properties such as the adhesion energy and the morphology of the adsorbed layer will ultimately determine their behavior under shear.

As mentioned before, mica is generally an inert material. Nevertheless, it can be chemically modified by water vapor plasma treatment, promoting the formation of reactive surface silanol groups and its reactivity to different species, e.g., chlorosilanes, as suggested by Parker and coworkers [94, 95]. In this way, molecularly smooth hydrophobic surfaces can be prepared, given that the chemical structure is modified without increasing the roughness of the substrates. Mica surfaces treated by this procedure have been used in many SFA studies. Kessel and Granick modified this procedure to induce the self-assembly of alkoxysilanes on mica, showing the formation of strongly bound monolayers [96]. Giasson and coworkers have explored a large number of systems based on this method of chemical modification [97, 98, 99].

Several groups have proposed modifying mica surfaces simply by depositing thin films of different materials, including metals and dielectrics. To be able to monitor the geometry of the contact region by *MBI*, it is important for the deposited layers not to be completely opaque. This does not impose a serious limitation for sufficiently thin films. The interpretation of the *FECO* becomes more involved because of the larger number of optical layers in the optical path of the white light, but information about the thickness and optical properties of the confined film can nonetheless be extracted. Different algorithms that are adequate for the modified experimental conditions have been described in recent years [100, 101].

Studies of mica modification by deposition of many different materials have been reported. Silver [100, 102], gold [100, 103], platinum [103, 104], and silica [104, 105], are only a few of a long list of materials investigated. Horn et al. grew single aluminum oxide crystals [106] by vapor phase condensation. The tribological behavior of these surfaces was later investigated by Berman et al. [107] Vigil et al. deposited smooth layers of amorphous silica on mica and studied the behavior under compression and shear of the resulting surfaces [105]. They found that oscillatory structural forces were absent in the interaction between the surfaces. In addition, they observed long-time-dependent adhesion and friction of the surfaces in the presence of water. McGuiggan et al. deposited amorphous carbon by magnetron sputtering on mica and used these surfaces in the SFA-nanotribometer [108]. They found the friction force to be proportional to the area of contact between the surfaces, and the measured shear stress to decrease strongly with increasing relative humidity. Hirz et al. sputtered thin films of zirconia and alumina on mica and investigated the behavior of these surfaces when lubricated with a linear perfluoropolyether [109]. They showed that these metal oxides formed smooth films on mica susceptible to being used as alternative substrates in SFA experiments.

Other groups have proposed simplifying the method of substrate preparation by eliminating the use of mica. A method of preparing silica surfaces for use in SFAs was proposed by Horn et al., although its use has not become widespread [110, 111]. Golan et al. proposed depositing a thin layer of silicon nitride on rigid silica disks previously coated with a reflective layer to replace mica substrates [112]. They also reported a succinct tribological study of this generic substrate.

In general, the abovementioned surface modification processes may alter the smoothness of the surfaces to some degree, complicating the system's geometry and changing it from a single-asperity to a multiple-asperity contact. However, they allow the investigation of surfaces of interest in many different fields, extending the range of applications of SFAs. In addition, in most cases, the roughness of the deposited layers can be controlled and/or modified to a certain extent, allowing the investigation of the effect of surface roughness on friction, an important field of research on its own. An interesting method to produce ultrasmooth metallic surfaces by template stripping [113] (using mica as the template surface) has been successfully applied in SFA experiments with gold [114, 115], platinum [85], and aluminum surfaces [116].

4 Case Study: Long-Range Interaction, Compliance, and Elastohydrodynamic Lubrication

To illustrate the full potential of the SFA technique for nanotribology studies, some experimental results obtained with mica coated with poly(N-isopropylacrylamide) cationic microgels are described in this section. These thermoresponsive microgels, which deswell upon heating, adsorb on negatively charged mica surfaces, forming a dense coating layer. They can also be chemically grafted onto mica, following the plasma-based mica activation method described above.

The interaction forces in the normal direction between microgel-coated surfaces are presented in Fig. 4a. As described above, the measured forces are normalized by the radius of curvature of the mica surfaces. These forces are dominated by long-range repulsion, which clearly demonstrates the thermosensitive character of the coating. By decreasing the temperature, the microgels are increasingly swollen, concomitantly increasing the range of the repulsive force. One main strength of the interferometric SFA technique is the possibility of monitoring the contact geometry at all times. An example of this is presented in Fig. 4b, where the apparent size of the contact region increases noticeably under normal load. Thus, it is questionable to apply the normalization of the force using the nondeformed radius. Compression-stiffening of microgels confined in the contact region makes it increasingly difficult to reduce the gap between the surfaces when load is applied. Under these conditions, conventional contact mechanic models (e.g., Hertz or JKR) are no longer suitable [27].

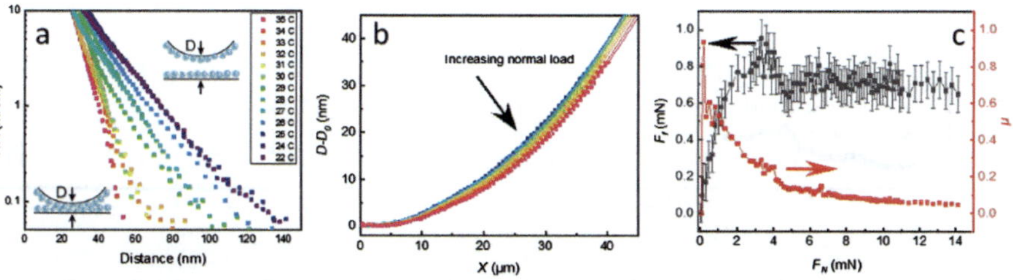

Fig. 4 **a** Interaction forces (normalized by the non-deformed radius of curvature of the surfaces, R) versus surface separation distance between mica surfaces coated with a chemically grafted layer of pNIPAM microgels immersed in a solution of 1 mM $NaNO_3$ at different temperatures, as indicated. **b** Geometry of the interacting surfaces under compression: surface separation shifted by the point of closest approach, D_0. $T = 22$ °C. Measured (points) and fitted (lines) data for different applied normal loads, between 0.03 and 1.81 mN. **c** Load-dependence of the friction force F_f between mica surfaces coated with a chemically grafted layer of pNIPAM microgels

The strong distance dependence of the normal force on surface separation has significant implications for friction forces. Typical results for friction F_f versus applied load F_N are presented in Fig. 4c. As can be observed in the figure, F_f quickly increases with L at low load values but remains practically unchanged in the higher load region. This unusual behavior is a consequence of the compliant surfaces being kept apart due to surface forces, which limits the contribution of microparticles distant from the contact apex to global friction, as we have extensively described before [117].

Additional information can be gathered by monitoring the geometry of the rubbing surfaces in real time. We have observed the emergence of velocity dependent, shear-induced normal forces, large enough to substantially separate the surfaces compared with the static equilibrium condition. This dynamic repulsive force of elastohydrodynamic origin can drive the system from boundary conditions to hydrodynamic lubrication, significantly reducing friction and wear. An example of this scenario is presented in Fig. 5a. As has been recently described, the dynamic lift force is determined by the viscoelastic properties of the surfaces. This is the reason for the strong influence of temperature on the measured lift.

Theoretical descriptions of elastohydrodynamic lubrication generally assume constant surface separation [118, 119], which does not describe what is usually observed in experiments: the emergent lift force pushes the surfaces apart. However, this scenario of constant separation can be reproduced in the interferometric SFA by measuring the actual separation between the surfaces while shearing and using a feedback loop to control it. In this way, the lift force can be accurately determined, and its dependence on driving speed and separation can be investigated and compared with theoretical models, as has been recently reported [120].

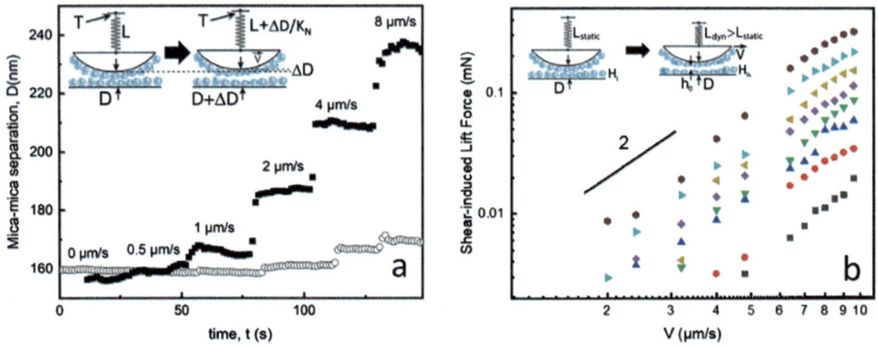

Fig. 5 **a** Variation in the separation between microgel-coated mica surfaces under shear, due to the elastohydrodynamic lift force. Black squares: $T = 35$ °C. Open circles: $T = 25$ °C. The driving velocity V was systematically increased, as indicated. **b** Lift force versus driving speed at different constant mica-mica separations, controlled by a feedback loop acting on the applied load. Black squares, $D = 300$ nm; red circles, $D = 250$ nm, blue triangles $D = 240$ nm, green inverted triangles D = 210, purple rhombi D = 190 nm, ochre triangles $D = 180$ nm, cyan triangles $D = 160$ nm, brown hexagons $D = 130$ nm. $T = 35$ °C

This example illustrates the capabilities of the SFA nanotribometer. It shows how the measurement of parameters pertinent to the friction problem, in particular the capability to monitor the geometry of the rubbing surfaces, greatly improves the understanding of the phenomena involved, allowing a quantitative comparison of the behavior of the system with theoretical models.

References

1. G.A. Tomlinson, Molecular cohesion. Philos. Mag. **6**(37), 695–712 (1928). https://doi.org/10.1080/14786441008564646
2. J.T.G. Overbeek, M.J. Sparnaay, Classical coagulation. London-van Der Waals attraction between macroscopic objects. Discuss. Faraday Soc. **18**(1), 12 (1954). https://doi.org/10.1039/df9541800012
3. Y.I.I. Rabinovich, B.V.V. Derjaguin, N.V.V. Churaev, Direct measurements of long-range surface forces in gas and liquid media. Adv. Colloid Interface Sci. **16**(1), 63–78 (1982). https://doi.org/10.1016/0001-8686(82)85007-0
4. D. Tabor, R.H.S. Winterton, The direct measurement of normal and retarded van Der Waals forces. Proc. R. Soc. A Math. Phys. Eng. Sci. **312**(1511), 435–450 (1969). https://doi.org/10.1098/rspa.1969.0169
5. J.N. Israelachvili, D. Tabor, The measurement of Van Der Waals dispersion forces in the range 1.5 to 130 nm. Proc. R. Soc. A Math. Phys. Eng. Sci. **331**(1584), 19–38 (1972). https://doi.org/10.1098/rspa.1972.0162
6. D. Dowson, *History of Tribology*, 2nd edn. (Professional Engineering Publishing Limited, London, 1998)
7. J.N. Israelachvili, D. Tabor, The shear properties of molecular films. Wear **24**(3), 386–390 (1973). https://doi.org/10.1016/0043-1648(73)90167-1
8. B.J. Briscoe, D.C.B. Evans, The shear properties of Langmuir-Blodgett layers. Proc. R. Soc. A **380**(1779), 389–407 (1982). https://doi.org/10.1098/rspa.1982.0048
9. A.I. Bailey, J.S. Courtney-Pratt, The area of real contact and the shear strength of monomolecular layers of a boundary lubricant. Proc. R. Soc. A **227**(1171), 500–515 (1955). https://doi.org/10.1098/rspa.1955.0026
10. B. Derjaguin, Untersuchungen über Die Reibung und Adhäsion, IV. Kolloid-Zeitschrift **69**(2), 155–164 (1934). https://doi.org/10.1007/BF01433225
11. S. Tolansky, *Multiple Beam Interferometry of Surfaces and Films* (University Press, Oxford, 1948)
12. J.N. Israelachvili, Thin film studies using multiple-beam interferometry. J. Colloid Interface Sci. **44**(2), 259–272 (1973). https://doi.org/10.1016/0021-9797(73)90218-X
13. R.G. Horn, D.T. Smith, Analytic solution for the three-layer multiple beam interferometer. Appl. Opt. **30**(1), 59–65 (1991)
14. C. Müller, P. Mächtle, C.A. Helm, Enhanced absorption within a cavity. A study of thin dye layers with the surface forces apparatus. J. Phys. Chem. **98** (43), 11119–11125 (1994). https://doi.org/10.1021/j100094a021
15. M. Heuberger, The extended surface forces apparatus. Part I. Fast spectral correlation interferometry. Rev. Sci. Instrum. **72**(3), 1700–1707 (2001). https://doi.org/10.1063/1.1347978
16. M. Heuberger, G. Luengo, J. Israelachvili, Topographic information from multiple beam interferometry in the surface forces apparatus. Langmuir **13**(14), 3839–3848 (1997). https://doi.org/10.1021/la960942a
17. J.N. Israelachvili, G.E. Adams, Measurement of forces between two mica surfaces in aqueous electrolyte solutions in the range 0–100 nm. J. Chem. Soc. Faraday Trans. **1**(74), 975–1001 (1978). https://doi.org/10.1039/f19787400975

18. J. Israelachvili, Direct measurements of forces between surfaces in liquids at the molecular level. Proc. Natl. Acad. Sci. **84**(July), 4722–4724 (1987)
19. J.N. Israelachvili, P.M. McGuiggan, Adhesion and short-range forces between surfaces. Part I: New apparatus for surface force measurements. J. Mater. Res. **5**(10), 2223–2231 (1990). https://doi.org/10.1557/JMR.1990.2223
20. J.L. Parker, H.K. Christenson, B.W. Ninham, Device for measuring the force and separation between two surfaces down to molecular separations. Rev. Sci. Instrum. **60**(10), 3135–3138 (1989). https://doi.org/10.1063/1.1140544
21. J. Israelachvili, Y. Min, M. Akbulut, A. Alig, G. Carver, W. Greene, K. Kristiansen, E. Meyer, N. Pesika, K. Rosenberg, H. Zeng, Recent advances in the surface forces apparatus (SFA) technique. Reports Prog. Phys. **73**(3), 036601 (2010). https://doi.org/10.1088/0034-4885/73/3/036601
22. K. Kristiansen, S.H. Donaldson, Z.J. Berkson, J. Scott, R. Su, X. Banquy, D.W. Lee, H.B. De Aguiar, J.D. McGraw, G.D. Degen, J.N. Israelachvili, Multimodal miniature surface forces apparatus (MSFA) for interfacial science measurements. Langmuir **35**(48), 15500–15514 (2019). https://doi.org/10.1021/acs.langmuir.9b01808
23. J. Parker, A. Stewart, Force feedback measurements of surface forces. Prog. Colloid Polym. Sci. **88**, 162–168 (1992). https://doi.org/10.1007/BFb0114431
24. A.M. Stewart, J.L. Parker, Force feedback surface force apparatus: principles of operation. Rev. Sci. Instrum. **63**(12), 5626–5633 (1992). https://doi.org/10.1063/1.1143392
25. A. Tonck, J.M. Georges, J.L. Loubet, Measurements of intermolecular forces and the rheology of dodecane between alumina surfaces. J. Colloid Interface Sci. **126**(1), 150–163 (1988). https://doi.org/10.1016/0021-9797(88)90109-9
26. K.G. Lodge, Techniques for the measurement of forces between solids. Adv. Colloid Interface Sci. **19**(1–2), 27–73 (1983). https://doi.org/10.1016/0001-8686(83)80003-7
27. K.L. Johnson, *Contact Mechanics* (Cambridge University Press, Cambridge, 1985)
28. A.M. Homola, J.N. Israelachvili, M.L. Gee, P.M. McGuiggan, Measurements of and relation between the adhesion and friction of two surfaces separated by molecularly thin liquid films. J. Tribol. **111**(4), 675–682 (1989). https://doi.org/10.1115/1.3261994
29. G. Luengo, F.-J. Schmitt, R. Hill, J. Israelachvili, Thin film rheology and tribology of confined polymer melts: contrasts with bulk properties. Macromolecules **30**(8), 2482–2494 (1997). https://doi.org/10.1021/ma9519122
30. M.L. Gee, P.M. McGuiggan, J.N. Israelachvili, A.M. Homola, Liquid to solidlike transitions of molecularly thin films under shear. J. Chem. Phys. **93**(3), 1895–1906 (1990). https://doi.org/10.1063/1.459067
31. J. Israelachvili, P. McGuiggan, M. Gee, A. Homola, M. Robbins, P. Thompson, Liquid dynamics in molecularly thin films. J. Phys. Condens. Matter **2**(S), SA89–SA98 (1990). https://doi.org/10.1088/0953-8984/2/S/009
32. C. Drummond, J. Israelachvili, Dynamic phase transitions in confined lubricant fluids under shear. Phys. Rev. E **63**(4), 041506 (2001). https://doi.org/10.1103/PhysRevE.63.041506
33. G. Luengo, J. Israelachvili, A. Dhinojwala, S. Granick, Generalized effects in confined fluids: new friction map for boundary lubrication. Wear **200**(1–2), 328–335 (1996). https://doi.org/10.1016/S0043-1648(96)07248-1
34. P.A. Schorr, T.C.B. Kwan, S.M. Kilbey, E.S.G. Shaqfeh, M. Tirrell, Shear forces between tethered polymer chains as a function of compression, sliding velocity, and solvent quality. Macromolecules **36**(2), 389–398 (2003). https://doi.org/10.1021/ma011207v
35. G. Luengo, M. Heuberger, J. Israelachvili, Tribology of shearing polymer surfaces. 2. Polymer (PnBMA) sliding on mica. J. Phys. Chem. B **104**(33), 7944–7950 (2000). https://doi.org/10.1021/jp0005773
36. H. Yoshizawa, Y.L. Chen, J. Israelachvili, Recent advances in molecular level understanding of adhesion, friction and lubrication. Wear **168**(1–2), 161–166 (1993). https://doi.org/10.1016/0043-1648(93)90212-5
37. H. Yoshizawa, Y.L. Chen, J. Israelachvili, Fundamental mechanisms of interfacial friction. 1. Relation between adhesion and friction. J. Phys. Chem. **97**(16), 4128–4140 (1993). https://doi.org/10.1021/j100118a033

38. C. Drummond, J. Israelachvili, P. Richetti, Friction between two weakly adhering boundary lubricated surfaces in water. Phys. Rev. E **67**(6), 066110 (2003). https://doi.org/10.1103/PhysRevE.67.066110
39. J. Van Alsten, S. Granick, Molecular tribometry of ultrathin liquid films. Phys. Rev. Lett. **61**(22), 2570–2573 (1988). https://doi.org/10.1103/PhysRevLett.61.2570
40. J. Peachey, J. Van Alsten, S. Granick, Design of an apparatus to measure the shear response of ultrathin liquid films. Rev. Sci. Instrum. **62**(2), 463–473 (1991). https://doi.org/10.1063/1.1142089
41. H. Hu, G.A. Carson, S. Granick, Relaxation time of confined liquids under shear. Phys. Rev. Lett. **66**(21), 2758–2761 (1991). https://doi.org/10.1103/PhysRevLett.66.2758
42. Y. Zhu, S. Granick, Viscosity of interfacial water. Phys. Rev. Lett. **87**(9), 096104 (2001). https://doi.org/10.1103/PhysRevLett.87.096104
43. A.L. Demirel, S. Granick, Origins of solidification when a simple molecular fluid is confined between two plates. J. Chem. Phys. **115**(3), 1498–1512 (2001). https://doi.org/10.1063/1.1380207
44. S. Granick, H.-W. Hu, Nanorheology of confined polymer melts. 1. Linear shear response at strongly adsorbing surfaces. Langmuir **10**(10), 3857–3866 (1994). https://doi.org/10.1021/la00022a076
45. S. Granick, H.-W. Hu, G.A. Carson, Nanorheology of confined polymer melts. 2. Nonlinear shear response at strongly adsorbing surfaces. Langmuir **10**(10), 3867–3873 (1994). https://doi.org/10.1021/la00022a077
46. J. Van Alsten, S. Granick, Shear rheology in a confined geometry: polysiloxane melts. Macromolecules **23**(22), 4856–4862 (1990). https://doi.org/10.1021/ma00224a014
47. Y. Zhu, S. Granick, Biolubrication: hyaluronic acid and the influence on its interfacial viscosity of an antiinflammatory drug. Macromolecules **36**(4), 973–976 (2003). https://doi.org/10.1021/ma025988r
48. M. Ruths, S.A. Sukhishvili, S. Granick, Static and dynamic forces between adsorbed polyelectrolyte layers (quaternized poly-4-vinylpyridine). J. Phys. Chem. B **105**(26), 6202–6210 (2001). https://doi.org/10.1021/jp010182z
49. J. Klein, D. Perahia, S. Warburg, Forces between polymer-bearing surfaces undergoing shear. Nature **352**(6331), 143–145 (1991). https://doi.org/10.1038/352143a0
50. U. Raviv, R. Tadmor, J. Klein, Shear and frictional interactions between adsorbed polymer layers in a good solvent. J. Phys. Chem. B **105**(34), 8125–8134 (2001). https://doi.org/10.1021/jp0041860
51. R. Tadmor, J. Janik, J. Klein, Sliding friction with polymer brushes. Phys. Rev. Lett. **91**(11), 115503 (2003). https://doi.org/10.1103/PhysRevLett.91.115503
52. U. Raviv, S. Giasson, N. Kampf, J.-F. Gohy, R. Jérôme, J. Klein, Lubrication by charged polymers. Nature **425**(6954), 163–165 (2003). https://doi.org/10.1038/nature01970
53. E. Eiser, J. Klein, T. Witten, L. Fetters, Shear of telechelic brushes. Phys. Rev. Lett. **82**(25), 5076–5079 (1999). https://doi.org/10.1103/PhysRevLett.82.5076
54. J. Klein, E. Kumacheva, D. Perahia, D. Mahalu, S. Warburg, Interfacial sliding of polymer-bearing surfaces. Faraday Discuss. **98**, 173–188 (1994). https://doi.org/10.1039/fd9949800173
55. J. Klein, E. Kumacheva, Simple liquids confined to molecularly thin layers. I. Confinement-induced liquid-to-solid phase transitions. J. Chem. Phys. **108**(16), 6996–7009 (1998). https://doi.org/10.1063/1.476114.
56. U. Raviv, P. Laurat, J. Klein, Fluidity of water confined to subnanometre films. Nature **413**(6851), 51–54 (2001). https://doi.org/10.1038/35092523
57. R.A. Quon, J.M. Levins, T.K. Vanderlick, High-precision automated measurement of surface separation for the surface forces apparatus. J. Colloid Interface Sci. **171**(2), 474–482 (1995). https://doi.org/10.1006/jcis.1995.1205
58. T. Grünewald, C.A. Helm, Computer-controlled experiments in the surface forces apparatus with a CCD-spectrograph. Langmuir **12**(16), 3885–3890 (1996). https://doi.org/10.1021/la960002q

59. M. Zäch, J. Vanicek, M. Heuberger, The extended surface forces apparatus. Part III. High-speed interferometric distance measurement. Rev. Sci. Instrum. **74**(1), 260–266 (2003). https://doi.org/10.1063/1.1527199
60. D.F. Kienle, T.L. Kuhl, Analyzing refractive index profiles of confined fluids by interferometry. Anal. Chem. **86**(23), 11860–11867 (2014). https://doi.org/10.1021/ac503469x
61. L. Qian, G. Luengo, D. Douillet, M. Charlot, X. Dollat, E. Perez, New two-dimensional friction force apparatus design for measuring shear forces at the nanometer scale. Rev. Sci. Instrum. **72**(11), 4171 (2001). https://doi.org/10.1063/1.1412860
62. K. Kristiansen, X. Banquy, H. Zeng, E. Charrault, S. Giasson, J. Israelachvili, Measurements of anisotropic (off-axis) friction-induced motion. Adv. Mater. **24**(38), 5236–5241 (2012). https://doi.org/10.1002/adma.201201566
63. R. Jin, I.-C. Liao, C. Cazeneuve, J.C. Chang, M. Ruths, G.S. Luengo, Effects of imprinted 3D surface patterning on localized changes in the tribology of human stratum corneum. Langmuir **35**(48), 15573–15584 (2019). https://doi.org/10.1021/acs.langmuir.9b01974
64. D.D. Lowrey, K. Tasaka, J.H. Kindt, X. Banquy, N. Belman, Y. Min, N.S. Pesika, G. Mordukhovich, J.N. Israelachvili, High-speed friction measurements using a modified surface forces apparatus. Tribol. Lett. **42**(1), 117–127 (2011). https://doi.org/10.1007/s11249-011-9746-1
65. J.N. Israelachvili, S.J. Kott, L.J. Fetters, Measurements of dynamic interactions in thin films of polymer melts: the transition from simple to complex behavior. J. Polym. Sci. Part B Polym. Phys. **27**(3), 489–502 (1989). https://doi.org/10.1002/polb.1989.090270302
66. A. Dhinojwala, S. Granick, New approaches to measure interfacial rheology of confined fluids. J. Chem. Soc. Faraday Trans. **92**(4), 619 (1996). https://doi.org/10.1039/ft9969200619
67. J.L. Parker, A novel method for measuring the force between two surfaces in a surface force apparatus. Langmuir **8**(2), 551–556 (1992). https://doi.org/10.1021/la00038a040
68. P. Frantz, N. Agrait, M. Salmeron, Use of capacitance to measure surface forces. 1. Measuring distance of separation with enhanced spatial and time resolution. Langmuir **12**(13), 3289–3294 (1996). https://doi.org/10.1021/la960014b
69. A.M. Stewart, Capacitance dilatometry attachment for a surface-force apparatus. Meas. Sci. Technol. **11**(3), 298–304 (2000). https://doi.org/10.1088/0957-0233/11/3/319
70. F. Restagno, J. Crassous, E. Charlaix, C. Cottin-Bizonne, M. Monchanin, A new surface forces apparatus for nanorheology. Rev. Sci. Instrum. **73**(6), 2292 (2002). https://doi.org/10.1063/1.1476719
71. L. Garcia, C. Barraud, C. Picard, J. Giraud, E. Charlaix, B. Cross, A micro-nano-rheometer for the mechanics of soft matter at interfaces. Rev. Sci. Instrum. **87**(11) (2016). https://doi.org/10.1063/1.4967713
72. S.H.J. Idziak, I. Koltover, J.N. Israelachvili, C.R. Safinya, Structure in a confined smectic liquid crystal with competing surface and sample elasticities. Phys. Rev. Lett. **76**(9), 1477–1480 (1996). https://doi.org/10.1103/PhysRevLett.76.1477
73. S.H. Idziak, C.R. Safinya, R.S. Hill, K.E. Kraiser, M. Ruths, H.E. Warriner, S. Steinberg, K.S. Liang, J.N. Israelachvili, The X-ray surface forces apparatus: structure of a thin smectic liquid crystal film under confinement. Science **264**(5167), 1915–1918 (1994). https://doi.org/10.1126/science.264.5167.1915
74. Y. Golan, A. Martin-Herranz, Y. Li, C. Safinya, J. Israelachvili, Direct observation of shear-induced orientational phase coexistence in a lyotropic system using a modified X-ray surface forces apparatus. Phys. Rev. Lett. **86**(7), 1263–1266 (2001). https://doi.org/10.1103/PhysRevLett.86.1263
75. O.H. Seeck, H. Kim, D.R. Lee, D. Shu, I.D. Kaendler, J.K. Basu, S.K. Sinha, Observation of thickness quantization in liquid films confined to molecular dimension. Europhys. Lett. **60**(3), 376–382 (2002). https://doi.org/10.1209/epl/i2002-00274-6
76. H. Weiss, H.W. Cheng, J. Mars, H. Li, C. Merola, F.U. Renner, V. Honkimäki, M. Valtiner, M. Mezger, Structure and dynamics of confined liquids: challenges and perspectives for the X-ray surface forces apparatus. Langmuir **35**(51), 16679–16692 (2019). https://doi.org/10.1021/acs.langmuir.9b01215

77. P. Kékicheff, J. Iss, P. Fontaine, A. Johner, Direct measurement of lateral correlations under controlled nanoconfinement. Phys. Rev. Lett. **120**(11), 118001 (2018). https://doi.org/10.1103/PhysRevLett.120.118001
78. P. Mächtle, C. Müller, C.A. Helm, A thin absorbing layer at the center of a Fabry-Pérot interferometer. J. Phys. II **4**(3), 481–500 (1994). https://doi.org/10.1051/jp2:1994139
79. P. Frantz, F. Wolf, X.-D. Xiao, Y. Chen, S. Bosch, M. Salmeron, Design of surface forces apparatus for tribology studies combined with nonlinear optical spectroscopy. Rev. Sci. Instrum. **68**(6), 2499–2504 (1997). https://doi.org/10.1063/1.1148148
80. A. Mukhopadhyay, J. Zhao, S.C. Bae, S. Granick, An integrated platform for surface forces measurements and fluorescence correlation spectroscopy. Rev. Sci. Instrum. **74**(6), 3067–3072 (2003). https://doi.org/10.1063/1.1570947
81. S.C. Bae, H. Lee, Z. Lin, S. Granick, Chemical imaging in a surface forces apparatus: confocal Raman spectroscopy of confined poly(dimethylsiloxane). Langmuir **21**(13), 5685–5688 (2005). https://doi.org/10.1021/la050233+
82. S.C. Bae, Z. Lin, S. Granick, Conjugated polymers confined and sheared: photoluminescence and absorption dichroism in a surface forces apparatus. Macromolecules **38**(22), 9275–9279 (2005). https://doi.org/10.1021/ma0510035
83. H.B. De Aguiar, J.D. McGraw, S.H. Donaldson, Interface-sensitive Raman microspectroscopy of water via confinement with a multimodal miniature surface forces apparatus. Langmuir **35**(48), 15543–15551 (2019). https://doi.org/10.1021/acs.langmuir.9b01889
84. J. Fréchette, T.K. Vanderlick, Double layer forces over large potential ranges as measured in an electrochemical surface forces apparatus. Langmuir **17**(24), 7620–7627 (2001). https://doi.org/10.1021/la011087k
85. S. Fujii, M. Kasuya, K. Kurihara, Characterization of platinum electrode surfaces by electrochemical surface forces measurement. J. Phys. Chem. C **121**(47), 26406–26413 (2017). https://doi.org/10.1021/acs.jpcc.7b09301
86. H. Zeng, Y. Tian, T.H. Anderson, M. Tirrell, J.N. Israelachvili, New SFA techniques for studying surface forces and thin film patterns induced by electric fields. Langmuir **24**(4), 1173–1182 (2008). https://doi.org/10.1021/la7017242
87. R. Tivony, D.B. Yaakov, G. Silbert, J. Klein, Direct observation of confinement-induced charge inversion at a metal surface. Langmuir **31**(47), 12845–12849 (2015). https://doi.org/10.1021/acs.langmuir.5b03326
88. C. Drummond, Electric-field-induced friction reduction and control. Phys. Rev. Lett. **109**(15), 154302 (2012). https://doi.org/10.1103/PhysRevLett.109.154302
89. P. Richetti, C. Drummond, J. Israelachvili, M. In, R. Zana, Inverted stick-slip friction. Europhys. Lett. **55**(5), 653–659 (2001). https://doi.org/10.1209/epl/i2001-00109-0
90. M. Ruths, N.A. Alcantar, J.N. Israelachvili, Boundary friction of aromatic silane self-assembled monolayers measured with the surface forces apparatus and friction force microscopy. J. Phys. Chem. B **107**(40), 11149–11157 (2003). https://doi.org/10.1021/jp0353946
91. C.A. Helm, J.N. Israelachvili, P.M. McGuiggan, Role of hydrophobic forces in bilayer adhesion and fusion. Biochemistry **31**(6), 1794–1805 (1992). https://doi.org/10.1021/bi00121a030
92. S. Yamada, J. Israelachvili, Friction and adhesion hysteresis of fluorocarbon surfactant monolayer-coated surfaces measured with the surface forces apparatus. J. Phys. Chem. B **102**(1), 234–244 (1998). https://doi.org/10.1021/jp972875e
93. G.G. Warr, Surfactant adsorbed layer structure at solid/solution interfaces: impact and implications of AFM imaging studies. Curr. Opin. Colloid Interface Sci. **5**(1–2), 88–94 (2000). https://doi.org/10.1016/S1359-0294(00)00038-8
94. J.L. Parker, D.L. Cho, P.M. Claesson, Plasma modification of mica: forces between fluorocarbon surfaces in water and a nonpolar liquid. J. Phys. Chem. **93**(16), 6121–6125 (1989). https://doi.org/10.1021/j100353a034
95. J.L. Parker, P.M. Claesson, D.L. Cho, A. Ahlberg, J. Tidblad, E. Blomberg, Plasma modification of mica. J. Colloid Interface Sci. **134**(2), 449–458 (1990). https://doi.org/10.1016/0021-9797(90)90155-H

96. C.R. Kessel, S. Granick, Formation and characterization of a highly ordered and well-anchored alkylsilane monolayer on mica by self-assembly. Langmuir **7**(3), 532–538 (1991). https://doi.org/10.1021/la00051a020
97. B. Liberelle, X. Banquy, S. Giasson, Stability of silanols and grafted alkylsilane monolayers on plasma-activated mica surfaces. Langmuir **24**(7), 3280–3288 (2008). https://doi.org/10.1021/la703522u
98. B. Liberelle, S. Giasson, Friction and normal interaction forces between irreversibly attached weakly charged polymer brushes. Langmuir **24**(4), 1550–1559 (2008). https://doi.org/10.1021/la702367f
99. C.S. Wang, R. Xie, S. Liu, S. Giasson, Tribological behavior of surface-immobilized novel biomimicking multihierarchical polymers: the role of structure and surface attachment. Langmuir **35**(48), 15592–15604 (2019). https://doi.org/10.1021/acs.langmuir.9b02018
100. J.M. Levins, T.K. Vanderlick, Extended spectral analysis of multiple beam interferometry: a technique to study metallic films in the surface forces apparatus. Langmuir **10**(7), 2389–2394 (1994). https://doi.org/10.1021/la00019a059
101. M.T. Clarkson, Multiple-beam interferometry with thin metal films and unsymmetrical systems. J. Phys. D Appl. Phys. **22**(4), 475–482 (1989). https://doi.org/10.1088/0022-3727/22/4/001
102. J.L. Parker, H.K. Christenson, Measurements of the forces between a metal surface and mica across liquids. J. Chem. Phys. **88**(12), 8013–8014 (1988). https://doi.org/10.1063/1.454260
103. C.P. Smith, M. Maeda, L. Atanasoska, H.S. White, D.J. McClure, Ultrathin platinum films on mica and the measurement of forces at the platinum/water interface. J. Phys. Chem. **92**(1), 199–205 (1988). https://doi.org/10.1021/j100312a044
104. N.A. Alcantar, C. Park, J.-M. Pan, J.N. Israelachvili, Adhesion and coalescence of ductile metal surfaces and nanoparticles. Acta Mater. **51**(1), 31–47 (2003). https://doi.org/10.1016/S1359-6454(02)00225-2
105. G. Vigil, Z. Xu, S. Steinberg, J. Israelachvili, Interactions of silica surfaces. J. Colloid Interface Sci. **165**(2), 367–385 (1994). https://doi.org/10.1006/jcis.1994.1242
106. R.G. Horn, D.R. Clarke, M.T. Clarkson, Direct measurement of surface forces between sapphire crystals in aqueous solutions. J. Mater. Res. **3**(03), 413–416 (1988). https://doi.org/10.1557/JMR.1988.0413
107. A. Berman, S. Steinberg, S. Campbell, A. Ulman, J. Israelachvili, Controlled microtribology of a metal oxide surface. Tribol. Lett. **4**(1), 43–48 (1998). https://doi.org/10.1023/A:1019134617374
108. P.M. McGuiggan, S.M. Hsu, W. Fong, D. Bogy, C.S. Bhatia, Friction measurements of ultra-thin carbon overcoats in air. J. Tribol. **124**(2), 239–244 (2002). https://doi.org/10.1115/1.1387035
109. S.J. Hirz, A.M. Homola, G. Hadziioannou, C.W. Frank, Effect of substrate on shearing properties of ultrathin polymer films. Langmuir **8**(1), 328–333 (1992). https://doi.org/10.1021/la00037a060
110. R.G. Horn, D.T. Smith, W. Haller, Surface forces and viscosity of water measured between silica sheets. Chem. Phys. Lett. **162**(4–5), 404–408 (1989). https://doi.org/10.1016/0009-2614(89)87066-6
111. M. Ruths, D. Johannsmann, J. Rühe, W. Knoll, Repulsive forces and relaxation on compression of entangled, polydisperse polystyrene brushes. Macromolecules **33**(10), 3860–3870 (2000). https://doi.org/10.1021/ma9920994
112. Y. Golan, N.A. Alcantar, T.L. Kuhl, J. Israelachvili, Generic substrate for the surface forces apparatus: deposition and characterization of silicon nitride surfaces. Langmuir **16**(17), 6955–6960 (2000). https://doi.org/10.1021/la000125h
113. M. Hegner, P. Wagner, G. Semenza, Ultralarge atomically flat template-stripped Au surfaces for scanning probe microscopy. Surf. Sci. **291**(1–2), 39–46 (1993). https://doi.org/10.1016/0039-6028(93)91474-4
114. L. Chai, J. Klein, Large area, molecularly smooth (0.2 nm rms) gold films for surface forces and other studies. Langmuir **23**(14), 7777–7783 (2007). https://doi.org/10.1021/la063738o

115. M. Valtiner, X. Banquy, K. Kristiansen, G.W. Greene, J.N. Israelachvili, The electrochemical surface forces apparatus: The effect of surface roughness, electrostatic surface potentials, and anodic oxide growth on interaction forces, and friction between dissimilar surfaces in aqueous solutions. Langmuir **28**(36), 13080–13093 (2012). https://doi.org/10.1021/la3018216
116. C.D. Van Engers, Z.D. Lamberty, P.M. McGuiggan, J. Frechette, Template-stripped ultra-smooth aluminum films (0.2 nm RMS) for the surface forces apparatus. Langmuir **37**(21), 6556–6565 (2021). https://doi.org/10.1021/acs.langmuir.1c00899
117. P. Vialar, P. Merzeau, E. Barthel, S. Giasson, C. Drummond, Interaction between compliant surfaces: how soft surfaces can reduce friction. Langmuir **35**(48), 15723–15728 (2019). https://doi.org/10.1021/acs.langmuir.9b02384
118. J.M. Skotheim, L. Mahadevan, Soft lubrication: the elastohydrodynamics of nonconforming and conforming contacts. Phys. Fluids **17**(9), 092101 (2005). https://doi.org/10.1063/1.1985467
119. A. Pandey, S. Karpitschka, C.H. Venner, J.H. Snoeijer, Lubrication of soft viscoelastic solids. J. Fluid Mech. **799**, 433–447 (2016). https://doi.org/10.1017/jfm.2016.375
120. P. Vialar, P. Merzeau, S. Giasson, C. Drummond, Compliant surfaces under shear: elastohydrodynamic lift force. Langmuir **35**(48), 15605–15613 (2019). https://doi.org/10.1021/acs.langmuir.9b02019

Ultrasonic Atomic Force Microscopies in Nanotribology

Chengfu Ma and Walter Arnold

Abstract In this chapter ultrasonic atomic force microscopy (AFM) techniques are discussed, which are dynamic AFMs working in contact mode that combine the excitation and detection of ultrasonic vibrations. Ultrasonic AFMs are widely used for quantitative mechanical property measurements and for non-destructive subsurface imaging. Here, we concentrate on the applications of ultrasonic AFMs in nanotribology studies. We will first introduce the working principles of ultrasonic AFMs, and then describe their applications in measuring surface properties and friction. Finally, we will summarize the use of ultrasonic AFMs to study and induce friction reduction and wear elimination.

1 Introduction

Since its invention in 1986 [1], atomic force microscopy (AFM) has been developed from a primarily topography imaging tool into instruments capable of measuring many physical properties of materials and components [2–5]. Among such instruments are ultrasonic AFM techniques, which comprise a number of methods that were developed in order to realize acoustic imaging with AFM [6–9]. According to their different schemes of exciting and detecting ultrasonic vibrations, ultrasonic AFMs are termed differently in the literature. These include ultrasonic force microscopy (UFM) [7], atomic force acoustic microscopy (AFAM) [10], ultrasonic atomic force microscopy (UAFM) [11], and heterodyne force microscopy (HFM) [9] which is also termed as scanning near-field ultrasound holography (SNFUH) elsewhere [12]. In order to improve the detection sensitivity, AFAM and UAFM are often operated at

C. Ma (✉)
State Key Laboratory of Solid Lubrication, Lanzhou Institute of Chemical Physics, Chinese Academy of Sciences, Lanzhou 730000, China
e-mail: chfuma@ustc.edu.cn

W. Arnold
Department of Materials Science and Engineering, Saarland University, 66123 Saarbrücken, Germany
e-mail: arnoldwk@ieee.org

E. Gnecco and E. Meyer (eds.), *Fundamentals of Friction and Wear on the Nanoscale*, NanoScience and Technology, https://doi.org/10.1007/978-3-031-63065-1_3

one of the cantilever resonances, while the tip is in contact with the sample. Therefore, they are also known as contact-resonance AFM (CR-AFM) [13].

Ultrasonic AFMs are widely used for the mapping of mechanical properties of sample surfaces [14–17]. In particular, due to the sensitivity of the contact resonances of the AFM cantilever to the contact stiffness and the contact damping between the tip and the sample, the elastic and viscoelastic properties of the material can quantitatively be measured [18–20]. Theoretical models describing the cantilever's contact-resonance characteristics have been developed for this purpose [21, 22]. In addition, ultrasonic AFMs also attract attention for their non-destructive subsurface imaging capabilities at the nanoscale. Imaging of embedded inclusions [12, 23], voids [24–26], and other structures [27, 28] has been demonstrated on well-prepared model samples. Their applications on detecting subsurface defects in materials [29], internal voids in integrated circuits [12], adhesive variations in interfaces [13], and subcellular structures [12] have been shown to be promising. Although the origin of subsurface imaging contrast in ultrasonic AFMs is still a matter of controversy, more evidence supports contact stiffness variations induced by subsurface features as the key subsurface contrast mechanism [30], instead of ultrasonic scattering. This especially holds for ultrasonic AFMs that work in frequency ranges over several tens of kHz to a few MHz. For quantitative mechanical property measurements and subsurface imaging with ultrasonic AFMs, we refer the reader to review articles in Refs. [21, 30, 31].

Although ultrasonic AFMs are mainly studied for surface and subsurface mechanical imaging as discussed above, they have also attracted extensive research interests from the nanotribology community. By laterally modulating the contact between the tip and the sample [32, 33], especially at the AFM cantilever's torsional resonances [34, 35], ultrasonic AFMs have been employed for friction measurements. This brings advantages for studying friction at much larger velocities (~ 1 mm/s) than in conventional friction force microscopy (FFM) (~ 100 μm/s) [33, 34], and allows friction measurement less affected by the surface topography [33, 34]. What inspires the research enthusiasm of using ultrasonic AFMs in nanotribology even more is the observation of an ultrasound induced lubricity effect [36]. Friction between the AFM tip and the sample has been shown both at the nano- [36–41] and atomic scales [40, 42–45] to be significantly reduced or, has even totally vanished by applying ultrasonic vibrations. Such an effect is therefore also termed as sono-lubricity [38] or dynamic superlubricity [43, 46]. Furthermore, wear at the nanoscale has been shown to be reduced by ultrasonic modulation, as evidenced by the elimination of wear of a tip sliding on a polymer [47] and by the suppression of the formation of periodic ripples on polymer surfaces induced by scanning AFM tips [48, 49].

In this chapter, we will describe the applications of ultrasonic AFMs in nanotribology studies. We will first introduce in Sect. 2 the principles of operations of ultrasonic AFMs, and then describe in Sect. 3 their applications in friction measurement. In Sects. 4 and 5, we will summarize the studies respectively on ultrasonic vibration induced friction reduction and wear elimination.

2 Ultrasonic Atomic Force Microscopy Techniques

According to their excitation and detection schemes, a number of ultrasonic AFM techniques have been developed and termed differently as mentioned above. In this section, we first describe the working principles of conventional ultrasonic AFM methods that employ the flexural (bending) vibrations of the cantilever for either qualitative or quantitative mechanical characterization of samples and for subsurface imaging. Then, we introduce ultrasonic AFMs that modulate the tip-sample contact laterally in-plane with ultrasonic vibrations, especially those variants that detect the ultrasonic vibrations by the cantilever torsional resonances.

As can be seen in Fig. 1a, ultrasonic AFMs are generally implemented using the contact mode. In this mode, the tip is brought into contact with the sample surface and probes it in a raster scanning manner in the x–y plane. The lateral scanning is realized with the help of piezoelectric scanners which drives either the cantilever or the sample. The tip load or the contact force is commonly quantified by measuring the vertical deflection of the cantilever by the optical lever method with a four-quadrant photodiode detector and by knowing the spring constant of the cantilever. During scanning, the contact force is kept constant at a setpoint value via a feedback loop by adjusting the z-scanner that drives the cantilever support to approach or to retract from the sample surface. In such a way, the topography of the sample surface is mapped from the feedback signal at each scanning position. Besides monitoring the flexural deflection of the cantilever, the torsion of the cantilever can also be measured with the photodiode detector by reading the laser spot's lateral displacement. This led to the implementation in most commercial AFMs [50, 51] of an important AFM mode, the lateral force microscopy (LFM) or friction force microscopy (FFM) [52]. In FFM, the in-plane friction force acting between the tip and the sample is mapped simultaneously with the topography imaging mode by measuring the torsion of the AFM cantilever. While the tip can be scanned in any direction for topography imaging, in FFM it slides over the sample surface in the direction that is perpendicular to the long axis of the cantilever. By using FFM, a variety of friction phenomena has been investigated in different environments on a wide range of surfaces at the nano- and even atomic scales [50–55].

While the contact mode AFM, including FFM, is a quasi-static method, ultrasonic AFMs are dynamic modes. As can be seen in Fig. 1a, ultrasonic excitations are applied to modulate the tip-sample contact either via the cantilever or the sample. The direct excitation of the cantilever can be achieved by the piezo shaker in the cantilever holder installed in most commercial AFMs for intermittent-contact imaging. The excitation can also be applied by attaching an external transducer beneath the sample. Both the piezo shaker and the transducer typically have center frequencies of a few MHz, and are excited by built-in or external function generators. The techniques are often known as UAFM (cantilever excitation) [11] or AFAM (sample excitation) [10]. The ultrasonic waves transmit to the tip-sample contact area and force the cantilever to vibrate, as schematically illustrated in Fig. 1b. The cantilever oscillation is then detected by the photodiode, and analyzed by a lock-in amplifier which outputs the

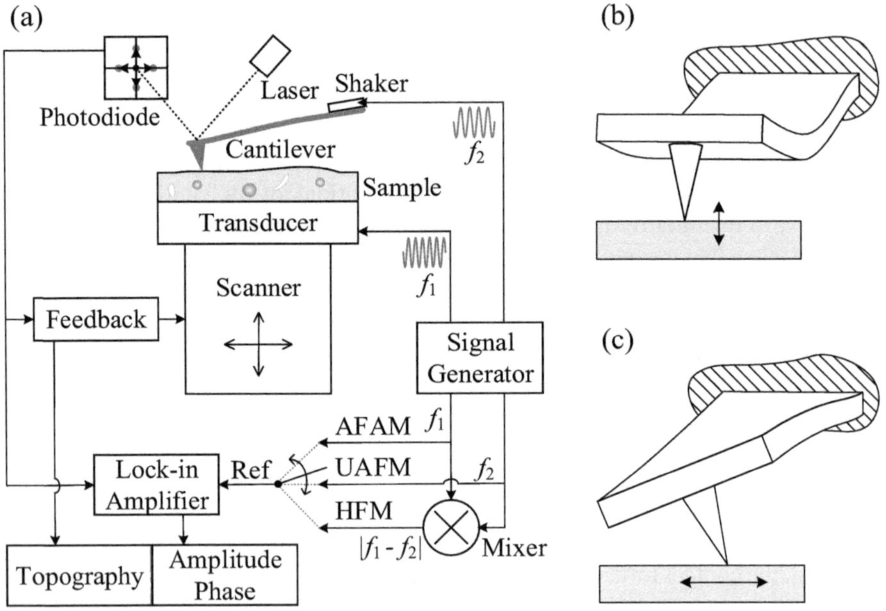

Fig. 1 **a** Principle of the setup for ultrasonic AFMs; **b**, **c** schematic illustrations of respectively the flexural and torsional oscillations of AFM cantilevers induced by vertical and lateral excitations, while the tips are coupled to sample surfaces. (**a**) is reproduced with permission from Ma et al., J. Appl. Phys. 128, 180901 (2020) [30] with minimal changes. Copyright 2020 AIP Publishing LLC

amplitude and phase signals. Since the working frequencies in ultrasonic AFMs are much higher than the frequency of the feedback loop for topography imaging, amplitude and phase images are obtained in ultrasonic AFMs simultaneously with the topography image. To improve the signal-to-noise ratio, UAFM and AFAM imaging are usually operated at frequencies close to the contact resonances (CRs) of the cantilever. This requires spectroscopy measurements in order to determine the CRs which can be done by a frequency sweep or by analyzing the thermal noise of the cantilever [56]. For example, Fig. 2a shows flexural resonance spectra of an AFM cantilever excited by thermal noise while its tip was respectively free and in contact with a silicon wafer. The cantilever's spring constant was calibrated to be approx. 5.4 N/m by using the thermal noise method [57]. A tip load of approx. 173 nN was applied while measuring on the silicon wafer. The first three flexural free resonance eigenmodes (FR^1, FR^2, FR^3) of the cantilever and its first flexural CR eigenmode (CR^1) on the wafer can be identified. The resonance of the cantilever is significantly shifted by stiffening due to the tip-sample contact. By fitting the frequency and quality factor Q values of the CRs, the elastic and viscoelastic properties of the sample can be quantified by employing a realistic CR model for the cantilever considering its geometry and the forces acting at its tip, as well as the contact mechanics for the tip-sample contact [21, 31]. Such quantitative measurements are also called CR-AFM [13]. However, unlike fast qualitative UAFM/AFAM amplitude and phase imaging

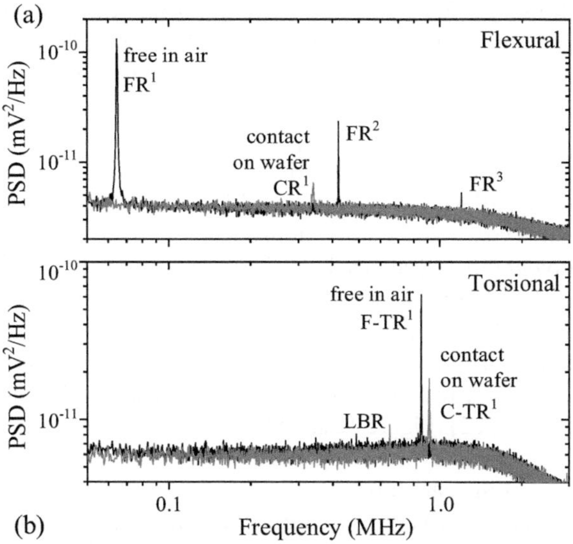

Fig. 2 **a** Flexural and **b** torsional resonance spectra of an AFM cantilever with respectively the tip being free in air and being coupled to a silicon wafer. The flexural and torsional power spectral density (PSD) spectra were obtained by analyzing the thermal noise of the cantilever in the vertical and lateral outputs of the photodiode

at fixed frequencies, quantitative CR-AFM spectroscopy measurement by frequency sweep is relatively slow. It is therefore not practical to obtain CR frequency and *Q* maps with many pixels by such a method. Instead, measurements are often carried out only at a few points of interest. To address this issue, resonance tracking [58] or band excitation methods [59] have been proposed which however need additional electronics.

As mentioned above, UAFM/AFAM are sensitive to the elastic and viscoelastic properties of sample surfaces [15–17] which can be quantified by CR measurements [18–20]. The underlying mechanism is that materials having different elastic and viscoelastic properties alter the boundary conditions in the tip-sample contact, i.e., the real and imaginary part (damping) of the contact stiffness [21, 31]. This leads to frequency shifts of the cantilever resonances and changes of their *Q* values, and in case of nonlinear contacts to deviations from Lorentz curves. This also applies when mechanically inhomogeneous features, including particles and voids, exist below the sample surface (subsurface). They alter the effective contact stiffnesses sensed by the AFM tip while indenting, and are therefore revealed [30].

Unlike UAFM and AFAM which work at one of the cantilever's CRs, UFM is operated away from the CR resonances, usually at higher frequencies of a few MHz. In this case the cantilever cannot follow the sample vibration (dynamically "frozen"), but indents the sample effectively. Therefore, instead of detecting the cantilever's CRs or amplitude and phase, UFM detects the additional cantilever deflection caused by the nonlinear tip-sample force-distance curve [7, 14, 60]. A further method that also exploits the nonlinear tip-sample interaction is HFM [9], which is also known as SNFUH [12]. In HFM, both the cantilever and the sample are excited by ultrasonic waves at slightly different frequencies (usually in the MHz range), as illustrated in Fig. 1a. The nonlinear tip-sample interaction produces an oscillatory signal at the difference frequency of the two ultrasonic waves which is called in analogy to radio technology the heterodyne signal. The amplitude and phase of the heterodyne signal are monitored with the lock-in amplifier by using the mixed frequency signal as reference.

In addition to out-of-plane vibrations, in-plane lateral vibrations can also be applied to modulate the tip-sample contact. This can be achieved by the sample, the cantilever or by both, just like the normal excitation methods described above. However, a shear-wave transducer [33, 61, 62] or a specially designed cantilever holder [35, 63] are needed to induce respectively the in-plane vibrations of the sample surface or the lateral oscillations of the AFM tip. According to the different polarizations of the in-plane excitations relative to the long axis of the cantilever as well as due to different excitation frequencies, different vibrational modes of the cantilever can be excited which include vertical bending [32], lateral bending [64], and torsion [33, 61, 62]. Figure 1c shows schematically the excitation of torsional vibrations of a cantilever by lateral displacements of the sample that are perpendicular to the long axis of the cantilever.

Except in very few cases where the in-plane modulations are detected from the vertical bending [32] or lateral bending [64] vibrations of the cantilever, the torsional [33, 61, 62] vibrations are commonly used and the corresponding method is often termed as torsional resonance (TR) AFM [34, 35, 63, 65]. This is due to the fact that the optical lever detection method used in most commercial AFMs is designed to detect the torsion of the cantilever in addition to its deflection. By monitoring the lateral and vertical movements of the laser spot on the photodiode as illustrated in Fig. 1a, the torsion and the deflection of the cantilever can be simultaneously measured. For example, in Fig. 2b we show the torsional resonance spectra of a cantilever that were obtained simultaneously with the flexural resonance spectra in Fig. 2a. The first free ($F\text{-}TR^1$) and contact TR ($C\text{-}TR^1$) on the silicon wafer can be identified. In addition, a small resonance peak (LBR) can be seen on the contact TR spectra which is probably due to a coupled mode with a lateral bending resonance caused by an asymmetric tip position [63].

Just like the out-of-plane ultrasonic AFM techniques, in-plane ultrasonic AFMs also work in two modes: in the spectroscopy mode (Fig. 2b) and the imaging mode. They too allow amplitude and phase imaging at a fixed operation frequency, as well as quantitative imaging of the TR frequencies and Q values. It should be mentioned that although we introduced the out-of-plane and the in-plane ultrasonic AFM techniques separately, it occurs commonly in experiments that out-of-plane ultrasonic vibrations influence the in-plane response (torsional) signal and vice versa. Moreover, the out-of-plane ultrasonic AFM measurements may exhibit coupling effects from in-plane vibrations or vice versa due to the coupling between different eigenmodes of the cantilever beam [33, 63].

3 In-Plane Surface Properties and Friction Measurement by Ultrasonic AFMs

The indentation modulus of a sample is determined when using vertical ultrasonic AFMs by evaluating the contact-resonance frequencies of the cantilever. When the tip is in contact with a sample, the out-of-plane cantilever oscillations exert a periodic normal force and hence modulates the indention depth of the tip. Quantitatively, the normal contact stiffness and contact damping (*Q*-value) can be quantified, which are related to the elastic and viscoelastic properties of the sample [21, 31]. The resulting amplitude and phase images reveal surface and subsurface mechanical heterogeneities of the sample with very high sensitivity. Analogously, lateral ultrasonic AFMs, which modulate the shear interaction between the tip and sample, allowing to quantify the in-plane mechanical properties of the sample. However, attention must be paid to the tip-sample contact condition while the modulation is applied. Depending on the applied lateral modulation amplitude and tip load, the AFM tip may stick to the contact point or slide locally over the sample [32–34, 61, 66–73]. In the two different regimes, either local viscoelastic or surface friction forces dominate the tip-sample interaction, leading to different cantilever responses and therefore to the possibility to characterize in-plane viscoelastic properties or surface friction.

In lateral ultrasonic AFMs, the tip, which is pinned to the sample surface, follows the in-plane motion due to the applied load. This leads to a torsional motion of the cantilever when the in-plane polarization of the sample vibration is perpendicular to its long axis [34, 61, 66, 71] as illustrated in Fig. 1c, respectively to buckling when the sample is vibrated parallel to its long axis [32, 73]. When the vibration amplitude is small, the in-plane tip-sample interaction resists the motion of the tip relative to the sample surface and the tip sticks to the contact point. However, with increasing vibration amplitude, the deformation in the cantilever by torsion or buckling increases and it gets more and more difficult to hold the tip. At one point where the restoring spring force of the cantilever surpasses a critical value, oscillatory slipping of the tip starts and friction determines the tip-sample contact forces. Before the regime of total sliding, micro-slip may occur between the tip-sample contact [74].

The stick-to-sliding transition of the tip-sample contact in ultrasonic AFMs can be identified from the cantilever's vibration amplitude [32, 33, 61, 67]. For example, as can be seen in Fig. 3a, Scherer et al. [61] measured the torsional vibration envelope of an AFM cantilever with its tip in contact with a sample surface that is oscillated in-plane and perpendicular to the length of the cantilever. Two silicon wafers, one dry and the other coated with the polar lubricant Z-Dol, were examined. It can be seen that while ramping the driving amplitude linearly at a frequency of 370 kHz, the cantilever's torsional amplitudes grew linearly at first for both samples. However, after the critical points marked as S1 and Z1 respectively for the two samples, the behavior changed, and the cantilever amplitude became almost independent of the surface amplitudes. The critical points indicate where the stick-to-sliding transition happens, yielding values of the static frictions.

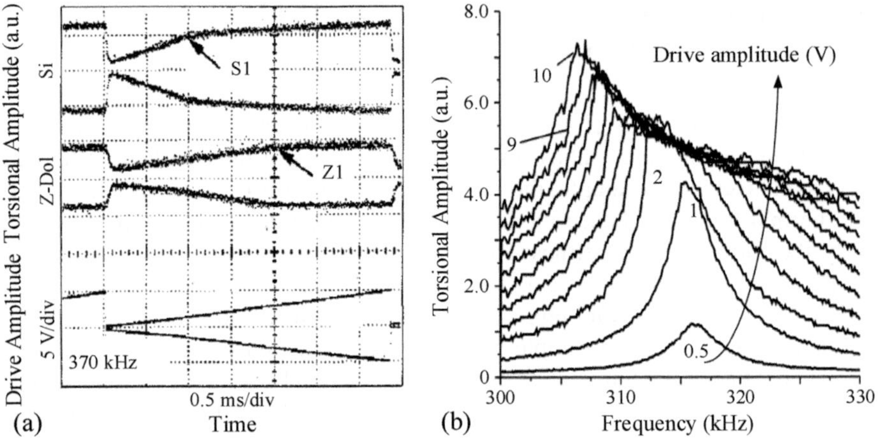

Fig. 3 **a** Torsional amplitudes of a cantilever with the tip in contact with a silicon substrate lubricated with Z-Dol as a function of the linearly increasing driving amplitude in comparison to a bare silicon substrate (Si). The frequency was 370 kHz; **b** Torsional resonance spectra of a cantilever measured for different driving amplitudes with its tip in contact with a fused silica sample. (**a**) is reproduced with permission from Scherer et al., IEEE Trans. Magn. 33(5), 4077–4079 (1997) [61] with minimal changes. Copyright 1997 IEEE. (**b**) is reproduced with permission from Reinstädtler et al., Tribol. Int. 38, 533–541 (2005) [72] with minimal changes. Copyright 2005 Elsevier Ltd.

Whether the tip sticks to or slides over the sample under an ultrasonic excitation can also be discerned from the nonlinear behavior of the cantilever's resonances [34, 70–73]. For instance, Fig. 3b shows the torsional resonance spectra of a cantilever under different excitation amplitudes measured by Reinstädtler et al. on a fused silica sample [72]. The excitations are applied from the sample's underside and are polarized in-plane perpendicular to the long axis of the cantilever. It can be observed that at low excitation amplitudes the resonances are Lorentzian. Under these conditions the AFM tip sticks to the sample and the tip-sample interaction is mainly determined by viscoelasticity. However, above a critical excitation amplitude, the tip starts to slide and a friction force develops at the tip-sample interface. This results in the deviations of the cantilever's resonances from the Lorentzian shape showing nonlinear plateaus. In fact, a similar nonlinear behavior was also observed in vertical ultrasonic AFMs [74], where large out-of-plane excitations led to in-plane stick-to-sliding transition of the tip in the direction of the length of the cantilever due to its buckling. Studies by both in-plane [35] and out-of-plane [74] ultrasonic AFMs show that by increasing the applied tip load, the critical amplitude to the onset of tip sliding increases. This is expected, considering that for larger applied tip loads, larger static friction forces are to be overcome to initiate sliding. Moreover, it should be mentioned that besides directly observing the shape of resonance curves, the nonlinearity induced by the stick-to-sliding of the tip can also be monitored from the appearance of higher harmonics in the cantilever oscillation [73].

By modulating the tip-sample contact in the sticking regime as described above, in-plane mechanical properties of the sample can be quantified [35, 65, 68, 69, 75, 76]. This involves the measurement of the lateral contact stiffness, which is defined as the amount of force divided by the lateral displacement. It can be measured by deducting the lateral stiffness of the cantilever from the total lateral stiffness of the system, where the latter can be measured in ultrasonic AFMs by the ratio of the lateral force amplitude to the lateral displacement amplitude after calibrating both of them [68, 69]. This can also be done by building the torsional resonance model of the cantilever which relates the lateral contact stiffness to the free and contact torsional resonance frequencies [35, 65, 75, 76]. Then, from the measured lateral contact stiffness, elastic properties of the sample can be measured such as the shear modulus. Furthermore, Mazeran and Loubet [69] estimated the Young's moduli of carbon fibers and the host epoxy matrix in a composite as shown in Fig. 4a by measuring the lateral contact stiffnesses from an in-plane modulation image. The lateral contact stiffness was obtained by multiplying the normal contact stiffness with a given value, here 4/5, which is based on the Derjaguin–Muller–Toporov (DMT) theory using a Poisson's ratio of 1/3. Hurley and Turner [75] developed a method to quantitatively measure Poisson's ratio, shear modulus, and Young's modulus by measuring the normal and lateral contact stiffnesses from the flexural and torsional cantilever CR frequencies. In the same way, by considering not only the CR frequencies, but also the Q values of the cantilever resonances, Rezaei and Turner [76] showed how to determine the indentation and shear loss tangents of the sample.

By applying a modulation amplitude large enough to induce sliding in the tip-sample contact, one can obtain images representing the frictional properties of sample surfaces [32–34, 66, 67, 69, 73]. For example, as shown in Fig. 4c, the in-plane modulation amplitude obtained at a low excitation amplitude (sticking regime) shows the contrast of the elastic properties of the sample, although not quantified here. The sample is a composite material composed of carbon fibers and an epoxy matrix, which have a nominal Young's moduli of 40 GPa and 3 GPa, respectively [69]. When imaging in the sliding regime by applying a larger excitation amplitude, the resulting amplitude image in Fig. 4d shows an opposite contrast which corresponds to the different friction properties of the two materials. This can be confirmed by the similar contrast of a quantitative FFM friction map as can be seen in Fig. 4b. Compared to conventional FFM, in-plane ultrasonic AFM allows friction measurements at tip-sample sliding velocities of the order of 1 mm/s, which is much higher than that of FFM (~ 100 μm/s) [33, 34]. In addition, the technique of friction imaging by in-plane ultrasonic AFM is independent of the scanning direction and is less sensitive to topography induced effects [32–34, 66, 69]. Indeed, a study done by Bhushan and Kasai [34] on a silicon ruler sample with a slope angle of $\sim 45°$ has shown that the topography-induced contribution in the friction force was 12% for the torsional resonance AFM mode, but was 85% for the conventional FFM mode.

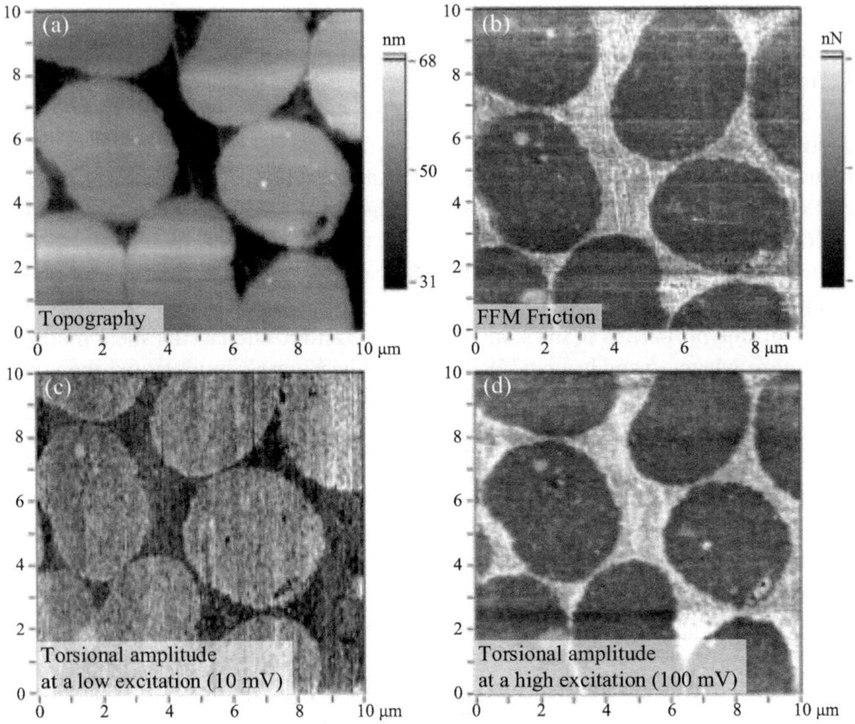

Fig. 4 **a** AFM topography and **b** FFM measured friction (trace-retrace) images of a composite sample made of carbon fibers and an epoxy matrix; **c**, **d** In-plane modulation images (cantilever torsional amplitude) obtained at a driving frequency of 37 kHz with an excitation amplitude of respectively 10 mV and 100 mV. Reproduced with permission from Mazeran and Loubet, Tribol. Lett. 7, 199–212 (1999) [69] with minimal changes. Copyright 1999 J.C. Baltzer AG, Science Publishers

4 Friction Reduction by Ultrasonic Vibration

Controlling friction, in particular reducing friction, is of great interest at all length scales. In the macroscopic world, one uses liquid or solid lubricants to reduce friction. In small-scale devices, such as in micro- and nanoelectromechanical systems (MEMS/NEMS), the much larger surface-to-volume ratios lead to the domination of surface forces. i.e., friction and adhesion over actuation and restoring forces [77]. This results in serious wear problems, which influence the reliable functioning of devices and their lifetimes. However, at this scale, conventional lubricants are not applicable and alternative ways to control friction are highly desirable [78]. Among such methods are inducing mechanical vibrations in the friction pairs, which has long been studied and has been widely used for applications at the macroscale [79–82]. Ultrasound induced lubricity has also been observed at the nanoscale with ultrasonic AFM [36, 60, 83] and has since sparked widespread research interests [37–43, 45, 47,

49, 84–88] due to its prospects in controlling friction and eliminating wear in micro-/nano-devices. Here, we concentrate on studies utilizing ultrasonic AFMs in order to explore the impact of ultrasonic excitations on friction at the nano- and atomic scale. Studies as to how the ultrasonic vibrations affect the wear will be summarized in the next Section.

The reduction of a cantilever's torsional amplitude by applying out-of-plane ultrasonic modulation in an ultrasonic AFM was first observed by Yamanaka et al. [60]. This phenomenon was later studied in detail by Dinelli et al. [36] by combining FFM and UFM. As schematically illustrated in Fig. 5a, the lateral (friction) force between the tip-sample was measured by using FFM [50–52] where the tip scans forward and backward perpendicularly to the long axis of the cantilever. During the scanning, the cantilever torsion is monitored and a so-called friction loop is generated whose half height is a measure of the friction force. Meanwhile, the sample was excited into ultrasonic vibrations at MHz and the cantilever's deflection response was monitored by using UFM. The effect of the ultrasonic vibration amplitude on the friction force and the cantilever deflection was then studied. A representative result is shown in Fig. 5b. Here, the measurements were carried out in an ambient environment on a silicon sample with a scanning speed of 50 nm/s by using a silicon–nitride cantilever with a spring constant of 0.05 N/m. Two tip loads of respectively 0 nN (F_1) and 2 nN (F_2) were applied which was below the adhesion (pull-off) forces between 5 and 20 nN. Starting from zero ultrasonic amplitude, it can be seen that with increasing amplitude the friction force decreases at first only slowly. Then, once a critical ultrasonic amplitude, a_{c1} or a_{c2}, is reached, a large cantilever deflection is observed due to the strong nonlinearity of the tip-sample force-distance curve. After that, the friction force reduces rapidly and almost zero friction is finally reached. Based on their observations, the authors attribute the sudden friction reduction above the critical amplitude to the break of the tip-sample contact for a part of the ultrasonic cycle. This is evidenced by the accompanying static shift of the cantilever deflection. Its dependence on the applied tip load can be derived from an analysis based on the force-distance curve, see below.

The ultrasound induced lubricity has since been explored at the micro-/nanoscale by many groups on different materials by using different excitation schemes [37–41, 47, 49, 84, 85]. For example, Behme and Hesjedal [37, 84] studied the effect of friction loss on GaAs wafers by using surface acoustic waves and Love waves having frequencies of 200 MHz to 3 GHz with their k-vectors perpendicular to the length of the cantilever. Love waves which possess only in-plane displacement, showed no effect. Thus, the friction reduction effect is only due to the out-of-plane oscillations of the surface waves. The negligible effect of in-plane excitations on friction was also observed by Cao et al. [41]. However, a study by Shi et al. [40] shows that the friction force on HOPG samples can also be lowered by applying in-plane excitations provided resonance conditions are obeyed given by the Tomlinson model [89]. Pfahl et al. [38] employed out-of-plane cantilever excitations both with UAFM and AFAM in order to study the effect of friction reduction at a few hundred kHz near the CR of the cantilever. The friction force was reduced in the same manner in both excitation schemes showing a universal behavior as a function of cantilever amplitude

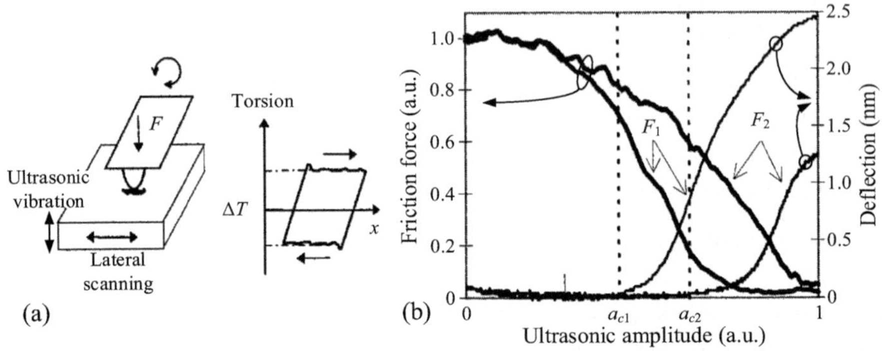

Fig. 5 **a** Schematic illustration for measuring friction in FFM while an out-of-plane ultrasonic excitation is applied to the sample; **b** measured dynamic friction and cantilever deflection as functions of ultrasonic amplitude on a silicon sample. The measurements were carried out at tip loads of respectively 0 nN (F_1) and 2 nN (F_2) with a sliding speed of 50 nm/s. The friction values are normalized. The ultrasonic amplitude was estimated to be at most between 10 and 20 Å. Reproduced with permission from Dinelli et al., Appl. Phys. Lett. 71(9), 1177 (1997) [36] with minimal changes. Copyright 1997 American Institute of Physics

regardless of the excitation frequency and the applied tip load. The investigations undertaken by Jeon et al. [85] and Shi et al. [40] demonstrate that friction reduction can also be achieved over frequencies of 1–100 kHz. From these studies, one can see that the ultrasound induced lubricity is a universal phenomenon over a wide range of frequencies. Some studies [36–38, 84] explain the ultrasonic vibration friction reduction by the non-linearity of the force-distance leading to a reduction of the average force exerted by the tip. This effect is called the mechanical diode effect. The diode effect leads to a reduction of adhesion and shear strength between the tip and the sample [39]. The diode effect has been examined in detail modeling the cantilever as a beam [90] and using an analytical expression of the force-distance curve by [91]. At low vibration amplitudes, Dinelli et al. [36] invoked another mechanism, namely that a viscoelastic layer develops a negative pressure and serves as a lubricant when losing its solid like behavior.

In atomic-scale friction, the ultrasonic vibration induced lubricity has also been demonstrated [40, 42, 45, 86–88]. A state of superlubricity, where the friction totally vanishes, was achieved by exciting the mechanical resonances of the sliding system [42]. This leads to the concept of dynamic superlubricity [43, 46, 92]. Socoliuc et al. modulated the tip-sample normal force through electro-capacitive actuation of the cantilever as illustrated in Fig. 6a. As can be seen in Fig. 6b, c, based on the Tomlinson model [89], a state of superlubricity could be achieved where the atomic-scale stick–slip instabilities are smoothed out when applying ultrasonic modulation. With increasing modulation amplitude, the average friction force dropped rapidly and vanished finally (Fig. 6d). The measurements were carried out in ultrahigh vacuum on an atomically flat NaCl surface at a tip load of 2.73 nN [42, 43]. An ac voltage with frequency 56.7 kHz was applied to excite the cantilever, which corresponds to

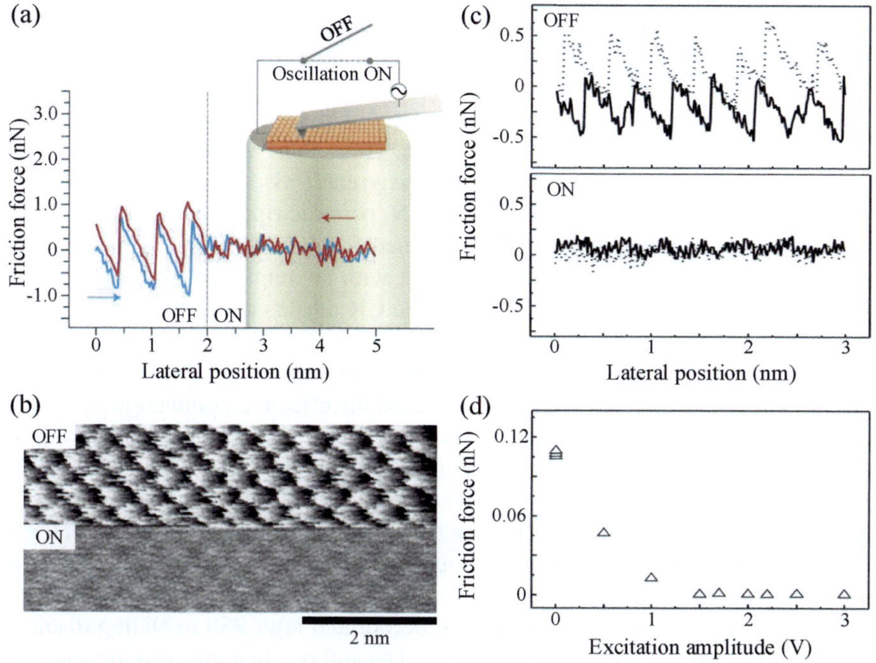

Fig. 6 **a** Principle of modulating the normal force between an AFM tip and a sample surface, as well as a recorded friction loop (forward and backward) showing atomic-scale stick–slip instabilities; **b** friction force map (backward) on a NaCl(100) surface at a normal load of 2.73 nN when the modulation is turned on and off; **c** friction loops on a NaCl(100) surface at a normal load of 2.73 nN without and with modulation; **d** measured friction force as a function of excitation amplitude. (**a**) is reproduced with permission from Carpick, Science 313(5784), 184–185 (2006) [77] with minimal changes. Copyright 2006 AAAS; (b) is reproduced with permission from Gnecco et al., Nanotechnology 20, 025501 (2009) [43] with minimal changes. Copyright 2009 IOP Publishing Ltd. (**c, d**) are reproduced with permission from Socoliuc et al., Science 313(5784), 207–210 (2006) [42] with minimal changes. Copyright 2006 AAAS

its first flexural CR [43]. Later, Gnecco et al. [43] have shown that such a dynamic superlubricity can also be realized under ambient conditions on other atomically flat surfaces. The work done by Roth et al. [45] further demonstrated that lateral vibrations can also lead to dynamic superlubricity.

5 Ultrasonic Vibration on Wear

Following the techniques of sono-lubricity or dynamic superlubricity as described above, the impact of mechanical oscillations on wear is also of much interest. A similar effect would provide a promising approach to reduce wear in small-scale devices, leading to an increase of their lifetime. This was indeed proven to be true by some studies [39, 47–49]. For example, Knoll et al. [48] and Pedraz et al. [49] have shown that the formation of ripples induced by scanning tips on polymer surfaces can be suppressed by mechanically modulating the tip-sample contact at ultrasonic frequencies. The formation of ripples can be frequently observed on polymer surfaces when scanning with an AFM in contact mode. This can be considered as a pre-stage of wear. For instance, while scanning a polystyrene (PS) surface in contact mode with a silicon tip under a tip load of 1.378 μN, periodic ripples with peak-to-valley amplitude of approximately 4 nm were formed for the first scanning (see Fig. 7). However, by exciting a mechanical vibration with an amplitude of 0.55 nm to the sample at 366 kHz, which corresponds to the CR of the cantilever, ripples are not formed, as can be seen in Fig. 7.

In addition, the impact of mechanical modulations on the abrasive wear were studied by Lantz et al. [47]. They demonstrated the effective elimination of wear of a tip sliding on a polymer surface over several hundred meters by modulating the tip–sample contact. Without any modulation, for a silicon tip used for writing and reading ultrahigh-density data on a polymer film in contact mode, it wears gradually with increasing sliding distance. This can be seen from the measured adhesion forces during the sliding process as shown in Fig. 8. The increase of adhesion indicates an enlarged tip apex, and thus the wear of the tip. This was confirmed by scanning electron microscopy (SEM) images of the tip before and after 750 m sliding distance. Here, a normal tip load of 5 nN was applied. The authors then repeated the measurements with another tip, however this time by applying a 500 kHz normal modulation to the tip-sample contact. From the results shown in Fig. 8, it can be seen that the adhesion remains essentially constant throughout the sliding process, indicating a

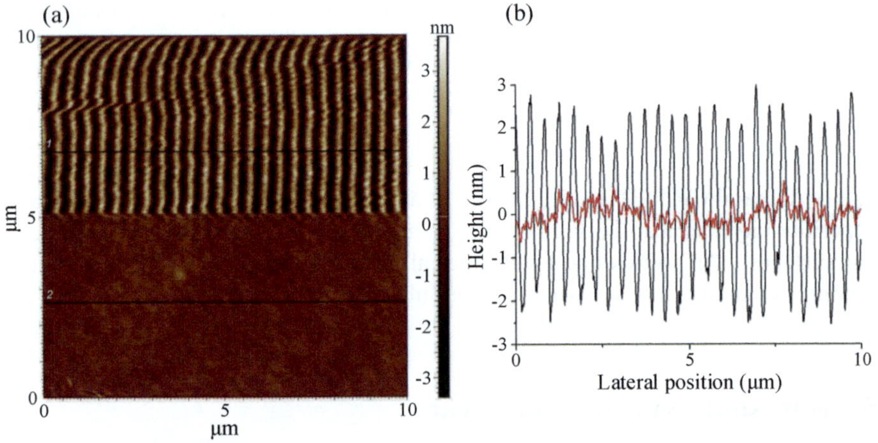

Fig. 7 **a** AFM topography image of a polystyrene surface while scanning in contact mode without (upper part) and with (lower part) excitation with a tip load of 1.378 μN; **b** height profiles along the scan lines in (**a**). An excitation amplitude of 0.55 nm was applied at 366 kHz while acquiring the lower part of the image. Reproduced with permission from Pedraz et al., ACS Nano 9(9), 8859–8868 (2015) [49] with minimal changes. Copyright 2015 American Chemical Society

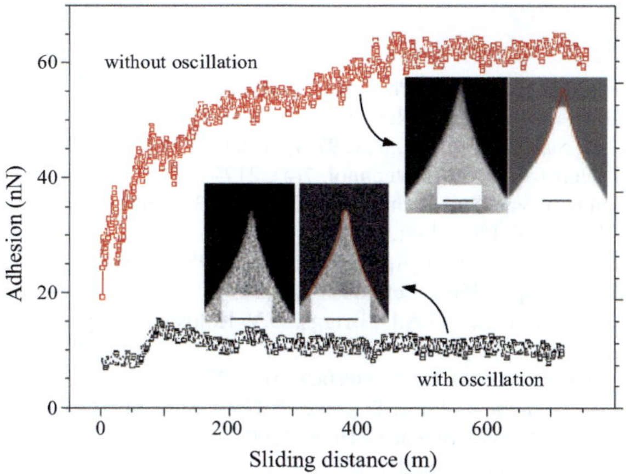

Fig. 8 Adhesion forces of AFM tips sliding on a polymer surface as a function of sliding distance, as well as SEM images of the tips before (left panels) and after (right panels) the scanning (insets; all scalebars: 100 nm). A tip load of 5 nN was applied for both measurements. A normal modulation at 500 kHz of the tip-sample contact was applied for the lower measurement. Reproduced with permission from Lantz et al., Nat. Nanotechnol. 4(9), 586–591 (2009) [47] with minimal changes. Copyright 2009 Macmillan Publishers Limited

stable tip. The SEM images of the tip taken before and after sliding showed negligible tip wear. With the same tip, it was demonstrated that data writing and reading at an areal density of 1 Tbit/inch^2 can be achieved with sufficient signal-to-noise ratio after the 750 m sliding.

6 Summary

In this chapter, we reviewed ultrasonic AFM techniques in nanotribology studies. First, the working principles and instruments of ultrasonic AFMs are introduced. Then their applications in measuring surface properties and friction are described. Finally, studies of friction reduction and wear elimination by using ultrasonic AFMs are reviewed. Besides these studies, there are also other applications of ultrasonic AFMs in nanotribology which we mention in short. They include, for instance, local adhesion hysteresis measurement [93, 94] as well as vibration assisted nanolithography [95, 96]. In summary, we showed that ultrasonic AFMs are important tools in nanotribology, not only for fundamental studies but also for practical applications.

Acknowledgements C.M. thanks for support by the National Natural Science Foundation of China (No. 52375556).

References

1. G. Binnig, C.F. Quate, C. Gerber, Phys. Rev. Lett. **56**(9), 930 (1986)
2. B. Cappella, G. Dietler, Surf. Sci. Rep. **34**(1–3), 1–104 (1999)
3. R.W. Carpick, M. Salmeron, Chem. Rev. **97**(4), 1163–1194 (1997)
4. R. Garcia, E.T. Herruzo, Nat. Nanotechnol. **7**(4), 217–226 (2012)
5. D. Sarid, V. Elings, J. Vac. Sci. Technol. B **9**(2), 431–437 (1991)
6. B. Cretin, F. Sthal, Appl. Phys. Lett. **62**, 829 (1993)
7. O. Kolosov, K. Yamanaka, Jpn. J. Appl. Phys. **32**, L1095 (1993)
8. U. Rabe, W. Arnold, Appl. Phys. Lett. **64**, 1493 (1994)
9. M.T. Cuberes, H.E. Assender, G.A.D. Briggs, O.V. Kolosov, J. Phys. D Appl. Phys. **33**, 2347 (2000)
10. U. Rabe, E. Kester, W. Arnold, Surf. Interface Anal. **27**, 386 (1999)
11. K. Yamanaka, A. Noguchi, T. Tsuji, T. Koike, T. Goto, Surf. Interface Anal. **27**, 600 (1999)
12. G.S. Shekhawat, V.P. Dravid, Science **310**, 89 (2005)
13. D.C. Hurley, M. Kopycinska-Mueller, E.D. Langlois, A.B. Kos, N. Barbosa III., Appl. Phys. Lett. **89**, 021911 (2006)
14. O.V. Kolosov, M.R. Castell, C.D. Marsh, G.A.D. Briggs, T.I. Kamins, R.S. Williams, Phys. Rev. Lett. **81**, 1046 (1998)
15. U. Rabe, M. Kopycinska, S. Hirsekorn, J.M. Saldana, G.A. Schneider, W. Arnold, J. Phys. D Appl. Phys. **35**, 2621 (2002)
16. H. Wagner, D. Bedorf, S. Küchemann, M. Schwabe, B. Zhang, W. Arnold, K. Samwer, Nat. Mater. **10**, 439 (2011)
17. X. Zhou, H. Miao, F. Li, Nanoscale **5**, 11885 (2013)
18. D.C. Hurley, M. Kopycinska-Mueller, A.B. Kos, R.H. Geiss, Meas. Sci. Technol. **16**, 2167 (2005)
19. M. Kopycinska-Müller, A. Clausner, K.B. Yeap, B. Köhler, N. Kuzeyeva, S. Mahajan, T. Savage, E. Zschech, K.J. Wolter, Ultramicroscopy **162**, 82 (2016)
20. J.P. Killgore, D.G. Yablon, A.H. Tsou, A. Gannepalli, P.A. Yuya, J.A. Turner, R. Proksch, D.C. Hurley, Langmuir **27**, 13983 (2011)
21. U. Rabe, M. Kopycinska-Müller, S. Hirsekorn, in *Acoustic Scanning Probe Microscopy*. ed. by F. Marinello, D. Passeri, E. Savio (Springer, Berlin, 2013), pp.123–153
22. P.A. Yuya, D.C. Hurley, J.A. Turner, J. Appl. Phys. **109**, 113528 (2011)
23. K. Kimura, K. Kobayashi, K. Matsushige, H. Yamada, Ultramicroscopy **133**, 41 (2013)
24. Z. Parlak, F.L. Degertekin, J. Appl. Phys. **103**, 114910 (2008)
25. C. Ma, Y. Chen, W. Arnold, J. Chu, J. Appl. Phys. **121**, 154301 (2017)
26. C. Ma, W. Wang, Y. Chen, W. Arnold, J. Chu, J. Appl. Phys. **126**, 124302 (2019)
27. S. Hu, C. Su, W. Arnold, J. Appl. Phys. **109**, 084324 (2011)
28. W. Wang, C. Ma, Y. Chen, L. Zheng, H. Liu, J. Chu, Beilstein J. Nanotech. **10**, 1636 (2019)
29. T. Tsuji, K. Yamanaka, Nanotechnology **12**, 301 (2001)
30. C. Ma, W. Arnold, J. Appl. Phys. **128**, 180901 (2020)
31. J.P. Killgore, F.W. DelRio, Macromolecules **51**, 6977 (2018)
32. T. Göddenhenrich, S. Müller, C. Heiden, Rev. Sci. Instrum. **65**(9), 2870–2873 (1994)
33. V. Scherer, W. Arnold, B. Bhushan, Surf. Interface Anal. **27**(5–6), 578–587 (1999)
34. B. Bhushan, T. Kasai, Nanotechnology **15**(8), 923 (2004)
35. M. Reinstädtler, T. Kasai, U. Rabe, B. Bhushan, W. Arnold, J. Phys. D Appl. Phys. **38**(18), R269 (2005)
36. F. Dinelli, S.K. Biswas, G.A.D. Briggs, O.V. Kolosov, Appl. Phys. Lett. **71**(9), 1177–1179 (1997)
37. G. Behme, T. Hesjedal, Appl. Phys. A **70**, 361–363 (2000)
38. V. Pfahl, C. Ma, W. Arnold, K. Samwer, J. Appl. Phys. **123**(3) (2018)
39. H.J. Sharahi, P. Egberts, S. Kim, Nanotechnology **30**(7), 075502 (2018)
40. S. Shi, D. Guo, J. Luo, Friction **9**, 1163–1174 (2021)

41. J. Cao, Q. Li, Friction **10**(10), 1650–1659 (2022)
42. A. Socoliuc, E. Gnecco, S. Maier, O. Pfeiffer, A. Baratoff, R. Bennewitz, E. Meyer, Science **313**(5784), 207–210 (2006)
43. E. Gnecco, A. Socoliuc, S. Maier, J. Gessler, T. Glatzel, A. Baratoff, E. Meyer, Nanotechnology **20**(2), 025501 (2009)
44. P. Steiner, R. Roth, E. Gnecco, T. Glatzel, A. Baratoff, E. Meyer, Nanotechnology **20**(49), 495701 (2009)
45. R. Roth, O.Y. Fajardo, J.J. Mazo, E. Meyer, E. Gnecco, Appl. Phys. Lett. **104**(8) (2014)
46. E. Gnecco, S. Maier, E. Meyer, J. Phys. Condens. Matter **20**, 354004 (2008)
47. M.A. Lantz, D. Wiesmann, B. Gotsmann, Nat. Nanotechnol. **4**(9), 586–591 (2009)
48. A. Knoll, H. Rothuizen, B. Gotsmann, U. Duerig, Nanotechnology **21**(18), 185701 (2010)
49. P. Pedraz, R. Wannemacher, E. Gnecco, ACS Nano **9**(9), 8859–8868 (2015)
50. R. Overney, E. Meyer, MRS Bull. **18**(5), 26–34 (1993)
51. J.A. Ruan, B. Bhushan, J. Appl. Phys. **76**(9), 5022–5035 (1994)
52. C.M. Mate, G.M. McClelland, R. Erlandsson, S. Chiang, Phys. Rev. Lett. **59**(17), 1942 (1987)
53. E. Gnecco, R. Bennewitz, T. Gyalog, E. Meyer, J. Phys. Condens. Matter **13**(31), R619 (2001)
54. M. Dienwiebel, G.S. Verhoeven, N. Pradeep, J.W. Frenken, J.A. Heimberg, H.W. Zandbergen, Phys. Rev. Lett. **92**(12), 126101 (2004)
55. C. Lee, Q. Li, W. Kalb, X.Z. Liu, H. Berger, R.W. Carpick, J. Hone, Science **328**(5974), 76–80 (2010)
56. C. Ma, C. Zhou, J. Peng, Y. Chen, W. Arnold, J. Chu, J. Appl. Phys. **129**(23), 234303 (2021)
57. J.L. Hutter, J. Bechhoefer, Rev. Sci. Instrum. **64**(7), 1868–1873 (1993)
58. B.J. Rodriguez, C. Callahan, S.V. Kalinin, R. Proksch, Nanotechnology **18**, 475504 (2007)
59. S. Jesse, S.V. Kalinin, R. Proksch, A.P. Baddorf, B.J. Rodriguez, Nanotechnology **18**, 435503 (2007)
60. K. Yamanaka, H. Ogiso, O. Kolosov, Jpn. J. Appl. Phys. **33**(5S), 3197 (1994)
61. V. Scherer, B. Bhushan, U. Rabe, W. Arnold, IEEE Trans. Magn. **33**(5), 4077–4079 (1997)
62. T. Kawagishi, A. Kato, Y. Hoshi, H. Kawakatsu, Ultramicroscopy **91**(1–4), 37–48 (2002)
63. C. Su, L. Huang, C.B. Prater, B. Bhushan, in *Applied Scanning Probe Methods V*. ed. by B. Bhushan, S. Kawata, H. Fuchs (Springer, Berlin, Heidelberg, 2007), pp.113–148
64. A. Caron, U. Rabe, M. Reinstädtler, J.A. Turner, W. Arnold, Appl. Phys. Lett. **85**(26), 6398–6400 (2004)
65. Y. Song, B. Bhushan, J. Appl. Phys. **97**(8), 083533 (2005)
66. K. Yamanaka, E. Tomita, Jpn. J. Appl. Phys. **34**(5S), 2879 (1995)
67. J. Colchero, M. Luna, A.M. Baro, Appl. Phys. Lett. **68**(20), 2896–2898 (1996)
68. R.W. Carpick, D.F. Ogletree, M. Salmeron, Appl. Phys. Lett. **70**(12), 1548–1550 (1997)
69. P.E. Mazeran, J.L. Loubet, Tribol. Lett. **7**, 199–212 (1999)
70. A. Spychalski-Merle, K. Krischker, T. Göddenhenrich, C. Heiden, Appl. Phys. Lett. **77**(4), 501–503 (2000)
71. M. Reinstädtler, U. Rabe, V. Scherer, U. Hartmann, A. Goldade, B. Bhushan, W. Arnold, Appl. Phys. Lett. **82**(16), 2604–2606 (2003)
72. M. Reinstädtler, U. Rabe, A. Goldade, B. Bhushan, W. Arnold, Tribol. Int. **38**(6–7), 533–541 (2005)
73. F. Mertens, T. Göddenhenrich, A. Schirmeisen, Appl. Phys. Lett. **104**(11) (2014)
74. C. Ma, V. Pfahl, Z. Wang, Y. Chen, J. Chu, M.K. Phani, A. Kumar, W. Arnold, K. Samwer, Appl. Phys. Lett. **113**(8) (2018)
75. D.C. Hurley, J.A. Turner, J. Appl. Phys. **102**(3) (2007)
76. E. Rezaei, J.A. Turner, Appl. Phys. Lett. **110**(10) (2017)
77. R.W. Carpick, Science **313**(5784), 184–185 (2006)
78. M. Urbakh, J. Klafter, D. Gourdon, J. Israelachvili, Nature **430**(6999), 525–528 (2004)
79. H.D. Fridman, P. Levesque, J. Appl. Phys. **30**(10), 1572–1575 (1959)
80. M. Heuberger, C. Drummond, J. Israelachvili, J. Phys. Chem. B **102**(26), 5038–5041 (1998)
81. V.C. Kumar, I.M. Hutchings, Tribol. Int. **37**(10), 833–840 (2004)
82. V.L. Popov, J. Starcevic, A.E. Filippov, Tribol. Lett. **39**, 25–30 (2010)

83. V. Scherer, W. Arnold, in *Micro/Nanotribology and Its Applications*, ed. by B. Bhushan. NATO ASI Series, vol 330 (Springer, Dordrecht, 1997), pp. 225–231
84. T. Hesjedal, G. Behme, IEEE Trans. Ultrason. Ferroelectr. Freq. Control **49**(3), 356–364 (2002)
85. S. Jeon, T. Thundat, Y. Braiman, Appl. Phys. Lett. **88**, 214102 (2006)
86. R. Guerra, A. Vanossi, M. Urbakh, Phys. Rev. E **78**, 036110 (2008)
87. O.Y. Fajardo, E. Gnecco, J.J. Mazo, Phys. Rev. B **89**, 075423 (2014)
88. X. Ma, X. Tan, D. Guo, S. Wen, Friction **11**, 1225–1238 (2023)
89. G.A. Tomlinson, The London, Edinburgh, and Dublin Philos. Mag. J. Sci. **7**(46), 905–939 (1929)
90. S. Hirsekorn, U. Rabe, W. Arnold, Nanotechnology **8**(2), 57 (1997)
91. N.A. Burnham, R.J. Colton, H.M. Pollock, Nanotechnology **4**(2), 64 (1993)
92. H. Hölscher, A. Schirmeisen, U.D. Schwarz, Philos. Trans. R. Soc. A **366**(1869), 1383–1404 (2008)
93. R. Szoszkiewicz, B.D. Huey, O.V. Kolosov, G.A.D. Briggs, G. Gremaud, A.J. Kulik, Appl. Surf. Sci. **210**(1–2), 54–60 (2003)
94. R. Szoszkiewicz, B. Bhushan, B.D. Huey, A.J. Kulik, G. Gremaud, J. Appl. Phys. **99**(1), 014310 (2006)
95. L. Zhang, J. Dong, P.H. Cohen, IEEE Trans. Nanotechnol. **12**(5), 743–750 (2013)
96. Y. Geng, Y. Yan, Y. Zhuang, Z. Hu, Appl. Surf. Sci. **356**, 348–354 (2015)

Atomic-Scale Friction on Crystal Surfaces in Ultra-High Vacuum

Yiming Song, Sabine Maier, Enrico Gnecco, and Ernst Meyer

Abstract This chapter reviews friction force microscopy investigations on single-asperity sliding contacts in ultra-high vacuum (UHV). The atomic-scale stick–slip observed under such conditions can be converted into a superlubric regime of motion by reducing the normal load and/or applying ultrasonic vibrations. Thermal vibrations and sliding direction (on a crystal surface) also influence the friction. The empirical Prandtl-Tomlinson (PT) model is introduced, which explains well the main experimental observations. The scenario is more complicated on two-dimensional (2D) materials, where the puckering effect explains the difference in friction observed on monolayers versus multiple layers. The moiré patterns formed on them are also influenced by elastic deformation, which can lead to significantly larger dissipation than that due to atomic stick–slip alone.

1 Introduction

Apart from playing a vital role in everyday life, sliding friction is a phenomenon of key importance in a variety of technological applications and different branches of science. Understanding friction at the atomic scale is important to minimize wear and energy losses in micro- and nano-scale devices involving sliding contacts. Sliding friction without wear corresponds to the energy transfer from macroscopic degrees of freedom arising from the relative motion of the bodies in contact to microscopic degrees of freedom such as phononic or electronic excitations. In this context,

Y. Song
Institute of Applied Physics, Justus Liebig University Giessen, Giessen, Germany

S. Maier
Department of Physics, University of Erlangen-Nuremberg, Erlangen, Germany

E. Gnecco
Marian Smoluchowski Institute of Physics, Jagiellonian University, Krakow, Poland

E. Meyer (✉)
Department of Physics, University of Basel, Basel, Switzerland
e-mail: ernst.meyer@unibas.ch

E. Gnecco and E. Meyer (eds.), *Fundamentals of Friction and Wear on the Nanoscale*,
NanoScience and Technology, https://doi.org/10.1007/978-3-031-63065-1_4

atomic force microscopy (AFM) can be successfully applied to study the friction in single asperity contacts. This led to substantial advancements in our understanding of the microscopic mechanisms of friction on various solid surfaces and different environmental conditions.

The first atomic-scale friction measurements were performed by Mate et al. using a tungsten wire tip sliding on the basal plane of a graphite surface [1]. The transverse force acting on the tip, as measured by optical interferometry, showed a saw-tooth-like modulation as a function of support position corresponding to the hexagonal symmetry of the surface lattice. This effect is due to elastic instabilities, causing the tip to slip between neighboring lattice sites. This motion is an example of so-called *stick–slip* and has been observed on a wide variety of materials on the atomic scale. If the force signal is measured in forward and backward directions, a hysteretic friction loop is observed. The area enclosed between the two curves corresponds to the energy dissipated in the sliding process. The force peaks reached at the instabilities define the static friction, whereas the kinetic friction acting on the tip (essentially in the slip phase) is equal to the average value of the lateral force signal, as proven by simple energy balance considerations [2]. An example of a friction map and a corresponding friction loop obtained on a KBr(100) surface is shown in Fig. 1. The transverse force in this measurement varies in the nano-newton range. In contrast, the first measurement on graphite revealed lateral forces in the micro-newton range [1], which was possibly due to the transfer of a graphite flake onto the tip causing an increased contact area.

In the following, we will describe different aspects of atomic stick–slip on single crystal surfaces and on long-range moiré superstructures formed by one or few monolayers of a second material grown on those surfaces. The experimental results are followed by a brief discussion of the PT model, which explains satisfactorily most

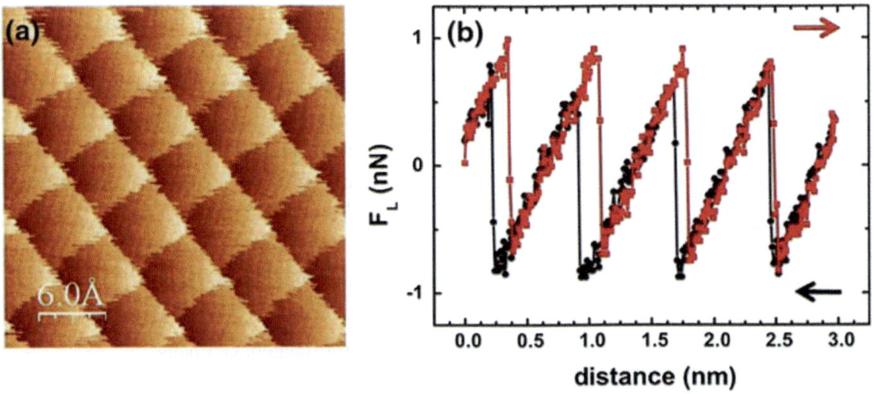

Fig. 1 **a** Friction force map and **b** corresponding friction loop along the (001)-direction on a KBr(100) surface. The sawtooth profiles are caused by elastic instabilities periodically reached in the so-called stick–slip motion of the probing tip, as explained in Sect. 2. Adapted from [3]

of the features observed on flat crystal surfaces. On 2D materials, additional mechanisms must be included to interpret peculiar features caused by moiré superstructures and weak interlayer interactions. To avoid issues with capillary condensation or airborne contaminants, the focus is limited to experiments performed in the UHV environment.

2 The Prandtl-Tomlinson Model

Friction experiments on atomically flat surfaces are well interpreted by the PT model [4, 5], proposed in the early 1920s. In spite of its simplicity, the PT model explains the basic feature observed in friction force microscopy measurements, i.e. the sawtooth variation of the friction force acting on a sliding tip. The PT model describes a point mass (corresponding to the tip apex) dragged along the surface by a rigid support connected via a spring k. Neglecting inertia, the total energy of the system is expressed by the superposition of the periodic tip–surface interaction energy, U_{eff}, and the elastic energy stored in the spring:

$$U_{tot}(\mathbf{r}, t) = U_{eff}(\mathbf{r}) + \frac{1}{2}k(\mathbf{r} - \mathbf{v}t)^2. \qquad (1)$$

In Eq. (1) the two-dimensional vector $\mathbf{r} \equiv (x, y)$ is the position of the tip on the surface, and $\boldsymbol{X} = \mathbf{v}t$ the position of the support, with $\mathbf{v}$ the scan velocity.

2.1 One-Dimensional Prandtl-Tomlinson Model

In one dimension, the total energy stored in the system simplifies to

$$U_{tot}(x, t) = -U_0 \cos\frac{2\pi x}{a} + \frac{1}{2}k(x - vt)^2. \qquad (2)$$

Assuming, as a first approximation, a sinusoidal interaction potential with the crystal lattice periodicity a and an amplitude U_0. $U_{tot}(x, t)$ has the form of a corrugated parabola, whose shape depends on the tip and support position as seen in Fig. 2a. The tip is located in the deepest minimum of U_{tot}. When the support starts moving over the surface, the lateral force increases almost linearly, according to the Hooke's law, and the minimum gets progressively flattened. At the position x^*, the equilibrium becomes unstable, and the tip suddenly slips into the next minimum in the potential profile. With the sinusoidal profile introduced in Eq. (2), the instability condition translates into

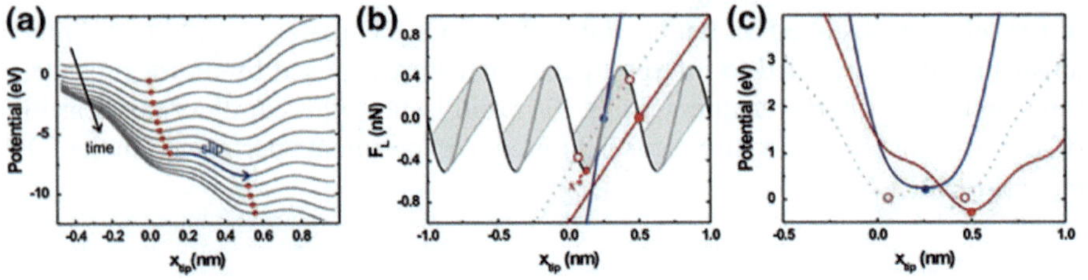

Fig. 2 **a** Time evolution of U_{tot} and corresponding tip position (*red dot*) capturing a slip event ($\eta = 3.2$). **b** Graphical solution of the equilibrium condition $\partial U_{tot}/\partial x = 0$ to illustrate the mechanism: The *straight lines* represent the spring term and the *curved lines* the periodic surface force. The *blue line* has a single intersection with the surface force for each value of x_{tip}, and thus a smooth sliding from one minimum to the next is possible ($\eta < 1$). The *red line* has double intersections in the *grey shaded* areas, indicating instable regions causing hysteresis between forward and backward scans ($\eta > 1$). **c** The corresponding U_{tot} for the different regimes in (**b**)

$$x^* = \frac{a}{2\pi} \arccos\left(-\frac{1}{\eta}\right) \text{ with } \eta = \frac{4\pi^2 U_0}{ka^2}. \tag{3}$$

The dimensionless parameter η quantifies the ratio between the strength of the surface potential and the elastic energy stored in the spring. It is clear from Eq. (3), that elastic instabilities are possible only if $\eta > 1$. In this case, a hysteretic friction loop resembling the results in Fig. 1b is observed if the support is moved forth and back along the x axis. The area of the loop represents the energy dissipated in the process. Note that the slip distance is usually equal to a, but it can be also a multiple of it if the tip oscillations accompanying the slip phase are underdamped and/or the parameter η exceeds well-defined critical values. This case is discussed in Sect. 3. For $\eta < 1$, the equilibrium is always stable, and the tip follows the support smoothly, with the lateral force profile oscillating according to the interaction potential.

In the PT model, the spring force $F_L = k(x_{tip} - vt)$ is represented by

$$F_L = -\frac{2\pi U_0}{a} \sin\frac{2\pi x_{tip}}{a}. \tag{4}$$

where the tip position x_{tip} is defined by the condition $\partial U_{tot}/\partial x = 0$. This force can be easily recorded by AFM, and in this context it is usually called lateral force. Its maximal absolute value, F_L^{max}, is reached if $x_{tip} = a/4$ and is related linearly to U_0 by

$$U_0 = \frac{aF_L^{max}}{2\pi}. \tag{5}$$

Hence, the amplitude of the interaction potential can be easily determined from the peak value of the lateral force, which approximately defines the *static* friction.

Another important parameter to consider in the framework of the PT model is the stiffness of the system. The effective spring constant k combines the compliance of tip and surface according to the relation [6]:

$$\frac{1}{k} = \frac{1}{k_{con}} + \frac{1}{k_{tip}} + \frac{1}{c_L}. \tag{6}$$

In Eq. (6), c_L is the spring constant associated with the lateral displacement due to torsional deformation of the cantilever supporting the tip (in standard AFM setups, see Chap. 1). k_{con} and k_{tip} are the stiffness values associated to the transverse deformation of contact and tip, respectively. There are different methods to measure the lateral spring constants c_L, e.g. the added mass method of Cleveland [7, 8] and the unloaded resonance technique of Sader [8, 9]. c_L can be also estimated numerically from the geometry and the elastic properties of the pinned cantilever [10]. Note that the lateral stiffness of high aspect ratio Si tips is usually comparable with the lateral spring constant of the cantilever, as observed by Lantz et al. using scanning transmission electron microscopy and finite element analysis [6]. Finally, the lateral contact stiffness k_{con} for an elastic contact between a sphere and a flat surface is $k_{con} = 8a_{con}G^*$ according to continuum mechanics [11], where a_{con} is the contact radius and the effective shear modulus G^* is defined by

$$\frac{1}{G^*} = \frac{2 - \nu_1^2}{G_1} + \frac{2 - \nu_2^2}{G_2}, \tag{7}$$

($G_{1,2}$ are the shear moduli of the sphere and the surface, respectively, and $\nu_{1,2}$ the corresponding Poisson numbers). According to the Hertz theory, the contact radius and, consequently, the contact stiffness follow a power-law as a function of the applied load. However, these results are valid (in the absence of adhesion) for macroscopic (continuum) contacts. In the case of atomic-scale stick–slip measured on an ionic crystal surface, Socoliuc et al. [12] have rather concluded that k_{con} is nearly constant, as shown in Fig. 3, whereas the amplitude of the interaction potential increases almost linearly with the applied load. Note that, in Fig. 3, the effective spring constant k is the measured quantity (as given by the slope of the stick phase in the friction loop (F_x vs. x curve) [13]. The value of k_{con} is obtained together with η according to the relation [14]

$$k = \left(1 + \frac{1}{\eta}\right) k_{exp}. \tag{8}$$

For a detailed discussion on the breakdown of continuum mechanics on the nanoscale, in relation to the contact stiffness, the reader is referred to the seminal paper by Luan and Robbins [15]. Among the different atomic structures of the tip apex examined, a spherical shape is still well described by continuum mechanics on the nanoscale, while stepped structures exhibit discontinuities in the contact area with increasing load.

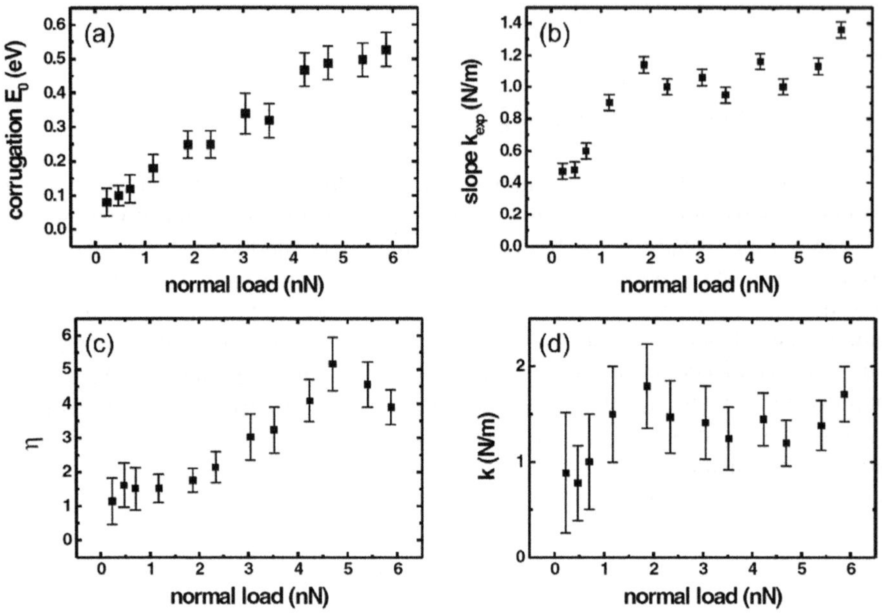

Fig. 3 **a** Energy corrugation E_0, **b** slope k_{exp} of the lateral force versus distance curve in the sticking part, **c** parameter η, and **d** effective lateral stiffness k of the contact, as a function of the normal load F_N acting on the tip. The quantities were determined from atomic stick–slip experiments on NaCl(001). From [12]

2.2 Extensions of the Prandtl-Tomlinson Model

Even if the one-dimensional PT model is sufficient to explain the basic mechanism of atomic-scale stick–slip (as revealed by the friction loops), the perpendicular (y) direction on the surface plane must be included to reproduce a standard friction map such as that shown in Fig. 1a. The upgrade would be straightforward, if the tip-surface interaction potential were a priori known [16]. However, this is not the case, since each tip has its own atomic structure. Nevertheless, the first Fourier components of $U_{eff}(x, y)$ are usually sufficient to understand the motion of the probing tip and the dependence of the static and kinetic friction on the scan direction. This is the case, e.g., for the square lattice of an ionic crystal [17] or the hexagonal lattice of graphite [18], which can be respectively described as follows:

$$U_{square}(x, y) = -U_0 \cos\frac{2\pi x}{a} \cos\frac{2\pi y}{a} \text{ and} \tag{9}$$

$$U_{hex}(x, y) = -\frac{U_0}{2.25}\left[2\cos\frac{2\pi x}{a}\cos\frac{2\pi y}{a\sqrt{3}} + \cos\frac{4\pi y}{a\sqrt{3}}\right]. \tag{10}$$

Extending the derivations in Sect. 2.2 to two dimensions, it is not difficult to determine how the tip moves in a two-dimensional 'zig-zag' fashion on these "energy landscapes" [19].

Other refinements are needed to explain important effects observed at finite temperatures, i.e. the distribution of static friction peaks [20] and the increase of the average static and kinetic friction force with the scan velocity v [21, 22]. A possible analytical solution of the PT model including thermal activation is discussed in Sect. 3. Numerically, the tip motion at finite temperatures can be estimated from the Langevin equation [23]

$$m\frac{d^2x}{dt^2} + m\gamma\frac{dx}{dt} + \frac{dU_{tot}}{dx} = \xi(t). \tag{11}$$

In Eq. (11), m *is* an effective mass of the system, and γ is a damping coefficient describing the coupling with phonon and possible electron oscillations in the substrate. The quantity $\xi(t)$ is a random noise term satisfying the fluctuation–dissipation relation $\langle\xi(t)\xi(t\prime)\rangle = 2m\gamma k_B T\delta(t - t\prime)$, in which k_B is the Boltzmann constant.

The PT model has been also extended to account for multiple springs. In the two-mass-two-spring model, one spring represents the cantilever and the other spring the microscopic contact [24]. More complex configurations require the use of the so-called Frenkel-Kontorova-Tomlinson (FKT) model. This model addresses the periodic sliding of two surfaces in contact, whereas the atoms of one surface are harmonically coupled with their nearest neighbors [25, 26]. The basic FK model was first proposed to describe dislocation in solids, and then used to address effects related to the commensurability and incommensurability of contacts as well as adsorbed films on surfaces [27]. In one dimension, one can think of a chain of particles of mass m connected by elastic springs and interacting with a periodic potential. In this case, the total potential takes the form

$$U_{tot} = \sum_n \left(-U_0 \cos\frac{2\pi x_n}{a} + \frac{1}{2}k(x_{n+1} - x_n - b)^2 \right). \tag{12}$$

where b is the distance between two particles in the chain. Static friction can be probed by adding an external force F_{ext} adiabatically increasing till sliding initiates. The ratio a/b is very important. For any (irrational) value of a/b there is a critical value η_c of the parameter η defined by Eq. (3) such that the static friction force vanishes when $\eta < \eta_c$. The parameter η_c takes its maximum value ($= 1$) when the ratio a/b is equal to the golden mean $\left(1 + \sqrt{5}\right)/2 = 1.618$ [28].

3 Experimental Observations of Atomic Stick–Slip

Since the first friction force measurements by Mate et al. [1], atomic-scale stick–slip has been experimentally observed on a variety of materials including insulators, metals, and semiconductors. As previously mentioned, this chapter focuses on atomic-scale friction measurements in UHV only. Due to their simple structure and comparatively easy and reproducible preparation, alkali halide crystals are typically used as model systems in UHV-FFM. Atomic stick–slip corresponding to the square symmetry given by Eq. (9) was observed on KBr [29], NaCl [12, 21], and KF [30]. It means that a strong interaction with the tip apex is achieved by only one of the two ionic species forming the crystal. Measurements on NaF(100) by Ishikawa et al. are the exception, as they could resolve both ions depending on the applied load [31].

Friction force experiments on clean metal surfaces prepared in UHV are more difficult to carry out due to the high surface energy of metals under clean conditions, which leads to an increased reactivity of surface atoms resulting in a junction formation under the tip apex [32]. For example, the formation of a neck of copper atoms at the tip was proposed to result in the observation of stick–slip type sliding over Cu(111) surfaces [33]. Friction force microscopy experiments at low loads performed on Cu(100) and Au(111) surfaces revealed a clear stick–slip modulation in the lateral force with almost zero dissipation [34]. For higher loads, significant friction occurred together with the onset of wear. The Cu(100) surface is easier worn than Cu(111), because it is not a crystallographic glide plane [35]. Even on the more ductile Pt(111) surface atomic stick–slip was observed using passivated tips [36].

3.1 Load Dependence: From Smooth Sliding and Stick–Slip to Wear

On the macroscopic scale, dry friction is governed by the Coulomb (or Amontons) equation $F_L = \mu F_N$, where the friction coefficient μ mainly depends on the material of the two bodies in motion and the interface roughness. In single asperity contacts, the dependence of the friction on applied load is usually non-linear. In nano-scale contacts, the load dependence of friction is often described by means of contact mechanics continuum theories [37], which, however, do not give information about the atomic interactions in the sliding contact region as they assume a spherical tip (see also end of Sect. 2.2). Fusco et al. suggested a power law dependence by simulating the dynamics of a tip scanning a rigid monolayer graphite surface in a PT-like model [38].

Experimentally, the amount of energy dissipated while sliding over an atomically flat surface can be controlled by the variation of load on the contact [12]. The mean friction force decreases with decreasing normal load and reaches zero before the probing tip jumps out of contact. The normal force in the experiment is represented in the PT model by the parameter η. If $\eta < 1$, elastic instabilities are suppressed

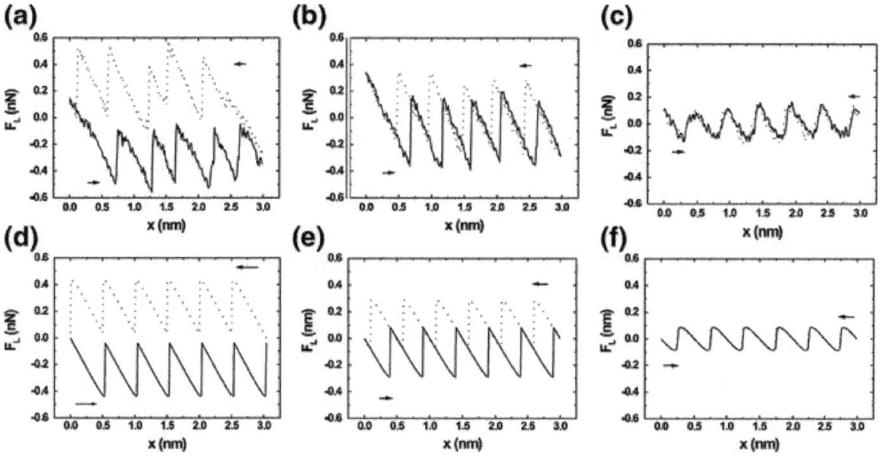

Fig. 4 Measurement of the lateral force sliding the tip forward and backward in (100) direction over a NaCl(001) surface, where the externally applied load was **a** $F_N = 4.7$nN, **b** $F_N = 3.3$nN, and **c** $F_N = -0.47$nN. **d–f** Corresponding numerical results from the Tomlinson model for **d** $\eta = 5$, **e** $\eta = 3$, and **f** $\eta = 1$. The stiffness was chosen as $k = 1$N/m and the lattice constant as $a = 0.5$nm. From [12]

and the tip smoothly follows the surface corrugation with negligible friction. This transition from stick–slip to smooth sliding was observed experimentally for friction loops recorded on NaCl(001) by Socoliuc et al. when a normal loading of about 1nN was applied, see Fig. 4.

There are other more feasible ways to achieve a state of ultra-low friction at the atomic scale. Dienwiebel et al. [39] observed vanishing friction while dragging a graphite flake out of registry over a graphite surface. The corrugation of the interaction potential between the graphite substrate and the graphite flake could be continuously varied by rotating the substrate with respect to the flake. This state of ultra-low friction is characteristic of incommensurable contacts. It is referred to as *superlubricity*, which goes back to a concept discussed by Hirano [27]. Friction can also be decreased by reducing the scan speed to a few nm/s or less, so that thermally activated jumps occur [40]. The thermally induced suppression of friction is called *thermolubricity*. Another way to switch friction on and off at the atomic scale was found by exciting the mechanical resonances of the sliding system perpendicular to the contact plane [41]. In this case, a similar transition from stick–slip to smooth sliding as in Fig. 4 was observed on NaCl, however by changing the excitation amplitude instead of the load. The state of *dynamic superlubricity* so-achieved could be also exploited to acquire lattice-resolved lateral force maps of crystal surfaces with no evidence of abrasive wear [42]. A significant decrease of friction has been also observed if AFM tips are shaken laterally instead of normally [43]. In this case, the excitation amplitude leads to effective values of the parameters in the PT model as described in [44].

In contrast, if the normal force acting in the contact region exceeds a critical value, fracture occurs and the sliding motion is accompanied by abrasive wear of at least one of the contacting surfaces. Quantitative information on the initial stages of abrasive wear on alkali halide surfaces could be obtained by comparing friction loops and topography images before and after scratching a KBr(001) surface [45]. With normal forces of few nN only a minor part about (30%) of the total energy dissipation during scratching the surface went into wear. If the tip is driven on distances of a few hundred nm or more, abrasive wear leads to the formation of wavy surface structures [46]. Please see Chap. 12 for similar effects on polymeric materials.

3.2 The Slip

The slip instability is accompanied by the dissipation of the elastic energy stored in the stick phase into phonon and electron excitations. In conventional AFM experiments, the slip occurs with a relaxation time well below the detection limit. However, recording the atomic-scale stick–slip with a bandwidth of few MHz has revealed a wide variation of slip durations up to several milliseconds, by far longer than expected for a relaxation process on the atomic scale [3]. These long slip events possibly result from a multiple contact established between tip and surface, as suggested by the correlation between irregular features in lateral force maps and long slip durations. This conclusion was also supported by comparing the experimental results with a multi-tip (multi-spring) simulation using a PT model including thermal activation.

3.3 Thermal Effects and Velocity Dependence

Coulomb's law of friction states that macroscopic friction is independent of the sliding velocity. However, atomic-scale friction measurements on mica [22], Cu(111) [33], NaCl [21], graphite [47], and silicon oxide [48] with driving velocities in the order of μm/s have shown a logarithmic dependence. The origin of this dependence can be also traced back to PT model with the inclusion of thermally activated slip. According to the reaction rate theory, slip only occurs when a variable energy barrier $\Delta E(t)$ defined by the interaction potential becomes zero. This is not the case at finite temperatures T. The probability that the tip remains stuck at a given minimum of the total potential, $p(t)$, decreases with time according to the equation [21]

$$\frac{dp(t)}{dt} = -f_0 \exp\left(-\frac{\Delta E(t)}{k_B T}\right) p(t). \tag{13}$$

In Eq. (13), f_0 denotes the resonance frequency in the potential well corresponding to the minimum. Assuming a linear dependence of the energy barrier $\Delta E = \lambda(F_L^* - F_L)$, on the increasing spring force F, the logarithmic dependence is obtained [21]

$$F(v) = \text{const.} + \frac{k_B T}{\lambda} \log \frac{v}{v_0}. \tag{14}$$

In Eq. (14), λ is in the order of the lattice constant. Thus, the slower the tip is driven, the higher the probability of a thermally activated early slip and hence a lower friction peak. Sang et al. [23] observed that barrier-hopping fluctuations occur preferentially when the tip is very close to athermal slip. In this case, the approximation $\Delta E \propto (\text{const.} - F_L)^{3/2}$ is more precise, and the master equation leads to

$$F(v) = \text{const.} - \left(\frac{k_B T}{\lambda}\right)^{2/3} \left(\log \frac{v_c}{v}\right)^{2/3}. \tag{15}$$

for velocities below the critical value

$$v_c = \frac{\pi \sqrt{2}}{2} \frac{f_0 k_B T}{ka}. \tag{16}$$

At high velocities, thermal vibrations become less effective, and the friction force tends to become independent of the velocity. Such a transition has been observed by Riedo et al. on mica in ambient conditions [22]. Worth mentioning is also the non-monotonic dependence of the friction force predicted by Reimann et al. with explicit inclusion of dissipation of inertia effects [49].

Thermal effects play an important role not only in the velocity dependence of nanoscale friction. Krylov and Frenken [40] found that they can lead to a variety of friction regimes not predicted by the athermal PT model. In particular, the range of ultra-low friction at low loads observed by Socoliuc et al. [12] on NaCl could be extended toward higher loads thanks to thermal vibrations. Thermal delocalization of the contact could also lead to extremely low friction independent of normal load, despite the observation of atomic stick–slip in the lateral force signal. As mentioned above, this concept of low friction owing to thermal fluctuations is referred to as thermolubricity.

With the further instrumental development of FFMs being operational with high resolution at cryogenic temperatures, the predictions of the PT model including thermal activation could be directly verified. Atomic stick–slip measurements on graphite and MoS_2 between 100 and 300K [47, 50] have indeed revealed an increase in friction with decreasing temperature in agreement with the aforementioned interpretation. Last but not least, another thermally activated effect is the increase of the contact area due to the formation of covalent bonds between tip and surface or the build-up of a solid neck caused by migration of surface atoms under the tip pressure. It explains the logarithmic increase of lateral stiffness with the time of contact, which was observed on SiO_2 and NaCl with a variable-temperature UHV-AFM [51, 52].

3.4 Maximal Lateral Force and Friction Anisotropy

Another important feature in the stick–slip motion is the peak value of the spring force, which is a direct measure of the surface-tip potential strength, see Eq. (5). The values of this peak are spread by thermal activation. A statistical analysis performed by Schirmeisen et al. on HOPG [23] revealed a site-dependent energy barrier of the surface potential (from the side to the center of the hollow-site rows) consistent with the thermally activated PT model. Moreover, a velocity-dependent statistical analysis of the slip lengths on HOPG by Evstigneev et al. [53] showed that a single-step reaction scenario is valid only at relatively high velocities, while at slower pulling speeds, a more complicated slip mechanism must be at work. Also in this case, a multiple bond formation of the tip-substrate contact may occur, with additional bonds taking finite time to establish, so that at fast pulling, new bonds do not have sufficient time to develop during a single stick phase. Regarding the dependence of the stick–slip on the scan direction, worth mentioning are also recent measurements performed by Konrad et al. on graphene and MoS_2 in UHV [54]. The results, which extend earlier observations in ambient conditions [55], agree in good agreement with the extension of the PT model presented in Sect. 2.2, with minor deviations possibly due to asymmetry of the tip apex [56].

3.5 Multiple Slips

As already mentioned, most of atomic stick–slip measurements in UHV reveal slip jump periodicities of one lattice spacing. Johnson and Woodhouse have predicted that under certain conditions, slips are not restricted to a single lattice constant and hopping over multiple spacings may occur (i.e. multiple slips) [57]. The transition between single and double slip mode and its dependence on contact damping, sliding velocity, and finite temperature has been studied in numeric simulations employing a generalized PT model [58, 59]. Medyanik et al. derived the threshold values of the parameter η (see Eq. 3) for the transition between different slip regimes from a simple analytic criterion based on the one-dimensional PT model in the quasi-static limit [60]. They found that the tip slides smoothly if $\eta < 1$, executes stick–slip motion with single slips for $1 < \eta < 4.6$ or, possibly exhibits double slips for $4.6 < \eta < 7.79$ or even larger multiple slips for $\eta > 7.79$. However, the threshold values are significantly modified by the damping coefficient γ and the temperature. A chaotic regime (with alternate jump lengths) is always expected if γ is low enough [61].

On this basis, it is not surprising that the transition from single to multiple slips occurs is observed at increasing load values. Figure 5 shows the experimental dependence of stick–slip behavior with load on an HOPG surface in ambient conditions. Superlubricity is observed only at the lowest applied load, whereas stick–slip instabilities with the periodicity of the HOPG lattice and multiples of it occur at higher

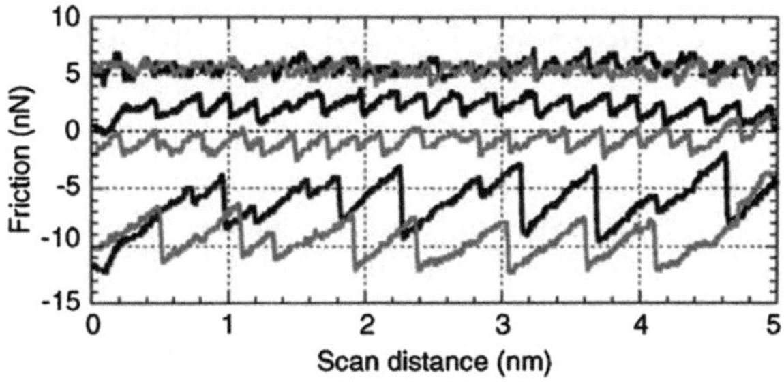

Fig. 5 Representative experimental friction loops with increasing load (offset for clarity) demonstrating the transitions from smooth sliding (*top*) to single (*middle*) and mostly double slips (*bottom*). From [60]

loads. Roth et al. [62] verified the occurrence of multiple slips also in UHV and used them to estimate the viscous damping accompanying the stick–slip motion on NaCl(001). A comparison with simulations based on the PT model suggested that a transition between underdamped and slightly overdamped tip oscillations at higher loads. In recent experiments on MoS_2/Au(111) multiple slips have been also related to a non-monotonic velocity dependence of atomic-scale friction with an abnormal weakening of the mean friction force within the range of 10–100 nm/s[63]. These results are corroborated by an analytical interpretation, where the multiple slips are associated to velocity, temperature and magnitude of the surface corrugation.

3.6 *Atomic Scale Friction at Step Edges*

A common observation in several friction force experiments on the micrometer-scale is the enhancement of friction at step edges related to an increased energy barrier at such locations, known as Schwöbel barrier [64–66]. Such a friction enhancement is strongly direction-dependent. While the friction increases linearly with the applied load in upward scans, it is load-independent for downward scans. This effect was experimentally observed on HOPG, MoS_2, and NaCl in ambient and UHV environments for monatomic up to about tenfold steps [64, 65].

Atomic stick–slip motion across a step has been measured on NaCl(001) and Ge(001) by Steiner et al. [67]. The lateral force was found to increase while scanning both up and down an edge, only if the AFM tip was blunt. However, with atomically sharp tips, the lateral force still increased upwards but decreased and even changed sign downwards, leading to almost overlapping lateral force profiles for both directions. The experimental results could be also interpreted within a modified PT including an additional asymmetric well simulating a sharp drop followed by a gradual recovery of the tip-surface interaction.

4 Atomic-Scale Friction of Reconstructed Surfaces

In this section, we transition from atomically flat to corrugated substrates and discuss how periodic long-range reconstructions, induced by the lattice mismatch between a 2D film and the substrate, affect the atomic friction process. Friction force microscopy in a low load regime was successfully applied to image surface reconstructions on metals, e.g. the Au(111) herringbone [34, 68], charge density wave induced superstructures [69] as well as on single-crystal semiconductors [70]. In order to resolve these surface modulations, the contact size in those experiments was significantly below the characteristic length of the superstructure and reconstruction, respectively (i.e. a few nm).

Recently, atomic friction experiments demonstrating stick–slip behavior have been performed on a variety of moiré superlattices. These structures include ionic crystal films deposited on various substrates [71, 72] and 2D material heterojunctions [73], all of which revealed intriguing variations in friction.

4.1 Amplitude Modulation

The lattice mismatch between two materials gives rise to long-range modulated superstructures at their interface. As an example, KBr film deposited on a NaCl(001) substrate gives rise to a (6 × 6) periodic pattern [74]. On such samples, the surface potential corrugation amplitude varies across the moiré superlattice, which strongly influences the atomic friction. Consequently, this leads to a distinct transition from a dissipative stick–slip motion to a smooth sliding regime aligned with the underlying superstructural pattern [71]. The experimental results by Maier et al. could be easily reproduced by introducing a spatially modulated tip-sample interaction in the PT model, i.e. by replacing. U_0 in Eq. (2) with

$$U_0\left(1 + \alpha \cos \frac{2\pi x}{b}\right). \tag{17}$$

In Eq. (17), α defines the normalized modulation amplitude and b the periodicity of the superstructure. As a consequence, the lateral force is also modulated along the underlying moiré superlattice.

4.2 Centerline Modulation

In the context of moiré superstructures on graphene grown on SiC(0001), a different type of modulation was experimentally observed [73]. The modulation is consistent with a moiré-related potential superimposed on the atomic corrugation potential

$$U = U_{eff} + \beta U_{sup}. \tag{18}$$

where β gives the ratio between the amplitudes of the moiré-related potential U_{sup} [75] and the atomic potential U_{eff}, as in Eq. (1). This so-called centerline modulation does not affect the width of the friction loop, and thus there is no change in the energy dissipated during frictional sliding (provided that load and velocity are low enough). Chan et al. [76] observed that the lateral force modulation critically changes on various graphene moiré patterns on Pt(111). MD simulations unveiled a comparable lateral force modulation caused by the moiré supercells, and related it to the geometric corrugation of graphene [77]. The distinction between these two forms of modulation can be ascribed to the distinctive physical and chemical properties inherent to the 2D layered materials. Indeed, the superposition of short-range and moiré-related potentials arises from the combination of weak interfacial van der Waals interactions between tip and surface, along with the high chemical inertness and strong intralayer covalent bonding, which is peculiar of 2D materials. The topic is further discussed in the next section.

5 Friction on 2D Materials

Recently, 2D materials have been of wide interest due to their extraordinary electronic, mechanical, optical, and chemical properties with respect to their bulk counterparts, making them suitable for applications in electronic and NEMS devices [78]. The most widely studied 2D material is graphene, whose nano- and microtribology properties are outlined in detail elsewhere in this book. Using AFM, the nano-scale frictional response of atomically thin sheets of supported graphene, molybdenum disulfide (MoS_2), niobium diselenide ($NbSe_2$), and hexagonal boron nitride have been investigated [73, 79, 80]. Both graphene and MoS_2 exhibited clear atomic stick–slip friction for exfoliated and grown films down to a single atomic sheet.

5.1 Layer Dependence of Friction

An intriguing question is the comparison of friction on a monolayer, two layers or multiple layers. The simple PT model would not predict a difference, but the experimental observation is showing that the first layer usually shows larger friction compared to the second or third layer. A continuum model based upon the puckering effect can easily explain those effects [79, 81]. Due to the limited bending stiffness of the 2D materials, the layers deform/pucker in the vicinity of the probing tip, as seen in Fig. 7a. This deformation leads to additional dissipation. Multiple layers have a larger bending stiffness compared to the monolayer, which explains the smaller friction on multiple layers compared to the monolayer (Fig. 6).

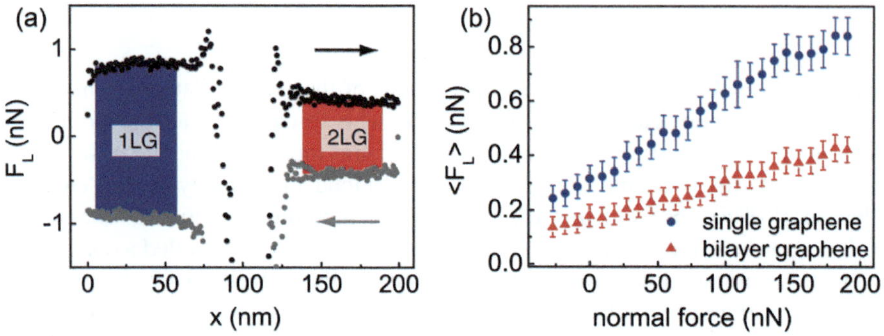

Fig. 6 **a** Friction of monolayer of graphene and a bilayer under UHV conditions. **b** Larger friction is observed on the monolayer compared to the bilayer. Adapted from [80]

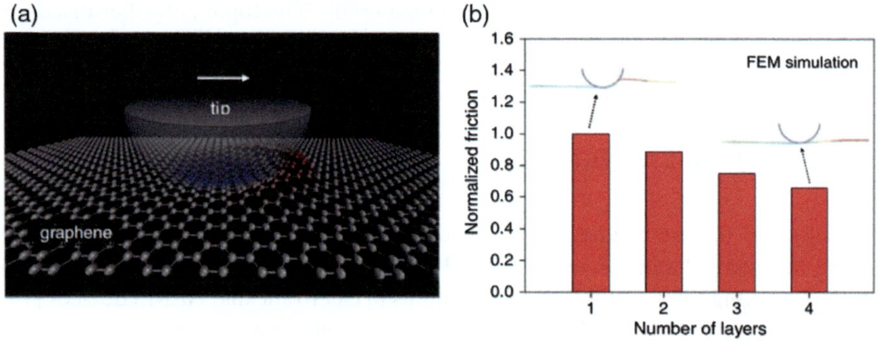

Fig. 7 **a** The puckering effect leads to additional friction compared to the atomic stick–slip. **b** Due to higher bending stiffness, multiple layers show less friction compared to monolayers. Adapted from [79]

5.2 Moiré-Scale Stick–Slip Dynamics

Finally, we consider the mechanical response of the moiré superstructure induced by the probing tip [82, 83]. The in-plane and out-of-plane deformations of the moiré supercell during sliding are important to the atomic friction process, as theoretically predicted by Anderson et al. [84]. Experimentally, more complex friction modulation phenomena are observed, which cannot be interpreted by the PT model assuming a stationary potential landscape. For instance, the friction measurements on a superlattice of graphene-coated Pt(111) surfaces have revealed two distinct regimes of friction depending on the sliding velocities: (i) the friction remains ultralow and nearly constant below some threshold, whereas (ii) it depends logarithmically on the velocity appears above it (see Fig. 8) [85]. In addition, the interfacial twist angle, which results in varying periodicities of moiré superstructures, significantly influences the threshold velocity. A larger superlattice dimension, in particular, results in

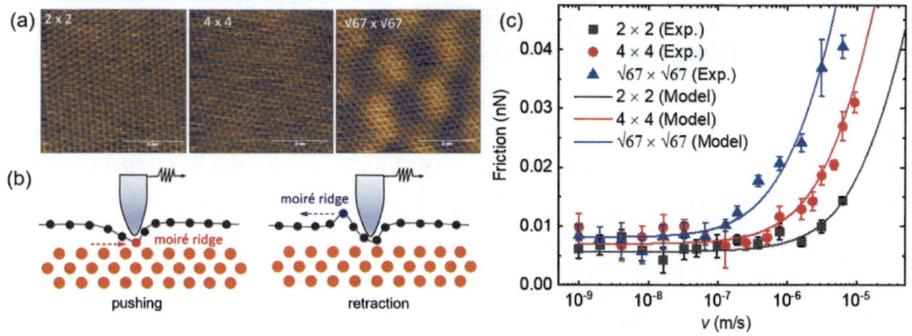

Fig. 8 **a** Topographic maps of graphene/Pt(111) with different moiré lattices. **b** Schematics of energy dissipation mechanism. **c** Velocity dependence of moiré friction demonstrating a transition from nearly constant ultralow values at lower speeds to logarithmic scaling at higher speeds depending on moiré superlattice dimensions. Adapted from [85]

a lower transition velocity. In addition to the well-known atomic stick–slip dynamics, larger-scale higher-friction patterns with a periodicity that matches the moiré superlattice are observed in regime (ii). At higher velocities and larger normal force, friction due to the moiré starts to dominate the energy dissipation process [85, 86]. Because of the significant extension of the elastic deformations during the moiré-level stick phase, a greater amount of energy can be dissipated in the subsequent slip phase. This explains why the friction stemming from the moiré pattern prevails on the atomic friction. This new emerging energy dissipation route due to the elastic deformation and relaxation of moiré ridge induced by the sliding tip has been also verified via atomic MD simulations.

Additionally, moiré dominated non-linear frictional responses to normal loads have been observed on various graphene superlattices [86, 87] in UHV and ambient environment. These findings could be explained in the framework of dissipative moiré-scale stick – slip dynamics due to tip-ridge interactions, as well. In this case, the transition of the frictional load dependence from a superlubric state to a dissipative regime could be achieved by increasing the normal force above a threshold value depending on the moiré pattern (Fig. 9).

Variations in the atomic thin film/substrate interactions provide an alternative solution for a subtly intrinsic control of moiré friction. Song et al. investigated MoS_2 films prepared on Au(111) surfaces under UHV conditions [88] where (this being also the case for as in the case of a Au-coated AFM tip on MoS_2 [89], the atomic layer strongly adheres to the substrate). Friction force measurements for the MoS_2/Au(111) system appeared to be unaffected by the applied load when a relatively large moiré size of 3.3 nm was present [88], in contrast to the previously reported observations of friction at the ridge of the moiré pattern leading to a highly dissipative regime. The puckering effect was not observable when analyzing the corresponding lateral force trace and retrace. This can be ascribed to stabilized atomic layer with moiré superlattice due to the strong adhesion between MoS_2 and the gold surface. Note that, in this case, the preparation and friction measurements of the MoS_2 surface

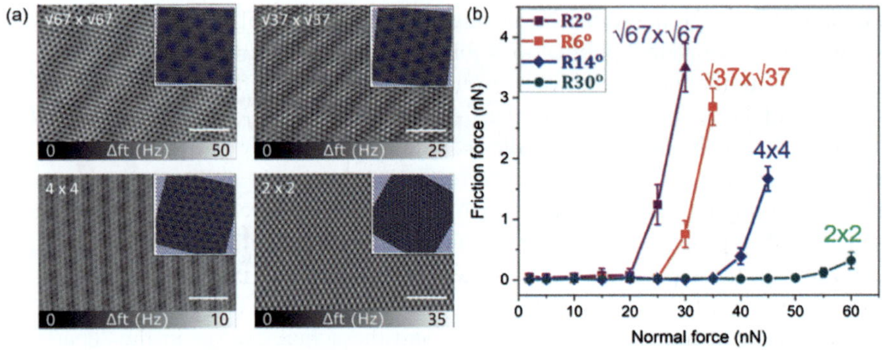

Fig. 9 **a** Friction force maps on graphene/Pt(111) with different moiré lattices. **b** Friction force versus normal force for the different moiré periodicities. The threshold force depends on the size of the moiré lattice. Adapted from [86]

under UHV conditions were very important to exclude potential contaminations in the experimental findings.

6 Conclusions

In conclusion, we have introduced the PT model for atomic-scale sliding friction and showed how it can reproduce the main results of AFM-based friction measurements on crystal surfaces in UHV. These results include the dependence of friction on load, and specifically the transition from smooth sliding to stick–slip, as well as on velocity, temperature and scan direction. The PT model can also reproduce the reduction of friction caused by ultrasonic vibrations, the occurrence of multiple slip events, of stick–slip across step edges and on reconstructed surfaces, where long-ranged height modulations significantly modulate the atomic stick–slip. However, this is not the case for atomic-scale friction measurements on 2D materials, where peculiar mechanisms, such as the puckering effect and elastic deformations of ridges play a dominant role in the sliding motion of sharp nanotips.

References

1. C.M. Mate, G.M. McClelland, R. Erlandsson, S. Chiang, Phys. Rev. Lett. **59**, 1942 (1987)
2. C. Fusco, A. Fasolino, Phys. Rev. B **71**, 045413 (2005)
3. S. Maier, Y. Sang, T. Filleter, M. Grant, R. Bennewitz, E. Gnecco, E. Meyer, Phys. Rev. B **72**(9), 245418 (2005)
4. L. Prandtl, ZAMM - Journal of Applied Mathematics and Mechanics / Zeitschrift für Angewandte Mathematik und Mechanik **8**, 85 (1928)
5. G.A. Tomlinson, Philos. Mag. **7**, 905 (1929)

6. M. A. Lantz, S. J. OShea, A. C. F. Hoole, and M. E. Welland, Appl. Phys. Lett. **70**, 970 (1997).
7. J.P. Cleveland, S. Manne, D. Bocek, P.K. Hansma, Rev. Sci. Instrum. **64**, 403 (1993)
8. C.P. Green, H. Lioe, J.P. Cleveland, R. Proksch, P. Mulvaney, J.E. Sader, Rev. Sci. Instrum. **75**, 1988 (2004)
9. J.E. Sader, I. Larson, P. Mulvaney, L.R. White, Rev. Sci. Instrum. **66**, 3789 (1995)
10. T.R. Albrecht, S. Akamine, T.E. Carver, C.F. Quate, J. Vac. Sci. Technol. A **8**, 3386 (1990)
11. K.L. Johnson, *Contact Mechanics* (Cambridge University Press, Cambridge, 1985)
12. A. Socoliuc, R. Bennewitz, E. Gnecco, E. Meyer, Phys. Rev. Lett. **92**, 134301 (2004)
13. R.W. Carpick, D.F. Ogletree, M. Salmeron, Appl. Phys. Lett. **70**, 1548 (1997)
14. E. Gnecco, R. Bennewitz, T. Gyalog, and E. Meyer, J. Phys.-Condens. Matter **13**, R619 (2001).
15. B. Luan, M.O. Robbins, Nature **435**, 929 (2005)
16. T. Gyalog, M. Bammerlin, R. Luthi, E. Meyer, H. Thomas, Europhys. Lett. **31**, 269 (1995)
17. P. Steiner, R. Roth, E. Gnecco, A. Baratoff, S. Maier, T. Glatzel, E. Meyer, Phys. Rev. B **79**, 045414 (2009)
18. H. Hölscher, U.D. Schwarz, O. Zworner, R. Wiesendanger, Phys. Rev. B **57**, 2477 (1998)
19. E. Gnecco, Europhys. Lett. **91**, 66008 (2010)
20. A. Schirmeisen, L. Jansen, H. Fuchs, Phys. Rev. B **71**, 245403 (2005)
21. E. Gnecco, R. Bennewitz, T. Gyalog, C. Loppacher, M. Bammerlin, E. Meyer, H. Guntherodt, Phys. Rev. Lett. **84**, 1172 (2000)
22. E. Riedo, E. Gnecco, R. Bennewitz, E. Meyer, H. Brune, Phys. Rev. Lett. **91**, 084502 (2003)
23. Y. Sang, M. Dube, M. Grant, Phys. Rev. Lett. **87**, 174301 (2001)
24. D.G. Abel, S.Y. Krylov, J.W.M. Frenken, Phys. Rev. Lett. **99**, 166102 (2007)
25. T. Gyalog, H. Thomas, Europhys. Lett. **37**, 195 (1997)
26. M. Weiss, F.J. Elmer, Phys. Rev. B **53**, 7539 (1996)
27. M. Hirano, Wear **254**, 932 (2003)
28. O.M. Braun, Y.S. Kivshar, *The Frenkel-Kontorova Model: Concepts, Methods, and Applications* (Springer-Verlag, Berlin, 2004)
29. R. Lüthi et al., J. Vac. Sci. Technol. B **14**, 1280 (1996)
30. R.W. Carpick, Q. Dai, D.F. Ogletree, M. Salmeron, Tribol. Lett. **5**, 91 (1998)
31. M. Ishikawa, S. Okita, N. Minami, K. Miura, Surf. Sci. **445**, 488 (2000)
32. U. Landman, W.D. Luedtke, N.A. Burnham, R.J. Colton, Science **248**, 454 (1990)
33. R. Bennewitz, T. Gyalog, M. Guggisberg, M. Bammerlin, E. Meyer, H.J. Guntherodt, Phys. Rev. B **60**, R11301 (1999)
34. N.N. Gosvami, T. Filleter, P. Egberts, R. Bennewitz, Tribol. Lett. **39**, 19 (2010)
35. R. Bennewitz, E. Gnecco, T. Gyalog, E. Meyer, Tribol. Lett. **10**, 51 (2001)
36. M. Enachescu, R.W. Carpick, D.F. Ogletree, M. Salmeron, J. Appl. Phys. **95**, 7694 (2004)
37. M. Enachescu, R.J.A. van den Oetelaar, R.W. Carpick, D.F. Ogletree, C.F.J. Flipse, M. Salmeron, Phys. Rev. Lett. **81**, 1877 (1998)
38. C. Fusco, A. Fasolino, Appl. Phys. Lett. **84**, 699 (2004)
39. M. Dienwiebel, G.S. Verhoeven, N. Pradeep, J.W. Frenken, J.A. Heimberg, H.W. Zandbergen, Phys. Rev. Lett. **92**, 126101 (2004)
40. S.Y. Krylov, K.B. Jinesh, H. Valk, M. Dienwiebel, J.W.M. Frenken, Phys. Rev. E **71**, 065101 (2005)
41. A. Socoliuc, E. Gnecco, S. Maier, O. Pfeiffer, A. Baratoff, R. Bennewitz, E. Meyer, Science **313**, 207 (2006)
42. E. Gnecco, A. Socoliuc, S. Maier, J. Gessler, T. Glatzel, A. Baratoff, E. Meyer, Nanotechnology **20**, 025501 (2009)
43. R. Roth, O.Y. Fajardo, J.J. Mazo, E. Meyer, E. Gnecco, Appl. Phys. Lett. **104**, 083103 (2014)
44. O.Y. Fajardo, E. Gnecco, J.J. Mazo, Phys. Rev. B **89**, 075423 (2014)
45. E. Gnecco, R. Bennewitz, E. Meyer, Phys. Rev. Lett. **88**, 215501 (2002)
46. A. Socoliuc, E. Gnecco, R. Bennewitz, E. Meyer, Phys. Rev. B **68**, 115416 (2003)
47. L. Jansen, H. Holscher, H. Fuchs, A. Schirmeisen, Phys. Rev. Lett. **104**, 256101 (2010)
48. A. Schirmeisen, L. Jansen, H. Holscher, H. Fuchs, Appl. Phys. Lett. **88**, 123108 (2006)
49. P. Reimann, M. Evstigneev, Phys. Rev. Lett. **93**(230802), 230802 (2004)

50. X.Y. Zhao, S.R. Phillpot, W.G. Sawyer, S.B. Sinnott, S.S. Perry, Phys. Rev. Lett. **102**, 186102 (2009)
51. M. Vorholzer, J.G. Vilhena, R. Perez, E. Gnecco, D. Dietzel, A. Schirmeisen, Phys. Rev. X **9**, 041045 (2019)
52. J.J. Mazo, D. Dietzel, A. Schirmeisen, J.G. Vilhena, E. Gnecco, Phys. Rev. Lett. **118**(246101), 246101 (2017)
53. M. Evstigneev, A. Schirmeisen, L. Jansen, H. Fuchs, P. Reimann, Phys. Rev. Lett. **97**, 240601 (2006)
54. J. Konrad, E. Gnecco, D. Dietzel, A. Schirmeisen, Phys. Rev. B **107**, 195438 (2023)
55. S. G. Balakrishna, A. S. de Wijn, and R. Bennewitz, Physical Review B **89**, 245440(8) (2014).
56. M.R. Vazirisereshk, K. Hasz, R.W. Carpick, A. Martini, J. Phys. Chem. Lett. **11**, 6900 (2020)
57. K.L. Johnson, J. Woodhouse, Tribol. Lett. **5**, 155 (1998)
58. J. Nakamura, S. Wakunami, A. Natori, Phys. Rev. B **72**(6), 235415 (2005)
59. Z. Tshiprut, S. Zelner, M. Urbakh, Phys. Rev. Lett. **102**, 136102 (2009)
60. S.N. Medyanik, W.K. Liu, I.H. Sung, R.W. Carpick, Phys. Rev. Lett. **97**, 136106 (2006)
61. E. Gnecco, R. Roth, A. Baratoff, Phys. Rev. B **86**, 035443 (2012)
62. R. Roth, T. Glatzel, P. Steiner, E. Gnecco, A. Baratoff, E. Meyer, Tribol. Lett. **39**, 63 (2010)
63. Y.M. Song, J. Wang, A. Hinaut, S. Scherb, S.Y. Huang, T. Glatzel, E. Tosatti, E. Meyer, Phys. Rev. Lett., in press (2024)
64. H. Hölscher, D. Ebeling, U.D. Schwarz, Phys. Rev. Lett. **101**, 246105 (2008)
65. T. Müller, M. Lohrmann, T. Kasser, O. Marti, J. Mlynek, G. Krausch, Phys. Rev. Lett. **79**, 5066 (1997)
66. E. Meyer, R. Luthi, L. Howald, M. Bammerlin, M. Guggisberg, H.J. Guntherodt, J. Vac. Sci. Technol. B **14**, 1285 (1996)
67. P. Steiner, E. Gnecco, F. Krok, J. Budzioch, L. Walczak, J. Konior, M. Szymonski, E. Meyer, Phys. Rev. Lett. **106**, 186104 (2011)
68. Q.Y. Li, Y.L. Dong, A. Martini, R.W. Carpick, Tribol. Lett. **43**, 369 (2011)
69. E. Panizon, T. Marx, D. Dietzel, F. Pellegrini, G.E. Santoro, A. Schirmeisen, E. Tosatti, New J. Phys. **20**, 023033 (2018)
70. M. Goryl, J. Budzioch, F. Krok, M. Wojtaszek, M. Kolmer, L. Walczak, J. Konior, E. Gnecco, M. Szymonski, Phys. Rev. B **85**, 085308 (2012)
71. S. Maier, E. Gnecco, A. Baratoff, R. Bennewitz, E. Meyer, Phys. Rev. B **78**(5), 045432 (2008)
72. T. Filleter, W. Paul, R. Bennewitz, Phys. Rev. B **77**, 035430 (2008)
73. T. Filleter, R. Bennewitz, Phys. Rev. B **81**, 155412 (2010)
74. S. Maier, O. Pfeiffer, T. Glatzel, E. Meyer, T. Filleter, R. Bennewitz, Phys. Rev. B **75**, 195408 (2007)
75. P. Steiner, E. Gnecco, T. Filleter, N.N. Gosvami, S. Maier, E. Meyer, R. Bennewitz, Tribol. Lett. **39**, 321 (2010)
76. N. Chan, S.G. Balakrishna, A. Klemenz, M. Moseler, P. Egberts, R. Bennewitz, Carbon **113**, 132 (2017)
77. J. Liu, S. Zhang, Q.Y. Li, X.Q. Feng, Z.F. Di, C. Ye, Y.L. Dong, Carbon **125**, 76 (2017)
78. A.K. Geim, Science **324**, 1530 (2009)
79. C. Lee, Q. Li, W. Kalb, X.Z. Liu, H. Berger, R.W. Carpick, J. Hone, Science **328**, 76 (2010)
80. T. Filleter, J.L. McChesney, A. Bostwick, E. Rotenberg, K.V. Emtsev, T. Seyller, K. Horn, R. Bennewitz, Phys. Rev. Lett. **102**, 086102 (2009)
81. S. Li, Q. Li, R.W. Carpick, P. Gumbsch, X.Z. Liu, X. Ding, J. Sun, J. Li, Nature **539**, 541 (2016)
82. S. Koch et al., ACS Nano **7**, 2927 (2013)
83. M. Yankowitz, K. Watanabe, T. Taniguchi, P. San-Jose, B.J. LeRoy, Nat. Commun. **7**, 13168 (2016)
84. D. Andersson, A.S. de Wijn, Nat. Commun. **11**, 420 (2020)
85. Y.M. Song, X. Gao, A. Hinaut, S. Scherb, S.Y. Huang, T. Glatzel, O. Hod, M. Urbakh, E. Meyer, Nano Lett. **22**, 9529 (2022)
86. Z. Liu, J.G. Vilhena, A. Hinaut, S. Scherb, F. Luo, J. Zhang, T. Glatzel, E. Gnecco, E. Meyer, Nano Lett. **23**, 4693 (2023)

87. S. Zhang, Q.Z. Yao, L.X. Chen, C.X. Jiang, T.B. Ma, H.M. Wang, X.Q. Feng, Q.Y. Li, Phys. Rev. Lett. **128**, 226101 (2022)
88. Y.M. Song, A. Hinaut, S. Scherb, Y. Pellmont, R. Pawlak, S.Y. Huang, Z. Liu, T. Glatzel, E. Meyer, Appl. Surf. Sci. **601**, 154230 (2022)
89. L. Agmon, R. Almog, D. Gaspar, G. Voscoboynik, M. Choudhary, J. Jopp, Z. Klausner, A. Ya'akobovitz, R. Berkovich, Tribol. Int.. Int. **174**, 107734 (2022)

The (Grain) Boundaries of Structural Superlubricity

Oded Hod, Michael Urbakh, and Diana Berman

Abstract Structural superlubricity, a state of ultra-low friction and wear arising from incommensurability between contacting surfaces, is an intriguing physical phenomenon that holds promise for the significant reduction of energy loss and material damage in mechanical systems. One of the most prominent realizations of superlubric motion is demonstrated for nano- and micro-scale heterogeneous layered material contacts and their twisted homogeneous counterparts. On the route to scaling up superlubricity stand a few obstacles. In this chapter, we focus on the effect of grain boundaries, which inevitably emerge in large-scale layered material contacts, on their frictional properties. New frictional mechanisms associated with grain boundaries, such as shear induced buckling and unbuckling of corrugated dislocations and moiré superstructure scattering, are discussed. These, in turn, are characterized by unique frictional behavior, including nonmonotonic dependence on normal load, sliding velocity, and temperature that can be harnessed to restore structural superlubricity at increasing length-scales.

1 Introduction

In the past two decades, significant progress in comprehending the atomic mechanisms underlying frictional processes have led to the observation of ultralow friction (with kinetic friction coefficients below 10^{-3}) at nano- and micro-scale interfaces of two-dimensional (2D) materials [1–5]. This intriguing phenomenon, often referred

O. Hod · M. Urbakh (✉)
Department of Physical Chemistry, School of Chemistry, The Raymond and Beverly Sackler Faculty of Exact Sciences and The Sackler Center for Computational Molecular and Materials Science, Tel Aviv University, Tel Aviv, Israel
e-mail: urbakh@tauex.tau.ac.il

O. Hod
e-mail: odedhod@tauex.tau.ac.il

D. Berman (✉)
Department of Materials Science and Engineering, University of North Texas, Denton, USA
e-mail: Diana.Berman@unt.edu

E. Gnecco and E. Meyer (eds.), *Fundamentals of Friction and Wear on the Nanoscale*, NanoScience and Technology, https://doi.org/10.1007/978-3-031-63065-1_5

to as structural superlubricity [5–7], has been theoretically predicted and experimentally demonstrated in various contacts of pristine single-crystal layered materials, including nano- and micro-scale graphitic structures [8–10], graphene/*h*-BN heterojunctions [11–14], homogenous MoS_2 interfaces [15–17], graphene/MoS_2 heterojunctions [18], as well as other crystalline interfaces [19, 20]. The common feature typical to contacts of layered materials is their anisotropic structure, consisting of covalently bonded layers coupled via weaker dispersive interactions. A necessary condition to obtain superlubric behavior in such contacts is incommensurability between the rigid crystalline networks of the contacting surfaces. For homogeneous layered material junctions this condition can be fulfilled when the contacting surfaces are laterally twisted with respect to each other, thus avoiding frictional lock-in, observed at the commensurate aligned configuration [8]. Such lock-in is completely circumvented in heterojunctions that possess mismatching lattice periodicities, providing superlubric behavior that is robust against interfacial reorientation [11–13].

Scaling up structural superlubricity towards the meso- and macroscales, bears great technological potential for the reduction of energy loss and material wear in (electro-)mechanical devices [20, 21]. Within the realm of layered material interfaces, this endeavor is challenged by a number of effects that may introduce additional routes for frictional energy dissipation. The latter include surface and edge elasticity, structural defects, interlayer binding, intercalation of contaminants, and corrugated grain boundaries (GBs). In this Chapter, we review recent advances regarding the understanding of the effect of GB dynamics on the frictional properties of polycrystalline layered interfaces.

2 Structural Superlubricity of Pristine Layered Interfaces

Polycrystalline layered material contacts consist of a set of randomly oriented pristine regions (grains) separated by localized GBs. Therefore, prior to discussing the effects of GBs, one should first recall the frictional properties of pristine interfaces. In the early 1990s, Hirano and Sokoloff conducted computational investigations on simple model systems that predicted the occurrence of superlubric sliding regimes in incommensurate, defect-free, atomically smooth, and highly-ordered solid crystalline interfaces [22, 23].

Their idea centered around the concept that when the lattices of two ordered material surfaces are in perfect registry and aligned in the direction of sliding, atomic-level interlocking leads to relatively high interfacial friction (see Fig. 1) [6, 7, 24–26]. Microscopically, this can be understood by the fact that the corrugation of the sliding energy landscape is dominated by Pauli repulsion variations between the electron clouds surrounding atoms on adjacent layers as they cross each other. In the commensurate configuration, therefore, sliding involves the simultaneous crossing of many energy barriers, each of relatively high magnitude. Conversely, if the lattices mismatch, either intrinsically or via relative lateral rotation, an incommensurate

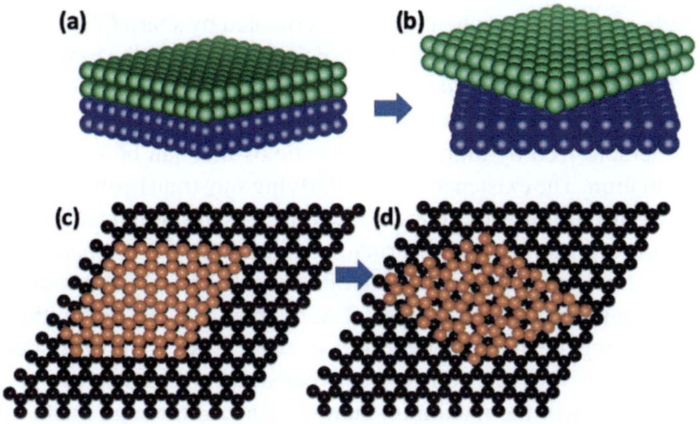

Fig. 1 A schematic representation of aligned commensurate (**a**, **c**) and twisted incommensurate (**b**, **d**) bulk cubic (**a**, **b**) and layered hexagonal (**c**, **d**) contacts. Reproduced with permission from Ref. [1]

state forms, where sliding involves the effective cancellation of lateral forces and a significant reduction of friction.

This prediction was followed by the development of novel measurement techniques that allowed for the experimental verification of superlubricity at incommensurate rigid contacts [1, 2, 8, 27]. Through the application of atomic force microscopy (AFM) and friction force microscopy (FFM), kinetic friction coefficients ranging from virtually zero (within experimental error bars) to 5×10^{-3} have been reported for nanoscale graphitic contacts [28–30]. These studies revealed that friction in pristine rigid crystalline interfaces varies from relatively high to vanishingly low when the interfacial configuration changes from commensurate to incommensurate.

3 Effect of Grain-Boundaries on the Friction of Layered Interfaces

At increasing contact dimensions, the interfacing surfaces often exhibit a mosaic of randomly oriented grains, separated by GBs that may form dislocation chains [31]. The scattered grain orientations can prevent frictional lock-in due to reduced overall interfacial commensurability, thus supporting superlubric behaviour [32, 33]. Nonetheless, the GBs themselves may induce large out-of-plane surface deformations [34–36] that can degrade or even completely eliminate superlubricity and induce wear [37].

A prototypical example of such polycrystalline two-dimensional (2D) surfaces is polycrystalline graphene (PolyGr), which is composed of randomly shaped and oriented single crystalline graphene patches separated by sharp GBs. The latter typically appear as arrays of pentagon-heptagon pairs (see Fig. 2), the exact configuration of which is determined by the relative orientation of the adjacent grains (characterized by the misfit angle θ between the corresponding lattice vectors). These topological defects are characterized by high local stress fields that can be relaxed via vertical surface deformation. The existence of an underlying substrate further limits the occurrence of downward protrusions, thus the local stresses developing at GBs are usually relieved via upward out-of-plan deformation. Figure 2b shows that at small misfit angles, highly corrugated GBs, consisting of a series of individual upward protruding bumps with a typical bump height of ~ 2.2 Å, appear [35, 36, 38]. In the range $0° < \theta < 12.5°$, increasing the misfit angle results in an increase of the dislocation density. This, in turn, induces stress cancellation between neighboring dislocations leading to a significant decrease of the maximum bump height. At the high misfit angle range ($21.8° < \theta < 30°$) the GBs form continuous dislocation chains that can be either corrugated or flat. Such GB deformations have been experimentally characterized with both scanning tunneling microscopy (STM) and AFM [35, 36].

Before studying the intricate frictional behavior of the entire mosaic structure [40, 41], it is instructive to first consider the effects of individual GBs, elucidate energy dissipation mechanisms induced by collective dynamical effects of different topological defects along the GB under shear, and identify their manifestations on the frictional characteristic. One of the earliest experimental investigations of single GB friction in layered materials was presented by the group of Bennewitz [42]. In that study, the lateral force acting on an AFM tip was recorded while crossing a GB of polycrystalline graphene grown atop a Pt(111) surface. Figure 3 presents a lateral force map (top) and a horizontal line cut through that map (bottom), both clearly demonstrating moiré patterns induced force variations at the grain regions

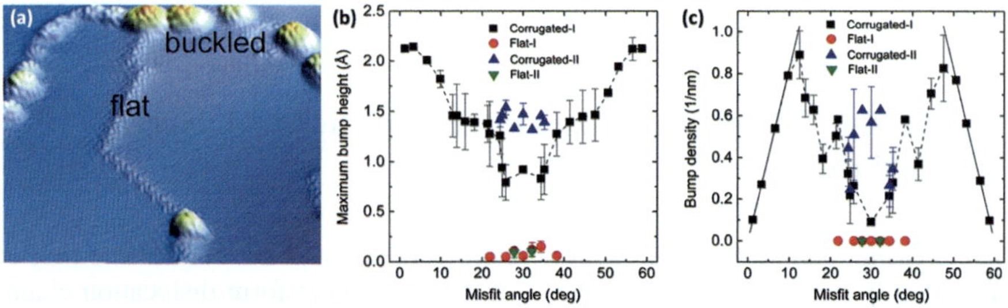

Fig. 2 **a** STM image of epitaxial graphene grown on a SiC($000\bar{1}$) substrate that exhibits buckled and flat grain boundaries separating single-crystalline domains with different lattice orientations. Reproduced with permission from Ref. [36]. Graphene GB **b** Maximum bump height; and **c** bump density calculated as a function of misfit angle, θ. Based on their corrugation and the types of dislocations, the GBs are categorized in four groups: Corrugated-I (black squares), Corrugated-II (blue up facing triangles), Flat-I (red circles), and Flat-II (green down facing triangles). The black solid lines in panel (**c**) indicate the theoretical prediction [34, 39]. The black dashed lines in all panels are plotted as guide for the eye. For flat GBs, the bump density is presented as zero and the maximum out-of-plane corrugation is shown in (**b**). Reproduced with permission from Ref. [38]

and enhanced tip drag at the corrugated GB region. Notably, a larger misfit between the contacting grains resulted in increased lateral force across the GB region.

The underlying mechanism of the friction enhancement was revealed via elaborate atomistic molecular dynamics simulations, which demonstrated that at low and moderate normal loads the frictional energy dissipation is dominated by shear induced buckling and unbuckling of GB dislocations [43]. When considering low temperature scenarios, at low normal load the buckling barrier is high and the contribution of this energy dissipation route is relatively low. As the normal load increases,

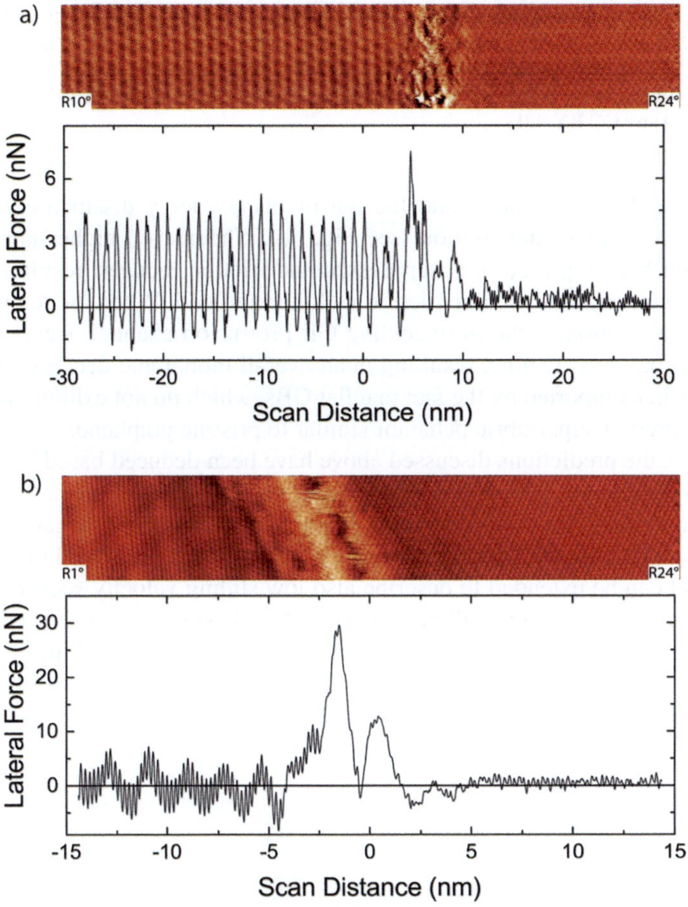

Fig. 3 **a** Lateral force map (top sub-panel) and a typical cross section (bottom sub-panel) of a GB separating two graphene domains, one rotated by 10° (R10°) and the other by 24° (R24°) with respect to the underlying Pt (111) substrate lattice vector. The lateral force profile is obtained as an average of 50 adjacent lines along the horizontal direction corresponding to a spatial width of 1.41 nm. **b** Same as panel (**a**) but for an R1°/ R24° graphene GB and lateral force averaging over a width of 0.70 nm. Reproduced with permission from Ref. [42]

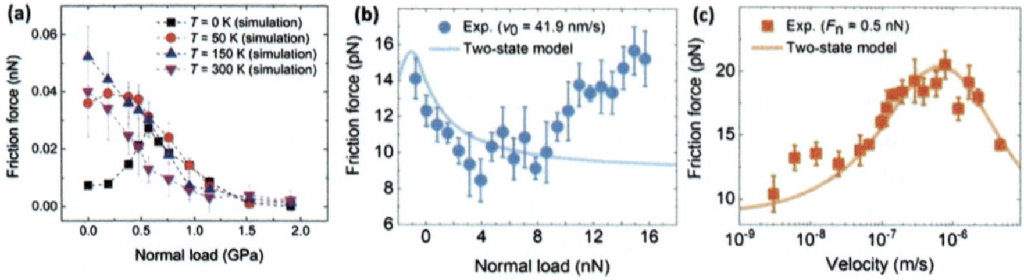

Fig. 4 **a** Molecular dynamics simulation results of normal load dependence of friction between pristine and polycrystalline graphene (made from patches oreiented at 0° and 8°) layers at representative temperatures of 0, 50, 150, and 300 K. Reproduced with permission from Ref. [43] by CC BY 4.0. **b, c** normal load and velocity dependence of friction, measured at a sliding velocity of 41.9 nm/s (blue circles) or a normal load of 0.5 nN (orange squares), respectively. Results of a phenomenological two-state model are presented by the solid lines. Reproduced with permission from Ref. [44] by CC BY 4.0

the buckling barrier reduces, and the out of plane energy dissipation channel is activated, thus enhancing friction (see Fig. 4a). Under even higher normal loads vertical (un)buckling motion is suppressed and friction reduces, resulting in atypical nonmonotonic variation of friction with normal load. When the temperature is increased, activation of the (un)buckling GB protrusion channel occurs already at the low normal load regime, resulting in an overall monotonic decrease of friction. This is further supported by the fact that flat GBs, which do not exhibit out-of-place buckling, present superlubric behavior similar to pristine graphene.

Notably, the predictions discussed above have been deduced based on molecular dynamics simulations that can only treat sliding velocities, which are higher by orders of magnitude than those considered in experiments. To fill in this gap, a two-state phenomenological model has been developed that could capture well the simulation results and can be extended to describe also low sliding velocity scenarios. Recent experiments on the polycrystalline-graphene/Pt(111) system [44] showed excellent agreement (at low to moderate normal loads, where the (un)buckling mechanism dominates) with the predictions of this phenomenological model both for the load and the velocity dependence of friction (see Fig. 4b, c).

At this point, it is evident that the contribution of buckling/unbuckling dynamics of GB protrusions to the friction of layered interfaces is based on a bistability mechanism that is not unique to the case of layered material interfaces. Nonetheless, a signature feature of layered material interfaces is the appearance of moiré superstructures. Upon shear motion, these superstructures propagate at a collective speed that may considerably exceed the actual sliding velocity. In the presence of GBs, the moiré patterns on both sides of the dividing line possess different periodicities and corrugation. Therefore, under shear motion GBs serve as impenetrable borders at which superstructure back-scattering dynamics occurs. This was recently demonstrated computationally for polycrystalline graphene/*h*-BN interfaces [45], where apart from the GB-localized buckling/unbuckling energy dissipation mechanism signatures, a delocalized moiré GB scattering friction route kicks-in at the high normal load regime.

We note that in a recent experiment, it was also demonstrated that GB friction can be modulated by an external electrical gate. Using AFM with a platinum tip sliding over GBs of a polycrystalline MoS_2 surface, Jeong et al. [46] demonstrated that charge accumulation at the GB region upon doping increases the attractive interaction between the tip and the surface, thus increasing friction. If a similar effect can be designed to enhance repulsion, friction reduction is expected to occur, thus establishing electrostatic control over friction.

4 Towards Macroscale Observations of Superlubricity Across Grain Boundaries

Underlying the above-discussed phenomenological two-state model for GB protrusion buckling/unbuckling the important assumption is that buckling events occurring at different dislocation centers are independent. The good agreement that this model demonstrates against fully atomistic simulation of various corrugated GBs supports this assumption and indicates that collective GB effects are of minor importance to polycrystalline interfacial friction. This, in turn, suggests that the frictional dissipation of corrugated GBs is proportional to their overall length that grows linearly with contact area, S_{tot}. To demonstrate this, we note that the rim of a single grain, namely the length of its GB, L_{grain}, is proportional to the square root of its surface area. Marking by $P(S)$ the probability density function to obtain a grain of area S, the average grain area is given by $\overline{S}_{grain} = \int P(S)S\mathrm{d}S$ and the corresponding average boundary length per grain is $\overline{L}_{grain} \propto \int P(S)\sqrt{S}\mathrm{d}S$. Given that the average number of grains is $N_{grain} = S_{tot}/\overline{S}_{grain}$, the average overall GB length is $L_{tot} \propto \zeta N_{grain}\overline{L}_{grain} = \zeta\left(\overline{L}_{grain}/\overline{S}_{grain}\right)S_{tot}$, which is proportional to S_{tot}. Here, ζ is a constant pre-factor of $O(1)$ accounting for overlapping boundaries of neighboring grains.

Considering that the friction coefficient is defined as the ratio between the friction and normal forces, $\mu = F_f/F_N$, where $F_f \propto L_{tot}$ and $F_N = S_{tot}\sigma$, where σ is the normal pressure, the GB contribution to the friction coefficient is expected to be proportional to the GB length per unit area: $\mu \propto L_{tot}/S_{tot} \propto \zeta\left(\overline{L}_{grain}/\overline{S}_{grain}\right)$.

Notably, the GB friction force scaling with contact area is stronger than the sublinear scaling found for pristine incommensurate layered contacts characterizing the grain surface area [3–5]. Therefore, polycrystalline layered junctions are expected to present higher friction than their pristine counterparts, despite the inherently enhanced interfacial incommensurability. Nevertheless, the predicted nonmonotonic GB frictional dependence on normal load leads to negative differential friction coefficients at the high normal load regime. Hence, contrary to common wisdom, enhancing the external load may actually reduce the dissipative effect of GBs and assist in restoring superlubricity. Additional gain can be obtained by considering multi-contact layered junctions [47–51]. For example, substrate pre-structuring may be used to divide the macro-scale flat contact into a set of independent nanoscale contacts. In this case, GB friction will scale linearly only with the actual contact area, thus eliminating the linear scaling with the nominal contact area. Such approaches may therefore open the way for achieving large-scale superlubricity by reducing the excess friction associated with each individual GB. In this respect, recent experiments on mixed MXene and MoS_2 tribofilms (see Fig. 5) indicated that when approaching the macroscale, several other frictional phenomena, e.g. chemistry/environment or wear effects, overshadow GB effects [52].

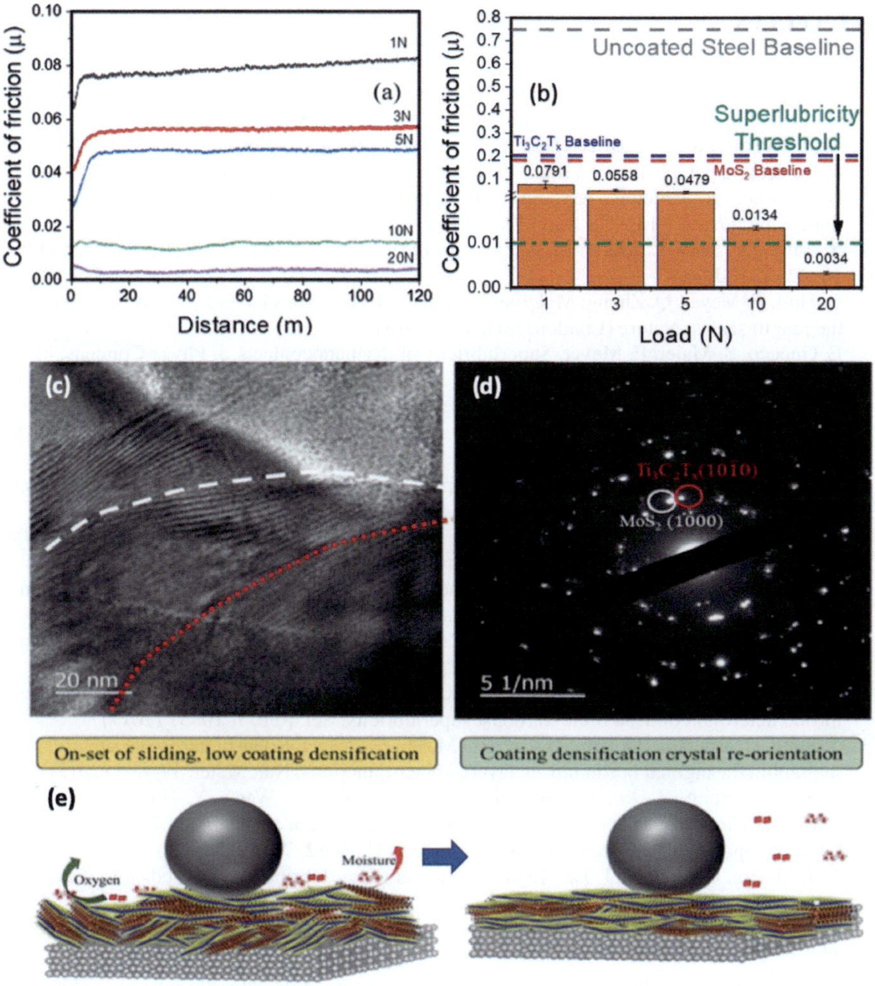

Fig. 5 **a** Coefficient of friction of MoS2-Mxene ($Ti_3C_2T_x$) solid lubricant coatings as a function of sliding distance under various normal loads for unidirectional sliding at a velocity of 0.1 m/s. **b** Summary of steady-state friction values as a function of normal load, compared to steel-on-steel, MoS_2, and Mxene reference values. **c, d** Bright field TEM and electron diffraction images, respectively, of the tribolayer at the end of the long-term frictional test. **e** Schematics of an in-operando mechanism consisting of coating densification and reorientation to obtain a superlubric coating. Adapted with permission from Ref. [52]

Acknowledgements "D.B. and M.U. acknowledge the support of the NSF (Award No. 2323452) and BSF (Award No. 2023614) under the NSF/BSF initiative. D.B. gratefully acknowledges support by the Fulbright U.S. Distinguished Scholar Program. O.H. is grateful for the generous financial support of the Heineman Chair in Physical Chemistry and the Tel Aviv University Center for Nanoscience and Nanotechnology".

References

1. A. Ayyagari, D. Berman, K.I. Alam, A. Erdemir, Progress in superlubricity across different media and material systems—a review. Front. Mech. Eng. **8** (2022)
2. M.Z. Baykara, M.R. Vazirisereshk, A. Martini, Emerging superlubricity: a review of the state of the art and perspectives on future research. Appl. Phys. Rev. **5**, 041102 (2018)
3. D. Berman, A. Erdemir, A.V. Sumant, Approaches for achieving superlubricity in two-dimensional materials. ACS Nano **12**, 2122–2137 (2018)
4. X. Sang, D. Liang, L. Xia, Y. Li, A review of recent advances in superlubricity. Tribol. Online **18**, 42–52 (2023)
5. O. Hod, E. Meyer, Q. Zheng, M. Urbakh, Structural superlubricity and ultralow friction across the length scales, Nature (London) 563, 485 (2018).
6. E. Gnecco, S. Maier, E. Meyer, Superlubricity of dry nanocontacts. J. Phys.: Condens. Matter **20**, 354004 (2008)
7. A. Socoliuc, R. Bennewitz, E. Gnecco, E. Meyer, Transition from stick-slip to continuous sliding in atomic friction: entering a new regime of ultralow friction. Phys. Rev. Lett. **92**, 134301 (2004)
8. M. Dienwiebel, G.S. Verhoeven, N. Pradeep, J.W. Frenken, J.A. Heimberg, H.W. Zandbergen, Superlubricity of graphite. Phys. Rev. Lett. **92**, 126101 (2004)
9. Z. Liu, J. Yang, F. Grey, J.Z. Liu, Y. Liu, Y. Wang, Y. Yang, Y. Cheng, Q. Zheng, Observation of microscale superlubricity in graphite. Phys. Rev. Lett. **108**, 205503 (2012)
10. Koren, E., Lörtscher, E., Rawlings, C., Knoll, A. W. & Duerig, U. Adhesion and friction in mesoscopic graphite contacts. Science 348, 679–683 (2015).
11. I. Leven, D. Krepel, O. Shemesh, O. Hod, Robust superlubricity in graphene/h-BN heterojunctions. J. Phys. Chem. Lett. **4**, 115–120 (2013)
12. D. Mandelli, I. Leven, O. Hod, M. Urbakh, Sliding friction of graphene/hexagonal–boron nitride heterojunctions: a route to robust superlubricity. Sci. Rep. **7**, 10851 (2017)
13. Y. Song, D. Mandelli, O. Hod, M. Urbakh, M. Ma, Q. Zheng, Robust microscale superlubricity in graphite/hexagonal boron nitride layered heterojunctions. Nat. Mater. **17**, 894 (2018)
14. Y. Liu, A. Song, Z. Xu, R. Zong, J. Zhang, W. Yang, R. Wang, Y. Hu, J. Luo, T. Ma, Interlayer friction and superlubricity in single-crystalline contact enabled by two-dimensional flake-wrapped atomic force microscope tips. ACS Nano **12**, 7638–7646 (2018)
15. J.M. Martin, C. Donnet, T. Le Mogne, T. Epicier, Superlubricity of molybdenum disulphide. Phys. Rev. B **48**, 10583–10586 (1993)
16. J.M. Martin, H. Pascal, C. Donnet, T. Le Mogne, J.L. Loubet, T. Epicier, Superlubricity of MoS_2: crystal orientation mechanisms. Surf. Coat. Technol. **68–69**, 427–432 (1994)
17. H. Li, J. Wang, S. Gao, Q. Chen, L. Peng, K. Liu, X. Wei, Superlubricity between MoS_2 Monolayers. Adv. Mater. **29**, 1701474 (2017)
18. L. Wang, X. Zhou, T. Ma, D. Liu, L. Gao, X. Li, J. Zhang, Y. Hu, H. Wang, Y. Dai, Superlubricity of a graphene/MoS_2 heterostructure: a combined experimental and DFT study. Nanoscale **9**, 10846–10853 (2017)
19. M. Marian, D. Berman, A. Rota, R.L. Jackson, A. Rosenkranz, Layered 2D nanomaterials to tailor friction and wear in machine elements—a review. Adv. Mater. Interfaces **9**, 2101622 (2022)
20. M. Hirano, K. Shinjo, R. Kaneko, Y. Murata, Observation of superlubricity by scanning tunneling microscopy. Phys. Rev. Lett. **78**, 1448–1451 (1997)
21. D. Berman, A. Erdemir, Achieving ultralow friction and wear by tribocatalysis: enabled by in-operando formation of nanocarbon films. ACS Nano **15**, 18865–18879 (2021)
22. K. Shinjo, M. Hirano, Dynamics of friction: superlubric state. Surf. Sci. **283**, 473–478 (1993)
23. M. Hirano, Superlubricity: a state of vanishing friction. Wear **254**, 932–940 (2003)
24. M.H. Müser, Structural lubricity: role of dimension and symmetry. EPL (Europhysics Letters) **66**, 97 (2004)
25. M. Dienwiebel, J.W. Frenken, *Experimental Observations of Superlubricity and Thermolubricity, Fundamentals of Friction and Wear* (Springer, 2007), pp. 159–176

26. A.E. Filippov, M. Dienwiebel, J.W. Frenken, J. Klafter, M. Urbakh, Torque and twist against superlubricity. Phys. Rev. Lett. **100**, 046102 (2008)
27. Z. Zheng, Z. Guo, W. Liu, J. Luo, Low friction of superslippery and superlubricity: a review. Friction **11**, 1121–1137 (2023)
28. C. Lee, Q. Li, W. Kalb, X.-Z. Liu, H. Berger, R.W. Carpick, J. Hone, Frictional characteristics of atomically thin sheets. Science **328**, 76–80 (2010)
29. D. Berman, A. Erdemir, A.V. Zinovev, A.V. Sumant, Nanoscale friction properties of graphene and graphene oxide. Diamond Relat. Mater. **54**, 91–96 (2015)
30. M. Hirano, K. Shinjo, Superlubricity and frictional anisotropy. Wear **168**, 121–125 (1993)
31. O.V. Yazyev, Y.P. Chen, Polycrystalline graphene and other two-dimensional materials. Nat. Nanotechnol. **9**, 755 (2014)
32. A.S. de Wijn, A. Fasolino, A. Filippov, M. Urbakh, Low friction and rotational dynamics of crystalline flakes in solid lubrication. EPL **95**, 66002 (2011)
33. S.-W. Liu, H.-P. Wang, Q. Xu, T.-B. Ma, G. Yu, C. Zhang, D. Geng, Z. Yu, S. Zhang, W. Wang, Y.-Z. Hu, H. Wang, J. Luo, Robust microscale superlubricity under high contact pressure enabled by graphene-coated microsphere. Nat. Commun. **8**, 14029 (2017)
34. O.V. Yazyev, S.G. Louie, Topological defects in graphene: dislocations and grain boundaries. Phys. Rev. B **81**, 195420 (2010)
35. J. Červenka, C.F.J. Flipse, Structural and electronic properties of grain boundaries in graphite: planes of periodically distributed point defects. Phys. Rev. B **79**, 195429 (2009)
36. Y. Tison, J. Lagoute, V. Repain, C. Chacon, Y. Girard, F. Joucken, R. Sporken, F. Gargiulo, O.V. Yazyev, S. Rousset, Grain boundaries in graphene on SiC($000\bar{1}$) substrate. Nano Lett. **14**, 6382–6386 (2014)
37. J. Zhang, X. Chen, Q. Xu, T. Ma, Y. Hu, H. Wang, A.K. Tieu, J. Luo, Effects of grain boundary on wear of graphene at the nanoscale: a molecular dynamics study. Carbon **143**, 578–586 (2019)
38. X. Gao, W. Ouyang, O. Hod, M. Urbakh, Mechanisms of frictional energy dissipation at graphene grain boundaries. Phys. Rev. B **103**, 045418 (2021)
39. J.P. Hirth, J. Lothe, *Theory of Dislocations* (1982)
40. A. Kavalur, W.K. Kim, Molecular dynamics study on friction of polycrystalline graphene. Comput. Mater. Sci. **137**, 346–361 (2017)
41. H. Li, W.K. Kim, Role of multigrain structure on friction of graphene layers. Comput. Mater. Sci. **165**, 23–33 (2019)
42. N. Chan, S.G. Balakrishna, A. Klemenz, M. Moseler, P. Egberts, R. Bennewitz, Contrast in nanoscale friction between rotational domains of graphene on Pt(111). Carbon **113**, 132–138 (2017)
43. X. Gao, W. Ouyang, M. Urbakh, O. Hod, Superlubric polycrystalline graphene interfaces. Nat. Commun. **12**, 5694 (2021)
44. O. Hod, Y. Song, X. Gao, R. Pawlak, S. Huang, A. Hinaut, T. Glatzel, M. Urbakh, E. Meyer, Non-Amontons frictional behaviors of grain boundaries at layered material interfaces (2023). https://doi.org/10.21203/rs.3.rs-3436204/v1
45. X. Gao, M.l Urbakh, O. Hod, Stick-slip dynamics of moir´e superstructures in polycrystalline 2D material interfaces, Phys. Rev. Lett. 129, 276101 (2022)
46. J.H. Jeong, Y. Jung, J.-U. Park, G.-H. Lee, Gate-tunable electrostatic friction of grain boundary in chemical-vapor-deposited MoS_2. Nano Lett. **23**, 3085–3089 (2023)
47. D. Berman, S.A. Deshmukh, S.K.R.S. Sankaranarayanan, A. Erdemir, A.V. Sumant, Macroscale superlubricity enabled by graphene nanoscroll formation. Science **348**, 1118–1122 (2015)
48. D. Berman, B. Narayanan, M.J. Cherukara, S.K.R.S. Sankaranarayanan, A. Erdemir, A. Zinovev, A.V. Sumant, Operando tribochemical formation of onion-like-carbon leads to macroscale superlubricity. Nat. Commun. **9**, 1164 (2018)
49. X. Yang, R. Li, Y. Wang, J. Zhang, Tunable, Wide-temperature and macroscale superlubricity enabled by nanoscale Van Der Waals heterojunction-to-homojunction transformation. Adv. Mater. 2303580 (2023)

50. X. Ren, X. Yang, G. Xie, F. He, R. Wang, C. Zhang, D. Guo, J. Luo, Superlubricity under ultrahigh contact pressure enabled by partially oxidized black phosphorus nanosheets, Npj 2D Mater. Appl. **5**, 44 (2021)
51. Liu, S.-W. et al. Robust microscale superlubricity under high contact pressure enabled by graphene-coated microsphere. Nat. Commun. 8, 14029 (2017).
52. A. Macknojia, A. Ayyagari, D. Zambrano, A. Rosenkranz, E.V. Shevchenko, D. Berman, Macroscale superlubricity induced by MXene/MoS_2 nanocomposites on rough steel surfaces under high contact stresses. ACS Nano **17**, 2421–2430 (2023)

Towards Application of Microscale Structural Superlubricity

Ming Ma and Quanshui Zheng

Abstract Structural superlubricity (SSL), a state of near-zero friction and no wear between contacting solid surfaces, offers a disruptive approach to minimizing friction and wear. Recent years have seen a surge in SSL research, expanding its focus from fundamental science to practical applications, where the realization of robust microscale SSL play a key role. This chapter summarizes the recent advancements in SSL, with a particular emphasis on aspects that promote practical applications. These include SSL electrical contacts, SSL-based generators, the stability of SSL systems, and the environmental impact of SSL.

1 Introduction

The relative sliding between two contacting solid surfaces is one of the most prevalent form of motion in the physical world. Friction, the force that resists an object's sliding against another object, and wear, the surface damage of one or all solid surfaces in contact subject to relative motion, are ancient and universal interface problems [1].

M. Ma (✉) · Q. Zheng
Center for Nano and Micro Mechanics, Tsinghua University, Beijing 100084, China
e-mail: maming16@tsinghua.edu.cn

Q. Zheng
e-mail: zhengqs@tsinghua.edu.cn

M. Ma
Department of Mechanical Engineering, and State Key Lab of Tribology in Advanced Equipment (SKLT), Tsinghua University, Beijing 10084, China

M. Ma · Q. Zheng
Institute of Superlubricity Technology, Research Institute of Tsinghua University in Shenzhen, Shenzhen 518057, China

Q. Zheng
Department of Engineering Mechanics, School of Aerospace Engineering, Tsinghua University, Beijing 100084, China

Tsinghua Shenzhen International Graduate School, Shenzhen 518057, China

E. Gnecco and E. Meyer (eds.), *Fundamentals of Friction and Wear on the Nanoscale*, NanoScience and Technology, https://doi.org/10.1007/978-3-031-63065-1_6

Friction and wear widely exists in many natural phenomena ranging from atomic and nanoscale to geologic scale.

At macro scale, various kinds of lubricants have been applied to reduce friction and wear, such as liquid lubricant, solid lubricant, grease, etc. However, as the scale decreases to micro and nanometer, reducing friction and wear with traditional lubricants may become inefficient as lubricants have high viscosity [2] and uneven distribution on the micro and nano meter scale surface [3]. Hence at this scale, direct contact between solids is inevitable, which significantly affects micro and nanoscale devices' design, functionality and longevity.

Yet, structural superlubricity (SSL), a state of no wear and nearly zero friction between two contacted solid surfaces [4, 5], poses the promise to significantly reduce fiction and wear at the micro and nano scale, which can hardly be achieved by traditional lubricants. SSL originates from the theoretical prediction that infinitely large and incommensurate contacts will lead to zero static friction [6]. Once sliding begins, the inter-surface stick–slip motion could be suppressed due to the offset of lateral interaction, hence resulting in vanishing dynamical friction [7, 8]. Depending on the junction geometry of the contacting materials, SSL can provide adjustable intrinsic lubrication, which remain robust under extreme conditions, such as vacuum, high pressure, high and low temperatures. The development of SSL provides a revolutionary paradigm shift for studying sliding contact interfaces, inspiring novel device design based on micro and nanometer relative contact sliding.

The history of SSL dates back to 1983, when Peyrard and Aubry first predicted a state of vanishing static friction in crystalline interfaces [6]. Almost a decade later, the term superlubricity was proposed, which refers to the prohibition of stick–slip motion through the elimination of a particular energy dissipation channel related to elastic instability [7]. In 2004, this phenomenon was observed experimentally for the first time [9], where the contact size was in the nanometer scale, and the experimental environment was ultra-high vacuum, which greatly limits the application of SSL.

With the development of high precision manipulation devices and fabrication of bulk pristine layered materials, from 2008 to 2013, Liu et al. [4], Zheng et al. [10] achieved microscale SSL under ambient conditions for the first time through the self-retracting motion of graphite mesa. To date, SSL has been achieved with different experimental systems, such as graphene-coated microsphere/graphene [11], gold islands on graphite [12], graphene nanoribbon on gold surfaces [13], graphene nanoscrolls [14], multi-walled carbon nanotubes [15], and so on. Among them, the SSL graphite mesa system holds the promise of an individual real contact area up to 100 μm^2 [4], sliding speeds up to 294 m/s, and sliding distances showing no wear up to 100 km [16]. The SSL based on graphite can be stably realized in both vacuum [9] and ambient conditions [4]. Recently, SSL between microscale graphite flakes and various two dimensional materials (graphene, hexagonal boron nitride) [17] and bulk phase materials [diamond-like-carbon (DLC), silicon (Si), silicon dioxide (SiO_2), aluminum oxide (Al_2O_3), hafnium dioxide (HfO_2)] have been observed [16]. Furthermore, a machine learning-assisted system is proposed to realize fully automated selective transfer and tribological performance measurement for SSL graphite mesa, supporting the realization of large-scale SSL [18].

As discussed above, the past decade has witnessed the rapid development of SSL, especially about the fundamental research on microscale SSL. In the following sections, we will focus on recent progress about which we have been working on during the last five years. This is centered on those investigations aiming at promoting the practical applications of SSL, including (1) SSL electrical contacts; (2) SSL-based generator; (3) stability issues of SSL; (4) environmental impact to SSL. For those whom are interested in an overview of SSL, the existing several nice reviews about SSL [5, 19–26] are strongly suggested.

2 Structural Superlubric Electrical Contacts

Sliding electrical contacts (SECs) are widely used in many devices to transmit current or signals, such as micro-motors [27], switches, relay [28–30], accelerometer [31], and storage in microelectromechanical system (MEMS) [32]. On macroscales, the performance of conductive slip rings [33] and pantographs [34] is also determined by SECs. There are three important indicators to measure SEC performance, namely critical current density, contact resistivity and friction coefficient.

The critical current density (CCD) of SECs is the maximum current that can maintain the normal operation of SEC under operating condition, which is an important indicator to measure the current carrying capacity of SEC. In various literature sources, three distinct definitions of CCD for friction pairs have been identified. The first pertains to the current density at which severe wear initiates within the friction pair [35–38]. Alternatively, CCD can refer to the point at which the constituent materials of the friction pair begin to melt, as per the second definition. The third type, which aligns with the definition utilized in this chapter, characterizes CCD as the critical current density at which interface defects start to manifest. Notably, the third standard represents the most rigorous criterion. SEC with high CCD has important value, which will benefit not only the maximum power can be transmitted, but also the miniaturization of device.

The contact resistivity of SEC is an important parameter to measure the energy transfer loss of SEC. SECs traditionally rely on low contact resistance to minimize energy loss and maximize energy transfer density [38, 39]. Elevated contact resistance results in substantial Joule heat generation at the sliding interface, leading to transmission energy losses and interface failure [39–41]. Thus, ensuring a low contact resistance at the interface is crucial for achieving high-energy density transmission while mitigating heat-related interface failures. Commonly employed strategies involve the implementation of noble metals such as copper, silver, platinum, and gold as coating layers between sliding interfaces to reduce contact resistance [39, 40, 42]. However, here we need to consider the third important parameter to measure SEC performance, the friction coefficient. A high friction coefficient usually indicates high friction and high wear, which will affect the contact stability and device lifespan of the SEC. Therefore, these noble metals often exhibit elevated friction levels, which contribute to wear and restrict device lifespan [38]. Moreover, wear debris generated during sliding can potentially trigger disastrous short circuits [43]. Consequently, some investigations have explored the application of two-dimensional materials, such as graphene, as lubricants to alleviate friction and wear issues at sliding interfaces. Nevertheless, the utilization of such materials may concurrently increase the contact resistance between interfaces [44, 45]. Some works can simultaneously achieve a decrease in friction coefficient and contact resistance by controlling the ratio of graphite lubricant to copper. Overall, the interface contact resistivity (4000 $\Omega\cdot\mu$m) and friction coefficient (0.255) are still at a relatively high level [46]. This means that traditional materials, whether metals or carbon-containing materials, will inevitably face the problem of high contact resistivity (low CCD value) or high friction coefficient, and the contradiction between the two cannot be easily solved.

Recently, SSL technology has been introduced to solve above contradictions. Considering that the actual contact area of the SSL system is approximately equal to the apparent contact area (as shown in Fig. 1) [16, 47] and the excellent thermal stability of graphite [48, 49], it is expected that SSL graphite contact is a good candidate to resolve the contraction between high CCD, low contact resistivity and low friction as well as wear for conventional SECs.

From recent studies, Wu et al. found that graphite/graphite contact in the SSL state can achieve CCD values up to 17.5 GA/m^2 under ambient conditions. Such a high value is 146 times larger than the existing CCD of SECs reported in the literature (0.12 GA/m^2 for Au/Cu friction pair [50]). The experimental optical microscope image and Raman mapping wear-free characterization results are shown in Fig. 2. Further research by the authors pointed out that, according to the theoretical and experimental results, the CCD converges to 6.69 GA/m^2 as the flake size increases, which remarks its valid on larger scales. Figure 2 illustrates the superior CCD performance of the graphite/graphite SSL electrical contacts in comparison to conventional SECs, highlighting the significant advancements in current carrying capability offered by SSL technology. At the same time, Typical characteristics of SSL, i.e., low coefficient of friction (<0.01) and negligible wear, is observed (as shown in Fig. 2). Therefore,

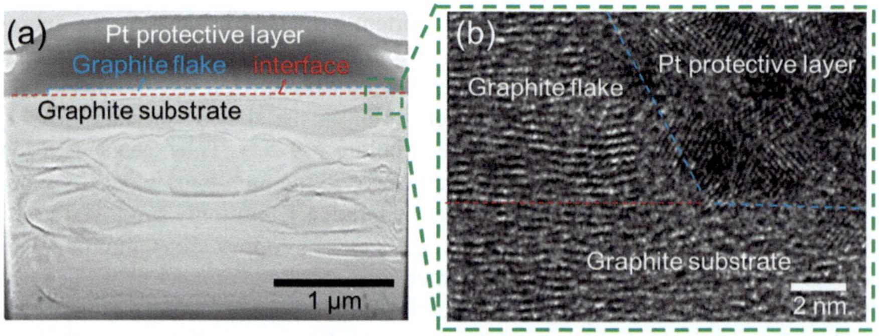

Fig. 1 Cross-section images of graphite/graphite interface via TEM. **a** TEM sample overall image. The red dashed line is the graphite/graphite interface. **b** Detail of the green dotted box in (**a**). The red dashed line is the graphite/graphite interface [47]

in addition to the CCD value, SSL SECs also have obvious advantages in friction coefficient compared to traditional SEC contacts, as shown in Fig. 3. Considering the zero-wear nature of SSL SECs, its stability and lifespan could be better than traditional SECs.

The high CCD of the SSL contacts comes from its high conductivity and thermal stability. The high conductivity of the interface is mainly determined by the actual contact area and the interlayer electron transfer barrier of the interface. The SSL SEC system has a two-dimensional interface with atomically smooth and fully contacted surfaces [16, 47], which has a significant advantage in the contact area for current transmission. In contrast, for traditional SECs like Au/Au [45], the contact area between conductive interfaces at the atomic level is often uneven, and the actual contact area is far smaller than the apparent contact area. Furthermore, the interlayer electron transfer barrier of the graphene-graphene interface is smaller than that of commonly employed materials in SEC. For example, the

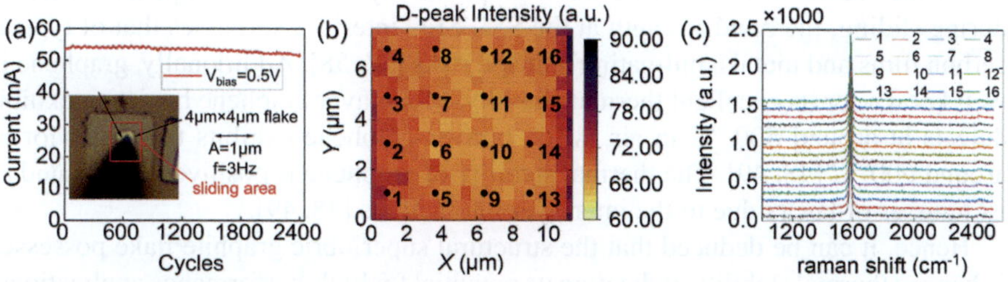

Fig. 2 CCD value testing and wear-free characterization of SSL contacts. **a** The transmitting current during more than 2000 cycles under the 0.5 V bias voltage. The amplitude and frequency of the cycle are 1 μm and 3 Hz. **b** Raman mapping of the sliding area. The brightness represents the signal intensity around the defecting peak (~ 1350 cm^{-1}). **c** Raman signals of different points in **b** with the Raman shift from 1000 to 2400 cm^{-1} [47]

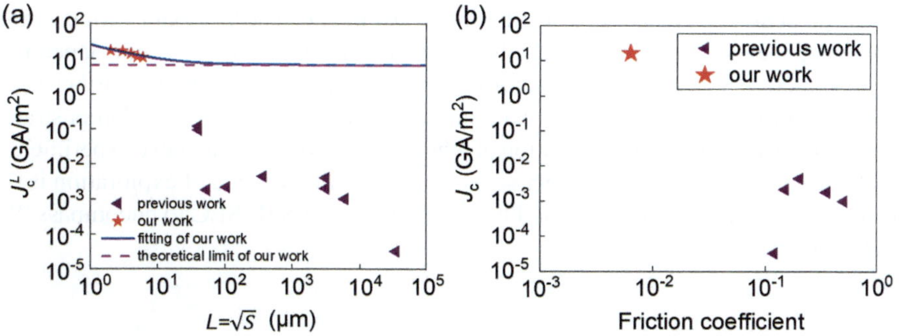

Fig. 3 The scaling relation of CCD and comparison with existing reports. **a** The relationship between CCD and the size of the contact. The red and purple points represent our and previous work [35–39, 50, 51], respectively. **b** The relationship between the CCD and the friction coefficient. The red and purple points represent our and previous work [36–39, 51], respectively [47]

contact resistivity of the graphite/graphite interface we measured using the four-wire method is $9.34 \times 10^{-10} \Omega \cdot m^2$. Such value agrees with existing measurement result $1.66 \times 10^{-10} \Omega \cdot m^2$ [52], and is 1/3 of copper-graphite electrical contact [53], a commonly used SEC in industry. It is worth mentioning that a metal surface can also form an atomically smooth surface after polishing, but the conductive sliding contact formed by the metal interface may encounter problems such as cold welding and excessive friction force [54], making it difficult to use as a sliding friction pair. Therefore, the core advantage of the SSL SEC interface is that it possesses a high actual contact area while maintaining an extremely low friction level, which is also an important characteristic of its underlying SSL technology [5, 9].

The SSL's low friction and wear-free properties from its defect-free 2D structure, requiring thermal stability during conductive sliding. Joule heating from conduction could incite carbon atom movement in the graphene, risking structural integrity. However, graphene's robust C–C bonds and high Young's modulus stabilize the hexagonal structure in the SSL SEC, protecting the interface [55, 56]. Additionally, while covalent bonding between graphene layers could disrupt the structure during sliding, the bond strength at the SSL SEC interface surpasses that of typical carbon films and metals, mitigating damage risks [57, 58]. Additionally, graphene is well known for its excellent thermal stability. Monolayer graphene begins to exhibit defects at around 500 °C in air, while bilayer graphene exhibits this behavior at around 600 °C [48, 49]. The thermal stability of graphene is positively correlated to its number of layers due to the interlayer interactions [48, 49].

Hence, it can be deduced that the structural superlubric graphite flake possesses enhanced thermal stability, indicating its potential for high-performance applications. According to the finite element simulation, the graphite interface temperature of steady state is about 946 °C which is consistent with the experimentally measured temperature at which defects start to appear in graphite [48, 49]. Altogether, the synergistic effects of relative low contact resistivity, excellent tribological properties, and high thermal stability of SSL graphite homojunction make it an ideal candidate for SEC.

At present, researches on SSL SEC have just begun. There remain various unresolved concerns pertaining to the robustness of SSL SEC. These include dependencies on factors like velocity, temperature, and atmospheric conditions. Furthermore, the feasibility of producing large-scale single-crystal graphite has been demonstrated [59], raising the challenging question of whether it is possible to achieve experimental realization of SSL SEC on a larger scale. Another valuable area of exploration is the extension of the 2D/2D homo- and hetero-junctions of SSL SEC to encompass 2D/3D contacts.

3 Power Generation Performance of SSL

With the rapid development of nanotechnology and microfabrication technology, miniaturized sensors and devices are constantly emerging in a vast number of applications in the fields of Internet of Things, sensor networks, big data, personal health systems, and artificial intelligence [60–64]. However, these sensors and devices have mostly been powered by line cords, or by batteries and external chargers, and thus have had limited applications [65, 66] particularly in independent, sustainable, maintenance-free operations of implantable biosensors, remote and mobile environmental sensors, nano-/micro-scale robots, and portable/wearable personal electronics [66–69]. The sustainability and the booming of sensors require small, wireless, portable, and sustainable power supplies.

A promising solution to these challenges is nanogenerators, which are proposed to collect weak and random mechanical energy into electricity. The common types of nanogenerators are piezoelectric nanogenerators (PENGs) [60, 70], triboelectric nanogenerators (TENGs) [71] and electret-based microgenerators (EBMGs) [72]. Recently, the Schottky generators (S-Gs) which generate direct current based on relatively sliding Schottky junctions were proposed. S-Gs have a simpler structure, higher current densities compared with common types of nanogenerators mentioned above. However, currently all reported S-Gs fail to simultaneously achieved sufficiently high current density and sufficiently long lifetime for real-world applications. The highest current densities in the range of 10^4–10^7 A·m^{-2} was achieved by friction of nanotips on semiconductor substrates [62, 73, 74], which refers to a very high normal pressure of 1–10 GPa. Meanwhile the corresponding currents decayed completely in the first 1–5 cycles [62, 74]. On the contrary, to achieve the long lifetime with the range of 3600–10,100 cycles [75, 76], a surface-to-surface contact sliding between the macroscopic MoS_2 or graphene was proposed and the corresponding normal pressure was of 0.05–5 MPa. The low normal pressure, however, generated exceedingly low current densities of around 0.033–1 A·m^{-2} [75, 76]. Even at this magnitude of low friction, notable wear was observed [75, 76], which is a main obstacle for sustainable use of S-Gs.

The fundamental challenge of increasing the efficiency of S-Gs is rooted in the conflict mechanism between high current density and friction-induced wear [62, 77–79]. However, another mechanism of generating current in a sliding Schottky joint, namely the mechanism of depletion layer establishment and destruction (DLED) [75, 80–82]. This mechanism was harnessed by using a specific Schottky joint, which is in a state of SSL to simultaneously achieve high current density of ~210 A·m^{-2} and long lifetime of at least 5000 cycles while maintaining stable high electrical current density (~119 A·m^{-2}). This new type of generator is a Schottky superlubric generator (S-SLG) which excludes the mechanism of friction-induced excitation in S-Gs.

3.1 *The Structure of a Typical Schottky Superlubric Generator (S-SLG)*

A typical S-SLG was fabricated by nanolithography and etching process. A square graphite flake (4 μm × 4 μm × 2.6 μm) made of highly ordered pyrolytic graphite (HOPG) with an Au film of 100 nm thickness was transferred onto a n-type silicon with a doping concentration of $N_D = 10^{15}$ cm^{-3} to form a Schottky junction. The measured work functions of the graphite flake and n-type silicon (n-Si) film are 4.77–4.82 and 4.36–4.42 eV respectively. Owing to the difference in work functions, electrons are automatically transferred between the graphite and n-Si contacted surfaces to form a depletion layer and a Schottky barrier at equilibrium. The current was generated by using a conductive atomic force microscope (AFM) tip (coated with Au) to induce sliding between the superlubric HOPG surface and the smooth n-Si surface (Fig. 4a, b). The measured currents in Fig. 4c show that the current during each sliding cycle remains nearly constant and gradually increases with the number of sliding cycles. The sliding life test with different speed shows a long lifetime (~5000 cycles) with stable high current density (~119 A·m^{-2}). In addition, the measured friction force (red line) declines [83] in the first 64 cycles and is stabilized in the subsequent thousands of slides at different sliding speeds (only slowly increasing with sliding speed) (Fig. 4d). The life time is much longer than that in experiment by sliding an AFM Au-tip with a radius of curvature of 35 nm and normal force 4.49 μN over the same n-Si film (Fig. 4e, f). Although a high current density (4.0 × 10^4 A·m^{-2}) was achieved, the current decayed very quickly within 30 cycles, which was much shorter than S-SLGs. The structural superlubric property of the system was verified by self-retracting motion (SRM) property [10].

3.2 *The Output Performance of S-SLG*

Two important factors to evaluate the performance of S-SLG are open-circuit voltage and short-circuit current. A typical output performance of S-SLG was depicted below. The positive direction of the current and voltage measurement was from the graphite flake to n-Si. The sustainability and stability of S-SLG was shown in Fig. 5a, as the open-circuit voltage Voc and short-circuit current Isc waveforms under a linear reciprocating motion with a speed of 4.3 μm/s and a normal force of $F_N = 250$ μN. The correlation of Voc, Isc and F_N was shown in Fig. 5a, which represented a weak correlation of normal pressure and Voc, Isc. The output current and voltage of the S-SLG were measured under different resistances (R) while keeping the normal force (250 μN) and speed (4.3 μm/s) constant.

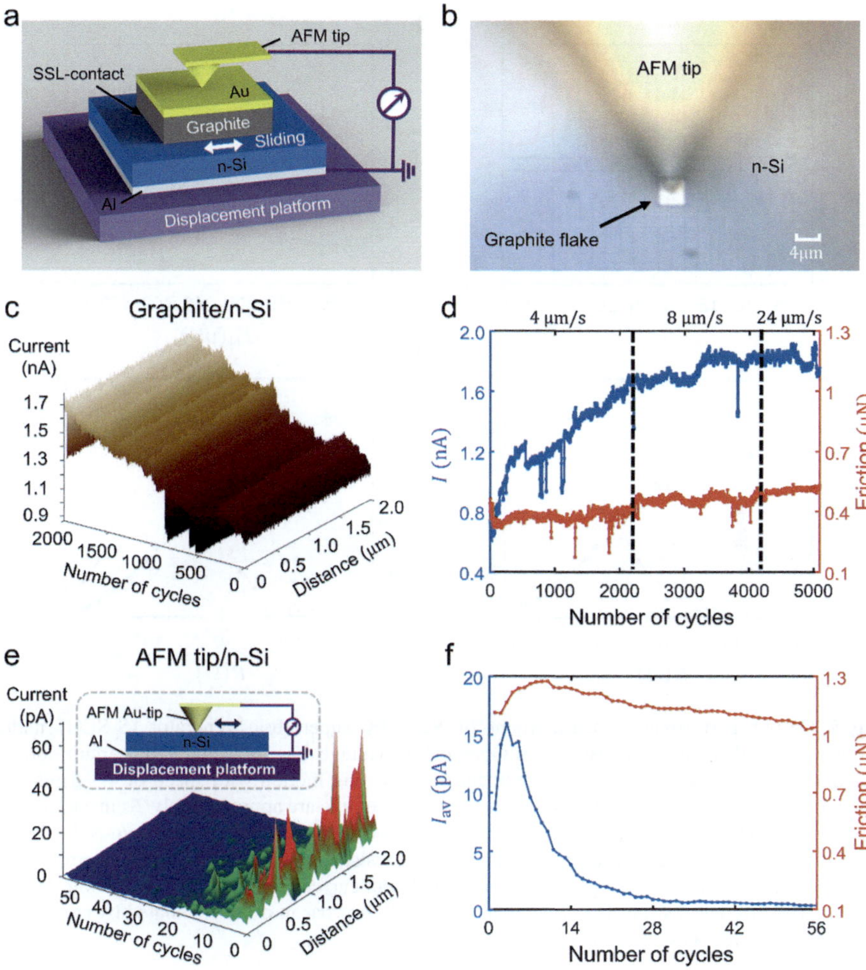

Fig. 4 Output current and friction measurement of Schottky superlubric generator (S-SLG) and comparison with an ordinary Schottky generator (S-G). **a** Structure of a graphite/n-Si S-SLG. **b** Optical microscopic image of a graphite/n-Si S-SLG. **c** Current maps in the first 2000 cycles of a graphite/n-Si S-SLG with speed of 4 μm/s under a controlled normal force of $F_N = 22.3$ μN. **d** Relationship between friction (red) and current I (blue) with sliding cycles of a graphite/n-Si S-SLG at different speeds. **e** Current maps in the first 56 cycles of an atomic force microscope (AFM) tip/n-Si ordinary S-G, and an illustration of the experimental setup. **f** Relationship between friction (red) and average current I_{av} (blue) with sliding cycles of an AFM tip/n-Si ordinary S-G [84]

3.3 The Mechanism of S-SLG

The friction in S-SLG is insufficient to excite electrons to generate measureable current in experiment. The friction measurement of graphite/n-Si S-SLG in Fig. 4d indicate the max friction force $f_{max} = 0.55$ μN. However, according to previous

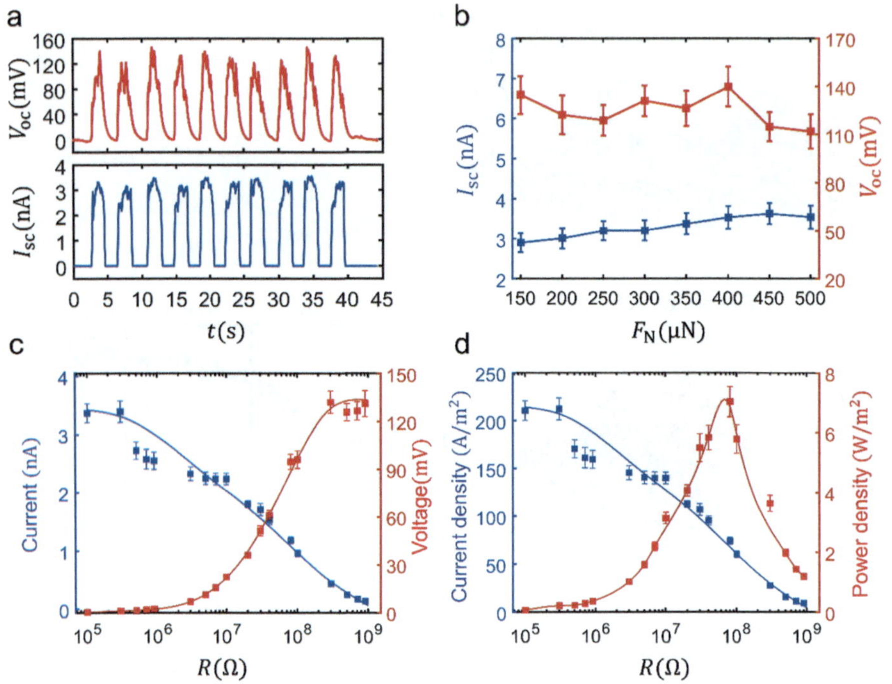

Fig. 5 Output performance measurement of Schottky superlubric generator (S-SLG) under different normal force and resistance. **a** Open-circuit voltage V_{oc} (red) and short-circuit current I_{sc} (blue) waveforms with time t under linear reciprocating motion with speed of 4.3 μm/s and normal force of $F_N = 250$ μN; the sliding time and pause time are approximately 2 s in each cycle. **b** Relationship between V_{oc} (red) and I_{sc} (blue) with normal force F_N of a graphite/n-Si S-SLG. **c** Output current (blue) and voltage (red) under different resistance R. Each data point and error bar in **b** and **c** are obtained by 4000 signal data points. **d** Output current density (blue) and power density (red) under different resistance R, each data point and error bar in **d** are calculated by combining the data points in **c** with the contact area $A = 16$ μm^2 [84]

SSL studies, friction during superlubric contact mainly occurs along edges through dangling bonds [85]. Therefore we could estimated the friction force of each dangling bond at the edges $f_0 \sim f_{max}a/4\,l = 8.5$ pN, where $a = 0.246$ nm and $l = 4$ μm are the lattice constant and side length of the two-dimensional graphite surface. The upper limit of friction excitation energy $\Delta E_f \sim f_0 b = 0.0287$ eV, where $b = 0.543$ nm is the lattice constant of silicon. However, the band gap of silicon $\Delta E_g = 1.12$ eV and the Schottky barrier height of graphite/n-Si interface $\Phi_B \sim 0.54$ eV are both much larger than friction energy $\Delta E_{f,}$ which indicates that it is difficult to excite electrons from friction to hop from valence band to conduction band of n-Si. It is also difficult for friction excited electrons to tunnel from graphite to the conduction band of n-Si through the Schottky barrier.

The mechanism of S-SLG remains unclear nowadays. The most reasonable model is the depletion layer establishment and destruction (DLED) model. When a metal is in contact with n-Si, electrons are transferred from n-Si to the metal [86]. Consequently, a positively charged region is formed in the n-Si, which is the depletion layer, and a built-in electric field is generated. In the DLED mechanism, sliding of the metal causes the head depletion layer to be established and the tail depletion layer to be destructed, thereby forming built-in electric field separation, which make drifting carriers form a DC current [75, 87]. However, to date, the DLED mechanism has not been directly proven experimentally or theoretically.

4 Stability Issue in SSL Systems

SSL essentially originates from the atomically flat, dangling-bond-free van der Waals (vdW) surfaces provided by 2D materials. By assembling these into 2D/2D or 2D/3D interfaces, extremely low friction and zero wear can be achieved [16, 17]. On such ultralow dissipation interfaces, even minute forces like surface energy are sufficient to overcome the interface resistance, inducing relative sliding motion. Zheng et al.'s [10], Liu et al. [4] pioneering experimental realization of microscale SSL was based on the self-retracting phenomenon in graphite interfaces, reflecting the extremely weak interactions and interlayer sliding freedom at SSL interfaces.

However, the current realization of SSL still relies on 2D material friction pairs with finite thicknesses, which are themselves composed of stacked vdW materials—have weak interlayer shear strength [88]. This inherent interlayer sliding freedom, while enabling superlubricity, also raises concerns about the structural stability and reliability of SSL systems [89], such as load-induced and shear-induced delamination [90, 91]. Herein, we focus on graphite, the most widely used 2D material for SSL, and discuss its structural stability challenges.

4.1 Graphite Delamination: A Roadblock for SSL Applications

Currently, the most critical stability issue hindering graphite's application in SSL is its widely reported delamination phenomenon [92]. Lu et al. [91] observed that graphite micropillars tend to split into multiple layers under shear, failing to maintain structural integrity. Peng et al. [90] also found that under compression, the internal graphene layers in graphite micropillars can buckle and become unstable. Since graphite friction pairs are inevitably subjected to shear and pressure in practical applications, this internal delamination instability directly determines their suitability for SSL. Even worse, when integrating graphite friction pairs into assembled SSL systems, the debris introduced by delamination could lead to catastrophic system failure. Not only can it introduce new sheared surfaces with defects, but the wear fragments generated by delamination can also trigger chain failure throughout the

SSL system. Overcoming such instabilities is crucial for transitioning SSL from an intriguing scientific concept to a reliable, practical technology across diverse applications.

4.2 Origin of Delamination in Graphite: Incommensurate Interfaces

Surprisingly, for graphite friction pairs without incommensurate interfaces, Liu et al. [93] revealed a remarkably high shear strength against delamination of 0.14 GPa, two orders of magnitude higher than that of incommensurate interfaces. This implies that commensurate-stacked graphite promises relatively robust performance, with delamination only occurring at incommensurate interfaces.

For graphite friction pairs containing incommensurate interfaces, Gongyang et al. [92] conducted in-depth investigations into the delamination issue. Figure 6b shows a typical example of the delamination of graphite mesas at the beginning of sliding on the DLC surface under ambient condition. Mechanically, the delamination instability originates from the competition between the friction force at the sliding interface and the internal resistance at the potential delamination interface. Delamination occurs, when the former overcome the latter. The internal resistance against delamination could be caused by the following three components: (i) the retracting force due to the cleavage energy, resulting in the self-retraction, (ii) the resistance near the edges caused by various bonds between neighboring surfaces, and (iii) other resistance which are intrinsically to the interfaces, e.g., the resistance caused by steps within the interface (as shown in Fig. 6a).

Lateral shear tests were further employed to decouple the three contributing factors to the internal resistance against delamination in graphite. A micro-force sensing probe fixed to the micro-manipulator was applied to shear the upper flake from the side. As shown in Fig. 6c, the static force measured afterwards (1.48 μN) is much smaller than that for the first measurement (19.2 μN), due to the change edge bonding strength at the delaminated interface. This indicates that for graphite containing incommensurate interfaces internally, the edge binding forces play a decisive role in impeding interlayer delamination.

4.3 Strategies Towards Robust Graphite

Based on their understanding of the interlayer delamination behavior in SSL graphite friction pairs, Zheng's group has explored three different approaches to enhance the structural stability of graphite: environmental control, chemical modification, and selective control.

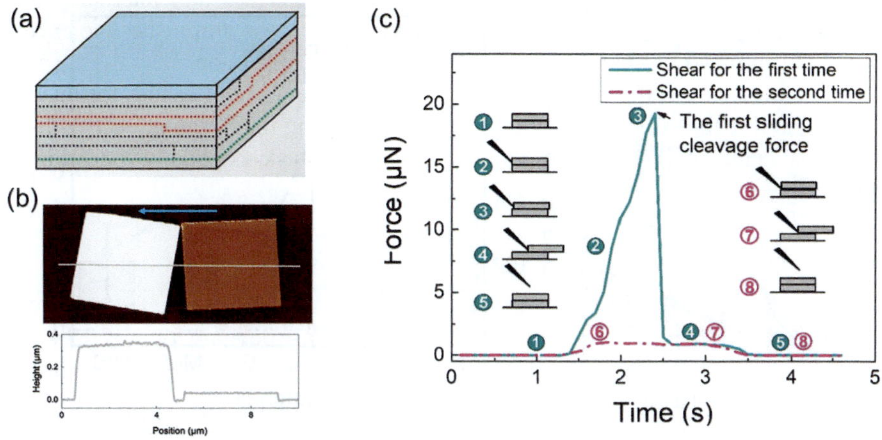

Fig. 6 **a** Schematic representation of the brick-stacking structure of graphite. **b** AFM height image of delaminated graphite friction pair on DLC substrate. **c** Typical measured lateral force–time curves for shearing graphite mesa during the first time (green solid line) and the second time (pink dashed dot line) [92]

Environmental Control Approach: Gongyang et al. showed that the delamination of graphite mesas on DLC can be eliminated after annealing at 150 °C and protection by nitrogen. Annealing alters the chemical structure of the graphite edges and removes adsorbed water molecules, significantly reducing friction at the graphite/DLC interface. In this case, the frictional force becomes insufficient to overcome the exfoliation resistance, thereby suppressing delamination. However, this method does not fundamentally address graphite's intrinsic delamination tendency and has limitations: (1) It imposes stringent environmental requirements on the SSL system to maintain low friction, limiting application flexibility. (2) The effects of edge alteration and adsorbate removal may diminish over time or with environmental changes, leading to delamination reoccurrence.

Edge Modification Approach: Hu [94] adopted an opposite strategy, irradiating the graphite edges with an argon ion beam to form covalent bonds between the interlayers, effectively stitching them together. This strategy enhances the edge binding forces, strengthening the internal resistance against delamination. Instead of reducing interface friction, this method targets the graphite friction pair itself. Experimental results demonstrated that after edge modification, graphite maintained structural integrity under applied shear stress.

However, concerns remain: (1) As the size increases, the effective constraints provided by edge modification may weaken, questioning its suitability for large-scale SSL graphite friction pairs. (2) The impact of modified edge regions on interface friction and transport properties lacks sufficient research, potentially degrading SSL performance.

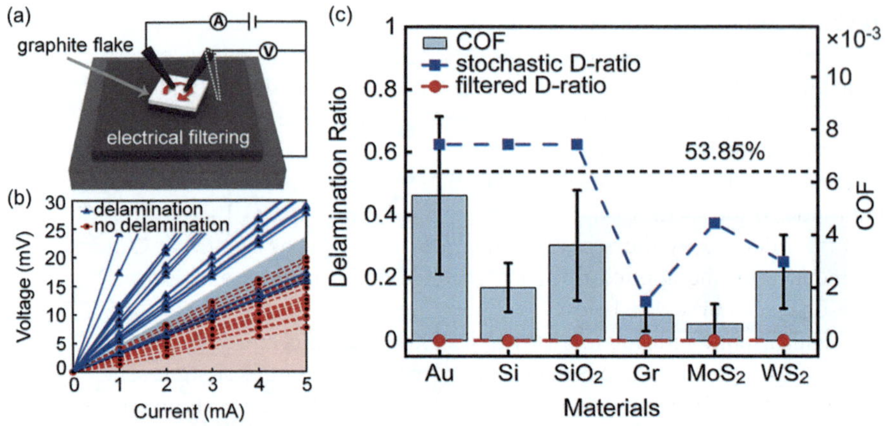

Fig. 7 **a** Schematic diagram of electrical measurement. **b** Filtering criteria for identifying graphite with and without delamination. **c** Comparing the statistical distribution (data points) of graphite layer thickness in flakes, with Gauss fit results (lines) representing the distribution probability for each set [95]

Selective Control Approach: For the incommensurate interfaces contained within SSL graphite friction pairs, they can indeed be viewed as potential delamination interfaces. In other words, if all internal incommensurate interfaces could be eliminated, it may be possible to obtain intrinsically robust graphite. It is based on this rationale that Chen et al. [95] proposed an electrical filtering method, aiming to identify robust graphite without internal incommensurate interfaces, as shown in Fig. 7a. Since the presence of incommensurate interfaces leads to increased electrical resistance, the existence of such potential delamination interfaces can be determined through electrical measurement, as depicted in Fig. 7b. Their experimental results confirm that all the filtered graphite exhibit delamination-free behavior after more than 7000 cycles of sliding on a series of 2D and 3D substrates, as shown in Fig. 7c.

While Chen et al.'s work provides a powerful approach to obtaining intrinsically stable graphite friction pairs, limitations in efficiency, scalability, and other aspects still need to be addressed before large-scale application. Automation could be a potential solution to overcome these challenges [18].

4.4 Discussions

Although the three aforementioned strategies aim to address graphite delamination instability, there remains a gap to true practical application. However, these efforts have provided valuable practical experiences and deeper mechanistic understandings. With advancements in materials science, more tangible and viable technical pathways will likely emerge to resolve the delamination problem. For instance, recent progress has enabled the growth of thick single-crystal graphite, introducing another potential route: directly using single-crystal graphite to prepare SSL friction pairs, potentially enabling bulk production of stable, delamination-free robust graphite. Through continuous innovations and efforts to overcome current challenges, we may ultimately transform this cutting-edge scientific concept into a reliable engineering practice, benefiting various fields.

5 Environmental Effects on SSL

Any surface exposed to ambient conditions will be contaminated by airborne adsorbed molecules, such as water and hydrocarbon [96, 97]. For tribo-pairs which are the most fundamental and common structure of mechanical devices, such airborne adsorbed molecules are an important source of third bodies, the substances which are confined between the two surfaces under relative sliding [98]. These airborne substances could not be completely eliminated even under high vacuum [99].

Third bodies have been found to play a key role in friction and wear [98, 100, 101]. For instance, analytic results and experiments in ultrahigh vacuum indicate that the static friction between two clean crystalline surfaces should almost vanish, yet their corresponding macroscopic objects always exhibit static friction in ambient condition [98, 102]. With molecular dynamics (MD) simulations, He et al. proposed that it is the third bodies, such as small hydrocarbon molecules adsorbed on the surfaces, that self-arranged to lock the two incommensurate contacting surfaces. This mechanism might explain the discrepancy above [98, 102]. Müser et al. [103] theoretically showed that an area-independent friction coefficient was obtained regardless of surface geometry when an adsorbed layer of mobile atoms was introduced between incommensurate surfaces.

Experimentally, by sliding a ta-C-coated AFM tip on HOPG, Hasz et al. [104] found a nonmonotonic dependence of friction on humidity, and attributed it to the changing quality of the contact between the water and the substrate. By pushing nanoparticles with the tip of an atomic force microscope, frictional resistance for contaminated incommensurate interface could increase by a factor of 40 for antimony islands on HOPG [105]. For a silica AFM tip sliding on a graphite basal plane in the presence of pentanol and phenol, Khajeh et al. found that both the mean value and standard deviation of the friction force with phenol were larger than those with pentanol. Mechanically driven chemical bonding between the tip and phenol molecules was believed to play a critical role [106]. Gallagher et al. [107] demonstrated that anisotropic friction domains with twofold symmetry and anisotropy exceeding 200% on graphene monolayers derive from self-assembly of environmental adsorbates into a highly regular superlattice of stripes with period 4–6 nm, not from periodic nanoscale ripples in the graphene [108]. The stripes and resulting frictional domains appear on monolayer and multilayer graphene on a variety of substrates, as well as on exfoliated flakes of hexagonal boron nitride. Furthermore, the stripe-superlattices are proved to be reproducibly and reversibly manipulated with sub-micrometer precision using a scanning probe microscope, allowing us to create arbitrary arrangements of frictional domains within a single flake [107]. Smith et al. investigated the hydrocarbon films confined between smooth mica surfaces using surface force apparatus (SFA), detecting the shear forces transmitted across nanoscale films of dodecane down to a single molecular layer. Either the commensurate surface alignment or presence of water at the interfaces leads to moderate or high friction. A single molecular layer of dodecane trapped between crystallographically misaligned dry surfaces is the origination for the friction coefficients down to 0.001 [109].

It is concluded that due to the universal presence and significant effects of the third bodies, there have been intensive studies either using AFM [101, 104–107] or SFA [109] experimentally or MD simulations theoretically [110–112]. These studies provide supports for investigating the effect of airborne third bodies on SSL. Since most SSL devices are operated in ambient conditions, it is crucial to understand the influence of airborne third bodies on the frictional properties [5]. Deng et al. report a running-in process on microscale SSL in graphite flake pairs assembled under ambient conditions, where airborne third bodies are inevitably present at the interfaces, as indicated in Fig. 8a. In this running-in process, the friction decreases with reciprocating motions (Fig. 8b), but no morphological or chemical changes can be observed. The underlying mechanism is attributed to the mechanical cleaning which describes the removal of contaminants confined between two surfaces through mechanical oscillations [83]. This mechanism is detailed introduced in Ma et al.'s work [113], which proposes that the diffusion coefficient of confined third bodies can be greatly enhanced under the influence of mechanical oscillations. This effect occurs due to bifurcations of particle trajectories caused by the reconstruction of the energy landscape during oscillations [113].

It is generally believed that SSL requires the contact interface being ultra-clean, so that cancellation of lateral forces can be guaranteed via direct contact between the two surfaces [114]. Experimentally, this is usually ensured by either a clean environment like ultrahigh vacuum [9, 13, 105, 115, 116] or mechanical cleaning of

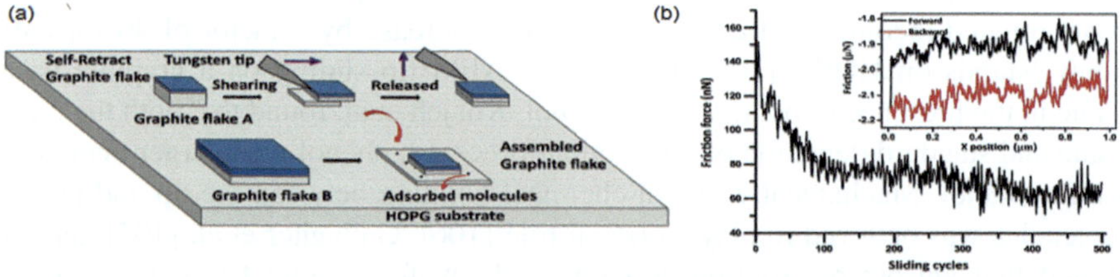

Fig. 8 **a** Illustration of the fabrication process of an assembled graphite flake system. Graphite flakes were sheared and then released with the help of a tungsten microtip. Deng et al. picked out these graphite flakes, which can retract back to their initial positions marked as graphite flake A and B. The top of the bigger graphite flake B was removed, and the top of the smaller graphite flake A was transferred onto the bottom of flake B, forming an assembled graphite flake [83]. **b** The friction force versus the number of sliding cycles for the assembled graphite flake. The inset shows a representative friction loop during the process [83]

the contact as introduced above [83, 113]. Although the observation of SSL under ambient conditions has been reported multiple times [4, 15, 117–119], most of these studies are attributed to the mechanical cleaning of the contact or there are no third bodies at the interface. Particularly, Cihan et al. reported that structural lubricity could exist in ambient conditions because the gold–graphite interface is expected to remain largely free from contaminant molecules [12]. Moreover, Dietzel et al. [105] suggested a link between the degree of surface contamination and the occurrence of superlubricity or "Amontons-like" friction, however, the impact of specific contamination types (hydrocarbons and water molecules) and contamination amount on SSL, and how these contaminants lead to the disappearance of SSL remain unambiguous. Recently, a study proposed by Wang et al. [120] directly proves that SSL can be maintained with an airborne third body-rich interface. In this work, after intentionally introducing airborne contaminants into microscale structural superlubric incommensurate graphite/graphite contacts, they find that the important features of SSL, the nearly-zero friction coefficient, ultralow friction, and symmetry in friction, are all well preserved. In the case of third body-free contact, a normal-force-independent SSL is always observed, which is caused by the weak interaction when the interface is in incommensurate contact, relying on the lattice misfit between two flat crystalline surfaces with high in-plane Young's modulus that leads to effective cancellation of the lateral forces during sliding motion [5]. For the contaminated samples with different exposure time under ambient conditions, i.e., the amounts of airborne third bodies are various, as shown in Fig. 9a, the friction force shows a negligible dependence on normal force as well, rather than increase linearly with the increase of the normal force as predicted by Robbins et al.'s theoretical work [98]. The friction coefficients (defined as the differential slope of friction versus normal force curve) are all on the order of 10^{-4}, which is a basic feature of SSL. Moreover, airborne third bodies are found to reduce the friction of a commensurate contact but increase the friction of an incommensurate contact, as shown in Fig. 9b. For incommensurate interface, the shear stress (friction per area) increases with exposure duration and transits to a plateau with exposure durations under ambient conditions at about 100 min. For commensurate interface, however, the shear stress decreases with exposure durations. Considering the magnitude of the changes in friction, obviously friction in commensurate state is more sensitive to the exposure duration. Generally, the friction for contaminant-free commensurate and incommensurate contacts is the largest and smallest respectively, with that for contaminated commensurate interface being larger than incommensurate in between as shown in Fig. 9b.

While abundant phenomena and interesting mechanisms have been revealed, investigation on the evolution of third bodies during sliding, e.g. changes in their quantity and morphology, remains exclusive [121]. This is important as in principle the physical properties are decided by the underlying structures [98]. For MD simulations, as their time scale is orders smaller than the experiments, the velocity used in it is much higher than that operated in experiments [122], leading the observed evolution questionable in terms of practical relevance [123, 124]. For experiments, the difficulties lie in the fact that third bodies are confined at the buried interface

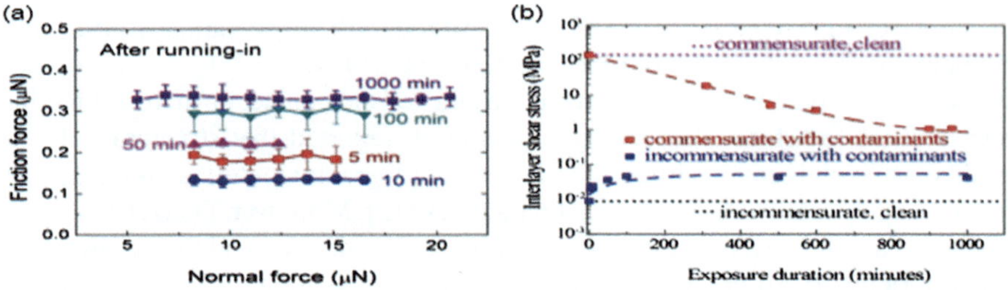

Fig. 9 **a** For samples with different durations of exposure, the measured dependence of friction force on normal load. The coefficient of friction, i.e. slope of the friction vs normal force curves, is $-6 \times 10^{-4}, 4 \times 10^{-4}, 6 \times 10^{-6}, 6 \times 10^{-4}, -2 \times 10^{-4}$, respectively, corresponding to exposure duration of 5, 10, 50, 100, and 1000 min. These coefficients of friction are practically indistinguishable from zero considering the precision of the current experimental system. **b** The interlayer shear stress of commensurate and incommensurate interfaces, which are either contaminant-free or contaminated [120]

[121] and the airborne ones are of nanometer in size [96, 97]. While ex situ measurements with an electron microscope could shed light on their changes [125–128], it is difficult to correlate them with the tribological properties since once the tribo-pairs are separated, the morphology, as well as the chemistry of the contacting surfaces and the third bodies, will inevitably be different from those in operation. For in situ measurement, direct observation is quite difficult due to the opacity of the substrates [129]. One relevant instrument is the surface force apparatus of which nanometer resolution on the thickness is possible [130]. However, the materials of the surfaces are limited to mica and the gap should be filled with transparent liquid, posing a stringent limit. More importantly, for most tribo-pairs wear often occurs. The generated wear debris will significantly increase the difficulty in studying the evolution of third bodies, as their physical evolution and tribo-chemical interaction are usually quite complicated, not to mention the accompanying complex changes in the contacting surfaces [131]. Overall, more studies are needed to address the above challenges, which are benefitting the design of structural superlubric devices operating under ambient conditions.

For the investigation on third bodies with structural superlubric systems, the absence of wear can eliminate the effects of wear debris, and the single crystalline structure of the contacting graphite surfaces provides an ideal model for theoretical study. Therefore, a structural superlubric system could be a good candidate for the investigation on the mechanism about the effect of third bodies.

6 Conclusion and Outlooks

To summarize, structural superlubricity (SSL), characterized by nearly zero friction and virtually no wear, holds revolutionary significance for both scientific researches and technological applications. Since the original success of extending nanoscale SSL to microscale, pioneers have ushered in a new era of research in the field. Today, SSL systems can robustly exist on microscale, at high speeds, in extreme temperatures, with diverse materials and interfaces, and across various enviornments. These outstanding properties enable SSL to play a central role and address fundamental issues in many important devices, as details in this chapter, such as SSL electrical contacts and SSL-based nanogenerators. Furthermore, SSL holds promise for applications in numerous fields including data storage, aerospace, and beyond [5].

There are also many intriguing topics and discoveries that we have not detailed here [21, 122], including the fundamental theories of SSL and the construction of large-scale SSL systems. Similarly, there are several questions that remain unanswered, such as performance of SSL under multi-physical fields. The study of friction has persisted for centuries and it continues. Now, the more pressing question is, are we ready to embrace the new world of "zero" friction and no wear brought forth by structural superlubricity?

References

1. A. Fall, B. Weber, M. Pakpour, N. Lenoir, N. Shahidzadeh, J. Fiscina, C. Wagner, D. Bonn, Sliding friction on wet and dry sand. Phys. Rev. Lett. **112**, 175502 (2014)
2. S. Granick, Motions and relaxations of confined liquids. Science **253**, 1374–1379 (1991)
3. S. Granick, Y. Zhu, H. Lee, Slippery questions about complex fluids flowing past solids. Nat. Mater. **2**, 221–227 (2003)
4. Z. Liu, J. Yang, F. Grey, J.Z. Liu, Y. Liu, Y. Wang, Y. Yang, Y. Cheng, Q. Zheng, Observation of microscale superlubricity in graphite. Phys. Rev. Lett. **108**, 205503 (2012)
5. O. Hod, E. Meyer, Q. Zheng, M. Urbakh, Structural superlubricity and ultralow friction across the length scales. Nature **563**, 485–492 (2018)
6. M. Peyrard, S. Aubry, Critical behaviour at the transition by breaking of analyticity in the discrete Frenkel-Kontorova model. J. Phys. C Solid State Phys. **16**, 1593–1608 (1983)
7. M. Hirano, K. Shinjo, R. Kaneko, Y. Murata, Anisotropy of frictional forces in muscovite mica. Phys. Rev. Lett. **67**, 2642–2645 (1991)
8. M. Hirano, K. Shinjo, Superlubricity and frictional anisotropy. Wear **168**, 121–125 (1993)
9. M. Dienwiebel, G.S. Verhoeven, N. Pradeep, J.W. Frenken, J.A. Heimberg, H.W. Zandbergen, Superlubricity of graphite. Phys. Rev. Lett. **92**, 126101 (2004)
10. Q. Zheng, B. Jiang, S. Liu, Y. Weng, L. Lu, Q. Xue, J. Zhu, Q. Jiang, S. Wang, L. Peng, Self-retracting motion of graphite microflakes. Phys. Rev. Lett. **100**, 067205 (2008)
11. S.W. Liu, H.P. Wang, Q. Xu, T.B. Ma, G. Yu, C. Zhang, D. Geng, Z. Yu, S. Zhang, W. Wang, Y.Z. Hu, H. Wang, J. Luo, Robust microscale superlubricity under high contact pressure enabled by graphene-coated microsphere. Nat. Commun. **8**, 14029 (2017)
12. E. Cihan, S. Ipek, E. Durgun, M.Z. Baykara, Structural lubricity under ambient conditions. Nat. Commun. **7**, 6 (2016)

13. S. Kawai, A. Benassi, E. Gnecco, H. Sode, R. Pawlak, X. Feng, K. Mullen, D. Passerone, C.A. Pignedoli, P. Ruffieux, R. Fasel, E. Meyer, Superlubricity of graphene nanoribbons on gold surfaces. Science **351**, 957–961 (2016)
14. D. Berman, S.A. Deshmukh, S.K. Sankaranarayanan, A. Erdemir, A.V. Sumant, Friction macroscale superlubricity enabled by graphene nanoscroll formation. Science **348**, 1118–1122 (2015)
15. R. Zhang, Z. Ning, Y. Zhang, Q. Zheng, Q. Chen, H. Xie, Q. Zhang, W. Qian, F. Wei, Superlubricity in centimetres-long double-walled carbon nanotubes under ambient conditions. Nat. Nanotechnol. **8**, 912–916 (2013)
16. D. Peng, J. Wang, H. Jiang, S. Zhao, Z. Wu, K. Tian, M. Ma, Q. Zheng, 100 km wear-free sliding achieved by microscale superlubric graphite/DLC heterojunctions under ambient conditions. Natl. Sci. Rev. **9**, nwab109 (2022)
17. Y. Song, D. Mandelli, O. Hod, M. Urbakh, M. Ma, Q. Zheng, Robust microscale superlubricity in graphite/hexagonal boron nitride layered heterojunctions. Nat. Mater. **17**, 894–899 (2018)
18. L. Chen, C. Lin, D. Shi, X. Huang, Q. Zheng, J. Nie, M. Ma, Fully automatic transfer and measurement system for structural superlubric materials. Nat. Commun. **14**, 6323 (2023)
19. J. Yuan, R. Yang, G.Y. Zhang, Structural superlubricity in 2D van der Waals heterojunctions. Nanotechnology **33**, 102002 (2021)
20. J. Luo, M. Liu, L. Ma, Origin of friction and the new frictionless technology-superlubricity: advancements and future outlook. Nano Energy **86**, 106092 (2021)
21. A. Vanossi, C. Bechinger, M. Urbakh, Structural lubricity in soft and hard matter systems. Nat. Commun. **11**, 4657 (2020)
22. Y. Song, C. Qu, M. Ma, Q. Zheng, Structural superlubricity based on crystalline materials. Small **16**, 1903018 (2020)
23. W. Zhai, K. Zhou, Nanomaterials in superlubricity. Adv. Funct. Mater. **29**, 1806395 (2019)
24. J.M. Martin, A. Erdemir, Superlubricity: friction's vanishing act. Phys. Today **71**, 40–46 (2018)
25. D. Berman, A. Erdemir, A.V. Sumant, Approaches for achieving superlubricity in two-dimensional materials. ACS Nano **12**, 2122–2137 (2018)
26. M.Z. Baykara, M.R. Vazirisereshk, A. Martini, Emerging superlubricity: a review of the state of the art and perspectives on future research. Appl. Phys. Rev. **5**, 041102 (2018)
27. T. Morita, Miniature piezoelectric motors. Sens. Actuat. A Phys. **103**, 291–300 (2003)
28. A. Basu, G.G. Adams, N.E. McGruer, A review of micro-contact physics, materials, and failure mechanisms in direct-contact RF MEMS switches. J. Micromech. Microeng. **26**, 104004 (2016)
29. O.Y. Loh, H.D. Espinosa, Nanoelectromechanical contact switches. Nat. Nanotechnol. **7**, 283–295 (2012)
30. B. Ma, Z. You, Y. Ruan, S. Chang, G. Zhang, Electrostatically actuated MEMS relay arrays for high-power applications. Microsyst. Technol. **22**, 911–920 (2015)
31. S. Arjun, Y. Navid, N. Khalil, A wide-range micromachined threshold accelerometer array and interface circuit. J. Micromech. Microeng. **11**, 118 (2001)
32. B. Bhushan, K.J. Kwak, Platinum-coated probes sliding at up to 100 mm s−1 against coated silicon wafers for AFM probe-based recording technology. Nanotechnology **18**, 345504 (2007)
33. M. Grandin, U. Wiklund, A wear tolerant slip-ring assembly, in *ICEC 2014; The 27th International Conference on Electrical Contacts* (2014), pp. 1–6
34. J. Ambrósio, J. Pombo, M. Pereira, P. Antunes, A. Mósca, Recent developments in pantograph-catenary interaction modelling and analysis. Int. J. Railway Technol. **1**, 249–278 (2012)
35. K.A. Aleutdinov, V.Y. Rubtsov, V.V. Fadin, M.I. Aleutdinova, Surface layer structure of AISI 1020 steel at different stages of dry sliding under electric current of high density. IOP Confer. Ser. Mater. Sci. Eng. **116**, 012022 (2016)
36. N. Argibay, J.A. Bares, J.H. Keith, G.R. Bourne, W.G. Sawyer, Copper–beryllium metal fiber brushes in high current density sliding electrical contacts. Wear **268**, 1230–1236 (2010)

37. Y.A. Wang, J.X. Li, Y. Yan, L.J. Qiao, Effect of electrical current on tribological behavior of copper-impregnated metallized carbon against a Cu–Cr–Zr alloy. Tribol. Int. **50**, 26–34 (2012)
38. J.A. Bares, N. Argibay, N. Mauntler, G.J. Dudder, S.S. Perry, G.R. Bourne, W.G. Sawyer, High current density copper-on-copper sliding electrical contacts at low sliding velocities. Wear **267**, 417–424 (2009)
39. N. Argibay, W.G. Sawyer, Low wear metal sliding electrical contacts at high current density. Wear **274–275**, 229–237 (2012)
40. M. Grandin, U. Wiklund, Influence of mechanical and electrical load on a copper/copper-graphite sliding electrical contact. Tribol. Int. **121**, 1–9 (2018)
41. A. Senouci, H. Zaidi, J. Frene, A. Bouchoucha, D. Paulmier, Damage of surfaces in sliding electrical contact copper/steel. Appl. Surf. Sci. **144–145**, 287–291 (1999)
42. C. Holzapfel, P. Heinbuch, S. Holl, Sliding electrical contacts: wear and electrical performance of noble metal contacts, in *Proceedings of the 2010 56th IEEE Holm Conference on Electrical Contacts* (2010), pp. 1–8
43. I.H. Sung, J.W. Kim, H.J. Noh, H. Jang, Effect of displacement and humidity on contact resistance of copper electrical contacts. Tribol. Int. **95**, 256–261 (2016)
44. B.H. Chudnovsky, Lubrication of electrical contacts, in *Proceedings of the Fifty-First IEEE Holm Conference on Electrical Contacts* (2005), pp. 107–114
45. M. Grandin, U. Wiklund, Wear phenomena and tribofilm formation of copper/copper-graphite sliding electrical contact materials. Wear **398–399**, 227–235 (2018)
46. K.H. Cho, U.S. Hong, K.S. Lee, H. Jang, Tribological properties and electrical signal transmission of copper–graphite composites. Tribol. Lett. **27**, 301–306 (2007)
47. T. Wu, W. Chen, L. Wangye, Y. Wang, Z. Wu, M. Ma, Q. Zheng, Ultrahigh critical current density across sliding electrical contacts in structural superlubric state. Phys. Rev. Lett. **132**, 096201 (2024)
48. F. Liu, M. Wang, Y. Chen, J. Gao, Thermal stability of graphene in inert atmosphere at high temperature. J. Solid State Chem. **276**, 100–103 (2019)
49. H.Y. Nan, Z.H. Ni, J. Wang, Z. Zafar, Z.X. Shi, Y.Y. Wang, The thermal stability of graphene in air investigated by Raman spectroscopy. J. Raman Spectrosc. **44**, 1018–1021 (2013)
50. S.H. Fan, Y.C. Chan, Current-carrying capacity of anisotropic-conductive film joints for the flip chip on flex applications. J. Electr. Mater. **32**, 101–108 (2003)
51. Z. He, Z. Ni, H. Wang, Z. Yang, W. Wei, X. Wang, L. Deng, G. Wu, Effect of magnetic field on the current-carrying friction and wear performance of C/Cu contact Pairs, in *Proceedings of the 2021 International Conference on Electrical Materials and Power Equipment (ICEMPE)* (2021), pp. 1–4
52. D. Dutta, A. Oz, O. Hod, E. Koren, The scaling laws of edge vs bulk interlayer conduction in mesoscale twisted graphitic interfaces. Nat. Commun. **11**, 4746 (2020)
53. B.D. Beake, A.J. Harris, T.W. Liskiewicz, J. Wagner, S.J. McMaster, S.R. Goodes, A. Neville, L. Zhang, Friction and electrical contact resistance in reciprocating nano-scale wear testing of metallic materials. Wear **121**, 474–475 (2021)
54. A.J.W. Moore, F.P. Bowden, Deformation of metals in static and in sliding contact. Proceed. Royal Soc. Lond. Ser. A Math. Phys. Sci. **195**, 231–244 (1997)
55. H.I. Rasool, C. Ophus, W.S. Klug, A. Zettl, J.K. Gimzewski, Measurement of the intrinsic strength of crystalline and polycrystalline graphene. Nat. Commun. **4**, 2811 (2013)
56. C. Lee, X. Wei, J.W. Kysar, J. Hone, Measurement of the elastic properties and intrinsic strength of monolayer graphene. Science **321**, 385–388 (2008)
57. P.L. de Andres, R. Ramírez, J.A. Vergés, Strong covalent bonding between two graphene layers. Phys. Rev. B **77**, 14 (2008)
58. J. Sun, Y. Zhang, Z. Lu, Q. Li, Q. Xue, S. Du, J. Pu, L. Wang, Superlubricity enabled by pressure-induced friction collapse. J. Phys. Chem. Lett. **9**, 2554–2559 (2018)
59. C. Liu, K. Liu, Monolayer mosaic heterostructures. Nat. Nanotechnol. **17**, 439–440 (2022)
60. Z.L. Wang, J.H. Song, Piezoelectric nanogenerators based on zinc oxide nanowire arrays. Science **312**, 242–246 (2006)

61. A. Li, Y. Zi, H. Guo, Z.L. Wang, F.M. Fernandez, Triboelectric nanogenerators for sensitive nano-coulomb molecular mass spectrometry. Nat. Nanotechnol. **12**, 481–487 (2017)
62. J. Liu, A. Goswami, K. Jiang, F. Khan, S. Kim, R. McGee, Z. Li, Z. Hu, J. Lee, T. Thundat, Direct-current triboelectricity generation by a sliding Schottky nanocontact on MoS2 multilayers. Nat. Nanotechnol. **13**, 112 (2018)
63. Y. Qin, X. Wang, Z.L. Wang, Microfibre-nanowire hybrid structure for energy scavenging. Nature **457**, 340–340 (2009)
64. R. Yang, Y. Qin, L. Dai, Z.L. Wang, Power generation with laterally packaged piezoelectric fine wires. Nat. Nanotechnol. **4**, 34–39 (2009)
65. H. Shao, C.-Y. Tsui, W.-H. Ki, The design of a micro power management system for applications using photovoltaic cells with the maximum output power control. IEEE Trans. Very Large Scale Integr. Syst. **17**, 1138–1142 (2009)
66. Z.L. Wang, W. Wu, Nanotechnology-enabled energy harvesting for self-powered micro-/nanosystems. Angew. Chem. Int. **51**, 11700–11721 (2012)
67. F.R. Fan, W. Tang, Z.L. Wang, Flexible Nanogenerators for energy harvesting and self-powered electronics. Adv. Mater. **28**, 4283–4305 (2016)
68. M.A.P. Mahmud, N. Huda, S.H. Farjana, M. Asadnia, C. Lang, Recent advances in nanogenerator-driven self-powered implantable biomedical devices. Adv. Energy Mater. **8**, 25 (2018)
69. A. Proto, M. Penhaker, S. Conforto, M. Schmid, Nanogenerators for human body energy harvesting. Trends Biotechnol. **35**, 610–624 (2017)
70. P.X. Gao, J. Song, J. Liu, Z.L. Wang, Nanowire piezoelectric nanogenerators on plastic substrates as flexible power sources for nanodevices. Adv. Mater. **19**, 67 (2007)
71. Z.L. Wang, Triboelectric nanogenerators as new energy technology for self-powered systems and as active mechanical and chemical sensors. ACS Nano **7**, 9533–9557 (2013)
72. C.C. Nguyen, D.C. Ranasinghe, S.F. Al-Sarawi, Analytical modeling and optimization of electret-based microgenerators under sinusoidal excitations. Microsyst. Technol. Micro-Nanosyst. Inform. Stor. Process. Syst. **23**, 5855–5865 (2017)
73. J. Liu, M. Miao, K. Jiang, F. Khan, A. Goswami, R. McGee, Z. Li, N. Lan, Z. Hu, J. Lee, K. Cadien, T. Thundat, Sustained electron tunneling at unbiased metal-insulator-semiconductor triboelectric contacts. Nano Energy **48**, 320–326 (2018)
74. J. Liu, Y. Zhang, J. Chen, R. Bao, K. Jiang, F. Khan, A. Goswami, Z. Li, F. Liu, K. Feng, J. Luo, T. Thundat, Separation and quantum tunneling of photo-generated carriers using a tribo-induced field. Matter **1**, 650–660 (2019)
75. S. Lin, Y. Lu, S. Feng, Z. Hao, Y. Yan, A high current density direct-current generator based on a moving van der Waals Schottky diode. Adv. Mater. **31**, 32–65 (2019)
76. Y. Lu, Z. Hao, S. Feng, R. Shen, Y. Yan, S. Lin, Direct-current generator based on dynamic PN junctions with the designed voltage output. Iscience **22**, 58 (2019)
77. R. Xu, Q. Zhang, J.Y. Wang, D. Liu, J. Wang, Z.L. Wang, Direct current triboelectric cell by sliding an n-type semiconductor on a p-type semiconductor. Nano Energy **66**, 1542 (2019)
78. Z. Zhang, D. Jiang, J. Zhao, G. Liu, T. Bu, C. Zhang, Z.L. Wang, Tribovoltaic effect on metal-semiconductor interface for direct-current low-impedance triboelectric nanogenerators. Adv. Energy Mater. **10**, 652 (2020)
79. M. Zheng, S. Lin, L. Xu, L. Zhu, Z.L. Wang, Scanning probing of the tribovoltaic effect at the sliding interface of two semiconductors. Adv. Mater. **32**, 1124 (2020)
80. S. Lin, R. Shen, T. Yao, Y. Lu, S. Feng, Z. Hao, H. Zheng, Y. Yan, E. Li, Surface states enhanced dynamic schottky diode generator with extremely high power density over 1000 W m^{-2}. Adv. Sci. **6**, 45–65 (2019)
81. Y. Lu, S. Feng, R. Shen, Y. Xu, Z. Hao, Y. Yan, H. Zheng, X. Yu, Q. Gao, P. Zhang, S. Lin, Tunable dynamic black phosphorus/insulator/Si heterojunction direct-current generator based on the hot electron transport. Research **2019**, 1542–5643 (2019)
82. Y. Lu, Q. Gao, X. Yu, H. Zheng, R. Shen, Z. Hao, Y. Yan, P. Zhang, Y. Wen, G. Yang, S. Lin, Interfacial built-in electric field-driven direct current generator based on dynamic silicon homojunction. Research **2020**, 16 (2020)

83. H. Deng, M. Ma, Y. Song, Q. He, Q. Zheng, Structural superlubricity in graphite flakes assembled under ambient conditions. Nanoscale **10**, 14314–14320 (2018)
84. X. Huang, X. Xiang, J. Nie, D. Peng, F. Yang, Z. Wu, H. Jiang, Z. Xu, Q. Zheng, Microscale Schottky superlubric generator with high direct-current density and ultralong life. Nat. Commun. **12**, 2268 (2021)
85. C. Qu, K. Wang, J. Wang, Q. Zheng, Origin of friction in superlubric graphite contacts. Phys. Rev. Lett. (2020)
86. Z. Zhang, J.T. Yates, Band bending in semiconductors: chemical and physical consequences at surfaces and interfaces. Chem. Rev. **112**, 5520–5551 (2012)
87. M.T. Angulo, C.H. Moog, Y.Y. Liu, A theoretical framework for controlling complex microbial communities. Nat. Commun. **10**, 12 (2019)
88. G. Wang, Z. Dai, Y. Wang, P. Tan, L. Liu, Z. Xu, Y. Wei, R. Huang, Z. Zhang, Measuring interlayer shear stress in bilayer graphene. Phys. Rev. Lett. **119**, 542 (2017)
89. Y. Liu, Y. Huang, X. Duan, Van der Waals integration before and beyond two-dimensional materials. Nature **567**, 323–333 (2019)
90. D. Peng, Z. Wu, D. Shi, C. Qu, H. Jiang, Y. Song, M. Ma, G. Aeppli, M. Urbakh, Q. Zheng, Load-induced dynamical transitions at graphene interfaces. Proceed. Natl. Acad. Sci. **117**, 12618–12623 (2020)
91. X. Lu, H. Huang, N.I. Nemchuk, R.S. Ruoff, Patterning of highly oriented pyrolytic graphite by oxygen plasma etching. Appl. Phys. Lett. **75**, 193–195 (1999)
92. Y. Gongyang, C. Qu, S. Zhang, M. Ma, Q. Zheng, Eliminating delamination of graphite sliding on diamond-like carbon. Carbon **132**, 444–450 (2018)
93. Z. Liu, S.-M. Zhang, J.-R. Yang, J.Z. Liu, Y.-L. Yang, Q.-S. Zheng, Interlayer shear strength of single crystalline graphite. Acta Mech. Sin. **28**, 978–982 (2012)
94. H. Hu, *Origin and Regulation of Friction in Superlubric Graphite Interface* (Tsinghua University, Beijing, 2022)
95. W. Chen, T. Wu, Y. Wang, Y. Wang, M. Ma, Q. Zheng, Z. Wu, Filtering robust graphite without incommensurate interfaces by electrical technique. ACS Appl. Mater. Interf. (2023). https://doi.org/10.1021/acsami.3c12234
96. Z.T. Li, A. Kozbial, N. Nioradze, D. Parobek, G.J. Shenoy, M. Salim, S. Amemiya, L. Li, H.T. Liu, Water protects graphitic surface from airborne hydrocarbon contamination. ACS Nano **10**, 349–359 (2016)
97. Z. Li, Y. Wang, A. Kozbial, G. Shenoy, F. Zhou, R. McGinley, P. Ireland, B. Morganstein, A. Kunkel, S.P. Surwade, L. Li, H. Liu, Effect of airborne contaminants on the wettability of supported graphene and graphite. Nat. Mater. **12**, 925–931 (2013)
98. G. He, M.H. Muser, M.O. Robbins, Adsorbed layers and the origin of static friction. Science **284**, 1650–1652 (1999)
99. D. Martinez-Martin, R. Longuinhos, J.G. Izquierdo, A. Marele, S.S. Alexandre, M. Jaafar, J.M. Gomez-Rodriguez, L. Banares, J.M. Soler, J. Gomez-Herrero, Atmospheric contaminants on graphitic surfaces. Carbon **61**, 33–39 (2013)
100. M. Urbakh, J. Klafter, D. Gourdon, J. Israelachvili, The nonlinear nature of friction. Nature **430**, 525–528 (2004)
101. H. Rokni, W. Lu, Direct measurements of interfacial adhesion in 2D materials and van der Waals heterostructures in ambient air. Nat. Commun. **11**, 87–92 (2020)
102. H. Gang, M.O. Robbins, Simulations of the static friction due to adsorbed molecules. Phys. Rev. B Condens. Matter Mater. Phys. **64**, 313 (2001)
103. M.H. Muser, L. Wenning, M.O. Robbins, Simple microscopic theory of Amontons's laws for static friction. Phys. Rev. Lett. **86**, 1295–1298 (2001)
104. K. Hasz, Z.J. Ye, A. Martini, R.W. Carpick, Experiments and simulations of the humidity dependence of friction between nanoasperities and graphite: the role of interfacial contact quality. Phys. Rev. Mater. **2**, 9 (2018)
105. D. Dietzel, C. Ritter, T. Mönninghoff, H. Fuchs, A. Schirmeisen, U.D. Schwarz, Frictional duality observed during nanoparticle sliding. Phys. Rev. Lett. **101**, 154 (2008)

106. A. Khajeh, Z. Chen, S.H. Kim, A. Martini, Effect of ambient chemistry on friction at the basal plane of graphite. ACS Appl. Mater. Interf. **11**, 40800–40807 (2019)
107. P. Gallagher, M. Lee, F. Amet, P. Maksymovych, J. Wang, S.P. Wang, X.B. Lu, G.Y. Zhang, K. Watanabe, T. Taniguchi, D. Goldhaber-Gordon, Switchable friction enabled by nanoscale self-assembly on graphene. Nat. Commun. **7**, 7 (2016)
108. J.S. Choi, J.S. Kim, I.S. Byun, D.H. Lee, M.J. Lee, B.H. Park, C. Lee, D. Yoon, H. Cheong, K.H. Lee, Y.W. Son, J.Y. Park, M. Salmeron, Friction anisotropy-driven domain imaging on exfoliated monolayer graphene. Science **333**, 607–610 (2011)
109. A.M. Smith, J.E. Hallett, S. Perkin, Solidification and superlubricity with molecular alkane films. Proceed. Natl. Acad. Sci. USA **116**, 25418–25423 (2019)
110. S.F. Cheng, B.Q. Luan, M.O. Robbins, Contact and friction of nanoasperities: effects of adsorbed monolayers. Phys. Rev. E **81**, 559 (2010)
111. W.G. Ouyang, A.S. de Wijn, M. Urbakh, Atomic-scale sliding friction on a contaminated surface. Nanoscale **10**, 6375–6381 (2018)
112. J.J. Li, W. Cao, J.F. Li, M. Ma, J.B. Luo, Molecular origin of superlubricity between graphene and a highly hydrophobic surface in water. J. Phys. Chem. Lett. **10**, 2978–2984 (2019)
113. M. Ma, I.M. Sokolov, W. Wang, A.E. Filippov, Q.S. Zheng, M. Urbakh, Diffusion through bifurcations in oscillating nano- and microscale contacts: fundamentals and applications. Phys. Rev. X **5**, 9 (2015)
114. M. Peyrard, S. Aubry, Critical-behavior at the transition by breaking of analyticity in the discrete frenkel-kontorova model. J. Phys. C Solid State Phys. **16**, 1593–1608 (1983)
115. X.F. Feng, S. Kwon, J.Y. Park, M. Salmeron, Superlubric sliding of graphene nanoflakes on graphene. ACS Nano **7**, 1718–1724 (2013)
116. D. Dietzel, M. Feldmann, U.D. Schwarz, H. Fuchs, A. Schirmeisen, Scaling laws of structural lubricity. Phys. Rev. Lett. **111**, 5 (2013)
117. E. Koren, E. Lortscher, C. Rawlings, A.W. Knoll, U. Duerig, Adhesion and friction in mesoscopic graphite contacts. Science **348**, 679–683 (2015)
118. C.Y. Qu, K.Q. Wang, J. Wang, Y.J. Gongyang, R.W. Carpick, M. Urbakh, Q.S. Zheng, Origin of friction in superlubric graphite contacts. Phys. Rev. Lett. **125**, 6 (2020)
119. C. Androulidakis, E.N. Koukaras, G. Paterakis, G. Trakakis, C. Galiotis, Tunable macroscale structural superlubricity in two-layer graphene via strain engineering. Nat. Commun. **11**, 11 (2020)
120. K.Q. Wang, Y.Q. He, W. Cao, J. Wang, C.Y. Qu, M.S. Chai, Y. Liu, Q.S. Zheng, M. Ma, Structural superlubricity with a contaminant-rich interface. J. Mech. Phys. Solids **169**, 118 (2022)
121. R.R. Chromik, Y.Y. Zhang, Nanomechanical testing of third bodies. Curr. Opin. Solid State Mater. Sci. **22**, 142–155 (2018)
122. J. Wang, A. Khosravi, A. Vanossi, E. Tosatti, Colloquium: sliding and pinning in structurally lubric 2D material interfaces. Rev. Mod. Phys. **96**, 011002–011027 (2024)
123. Z. Xu, H. Li, M. Ma, Accurate estimation of dynamical quantities for nonequilibrium nanoscale systems. Phys. Rev. E **107**, 632 (2023)
124. X.Z. Liu, Z.J. Ye, Y.L. Dong, P. Egberts, R.W. Carpick, A. Martini, Dynamics of atomic stick-slip friction examined with atomic force microscopy and atomistic simulations at overlapping speeds. Phys. Rev. Lett. **114**, 32–65 (2015)
125. P. Gallagher, Y.L. Li, K. Watanabe, T. Taniguchi, T.F. Heinz, D. Goldhaber-Gordon, Optical imaging and spectroscopic characterization of self-assembled environmental adsorbates on graphene. Nano Lett. **18**, 2603–2608 (2018)
126. A. Temiryazev, A. Frolov, M. Temiryazeva, Atomic-force microscopy study of self-assembled atmospheric contamination on graphene and graphite surfaces. Carbon **143**, 30–37 (2019)
127. W.F. Foster, K.M. Miyazawa, T.F. Fukuma, H.K. Kusumaatmaja, K. Voitchovsky, Self-assembly of small molecules at hydrophobic interfaces using group effect. Nanoscale **12**, 5452–5463 (2020)
128. A. Palinkas, G. Kalvin, P. Vancso, K. Kandrai, M. Szendro, G. Nemeth, M. Nemeth, A. Pekker, J.S. Pap, P. Petrik, K. Kamaras, L. Tapaszto, P. Nemes-Incze, The composition and structure

of the ubiquitous hydrocarbon contamination on van der Waals materials. Nat. Commun. **13**, 5498 (2022)
129. B.T. Liu, J. Wang, S.J. Zhao, C.Y. Qu, Y. Liu, L.R. Ma, Z.H. Zhang, K.H. Liu, Q.S. Zheng, M. Ma, Negative friction coefficient in microscale graphite/mica layered heterojunctions. Sci. Adv. **6**, 8 (2020)
130. T.Y. Han, W. Cao, Z. Xu, V. Adibnia, M. Olgiati, M. Valtiner, L.R. Ma, C.H. Zhang, M. Ma, J.B. Luo, X. Banquy, Hydration layer structure modulates superlubrication by trivalent La^{3+} electrolytes. Sci. Adv. **9**, 5487 (2023)
131. E. Gnecco, E. Meyer (eds.), *Fundamentals of Friction and Wear on the Nanoscale*, 2nd edn (2015), pp. 1–704

Molecular Tribology: Chemically Engineering Energy Dissipation at the Nanoscale

Oscar Gutiérrez-Varela, Rémy Pawlak, Giacomo Prampolini, Ernst Meyer, and Jose Guilherme Vilhena

Abstract Friction is a phenomenon which is present in our everyday life although we tend to remember it only when it is nearly absent such as when "slipping on ice". Its presence across disparate length scales (earthquakes, car engines down to molecular machines) reminds us of its ubiquity which endows friction of an utmost practical importance. Therefore, attempts to control it are almost as old as civilization and intrinsically tied to our technological progress. Interestingly, during the past decades we have witnessed a growing progress in miniaturization of devices down to the nanometer scale. "Special problems occur when things get small [...] and it might turn out to be advantages if we knew how to design for them", said Feynman when discussing the prospects of building "infinitesimal machinery". To achieve this goal, and to design efficient molecular nano-engines, it becomes imperative to unveil the non-equilibrium processes governing friction and energy dissipation at a molecular level. This chapter provides a comprehensive review of recent advancements in understanding nanoscale friction and the role of internal molecular degrees of freedom in controlling energy dissipation during friction. We discuss how recent advancements in experimental techniques, particularly those linked to Scanning Probe Microscopy, have significantly enhanced our comprehension of the mechanical characteristics of individual molecules and their influence on dissipation processes. We delve into

O. Gutiérrez-Varela · J. G. Vilhena (✉)
Instituto de Ciencia de Materiales de Madrid, Consejo Superior de Investigaciones Científicas, 28049 Madrid, Spain
e-mail: guilherme.vilhena@csic.es

O. Gutiérrez-Varela
e-mail: oscarj.gutierrez@uam.es

O. Gutiérrez-Varela
Departamento de Física Teórica de la Materia Condensada, Universidad Autónoma de Madrid, 28049 Madrid, Spain

R. Pawlak · E. Meyer
Department of Physics, University of Basel, Klingelbergstrasse 82, 4056 Basel, Switzerland

G. Prampolini
Istituto Di Chimica dei Composti Organo Metallici, Consiglio Nazionale delle Ricerche (ICCOM-CNR), 56124 Pisa, Italy

E. Gnecco and E. Meyer (eds.), *Fundamentals of Friction and Wear on the Nanoscale*, NanoScience and Technology, https://doi.org/10.1007/978-3-031-63065-1_7

how these internal degrees of freedom facilitate control over energy dissipation, unlocking various pathways to achieve different applications at the nanoscale, such as superlubric states through molecular flexibility. Furthermore, we analyze potential applications of the energy dissipation pathways in novel mechanisms for achieving controlled locomotion of molecular machines.

1 Introduction

Friction is an ubiquitous phenomenon across disparate length scales, from earthquake dynamics [1] to nanoscale contacts [2, 3]. Historical examples, such as wetting of the desert Saharan sands to ease the sliding of the building blocks of ancient Egyptian pyramids [4] and the development of what we know now as the first two laws of friction by Leonardo da Vinci for advancing his machine designs [5], highlight the connection between technological progress and friction understanding. More recently, advanced lubrication mechanisms for applications like car engines and space applications [6, 7] further demonstrate the growing need to control friction. Additionally, to fully unlock the potential of emerging nanomechanical devices like molecular machines [8, 9] and nanoarchitectures [10], calls for a better understanding of energy dissipation at such scales. However, the intricate nature of the many interactions among atoms and molecules participating, along with their diverse configurations during sliding, still presents a formidable challenge, leaving a gap in our understanding of the fundamental origins of friction.

In a renewed attempt to control friction by atomistic design and aiming to bridge the gap across different length scales, the fields of nanotribology and nanomanipulation were established [11–13]. Powered by major advances in nanotechnology, they allowed for a nanometer scale control of sliding interfaces, e.g. through Ultra-High-Vacuum (UHV) and Low-Temperature (LT) experiments, which resulted in major advances on our understanding of friction at an atomic scale [11–13]. One of the key to its advancement is the use of Atomic Force Microscopy (AFM), enabling precise measurements of nano-scale friction forces when scanning a surface using a sharp tip. However, relating the forces measured with AFM to specific energy dissipation processes require advanced atomistic numerical simulations, particularly Molecular Dynamics (MD). They provide an atomically detailed understanding of what is occurring at the interface—often inaccessible in most experiments. Likewise we have gained an atomic detailed insight into such properties [14–20], leading to significant progress in understanding nanotribological properties. Up to date, the main focus of nanotribology has been understanding and controlling friction one atom at-a-time, with the hope of bridging the gap across different length-scales, from atomic up to macroscopic contacts [12]. Nevertheless, nanotribology and nanomanipulation also presents a golden opportunity to study and unveil energy dissipation processes in a purely nanoscale context, without seeking a macroscopic counterparts, and instead to fully unleash the potential of nanoarchitectures and nanomachines [20, 21].

The pioneering works by Eigler and Schweizer [22], Eigler et al. [23] in the early 1990s marked the start of systematic atom-by-atom nanomanipulation using a Scanning Tunneling Microscope (STM). Subsequent experiments demonstrated the ability to move vertically and laterally individual atoms [24, 25], molecules [26, 27], and bonds [28, 29]. It took however over 15 years after the first atomic-manipulation by D. Eigler until were reported the first experiments in which one can manipulate molecules on virtually any kind of solid surface at LT and UHV conditions, and at the same time recording the time variations of mechanical forces as doing so [30–35]. In a very general way, the experiments consist in the deposition of the molecule onto a surface and then a tip of the Scanning Probe Microscopy (SPM) is brought close and subsequently manipulations are performed on it, either detaching it and/or dragging it on the surface. Then, signals are recorded from tip during manipulation to later relate them with the mechanical properties associated with the internal degrees of freedom of the molecule. MD simulations can be used to mimic these experiments and reproduce the results, and then obtain an atomistic description of the mechanical properties of the molecule by explicitly considering intramolecular bond behaviors like stretching, twisting, etc. [19, 20, 36].

The aforementioned experiments when synergistically combined with numerical simulations, revealed for the first time the importance of the internal degrees of freedom in the dynamics of complex molecular systems. For example, it has been shown how internal degrees of freedom of flexible polymeric chains give rise to a new form of superlubricity [20]. This and other developments ultimately paved the road to for a novel nanotribology branch seeking new phenomenon (energy dissipation mechanisms) without macroscale counterparts—note that in the macroscale, the action of molecular internal degrees of freedom can be averaged to a negligible contribution.

A systematic understanding of internal mechanical properties of single-molecules enabled through nanomanipulation techniques [19, 21, 37] has fueled a significant advancement in the field of nanotribology and nanomechanics. Among other examples, they have allowed to conceive novel molecular propulsion mechanisms, in which the motion is driven by the release of mechanical energy from particular degrees of freedom. For example, using a Scanning Tunneling Microscopy (STM) tip, it is possible to achieve directed rotations or translations of single molecules by mechanically rotating one or more of their chemical groups, creating with this molecular wheels [38, 39]. Furthermore, it has also been demonstrated that it is possible to activate the internal degrees of freedom of a molecule with light, by the rotation of one of its chemical groups after being irradiated [40].

Having unlocked the connection between internal molecular degrees of freedom and its nanomechanical properties, the logical step-forward is now to chemically design on-surface molecular mobility [41–43]. To optimize the on-surface diffusion and control the auto-assembly of molecular structures, it is essential to explore chemical moieties capable of adjusting both the grip and on-surface propulsion mechanisms [44]. Through them, we can either hinder or enhance the diffusion process in a controlled manner and thus provide a road-map for the bottom-up design of complex nanoarchitectures. However, such control calls for a better understanding of the energetics of on-surface diffusion.

In this chapter, we review both experimental and simulations of nanomanipulations mainly focusing on the importance of internal molecular degrees of freedom during such process. In this manner, we wish to convey the possibility of leveraging these internal properties of molecules in our favor when addressing nanotribological problems. Therefore, a comprehensive study of the mechanical and tribological properties of various molecular compounds deposited on a surface becomes essential. In Sect. 2, relevant experiments related to nanomanipulation are presented, followed by a brief discussion on experimental limits in the end of the section. Section 3 provides a general overview of the protocols employed in Molecular Dynamics to model nanomanipulation experiments, while Sect. 4 presents relevant results from Molecular Dynamics simulations. Finally, Sect. 5 offers a discussion of the addressed topic, along with its future perspectives.

2 Nanomanipulation Experiments

2.1 First Atom by Atom Nanomanipulation

In the early 1990s, Eigler and Schweizer [22], Eigler et al. [23] at IBM's laboratories pioneered the field of nanomanipulation. Their groundbreaking experiments began with the manipulation of Xenon (Xe) atoms deposited on a nickel surface in LT-UHV conditions [22]. Using a STM tip, they demonstrated the precise control of Xe atom placement by adjusting the tip's current, as seen in Fig. 1. The nanomanipulation process involves several key steps. As illustrated in Fig. 1a, initially, Xe atoms are randomly deposited onto the nickel surface. Subsequently, the surface is scanned using a STM. The STM tip is then precisely positioned on a selected Xe atom, and the current is increased to trap the atom. The trapped atom is moved along the surface to a predetermined location without detaching it. Throughout this nanomanipulation, the current is maintained at a constant level. Once the atom reaches the desired position, the current is lowered, resulting in the release of the atom. This process is repeated sequentially for each Xe atom. By precisely arranging these atoms, D. Eigler and collaborators were able to construct the letters of IBM.

Their work revealed that nanomanipulation of Xe atoms was height-dependent, independent of bias voltage sign or magnitude, suggesting a predominant role of Van der Waals interactions. Indeed, these types of interactions can enable the manipulation not only of individual atoms but also of more complex molecules such as polymeric chains or biomolecules like single-stranded DNA [19], as discussed later on. D. Eigler and collaborators observed different threshold tip heights for moving Xe atoms along the nickel rows and perpendicular to them, indicating a force dependence on the dragging direction. Furthermore, their experiments confirmed the stability of Xe

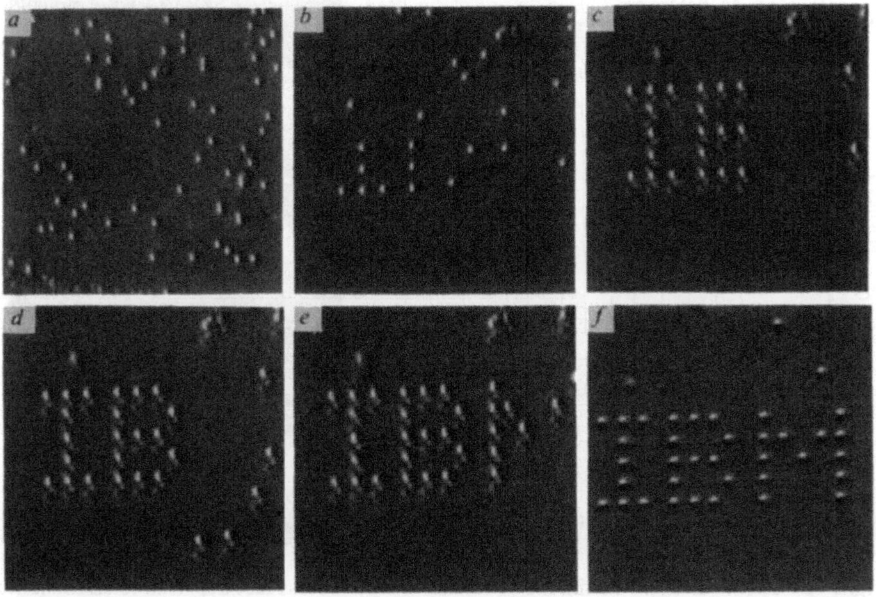

Fig. 1 Sequence of images taken with STM showing the creation of the IBM logo through the manipulation of Xe atoms. Each letter has dimensions of 50 Å from top to bottom. Reprinted from Eigler, D., Schweizer, E. Nature 344, 524–526 (1990), with permission from SNCSC

atom chains along nickel rows, maintaining a nearly uniform atom-to-atom distance, varying by only 0.2 Å.

In a subsequent experiment, conducted under the same laboratory conditions as their previous work [22], Eigler et al. [23] demonstrated their ability to detach and reposition Xe atoms from the nickel surface using a STM tip with varying current. The crucial aspect of trapping the Xe atom relies on a voltage switching strategy. The process begins with the Xe atom on the surface and the system in a low-conductance state (arrow A in Fig. 2a) with the tip initially at a certain negative potential and a height characterized by its tunneling-junction resistance. A positive potential is applied for a short time, resulting in a transient current spike (arrow B in Fig. 2a), and then the voltage is returned to its initial value. This procedure leads to the transfer of the Xe atom from the surface to the tip, establishing a high-conductance state (arrow C in Fig. 2a). Subsequently, a negative potential with a magnitude greater than the initial voltage is applied for a short time, producing a transient current spike (arrow D in Fig. 2a), and then the voltage is returned to its initial value. This re-establishes the low-conductance state by transferring the Xe atom from the tip to the surface.

To validate the correlation between these current variations and Xe atom trapping and release, imaging scans were performed at a higher tip position for each step. These scans visually confirmed that the Xe atom disappears and reappears during these voltage switches, as observed in Fig. 2b–d. Furthermore, these images demonstrated that the resolution increases when the tip has captured the Xe atom, as shown in

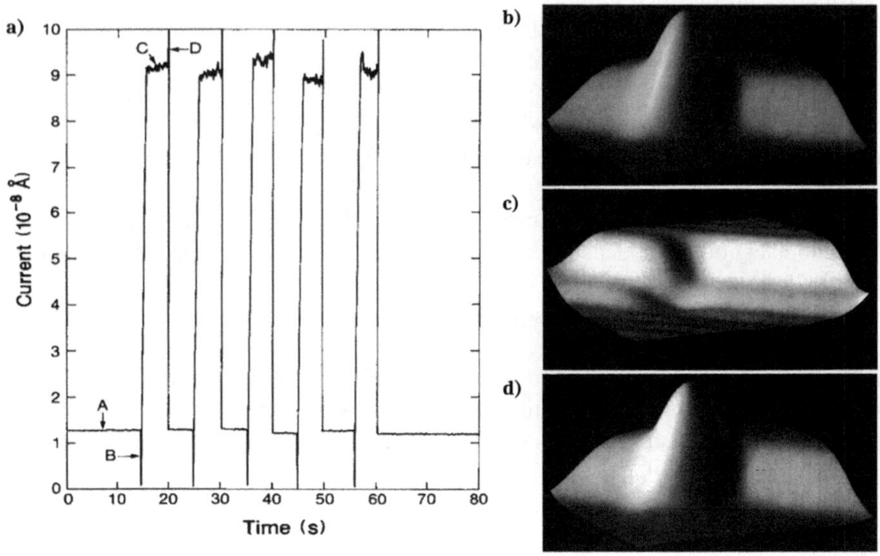

Fig. 2 **a** Current variations during the switching voltage process that traps and releases the Xe atom. The diagram shows the switch repeatedly alternating between the two conductance states. **b–d** A sequence of 25 × 25 Å images taken with STM showing **b** the Xe atom on the surface before trapping it, **c** the surface without the atom after trapping, and **d** the Xe atom after redepositing it onto the surface. In image **c** the surface structure is clearly visible, something that is not seen in images (**b**) and (**d**). Reprinted from Eigler, D., Lutz, C. and Rudge, W. Nature 352, 600–603 (1991), with permission from SNCSC

Fig. 2c. This discovery suggested the potential for enhancing STM resolution through tip functionalization, a concept later systematically confirmed with CO molecules [45].

These groundbreaking studies showcased the capability to manipulate atoms both laterally and vertically on a surface using SPM, sparking subsequent research in manipulating diverse molecules. Despite these advancements, the systematic measurement of forces during such atomically precise nanomanipulation remained elusive. Over fifteen years elapsed before were achieved the successful measurement of forces during such nanomanipulation conducted in UHV and LT. This breakthrough ultimately enabled a systematic study of the mechanical properties of various molecules with an unprecedented resolution.

2.2 *First Experiments in Measuring Forces During Nanomanipulation*

In the late 90 s, Rief et al. [46] were among the first to measure forces during manipulation of molecules using SPM in a liquid environment. They used force spectroscopy with a silicon cantilever to mechanically unfold proteins adsorbed onto a surface. In this methodology, a silicon cantilever, picks up the absorbed molecule using specific linkers like biotin. By retracting the tip, a force is applied to the molecule, and real-time measurement of cantilever deflection allows measuring the force as a function of the molecule-elongation. Despite undisputed merit of these pioneering works, such experiments had several limitations. Among other limitations [47], complex tip-functionalization schemes impedes having an image of the nanomanipulated object before/after this process and often results in the nanomanipulation of more than one object.

Measuring accurately the mechanical response of the various internal degrees of freedom of a molecule requires thorough isolation of the system. In typical ambient conditions, the molecule experiences stochastic collisions with its environment, introducing noise into measurements. This noise complicates the interpretation of the results, owing to the difficulty in separating the contributions from internal properties and those stemming from the environment. To mitigate this undesirable effect, precise control of the system is imperative. An effective approach involves operating under vacuum conditions and maintaining extremely low temperatures. These controlled conditions effectively eliminate unwanted environmental and thermal contributions, thus facilitating the correlation of measurements during manipulation with the activation of specific internal degrees of freedom of the molecule.

Besides a precise control of the atmosphere and system temperature, it is also necessary for the SPM to be sensitive enough to measure variations on the order of picometers (pm) and piconewtons (pN). This level of precision was achieved when resorting to a tuning fork sensor that allows tip oscillation amplitudes on the order of tens of pm, thereby achieving the necessary resolution for sub-molecular detail [47]. In this case, the typical nanomanipulation process begins with gently indenting the tip into the surface to sharpen its apex. Following this, the tip is brought close to one end of a selected molecule until a sudden increase in tunneling current occurs, indicating the establishment of a stable tip-molecule bond. Subsequently, the tip undergoes oscillations with amplitudes on the order of tens of pm and at a defined frequency f_0. Throughout the nanomanipulation process, the tip monitors variations in its resonance frequency.

A key aspect is that small oscillations of tuning fork sensors allows simplifying the relation between the normal force experienced by the tip with its oscillation frequency [47]. In particular, assuming low oscillation amplitudes, the relationship between the frequency shift (Δf) and the tip-sample stiffness k_{ts} can be approximated as:

$$\frac{\Delta f(z)}{f_0} = -\frac{1}{2k}\left\langle \frac{\partial f_{ts(z)}}{\partial z} \right\rangle \approx \frac{k_{ts}}{2k}$$

with z the tip-sample distance, f_0 the resonant frequency, k the sensor stiffness, $\langle dF_{ts}(z)/dz \rangle$ the averaged force gradient over the tip oscillation amplitudes and thus approximated by $k_{ts} \approx dF_{ts}/dz$, the tip-sample stiffness. Although not trivially connected, k_{ts} provides a detailed understanding on the dynamics of the system attached to the tip.

2.2.1 Polymerized Fluorene Chains

In 2014, Kawai et al. [31] manipulated long polymer chains adsorbed over an Au(111) in LT-UHV whilst simultaneously measuring for the first time k_{ts}, the normal force gradient. In their study, the researchers anchored the AFM tip to the edge of the polymer chain (imaged with an atomic detail before and after the manipulation process, as seen in Fig. 3a) and then retracted the tip. During this detachment process, they made a remarkable observation: a consistent periodicity in the force gradient (k_{ts}), as depicted in Fig. 3b. Remarkably, this periodicity is maintained throughout the detachment process, ultimately terminating with the full desorption of the chain. In particular, when lifting the tip a drop in the force gradient (k_{ts}) is observed every 0.91 nm in average, similar to the size of a fluorene monomer (0.845 nm). Then, the modulation of the force gradient can be correlated with the size of monomer composing the fluorene chain, leading to relate the drops with the detachment of a subunit.

As shown in Fig. 3b, as the tip is lifted, one observes an increase in the stiffness (k_{TS}) between the tip and the sample until it suddenly drops. Such stiffness increase results from an increase of normal force as the tip is being lifted. When the normal force is high enough to overcome the adhesion force of the subunit, the latter detaches, thus resulting in a modulation of the stiffness (k_{TS}) with the about the same periodicity as the monomer length. Given that the amplitudes in the force gradient are relatively similar throughout the entire process, these experiments also allowed concluding that the force required to detach each molecular subunit is similar regardless of its position in the chain. The periodic detachment of fluorene units emerged as a prominent mechanical feature of the polymer chains. This periodicity implied that fluorene chains exhibited a stepwise detachment behavior, with each periodic force gradient variation corresponding to the removal of a single fluorene unit from the surface.

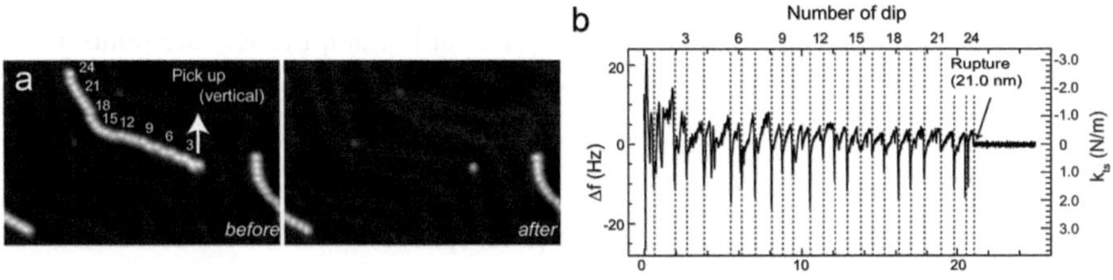

Fig. 3 **a** An STM image of the same area on the Au(111) surface before and after detachment of the polymeric chain. **b** Presents a graph of the frequency shift during detachment as a function of the tip height until the molecule is completely detached. Reprinted from S. Kawai et al., PNAS 111, 3968–3972 (2014)

2.2.2 Porphyrin Molecule

Subsequently, Pawlak et al. build upon such endeavor by a study which highlighted the critical role of internal molecular degrees of freedom of a single molecule during its lateral nanomanipulation. Such experiments consisted in vertical and lateral manipulation of a porphyrin molecule affixed to the apex of an AFM tip in LT-UHV over a Cu(111) surface [33]. An important finding in this study is the conspicuous hysteresis observed in the extracted normal force during both the approach and retraction phases, as shown in Fig. 4a. In the approach, initially, the magnitude of the force tends to increase smoothly with a linear-like behavior. Upon reaching the region delimited by the dashed black line in Fig. 4a, the slope of the force increases, causing its magnitude to grow rapidly and then maintains a relatively constant value. For the retraction, the changes are more evident. Initially, there is a plateau-like behavior with a force magnitude greater than that for the approach case. Then it undergoes a decrease in magnitude until reaching another plateau-like behavior. Finally, the force magnitude increases again before starting to decrease and converge with the force observed during the approach movement.

The sudden fluctuations in the normal force suggest that the molecule undergoes specific structural transformations as it interacts with the surface and is approached or retracted. More specific, this hysteresis is ascribed to conformational changes transpiring within the porphyrin molecule during manipulation, manifesting through shifts in the normal force at different distances from the surface, as shown in the inset

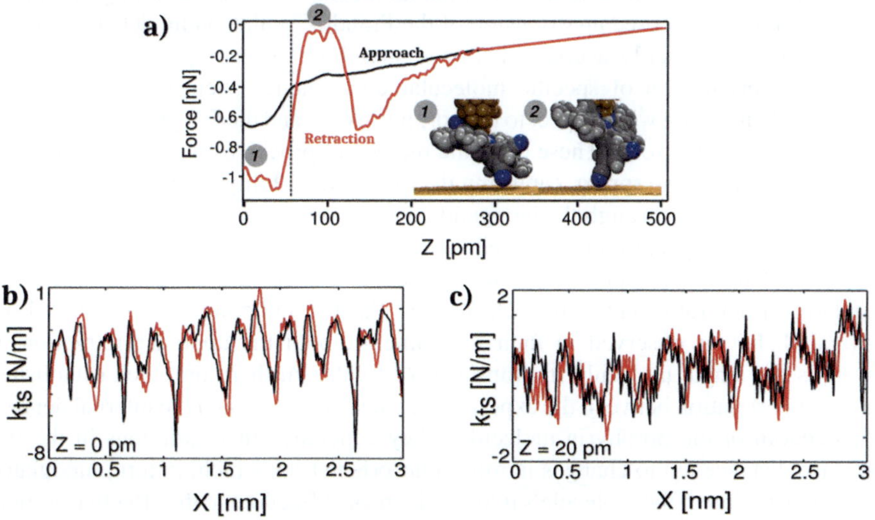

Fig. 4 **a** Normal force during the approach (black) and retraction (red) of the porphyrin molecule. The inset shows the different conformations during manipulation. **b** and **c** The stiffness during lateral manipulation considering a tip height of 0 and 20 pm, respectively. Reprinted (adapted) with permission from R. Pawlak et al., ACS Nano 2016, 10, 1, 713–722. Copyright 2016 American Chemical Society

of Fig. 4a. It is discerned that these transitions arose from the mechanical properties of the molecule acting as a "spring" connecting the tip apex to the copper surface.

This last point is of paramount importance. As introduced in the beginning of this Section, the critical measurement involves monitoring the frequency variation at the tip, directly tied to the gradient of the normal force. However, in a lateral manipulation, the force acts perpendicular to the normal force, so a direct relation between lateral and normal force is not trivial. This study succeeded in establishing a connection between them by the molecule situated between the tip and the surface, effectively acting as a spring connecting them.

In the course of lateral manipulation, atomic-scale friction patterns became evident, as illustrated in Fig. 4b and c. These friction patterns are associated with the interaction between the carbonitrile (CN) end groups of the molecule and the surface potential. Notably, the frictional response during lateral manipulations exhibits different behavior based on the tip height, a variability attributed to changes in the molecular conformation. As observed, the amplitude of the force gradient is larger at a lower heights (Fig. 4b) as compared to higher heights (Fig. 4c).

To interpret their findings, Pawlak and his colleagues made a key observation. They proposed that the differences in force measurements observed during the approach and retraction cycles, as well as the lateral scans, were fundamentally influenced by specific mechanical factors governing the behavior of the molecule. The dominant factors at play were identified as the σ-bond connecting the dicyanophenyl group facing the copper surface to the porphyrin core and the corresponding bond between the CN group facing the surface and the phenyl ring. These bonds were recognized as the most flexible within the molecular structure. Building upon this, the researchers proposed an extension of the Prandtl-Tomlinson model. This generalized model implicitly accounted for the stretching and compression of σ-bond and the reorientation of specific molecular components, such as the meso-(3,5-dicyanophenyl) ring with respect to the porphyrin core and the CN group with respect to the phenyl side group. These dynamic molecular processes were modeled as the action of an effective spring, capturing the molecule's ability to absorb and liberate mechanical energy through the elongation and compression of its bonds.

The critical aspect of the proposed model lies in the incorporation of the internal degrees of freedom within the molecule, accounting for the molecule's ability to undergo structural changes. In such manner, Pawlak and colleagues replicated the overall behavior observed in their experiments, both the approach-retraction as well as the single-molecule friction measurements. Furthermore, they established a direct correlation between the experimental measurements and the internal degrees of freedom of the porphyrin molecule. They concluded that variations in the tip-surface distance led to changes in molecular orientation, which, in turn, modulated the contribution of the molecule's internal degrees of freedom to the effective vertical and lateral stiffness. As a result, these changes affected the overall measured forces, as was evident in the experimental data.

2.2.3 Graphene Nanoribbons

In the study conducted by Kawai et al. [15], such LT-UHV nanomanipulation procedure was employed to address a fundamental problem in nanotribology—the asymptotically vanishing friction per unit area of well defined incommensurate contacts. In particular, they measured the friction, adhesion, and elastic properties of graphene nanoribbons (GNRs) deposited over a clean Au(111) surface. The nanomanipulation process involved anchoring one end of the GNR to the AFM tip, then lifting it to a specific height, and finally sliding sideways at a fixed distance from the surface. In this manner, the authors were able to measure the static friction required to displace the GNR of different sizes, whilst characterizing with an atomic precision the contact. Their findings, i.e. that the static friction per unit length decreased with the ribbon length, ultimately provided a unmatched experimental realization of structural-superlubricity—a state of ultra-low friction between incommensurate sliders. To achieve this, they mapped the frequency shift while scanning the tip laterally along a GNR (in the x-direction) at different constant tip-sample separations, as observed in Fig. 5a. Then, they estimated the energy landscape experienced by the tip by integrating two times the $\Delta f(z)$ sections extracted from Fig. 5a. Next, they differentiated the normal potential along x and multiplied it by a constant associated with the tuning fork's characteristics, obtaining the lateral force that it is associated with the static friction force that is in the order of tens of pN. They performed this procedure for different GNRs with varying lengths and managed to demonstrate a decreasing trend in static friction force as the GNR size increased, as presented in Fig. 5b.

GNRs are exceptionally stiff, which means they do not readily stretch or adapt to the lattice of the underlying gold substrate while sliding. This exceptional lateral stiffness is one of the important factors contributing to superlubricity as it favors incommensurability at interface. The surface characteristics of the gold substrate plays an important role in modulating the friction properties. The degree of commensurability between the GNRs and the substrate is altered as GNRs cross boundary regions between different surface structures. The presence of these boundary regions leads to variations in the friction profiles, with some areas experiencing higher friction forces, as observed in the frequency shift graphs in Fig. 5c. This local modulation of friction is more pronounced at small tip-surface separations. When considering high tip-surface separations, the bending of the suspended portion of the GNR can be considerably more pronounced during backward scanning, resulting in a delay in slip events. This distinction is aparent in the bottom of Fig. 5c, where, at a height of 5 nm, the frequency shift profile during backward movement differs notably from forward movement, displaying jumps with much larger spacing.

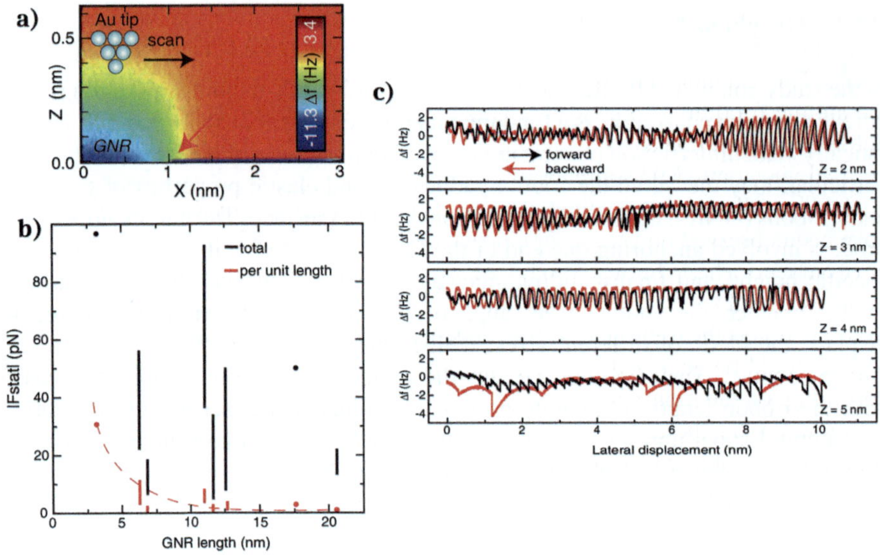

Fig. 5 **a** 2D Δf map along the longitudinal axis of the manipulated GNR. **b** Absolute value of the static friction force as a function of the GNR length (black) and per unit length (red). Dots are single measurements and the bars connect the largest and the smallest values measured. **c** Frequency shifts measured during lateral manipulation at different tip heights. From S. Kawai et al., Science 351, 6276, 957–961 (2016). Reprinted with permission from AAAS

2.3 *Study of Polymer Chains in Vacuum*

With the previous achievements, experiments were expanded to comprehensively study and analyze the mechanical characteristics of various molecular chains, elucidating the interplay between the mechanical attributes of the molecules and its tribological behavior. Substantial endeavors have been invested in comprehending the properties of polymeric chains, owing to their versatility in various scenarios and their widespread utility in diverse practical applications [17, 20, 37, 48].

2.3.1 Poly-pyrenylene Chains

Pawlak and his colleagues carried out experiments where the focus was on the analysis of mechanical properties of poly-pyrenylene chains when they are detached from a surface [48]. The researchers initiated their experiments by synthesizing extended poly-pyrenylene chains on a previously cleaned Au(111) surface. When scanning over the surface, they notably observed that the configuration of the molecules on the Au(111) surface displayed diverse curvatures, as seen in Fig. 6a, what is in contrast to the straight configuration observed in the case of GNRs [15]. Subsequently, the

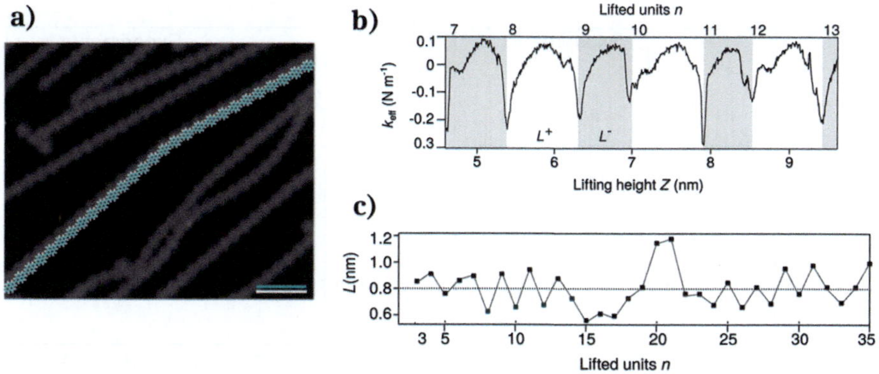

Fig. 6 **a** STM image of the polymeric molecules on the surface compared to a structure generated by MD. **b** Force gradient during the detachment of monomers $7 \leq n \leq 13$ showing the alternation in periodicity. **c** Spacing detachment as a function of unit *n*. Reprinted (adapted) with permission from R. Pawlak et al., Nano Letters 2020, 20, 1, 652–657. Copyright 2020 American Chemical Society

molecules were lifted from the gold surface by indenting the AFM tip at the end of the chain, conducted under LT-UHV conditions.

One of the most striking aspect is that when detaching the molecule from the surface, a periodic alternation in the spacing of the drop in the force gradient is observed, as shown in Fig. 6b and c. This indicates that the features for detaching monomers depend on whether their position in the chain is even or odd. More precisely, the results are associated with the fact that the force and energy required to detach the different monomers are different for the monomers in even positions (units 8, 10, and 12 in Fig. 6b) than in odd positions (units 7, 9, and 11 in Fig. 6b). However, to explain the behavior in more detail, an atomistic analysis of the problem is necessary. Poly-pyrenylene chains are known for their inherent flexibility, a consequence of carbon–carbon (C–C) single bonds. Indeed, data from molecular dynamics (MD) simulations, which will be elaborated in Sect. 4.1, revealed the presence of symmetry-broken conformations within the sliding physisorbed segment. The obtained results shed light on the intricate interplay between mechanical forces and the internal degrees of freedom inherent to these polymer chains.

In another study with a similar system, Vilhena et al. [20] examined these chains during lateral manipulations using AFM in LT-UHV conditions. In their experiments, the recorded force gradient exhibited periodic drops, as seen in Fig. 7, similar of those observed when polymeric chains detach from a surface. However, these profiles show differences in the spacing of periodicity when considering different tip heights. At low height (0.36 nm), a smaller spacing is observed (see Fig. 7a), compared to the greater spacing (~1 nm) observed at high height (0.51 nm) in Fig. 7b. Essentially, this shows that the dynamics of the portion of the poly-pyrenylene chain resting on the surface can be modulated by varying the tip height during manipulation. Another important finding is that the smaller spacing in the low-tip-height scenario (~0.22 nm)

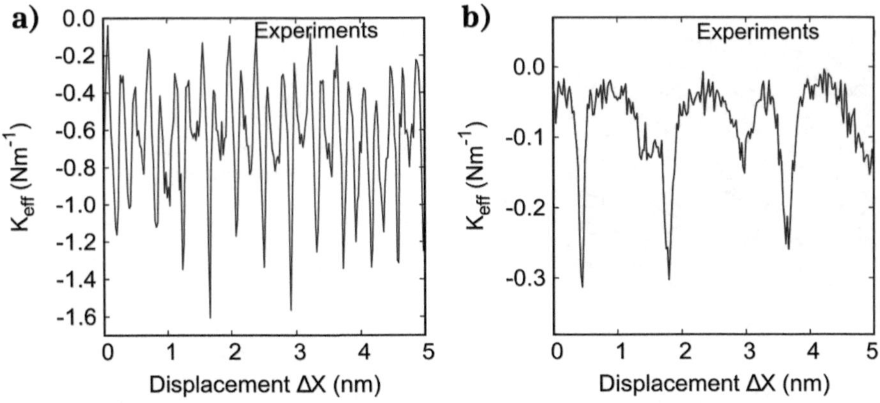

Fig. 7 Force gradient for lateral manipulation with **a** low and **b** high tip-sample separation. Reprinted with permission from J.G. Vilhena et al., Physical Review Letters 128, 216102 (2022). Copyright 2022 by the American Physical Society

is less than the periodicity of the distance between gold atoms (0.28 nm). This can be explained by considering a non-trivial trajectory of the chain over the crystalline structure while sliding, demonstrating side-winding motion of different types.

As previously shown in Fig. 6b and c, the features during the detachment of these polymeric chains depend on whether the monomer is in an even or odd position. This behavior, along with the dependence of the force gradient on height in lateral manipulation, allows considering a similar mechanism that gives rise to these different characteristics with origin in the internal properties of the molecule. To corroborate this point, MD simulations were conducted. The results will be discussed more extensively later on, however, we can anticipate the following: the differences in the profiles of the force gradient with respect to tip height are intimately linked to the geometric configuration of the poly-pyrenylene chain during the dragging process. Furthermore, the simulations allow the calculation of the static friction force, resulting in the order of a few tens of pN, demonstrating the superlubric character of these polymeric chains. This characteristic is associated to the molecular flexibility, a performance rivaling the superlubricity typically associated with rigid structures.

2.3.2 Very Long Polymer Chains

Lu et al. [37] conducted an investigation into the mechanical properties of very long polymeric chains, specifically poly-dimethylsiloxane (PDMS), polyoxymethylene (POM), and poly(methyl methacrylate) (PMMA), deposited onto quartz substrates under high-vacuum conditions (HV) and in a temperature range spanning from 153 to 300 K. They employed the Atomic Force Microscopy-based single-molecule force spectroscopy (AFM-SMFS) technique, by extending the chains and plotting the

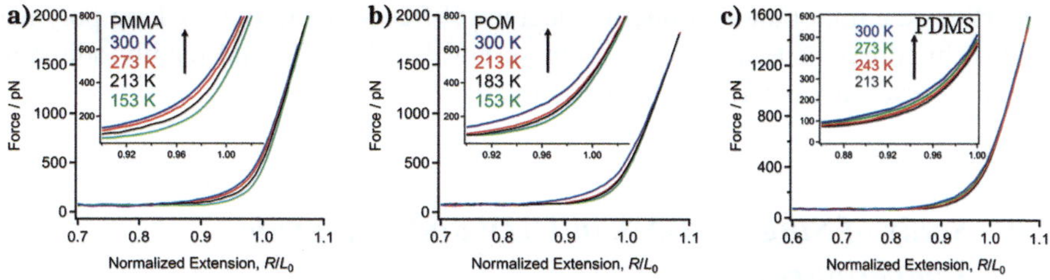

Fig. 8 Force as a function of the normalized extension for **a** PMMA, **b** POM, and **c** PDMS at different temperatures. L_O is larger than 100 nm. Reprinted (adapted) with permission from S. Lu et al., Macromolecules 2023, 56, 8, 3204–3212. Copyright 2023 American Chemical Society

resulting force against molecular extension. Note that at variance with the results shown thus far, in this study the authors use conventional silicon cantilevers. It is also worth highlighting that in this scenario, the detachment of polymers occurs with less precision. The absence of a prior image, like an STM image, depicting the polymers on the surface means that the tip is not systematically directed towards a specific point on the polymer for the detachment process. Then, the specific anchoring position remains unknown, contributing to a less controlled detachment procedure.

As shown in Fig. 8a–c, researchers identified three main force regimes during the elongation process: (i) the low-force, (ii) the middle-force, and (iii) the high-force regimes. To provide an explanation, the study applied the Freely Rotating Chain (FRC) model for polymers. This model conceptualizes the behavior of flexible polymers as a sequence of freely rotating units. However, the authors expanded upon this model by incorporating parameters derived from quantum mechanical (QM) calculations, giving rise to the modified QM-FRC model. The QM-FRC model, which incorporates the notion of temperature-independent Si–O and C–O bond rotations, explains the experimental data. A notable insight is that even as the temperature decreased, the internal rotations of covalent bonds persisted. This indicates that covalent bonds maintain their internal degrees of freedom, even at low temperatures, enabling the polymer chains to adapt to stretching forces.

The authors present a coherent summary of their observations focusing on PDMS chains, and confirm this with their theoretical model. During stretching, the PDMS chains suffer transition from a low-force regime, where Si–O bonds exist in gauche or trans states, to a middle-force regime, resulting in a decreased conformational entropy. This corresponds in Fig. 8c to the normalized extension that goes from 0.7 to ~0.9, depending on the temperature. Internal rotation of covalent bonds play an important role in conformational change and the associated decrement of entropy. The preference for the trans conformation, with lower energy and greater extension, became evident as the chains continued to stretch. However, in the high-force regime, corresponding to normalized extension larger than ~0.98 in Fig. 8c, enthalpic deformations occurred in the trans conformation with minimal internal rotations. In this case, the Si–O bonds, including their angle and length, undergo deformation due to the high force, and the molecule tends to remain stretched as a response. This finding underlines the importance of internal degrees of freedom within very long polymer chains (hundreds of nanometers), as these covalent bonds could readily adjust to external forces, irrespective of temperature.

2.4 *Study of Single ssDNA in UHV and Cryogenic Temperatures*

SPM methods offer the means to explore the mechanical and tribological properties of molecules beyond polymeric chains. One example is the work conducted by Pawlak et al. [19], where a comprehensive study on the mechanical and tribological properties of single-stranded DNA (ssDNA) is performed. The study encompass several stages to investigate the behavior of ssDNA oligomers on a Au(111) surface.

Initially the ssDNA molecules are deposited onto a clean Au(111) surface at room temperature. STM is then used to examine the Au(111) surface after ssDNA deposition observing solvent molecules surrounding the ssDNA chains. To eliminate the solvent, they anneal the system at different temperatures up to 500 K. Remarkably, even after these annealing steps, the ssDNA molecules consistently exhibited a folded configuration.

Following each annealing stage, the experimental setup is cooled to 5 K in which the samples are imaged. After removing most solvent molecules, one proceeds to lift the ssDNA molecules from the gold surface. During lifting, the measured stiffness exhibits a peculiar behavior. As the tip-sample separation increase, the stiffness decreases progressively, as shown in Fig. 9. This contrasts with typical polymer chains where stiffness maintains a mostly constant amplitude during lifting [31, 48], as seen in Figs. 3 and 6. Another observation is the presence of sharp dips in the measured stiffness occurring at 0.2–0.3 nm intervals, indicating the detachment of single nucleotides from the gold surface. Significantly, an abrupt drop in the frequency shift to zero is observed at a tip height of around 1.4 nm. This indicates that the ssDNA molecule is detaching from the tip well before the entire molecule is completely lifted, as its total length is about 4 nm.

In Fig. 9 we observe that at the beginning of the lifting process, we record a large stiffness (k). As the nucleotides composing the ssDNA detach, the stiffness gradually decays in a 1/Z fashion, being Z the the lifting height of the ssDNA. The atomic interpretation of these findings is provided by all-atom molecular dynamics simulations discussed in Sect. 4 at greater length. Essentially, in the initial stage the normal force felt by the tip is mostly the adhesion energy of the first nucleotide. Then, as it detaches the stiffness decays, as subsequent deformation process is now softened by the internal stiffness of the just detached nucleotide. Being much softer than the ssDNA-Gold interaction, this leads to a natural weakening of the contact stiffness. As more nucleotides are detached, the softer the contact becomes—analogously to springs in series, an additional spring (nucleotide) softens the collective stiffness.

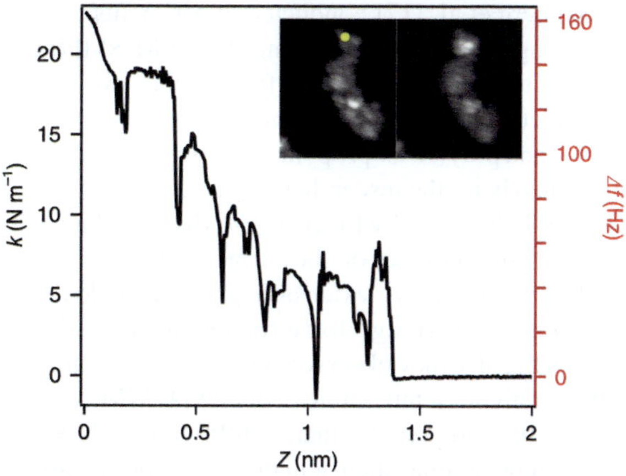

Fig. 9 Force gradient and frequency shift during the lifting of the ssDNA molecule. The inset shows an STM image of the molecule, and the yellow dot represents the position from which the tip lifts the molecule. Material from: R. Pawlak et al., Nature Communications 10, 685 (2019)

These experimental findings emphasize the importance of the internal degrees of freedom of the ssDNA in the detachment process. The mechanical properties of ssDNA are intricately linked to its flexibility of each nucleotide, in-turn governed by a delicate interplay of covalent and non-covalent forces [49]. Covalent bonds in the backbone provide structural integrity, while hydrogen bonds among nucleotide bases enable dynamic conformational changes. This interplay is essential for ssDNA's ability to adapt its configuration to its surroundings. Its flexibility is relevant for absorbing and releasing energy during mechanical manipulation, leading to sharp dips in stiffness as individual nucleotides detach from the surface. The strong adhesion and complex folding of ssDNA are evident, particularly in its tendency to keep folded configuration under high temperatures. Furthermore, the sequential base detachment, resembling the process of peeling off adhesive tape, shows a relatively high stiffness per stretched ssDNA repeat unit (i.e. nucleotide) with an approximate value of 30–35 N/m. This high force employed ensure that ssDNA is being entalpically deformed to extremes never explored before as all prior experiments failed to probe DNA nanomechanics in the nN regime.

2.5 Study of Polymer Chains in an Aqueous Medium

In previous experiments, vacuum conditions, often at temperatures close to absolute zero, were maintained to isolate the system. This ensured that measurements were not affected, allowing correlation with the internal mechanical properties of the studied molecule. While these conditions are crucial for measurement precision, for practical applications it is important to inspect the role of temperature and surrounding environment. To leverage this, it is essential to understand the role of internal and external factors in the mechanical response of the system.

Kolberg et al. [17] conducted experiments to study the mechanical properties of very long polymer chains using the AFM-SMFS technique in an aqueous environment at temperatures ranging from 278 to 318 K. They focused on the temperature-dependent elastic response of two water-soluble polymers, polyethylene glycol (PEG) and poly(N-isopropylacrylamide) (PNiPAM). Unlike most of the experiments presented here, the researchers utilized functionalized silicon nitride cantilevers, and similar to the study by Lu et al. [37] discussed in Sect. 2.3.2, there is also no previous image of the polymer on the surface, which diminishes experimental control.

The force versus extension graph revealed three force regimes: (i) low, (ii) medium, and (iii) high force regime, as noticed also by Lu et al. [37]. Notably, as seen in Fig. 10, they observed opposing temperature responses in which PEG became softer at higher temperatures, while PNiPAM became stiffer with increasing temperature. The study reveals that both PEG and PNiPAM initially exhibit entropic properties, representing their tendency to explore various configurations and maximize disorder. This behavior is prominent at low stretching forces, where the polymers act like entropic springs driven by the flexibility of their chains. A similar reasoning to that used by Lu et al. [37] can be followed for these systems in the low-force regime, where in Lu et al. case, the poly-dimethylsiloxane chain tended to have both trans and gauche states in the low-force regime, favoring the increase of its entropy. In this case, we can also consider that in the low force regime, polymeric chains tend to adopt a disorder configuration in such a way that maximizes their entropy.

However, as stretching force increases, the polymers response starts to be dominated by enthalpic energy terms. This is associated with interactions with their surroundings and the fact that its configuration begins to increase its energy in order to adapt to the external force and the medium, being more evident for PNiPAM.

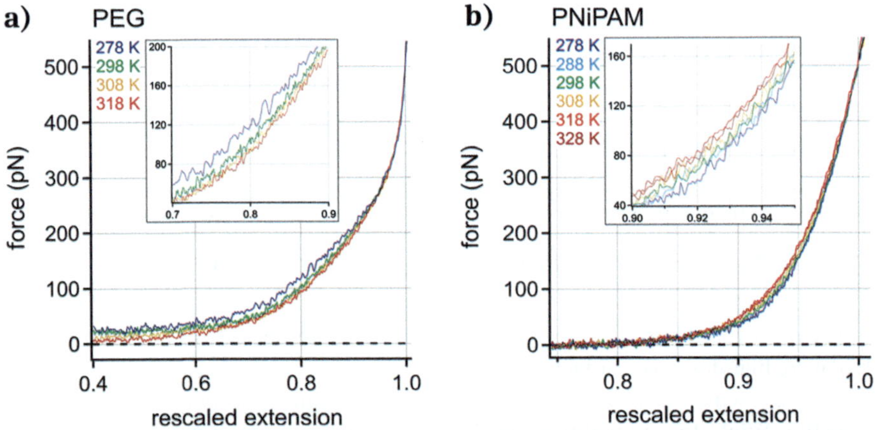

Fig. 10 Force versus rescaled extension at different temperatures and humidity conditions for the molecules of **a** PEG and **b** PNIPAM. Reprinted (adapted) with permission from A. Kolberg et al., J. Am. Chem. Soc. 2019, 141, 29, 11603–11613. Copyright 2019 American Chemical Society

For the PNiPAM, hydrogen bonds with water maintain its solubility at low temperatures. However, as temperature rises, these bonds become energetically unfavorable, leading to partial dehydration of the polymer and a shift to a more ordered and compact state. This transition to an energetically favored conformation makes PNiPAM stiffer and more resistant to deformation as temperature increases. The disruption of water bridges is a crucial factor in this shift from entropic to energetic behavior.

The study highlights the significance of hydration in macromolecular mechanics. Additionally, the authors demonstrate that both PEG and PNiPAM deviate from classical models like the Freely Jointed Chain (FJC) and Worm-Like Chain (WLC), especially at intermediate and high stretching forces, displaying superenergetic or mixed spring behaviors. By employing a T^n exponent model for the system's free energy vs stretching force, the authors categorize the polymer's temperature response as either energetic or entropic. They find that PEG shows a superenergetic response, whereas PNiPAM exhibits a mixed response, closely tied to alterations in hydrogen bonding with water and intramolecular interactions in the polymers, marking the difference between these polymeric chains.

2.6 Nanoscale Locomotion: The Power of Internal Degrees of Freedom

So far, the discussed studies have delved into the examination of tribological and mechanical properties of molecules on surfaces using SPM techniques. The studies underscore the importance of the molecule's internal degrees of freedom in elucidating its mechanical and dynamic attributes. Molecular flexibility allows for changes in different bonds controlling thus conformational changes. However, inducing such changes requires overcoming a potential energy barrier. By supplying the appropriate energy, such changes, e.g. the rotation of functional groups, can be induced. This characteristic, intimately connected with the molecule's internal degrees of freedom, can be designed in such manner that it is harnessed advantageously. For instance, the rotation of a functional group resembling a wheel can initiate controlled active molecular movement, leading to a directed nanoscale locomotion.

In a study conducted by Kudernal et al. [38], controlled molecular motion was demonstrated through the manipulation of internal degrees of freedom. The experiment achieved precise locomotion of meso-isomer molecules on a Cu(111) surface using a STM under LT-UHV conditions. The key to this controlled motion lies in manipulating the molecule's internal degrees of freedom, particularly focusing on controlled conformational isomerization. The molecule in question comprises four rotary motors (see Fig. 11a), functional units capable of undergoing controlled conformational changes triggered by sequential electronic and vibrational excitation.

The authors showcased the ability to induce controlled conformational changes in the molecule through electronic excitation using the STM tip. This mechanism

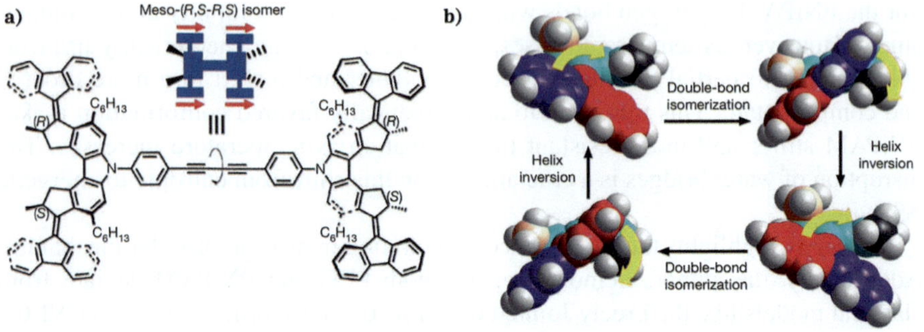

Fig. 11 **a** Representation of the meso-(R,S-R,S) isomer. **b** Schematic representation of the 360° rotation of the rotary motor. Reprinted from T. Kudernac et al., Nature 479, 208–211 (2011), with permission from SNCSC

involves double bond isomerization, followed by helix inversion of the functional groups, creating a paddlewheel-like motion that propels the entire molecule as seen in Fig. 11b. The molecule's precise design enables directed conformational changes, synchronized across all rotary motor units, and allowing the molecule to move in a coordinated linear or circular fashion. The authors emphasize the importance of chirality and geometry in the locomotion, stressing the crucial role of the molecule's orientation on the surface. When correctly oriented, the coordinated action of motor units produces controlled movement, while an incorrect conformational state leads to uncontrolled motion.

The study highlights the necessity for the molecule's absorption energy being lower than the energy released during isomerization so to facilitate locomotion. Indeed, despite the remarkable achievement of controlled molecular motion, it was not possible to determine the energy released during the isomerization, the forces involved in translational movement, and the adhesion energy between the molecule's isomers and the surface. Certainly, measuring these parameters is a complex task. Nonetheless, future endeavors in experimental, theoretical, and simulation work should be directed toward quantifying these aspects. This will pave the way for the design of more efficient molecular motors with active motion mechanisms.

2.7 *Experimental Limitations: The Gateway to Theoretical and Simulation Models*

Our discussion has portrayed various SPM experiments which successfully manipulate molecules on surfaces, emphasizing the crucial role of internal degrees of freedom in understanding mechanical and tribological properties. Manipulated molecules can absorb or release energy from these degrees of freedom, enabling the design of nanocontacts with specific characteristics. For instance, the flexibility of molecules can be leveraged to achieve superlubric states [20]. This release of energy can also be employed in designing controlled locomotion and molecular motors, as illustrated in the previous experimental example.

To design nanocontacts with specific traits, rigorous research on molecular compounds with relevant tribological and mechanical properties is essential. Chemical tuning can be applied to create molecules that either enable or restrict surface mobility, depending on the problem at hand. Understanding and quantifying parameters like adhesion energy, friction force, and activation energy of internal degrees of freedom are crucial for generating controlled molecular locomotion on surfaces. Additionally, most experiments must be performed under highly controlled conditions, such as UHV and temperatures close to absolute zero.

Experimental testing of numerous molecular compounds for potential use in nanotribological problems is time-consuming and resource-intensive. Technical limitations, such as the inability to precisely understand interface dynamics and challenges in achieving a fully atomistic description in experiments, prompt collaboration with theoretical and simulation approaches. Considering the impossibility of explaining nanocontact dynamics without explicitly considering internal degrees of freedom, various simulation techniques are explored. Coarse-grained models offer cost-effective consideration of some internal degrees of freedom but lack full atomistic detail. Performing fully atomistic simulations with quantum mechanics calculations is desirable but faces challenges in dealing with typical nanotribological systems' size and timescales.

Molecular Dynamics (MD) emerges as a valuable tool, providing a fully atomistic description and versatility for systems in equilibrium or non-equilibrium. MD allows the investigation of a broad range of nanotribological phenomena and offers a potential avenue for obtaining a detailed understanding with atomistic character of the problems under examination. Furthermore, MD allows for atomic-level interface analysis, often inaccessible in experiments, and cost-effective testing of molecular compounds before laboratory experiments. However, to calculate key physical parameters in nanotribology using MD, realistic systems for the simulations are necessary.

In the following sections, we present the fundamental principles that molecular systems must follow in MD to mimic the nanomanipulation experiments and, in doing so, complement the interpretation of experimental data. To validate our point, we present examples from different studies in which MD simulations successfully reproduce experiments and quantum mechanics results. Furthermore, these simulations go beyond by providing a fully atomistic interpretation of the experimental measurements.

3 Molecular Dynamics Approach

3.1 Main Steps for a Nanomanipulation Simulation

In essence, Molecular Dynamics involves solving numerically Newton's second law for each atom composing the system. The interactions, both intra- and intermolecular, are considered through a potential energy V. Thus, each atom i experiences a force $\boldsymbol{F}_i$ due to its interactions with other atoms:

$$m_i d^2 \boldsymbol{r}_i / dt^2 = F_i = -\nabla_i V. \quad (1)$$

m_i is the mass of atom i, $\boldsymbol{r}_i$ its vector position, and V is the potential energy of the system. The potential energy V is a sum of intra- and intermolecular interactions as follows:

$$V = V_{intra} + V_{inter}, \quad (2)$$

where V_{intra} is the sum of all intramolecular interactions (e.g. bond stretching, bending, and torsional degrees of freedom). As for V_{inter}, it encompasses all intermolecular non-bonded interactions as Van der Waals and Coulomb potential.

By numerically integrating Eq. (1) one obtains the velocities and positions of atom i for each instant of time [50–52]. It is possible to account for environmental effects, such as pressure and temperature, by using algorithms that include the numerical solution of the Langevin equation [53, 54] or through the extended Lagrangian methodology [55–58]. The time step employed is typically around ~ 1 fs, with the specific value determined by the higher frequencies of the atoms in the system. For a detailed explanation of the basic principles of Molecular Dynamics, we refer to [50, 51, 59].

For the problem at hand, i.e. modeling molecular nanotribology experiments, the system to be simulated is composed of the molecule adsorbed to a surface. The coordinates of the molecules and the surface can be obtained from a database [60] or constructed using software [61, 62]. Then, we first minimize the total energy of the system by relaxing its initial configuration. Then, the thermalization process at room temperature begins, either in an isochoric-isothermal (NVT) or isobaric-isothermal (NPT) ensemble. This process takes place for several nanoseconds until the temperature, density and energy stabilize.

In order to mimic annealing processes often used in experiments to synthesize the polymeric chains, the system's temperature is raised to several tens or hundreds of Kelvin above room temperature. This process is important as the final configuration of the molecule on the surface (before its manipulation) is strongly dependent on this step. Following this annealing, the temperature is gradually lowered to values close to absolute zero. Upon reaching the desired temperature (typically 4 K), the system is allowed to evolve at a constant temperature for several nanoseconds to relax. After these steps, the system is ready to enter the production stage, i.e. its manipulation.

To mimic the procedure by which the tip manipulates the molecule, the Steered Molecular Dynamics technique is often employed. This technique entails linking a group of atoms in the system by means of a spring to a virtual particle that is displaced at a constant velocity mimicking thus the motion of the AFM tip. The spring is characterized by its stiffness constant, typically in the order of ~0.2 N/m. It is important mentioning that the value this stiffness introduces a cut-off, i.e., this is the maximum stiffness one is capable of measuring since $1/k_{eff} = 1/k_{tip} + 1/k_{all}$ and if $k_{all} >> k_{tip}$ then $k_{eff} \approx k_{tip}$. Another point to mention is that the value of ~0.2 N/m is of an empirical nature, and although it describes the stiffness of the bond formed between the AFM tip and a broad range of manipulated polymers, its fundamental origins as of today remains unknown. During the simulation, information 'detected' by the virtual particle, namely the elongation/force as it is displaced at a constant velocity is saved with the desired frequency (~ps). It is through such information that we can establish a direct connection between experiments and simulation data. Additionally, relevant atomistic information of the system is saved, such as the molecule's geometry during lifting/sliding, to interpret the results with a full-atomic detail.

3.2 *Intra and Intermolecular Interactions*

In MD simulations, an accurate description of molecular and atomistic interactions is paramount for a proper system representation. The potential energy function, denoted as *V* in Eq. (2), plays a pivotal role in defining these interactions. Indeed, the precision of MD simulations strongly relies on our ability to correctly represent both intra- and intermolecular interactions. This is where the concept of the force field (FF) comes into play. A FF encompasses the total potential energy function, *V*, including all associated parameters used to describe a specific system. A well-constructed FF is essential to accurately model the behavior of the system, making it a cornerstone of MD simulations.

Many FFs are designed for transferability across diverse scenarios [63–65], like room temperatures or aqueous environments, using common mixing rules for molecule and atom crossed interactions. This proves beneficial for simple systems under typical conditions, eliminating the need for frequent FF parameterization. However, in nanomanipulation, where conditions differ significantly from standard environments with precise control in UHV and near absolute zero temperatures, conventional FFs may struggle. Nanomanipulation simulations require a fine balance between molecule-surface and intramolecular interactions. In these specific scenarios, conventional FFs often fall short of reproducing experimental outcomes even qualitatively.

Tailored FFs, designed for specific system characteristics, offer an important solution in nanomanipulation simulations. Especially effective in capturing the fine molecule-specific events on the pm and pN scales, these customized FFs provide a methodology for modeling intricate dynamics at cryogenic temperatures.

3.2.1 Intramolecular Interactions

Cacelli and Prampolini introduced an innovative methodology for precise intramolecular interaction parameterization sourced from preliminary Quantum Mechanically (QM) calculations—a procedure known as Quantum Mechanically Derived Force Field (QMD-FF) [66]. This approach accuracy stems from expressing the FF as a linear combination of functions of redundant internal coordinates, with data derived from density functional theory (DFT) calculations. It employs the energy and its first and second geometrical derivatives, computed at a chosen QM level for a number of relevant conformations of the target molecule, in its ground or excited state, and minimizes the differences of the collected QM descriptors with their classical counterparts, computed as a function of the FF parameters. Such a process allows the user to entail great part of the QM accuracy onto the QMD-FF description, paving the way (i) to the reliable description of peculiar structures not considered within general-purpose FFs, (ii) to overcome structural mismatch issues and produce FFs also for excited states, and (iii) to a balanced representation of flexible and stiff internal degrees of freedom, possibly accounting for their coupling, even when driven by electronic effects. This methodology has been successfully applied in various scenarios, ranging from transition metal compounds [67, 68] to simulations of electronic absorption or emission spectra [69–71].

Cerezo et al. [72] extended the previous method by introducing an innovative model for intramolecular interactions. This model explicitly links flexible dihedral angles with the stiffness of internal structural characteristics, going beyond traditional steric interactions to include conjugation effects and changes in bond orders induced by conformational variations. Non-bonded intramolecular potential terms are introduced for complex interactions between different internal coordinates (ICs), allowing a more comprehensive representation of molecular behavior. The approach introduces distinct coupling terms for specific pairs of ICs, offering a more accurate description of how dihedral angles influence other ICs' equilibrium values. The parameterization method includes external fitting (explicit coupling terms) and global fitting (all FF parameters) using quantum mechanics calculations for heightened accuracy in modeling molecular behavior. These methods find versatility in applications ranging from liquid crystals to polymers and organic electronics, offering a comprehensive and robust approach to molecular modeling.

3.2.2 Intermolecular Interactions

Iori et al. [73] developed GolP, a classical atomistic force field designed for modeling the interaction between organic-molecules and gold surfaces. Developed through a combination of quantum mechanical calculations and experimental data, GolP is compatible with popular bio-oriented molecular dynamics simulation software like GROMACS and NAMD. It considers various components contributing to the interaction energy, including image-charge interactions, chemisorption, Van der Waals interactions, and interactions specific to conjugated systems.

Handling charge-image interactions involves simulating metal polarization effects using a code-independent approach, where gold atoms are replaced with dipoles representing electronic metal polarization. Chemisorption energy is addressed through a systematic approach starting with quantum mechanics calculations for small molecules, identifying chemisorption characteristics and parameterizing interactions. Van der Waals interactions use a 12–6 Lennard–Jones (LJ) potential, with parameters determined by fitting to experimental data. A mixing rule based on post-Hartree Fock calculations determines generic gold-atom Van der Waals interactions. For conjugated systems, a modified LJ potential with virtual sites is employed, mimicking hollow sites where these systems adsorb. LJ parameters for virtual sites are defined based on experimental data, adjusted to align with both experimental results and quantum mechanics calculations, while real surface gold atoms remain free of LJ interactions.

As for non-metalic surfaces, such as 2D materials, the force field developed by Ouyang et al. [74, 75] stands out as the best suited to describe sliding processes. Their FFs focus on accurately modeling binding energy curves and sliding potential surfaces for various 2D materials (e.g. MoS_2, hBN). Notably, their FFs are meticulously benchmarked against DFT calculations, incorporating hybrid density functional approximation and nonlocal many-body dispersion treatment for long-range correlations. What sets this FF apart is its emphasis on capturing anisotropic interactions between different layers of materials, especially in 2D materials. Unlike standard isotropic FFs, this approach accounts for lateral interatomic distances between adjacent 2D layered materials.

Ouyang et al. [76] also address interactions between graphene, benzene molecules, and C_{60} fullerenes on a gold surface, emphasizing anisotropy. They introduce a Semi-Anisotropic Interfacial Potential (SAIP) designed to accurately represent interactions between gold and two-dimensional carbon allotropes. SAIP includes an isotropic term for long-range dispersive interactions and an anisotropic term for Pauli repulsion between graphene's π electrons and the gold surface's electron density. The parameters for SAIP are meticulously determined through fitting, ensuring precise interatomic interactions. SAIP accurately reproduces DFT calculations for interactions in various Au-C interfaces, including both flat and curved graphitic systems. However, the authors stress the importance of considering curvature in SAIP parameterization, especially for highly curved structures like C_{60} fullerenes. The distinct properties of highly curved systems require adjusting SAIP parameters for reactivity and interaction energies, highlighting the necessity of tailoring potentials to accurately represent each system's unique features for dependable material interface interaction modeling.

The presented methodologies for parameterizing tailored FFs highlight diverse approaches for constructing these FFs [77–86]. While reliance on experimental data for FF parametrization is ideal, its scarcity often makes this impossible. Quantum mechanics calculations provide essential data for accurate FF parametrization, allowing for specificity in models crucial for chemical engineering. Generic force fields like OPLS lack the precision needed to capture dynamics, even qualitatively. In this regard, to chemically engineer it is needed the specificity endowed by models whose accuracy is assured by supporting quantum mechanical data. The following section discusses studies using specific FFs to replicate nanomanipulation experiment results through MD simulations. These studies achieve a fully atomistic description of the investigated phenomena, showcasing the significance of accurate FFs in understanding and simulating complex molecular interactions.

4 Molecular Dynamics Evidence

4.1 Mechanical and Adhesion Properties of Polymers

4.1.1 Polymerized Fluorene Chains

In this section, we revisit the work of Kawai et al. [31], previously introduced in Sect. 2.2.1 but now with a focus on the theory/modeling proposed. The study focuses on the examination of a poly-fluorene chain lifted from a gold surface and considers an extended Frenkel-Kontorova (FK) model to explain the experimental results. The FK model considers a chain of masses connected by springs with constant k and on a periodic surface, with each mass symbolizing a fluorene monomer, and one end connected to a virtual particle simulating an AFM tip, as depicted in Fig. 12a. The model incorporates various interactions, including physisorption potential energy and a viscous force to address damping. The sum of forces acting on each mass allows for the derivation of equations of motion. Although the methodology diverges from traditional MD, employing larger masses and integration times, the underlying principles remain consistent—numerical integration of Newton's second law for particles interacting through a potential. It is noteworthy that the authors conducted thorough parameterization of potential energy's parameters, incorporating Density Functional Theory (DFT) calculations.

As seen in Fig. 12b, the model is able to replicate the experimental results. Furthermore, it allows to compute normal and lateral forces, as depicted in Fig. 12c, and to calculate the dissipation energy by relating the internal energy of the chain to its interaction energy with the surface, as shown in Fig. 12d. The lateral force tends to decrease as the chain is peeled off, as the lateral displacements of the lifted units gradually decrease towards the last unit connected to the tip. The dissipation energy corresponds entirely to the change in absorption energy, while elastic energy exhibits slight variations throughout the entire process, increasing and decreasing during the detachment of the different subunits similar to a stick–slip behavior. The model calculates a physisorption energy of 0.27 eV per monomer, that is reasonably close to a lower bound of 0.21 eV estimated for tilted methylated fluorene [31].

As observed in Fig. 12c, the normal force remains with relatively uniform amplitudes and oscillation periods throughout the lifting process, as long as the effective stiffness of the detached segment is greater than the stiffness of the AFM tip k_{tip}. This is always the case if the number of units is well below k/k_{tip}. The model helps

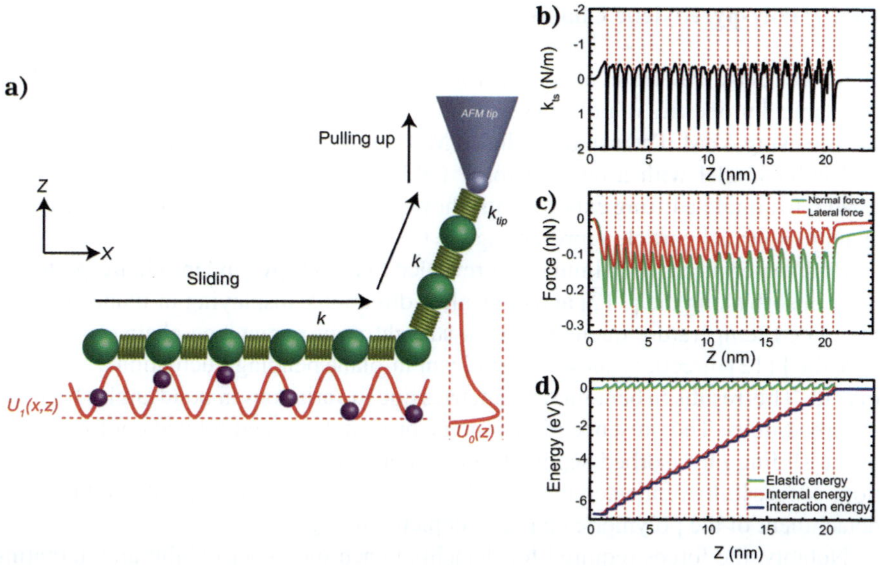

Fig. 12 **a** Schematic representation of the extended Frenkel-Kontorova model. **b** Calculated force gradient versus tip-surface distance. **c** Normal and lateral force versus tip-surface distance. **d** Interaction energy, elastic energy, and internal energy of the chain versus tip-surface distance. Reprinted from S. Kawai et al., PNAS 111, 3968–3972 (2014)

understand the reason behind this outcome. When a unit detaches, it alters the energy landscape influenced by the periodic potential corresponding to the substrate structure. The key is that the effective stiffness depends more on the properties of the detached segment than on the length of the chain or the remaining segment on the surface. This means that whether you are pulling a short or long segment, the effective stiffness of the detached portion remains relatively constant. This leads to the normal force remaining with a relatively similar behavior throughout the detachment process, resulting in uniform force patterns as individual units detach and encounter similar energy landscapes.

Despite the suitability of the model to qualitatively describe the experimental findings, it has limitations in representing all the important details, such as the actual structure of the polymeric chain with a fully atomistic detail. Namely, experimental observations reveal that the force modulation, while maintaining periodicity, tends to be more irregular (see Fig. 3b). These variations may be due to factors in the geometric structure of the chain that the model does not take into account, as it solely focuses on the elongation/compression between different monomers as part of the internal molecular structure.

4.1.2 Poly-pyrenylene Chains

Here, we revisit Pawlak et al.'s [48] work from Sect. 2.3.1, where they explored the mechanical properties of poly-pyrenylene chains by detaching them from a Au(111) surface using AFM. Their study included MD simulations, employing a full-atomistic molecular model with a tailored force field. Using Steered Molecular Dynamics (SMD) to simulate lifting experiments, they attached one end of the poly-pyrenylene chain to a virtual particle representing the tip.

Both experiments and simulations revealed that poly-pyrenylene chains preferentially absorb along the $\left[11\overline{2}\right]$ and equivalent directions, displaying in-plane bending. Simulated temperature quenching of a straight poly-pyrenylene chain, with reference the $\left[11\overline{2}\right]$ direction, showed significant in-plane bending fluctuations, as seen in Fig. 13. This results in bent final geometries due to variations in bond angles between adsorbed units. The presence of symmetry-broken, bent, and rotated conformations in the sliding physisorbed segment during detachment was observed. The simulations successfully replicated the periodic alternation of the force gradient during the detachment of the polymeric chain, as depicted in Fig. 14b.

Notably, the forces required for detaching each monomer exhibit an alternating value, characterized by a stick–slip pattern that can be seen in Fig. 14a. Thus, the periodic alternation in detachment distances is a result of differing force/energy costs associated with detaching odd or even units. The simulations reveal that internal degrees of freedom lead to diverse configurations during pulling, such as in-plane bending of C–C bonds. When lifting an odd unit, the energy cost is influenced by the arrangement of internal degrees of freedom, resulting in a specific torsional strain. Lifting an even unit involves a different arrangement, leading to a distinct energy landscape. Hence, the molecular structure responds with a wavelike in-plane bending deformation along the chain, influencing its overall conformation and desorption energy.

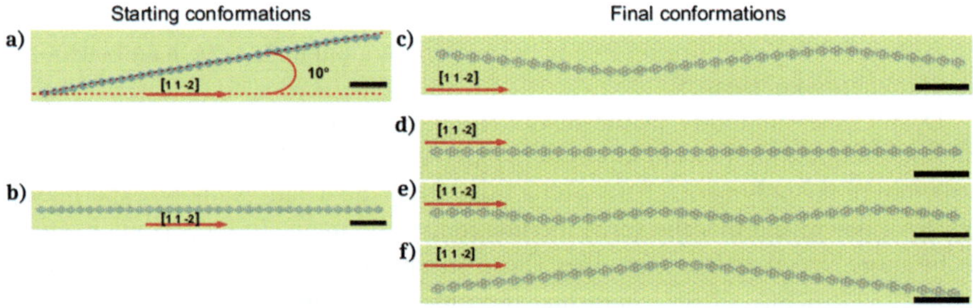

Fig. 13 Two different starting configurations **a** at an angle with $\left[11\overline{2}\right]$ and **b** aligned with $\left[11\overline{2}\right]$. **c–f** Are final configurations after quenching MD simulations. The final configuration **c** is for the initial configuration (**a**), while **d–f** are for the initial configuration (**b**). Reprinted (adapted) with permission from R. Pawlak et al., Nano Letters 2020, 20, 1, 652–657. Copyright 2020 American Chemical Society

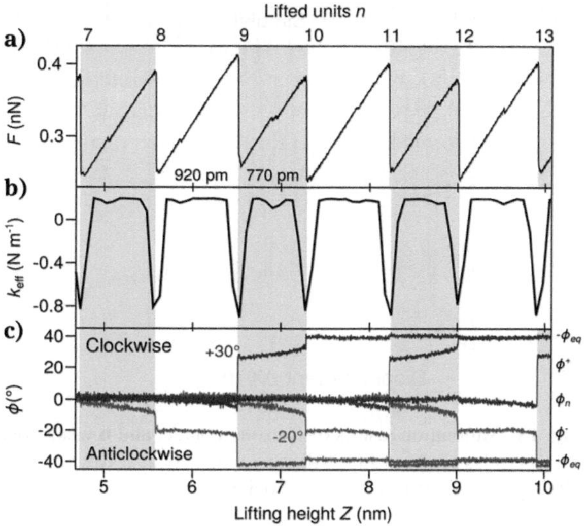

Fig. 14 **a** Computed normal force, **b** associated force gradient, and **c** angle ϕ between consecutive monomers planes. Reprinted (adapted) with permission from R. Pawlak et al., Nano Letters 2020, 20, 1, 652–657. Copyright 2020 American Chemical Society

To understand in more detail the periodic alternation, the authors focused on the dihedral angle between consecutive poly-pyrenylene monomers. On surface, the molecule maintains a flat configuration, but in the gas-phase, torsion between monomers is of ±40°. As monomers detach, the dihedral between consecutive units changes from 0°—imposed by surface adsorption—to more than 20°, so to approach the gas-phase equilibrium angle. As shown in Fig. 14c, the detached even and odd units exhibit torsion angles of +30° and −20°, respectively, approaching to the equilibrium value of the gas phase as the lifting continues. The lifted segment gradually straightens, inclined by at most 10°, reflecting the competition between the energy gained from twisting and the lingering attraction to the surface, causing steric hindrance.

Vilhena et al. [20] furthered the study by investigating lateral manipulations with varying tip-to-sample heights. The force gradient during lateral manipulation was studied using the same system at varied tip-surface distances, reproducing the experimental traces as seen in Fig. 15. The high height of the tip (0.51 nm—weak contact) led to increased spacing in the periodicity of the force gradient drop compared to the low tip height (0.36 nm—strong contact).

The simulations revealed a side-winding motion during sliding. In strong contact, similar to experiments, simulations demonstrate periodic force gradient with a spacing of ~0.22 nm, as seen in Fig. 15a, shorter than gold atom distances. Simulations indicate that this small spacing results from stick or slip events, inducing a nontrivial zigzag motion for each monomer, as shown in Fig. 16. The chain's twisting along its longitudinal axis and pyrenylene's zigzag movements explain the distance in force gradient periodicity, confirming molecular undulation during sliding. The chain's flexibility allow it to break down on-surface motion into small slip events, resulting in a smaller slip length than the distance between gold atoms.

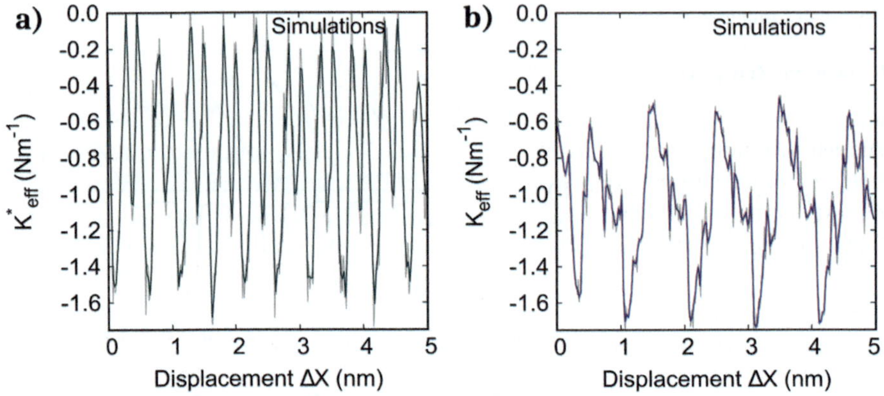

Fig. 15 Simulation results of **a** strong contact and **b** weak contact of the force gradient. Reprinted with permission from J.G. Vilhena et al., Physical Review Letters 128, 216102 (2022). Copyright 2022 by the American Physical Society

Fig. 16 Paracarbon atom trajectory of the 2nd and 5th units for the strong contact regime (S-U2, S-U5) and the 2nd, 3rd, and 8th units for the weak contact regime (W-U2, W-U3, W-U8). Reprinted with permission from J.G. Vilhena et al., Physical Review Letters 128, 216102 (2022). Copyright 2022 by the American Physical Society

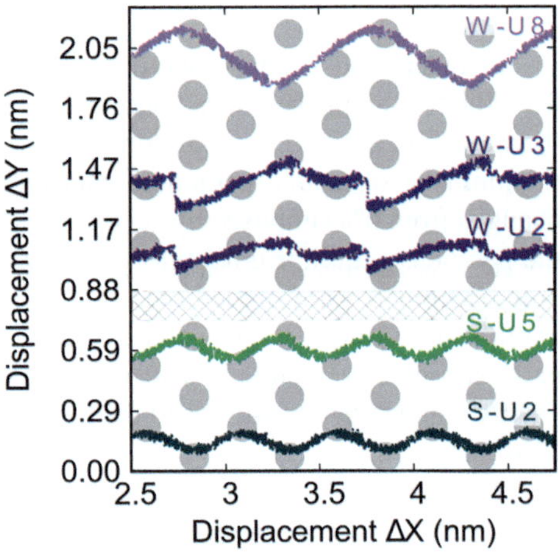

In weak contact, simulations observe a ~1 nm periodicity in the force gradient, as depicted in Fig. 15b, similar to experimental results. They also reveal that each monomer advance in a specific way due to torsion between consecutive units, favoring a 40° angle as in the gas-phase state. Lifting the second unit alters the angle between it and the unit that follows (unit 3), influencing torsion between units and changing the on-surface dynamics, as shown in Fig. 16. The weak-contact regime shows larger torsions/energy explored as compared to the strong-contact case, favoring off-axis motion to modulate torsion as it slides.

Despite high in-plane flexibility, when sliding the molecule backwards exhibits a surprising tail-forward sliding. Interestingly, side-winding motion persists even in backward sliding owing to the ultra-low static friction force of ~40 pN. Incommensurability between polycyclic-aromatic-hydrocarbon chains and Au(111) induced mechanical stress on the chain, which in turn deforms the chain to promote its registry with the surface. In this manner, the tension accumulated in the chain reduces diffusion energy barriers and facilitates controlled release or storage of strain. Comparisons with rigid molecules emphasize the advantages of flexible sideway-zigzag motion in reducing slip length (and thus energy barrier) and static-friction force. These studies collectively demonstrated how molecular flexibility contributes to particular behavior in the detachment process and unveiled a novel form of superlubricity at the nanoscale, i.e. the flexible-superlubricity, where internal degrees of freedom can be harnessed to store energy.

4.2 Sliding of Graphene Nanoribbons on Different Substrates

Ouyang et al. [74] conducted a computational study on the tribological properties of graphene nanoribbons (GNRs) on graphene (homogeneous interface) and hexagonal boron nitride (heterogeneous interface) substrates. Using MD simulations at 0 K, the study aims to uncover the relation between GNR mechanical properties and interfacial commensurability, shedding light on frictional behaviors. The research identifies a nuanced link between in-plane GNR elasticity and interfacial registry, resulting in snake-like movements and a non-linear friction force dependence on GNR length.

Figure 17a and b display the configurations of a GNR on different substrates after an energy optimization process. GNRs on graphene remained straight regardless of length, while on h-BN substrates GNRs $\geq$ 20.4 nm buckled. These different configurations can be modified dynamically during sliding. When pulling GNRs on graphene, short GNRs stayed straight with negligible structural deformation in lateral direction. However, GNRs larger than ~25 nm exhibit shear-induced buckling resulting in snake-like motion. In the case of h-BN substrate, there is a different behavior. Short GNRs exhibit in-plane bending and irregular motion, while GNRs larger than ~35.5 nm show shear-induced unbuckling followed by nearly unidirectional motion.

Figure 17c and d depict static friction's linear growth with GNR length until stabilization for both interfaces. Stabilization occurs at a shorter length for homogeneous interface compared to heterogeneous ones. This observation links length-dependent friction to the competition between elastic energy and interlayer registry matching in GNRs, due to finite in-plane stress presence. GNR configurations are influenced by internal degrees of freedom and interface type, with commensurability crucial for frictional outcomes. Misaligned contacts in the heterogeneous interface result in varying friction forces, including ultralow friction or dynamic transitions between low and high friction states. Local compensation of lattice vector mismatch through in-plane deformations forms moiré superstructures in the h-BN substrate, influencing GNR frictional dynamics.

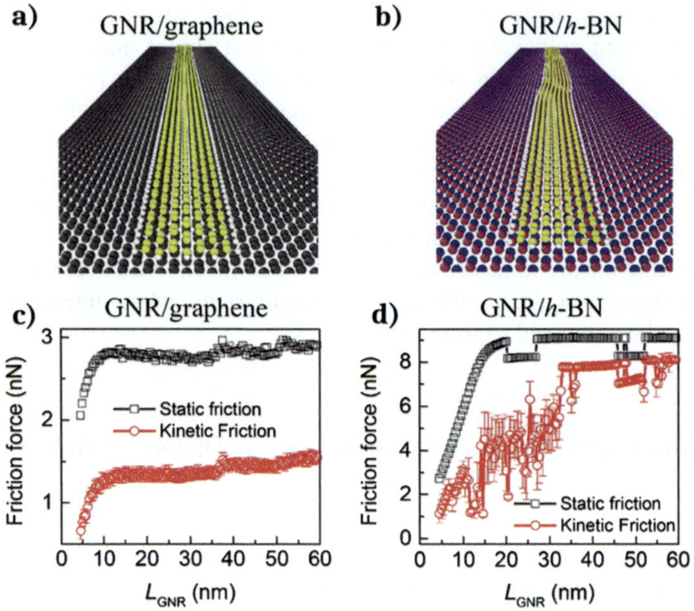

Fig. 17 Optimized geometries of a 36.76 nm long armchair GNR along the armchair axis of **a** graphene and **b** h-BN substrates. Static (black data) and kinetic friction of GNRs sliding over **c** graphene and **d** h-BN. Reprinted (adapted) with permission from W. Ouyang et al., Nano Letters 2018, 18, 9, 6009–6016. Copyright 2018 American Chemical Society

Fluctuating kinetic friction in the heterogeneous interface seen in Fig. 17d is explained by commensurability and moiré superstructures. The study introduce shear-induced dynamics, where the moiré pattern's periodicity influenced length-dependent static and kinetic friction for heterogeneous interface. Commensurability, influenced by GNR length and substrate interaction, led to a complex interplay of buckling and sliding dynamics. Higher friction in heterogeneous interface is attributed to differences in energy barriers during sliding. Internal degrees of freedom in GNRs allow zigzag motion in the homogeneous interface, reducing potential energy barriers and consequently the friction. While the heterogeneous interface, after initial unbuckling, the GNR's head deviation leading to more corrugated sliding paths and consequently experiencing a higher friction.

In Fig. 18, the lateral force for both interfaces is depicted. The aligned homogeneous interface displays a consistent double-peaked stick–slip pattern. The varying heights of the peaks indicate that slip initiation results from the pullout of leading-edge atoms at different potential depths along the sliding path. On the other hand, the aligned heterogeneous interface exhibits distinct force traces, featuring precursor partial slips before the occurrence of global sliding. These precursor events represent the gradual straightening of ribbons by eliminating buckled regions, starting from the leading edge. Once the graphene nanoribbon (GNR) is completely straightened, a global slip event takes place.

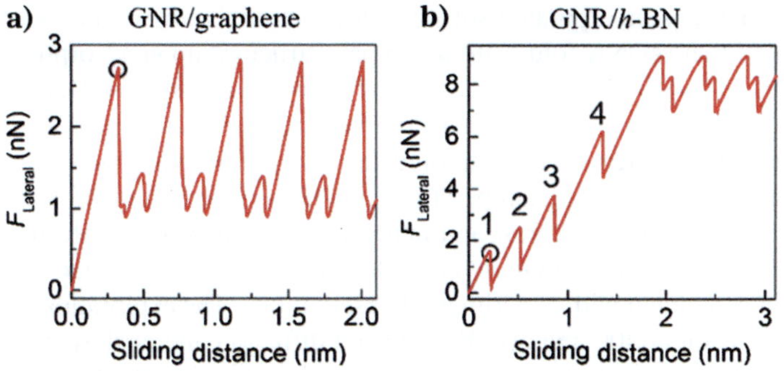

Fig. 18 Lateral forces during sliding of GNR of 59.4 nm on **a** graphene and **b** h-BN. Reprinted (adapted) with permission from W. Ouyang et al., Nano Letters 2018, 18, 9, 6009–6016. Copyright 2018 American Chemical Society

The study also considered temperature effects considering room temperature. While the qualitative dependence of static and kinetic friction forces on GNR length persisted at room temperature, thermolubricity effects were more pronounced in heterogeneous contacts. In this latter case, thermal fluctuations led to a significant reduction in kinetic friction forces, rendering them practically independent of GNR length.

When comparing the sliding dynamics of GNRs on graphene or h-BN with the previous case of poly-pyrenylene chains sliding on gold, we find that side-winding motion is more regular for poly-pyrenylene. For GNRs on graphene, side-winding motion occurs for longer lengths. However, when considering h-BN, the sliding dynamics are less regular, displaying bending and irregular motion for short GNRs and unbuckling behavior for longer GNRs. Moreover, in the case of poly-pyrenylene chains, static friction tends to decrease as their length increases. In contrast, for GNRs on graphene or h-BN, static friction initially increases with length before reaching a stable value. Furthermore, static friction proves to be lower in the case of poly-pyrenylene chains, in the order of tens of pN, compared to the GNR on graphene or on h-BN, with values in the nN range.

4.3 *Mechanical and Adhesion Properties of Single-Stranded DNA*

Now we revisit Sect. 2.4 to further the theoretical details on the nanomechanical and tribological properties of single-stranded DNA (ssDNA) deposited on a gold surface. The authors performed MD simulations in conjunction with experiments so to describe the process with atomically detail. The process was composed of three stages: (i) deposition of molecules onto the surface at room temperature, (ii) annealing at high temperatures, and (iii) cooling to 5 K for manipulation of ssDNA.

The simulations incorporated solvent effects through a hydrating water droplet, and unveiled that ssDNA folds into a compact structure upon adsorption as shown in Fig. 19a. The stability of the hydration layer surrounding ssDNA in simulations is consistent with experimental observations. Subsequently, dehydration and thermal annealing effects were explored in simulations. Various MD simulations were conducted at increasing temperatures, including evaporation of water at 450 K, removal of all water molecules, and annealing without water molecules. Despite different procedures, all simulations consistently demonstrated similar folded conformations, in agreement with experimental observations. The exploration of thermally assisted diffusion at 400 and 500 K indicated diffusion and coalescence occurring only at 500 K, as seen in Fig. 19b, matching with experiments. Figure 19b shows that even during diffusion, ssDNA maintained its folded structure.

When analyzing the force during the lifting procedure (see Fig. 20a) we observe that its peak-value decrease during the lifting is less pronounced than the corresponding gradient shown in Fig. 20b. The force gradient exhibits periodic jumps and a systematic decrease as nucleotides are sequentially detached. Unlike experiments, in simulations we can complete the detachment, providing insights into behavior inaccessible in experiments. Analysis of lifting stages revealed peeling, lifting, and detachment of cytosine bases, indicating strong adhesion between these bases and gold, consistent with experimental challenges. During the lifting of ssDNA, a distinctive characteristic is noted: the nucleotides that remain attached to the surface and do not slide, as illustrated in Fig. 20c. This stands in contrast to the behavior observed in poly-pyrenylene and poly-fluorene chains explored in Sects. 4.1.1 and 4.1.2. In those cases, as the molecules were lifted, the remaining part on the surface slides toward the vertical lifting axis as they detached and increased the tip height. This distinction highlights a difference in the weak adhesion of these polymeric chains to

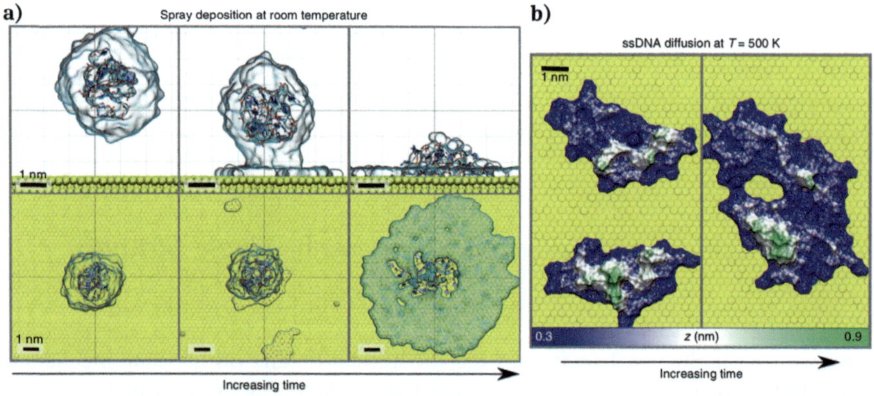

Fig. 19 MD simulations of the ssDNA absorption and diffusion process on the Au(111) surface. **a** Side and top views of the process by which the water droplet containing ssDNA is absorbed by the gold surface. **b** 500 ns simulation of two ssDNA oligomers assembled by diffusion at 500 K. Material from: R. Pawlak, et al. Nature Communications 10, 685 (2019)

the surface compared to ssDNA, what imposes experimental limits on the maximum detachment of ssDNA.

Considering a strong surface absorption, a spring-in-series model demonstrate a decreasing stiffness of the lifted ssDNA segment as it lengthened, as illustrated in Fig. 21. As more nucleotides detach, the stiffness of the detached part of the ssDNA tends to decrease. The stiffness per repeat distance in fully stretched ssDNA was calculated, revealing a value of 32.6 N/m. This confirmed that ssDNA is strongly absorbed to the surface, and fully stretched ssDNA is much more compliant than other polymeric systems, suggesting that the ssDNA backbone stiffness is dominated by bond angle bending. Calculating the stiffness of fully stretched ssDNA resulted in a value of 1.7 N/m, yielding an average tension in stretched ssDNA of ~2 nN for a 20-cytosine oligomer.

The nuanced understanding gained from MD simulations elucidated experimentally challenging aspects of ssDNA folding and adhesion on a gold surface. Despite challenges, the findings provided valuable insights into the mechanics of ssDNA at cryogenic temperatures, shedding light on its adhesion and mechanical response on

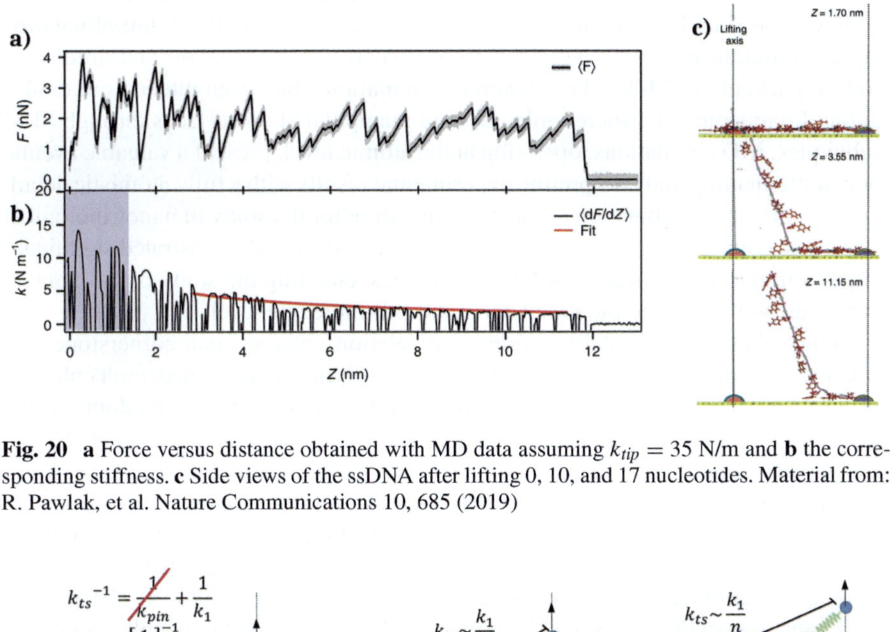

Fig. 20 **a** Force versus distance obtained with MD data assuming $k_{tip} = 35$ N/m and **b** the corresponding stiffness. **c** Side views of the ssDNA after lifting 0, 10, and 17 nucleotides. Material from: R. Pawlak, et al. Nature Communications 10, 685 (2019)

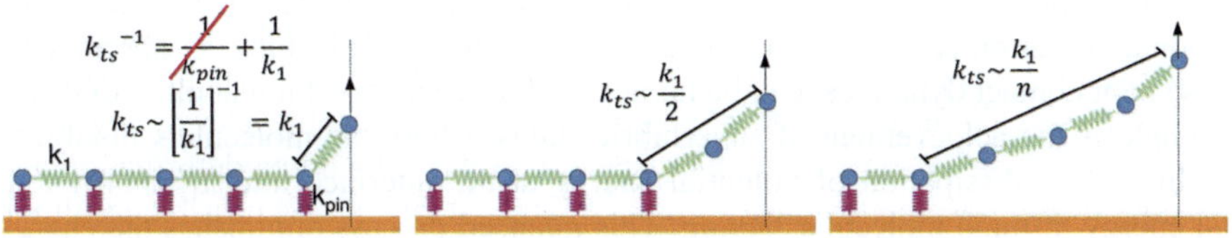

Fig. 21 Schematic representation of a series of springs model depicting the lifting of ssDNA. k_{pin} represents the stiffness of the adhesion force between the surface and ssDNA, which is considered greater than the stiffness between the detached nucleotides k_1

a gold surface. The combination of experimental and simulation approaches offered a comprehensive understanding of the interplay between structure and mechanical properties at the molecular level of ssDNA, with potential applications in nanodevices and biosensors.

5 Overview and Future Perspectives

In the pursuit of unraveling the intricacies of friction, an omnipresent force with ramifications from the macroscopic scale of tectonic plate movements to the molecular realm of surface diffusion, an important gap persists in our fundamental understanding of its origins. As we have shown, recent strides in experimental techniques, particularly those conducted under Ultra-High-Vacuum and Low-Temperature conditions using Scanning Probe Microscopy, notably Atomic Force Microscopy, have ushered in a renewed effort to delve into the atomistic underpinnings of friction.

These experimental techniques, carried out in meticulously controlled environments, have not only granted access to sub-nanometric precision in measurements but have also shed light on fundamental characteristics within the realm of nanotribology. Simultaneously, the synergy between experimental endeavors and simulation studies, particularly Molecular Dynamics simulations, has strenghtened our exploration of nanocontacts, where molecules are manipulated on surfaces through SPM techniques. MD simulations, operating at the atomic level, present a valuable avenue for complementing and interpreting experimental results with a fully atomistic detail. Furthermore, the combination of nanomanipulation for the study of nanotribological properties, coupled with the use of simulation methods, has demonstrated significant progress in the field of nanotribology-mechanics, enabling the study of mechanical properties with considerable precision at the sub-molecule level.

Amidst these commendable strides, a revelation emerges as a cornerstone: the dynamic of nanocontacts, particularly those involving manipulated molecules on surfaces, are intricately entwined with the internal degrees of freedom of the molecules themselves. These internal degrees of freedom, intimately connected to the mechanical properties of the nanocontact, spotlight the flexibility embedded within molecular bonds. Bonds, capable of stretching, shrinking, twisting, etc. harbor the ability to release or store potential energy—a pivotal characteristic that introduces innovative prospects for energy dissipation mechanisms in nanotribological and nanomanipulation scenarios. The inherent flexibility of molecules, as observed in these nanocontact dynamics, can be harnessed for groundbreaking applications. An example is the achievement of superlubric states, where the molecule's flexibility facilitates the dissipation of potential energy at the interface into internal potential energy through molecular flexing. This nuanced process diminishes the energy barrier for sliding, consequently leading to a reduction in friction.

Yet, as we stand at the frontier of this burgeoning field, a call echoes for intensified efforts to expand our current knowledge concerning the different energy dissipation mechanisms in nanocontacts. This imperative task extends beyond a mere academic pursuit; it is a needed prerequisite for the control and practical utilization of these phenomena. The forthcoming strides in understanding energy dissipation in nanocontacts have the capacity to propel emergent areas in nanosciences, spanning from the meticulous design of artificial molecular motors to the creation of intricate nanoarchitectures, among many others. However, to harness the full potential of these technological advancements, a meticulous understanding of the intricacies occurring at the molecular interface becomes imperative. For instance, in the realm of molecular design enabling active movement, considerations span from determining the adhesion energy between the molecule and the surface to comprehending the energy released during the isomerization process, ensuring it surpasses the adhesion energy. Equally crucial is an in-depth understanding of the conformational structure of the molecule, guaranteeing controlled locomotion. Additionally, the supply of appropriate energy to induce isomerization and subsequent locomotion necessitates careful consideration. Delving into the details of each of these requirements is very far from a trivial task.

This complex landscape underscores the necessity for a concerted effort, where experimental, theoretical, and simulation methodologies converge to synergistically advance our collective knowledge. While experimental observations ultimately hold the final say, theoretical and simulation contributions serve as indispensable tools to fill in the gaps that may elude experimental scrutiny.

An aspect demanding particular attention is the meticulous control of energy dissipation pathways during the sliding of nanocontacts, where manipulated molecules undergo deformations owing to their internal degrees of freedom. These molecular deformations necessitate potential energy to alter bond distances or angles, for example, thereby providing a rich reservoir for dissipation processes. The flexibility inherent in molecular structures becomes a potent means of energy dissipation, allowing precise tuning of the energy required for either facilitating sliding or inhibiting it. In order to achieve this, it is not only necessary to understand the internal mechanical properties of molecules but also the characteristics that the surface presents during the sliding of the nanocontact. This is because the nanotribological properties of the system result from a combination of both components: the surface and the sliding molecule. Therefore, it becomes important to explore and study systems that exhibit certain desired characteristics, either to facilitate the sliding of molecules on the surface or, conversely, to inhibit surface molecular motion.

Then, commensurability properties between manipulated molecules and the interacting surface are also important for determining sliding friction. Aligning the molecular structure with the surface lattice enhances commensurability, making sliding more difficult. Conversely, exploring systems with surface incommensurability properties is equally significant, since they encourage sliding. Designing systems with controlled commensurability or deliberate incommensurability properties serves various purposes. In applications seeking enhanced molecular mobility, designing systems with incommensurate characteristics proves advantageous. Conversely, for applications requiring precise control and inhibition of molecular motion, a commensurate system may be more fitting. Leveraging the flexibility of molecules to get both, commensurable or incommensurable nanocontacts, allows for strategic design. By selectively choosing or designing systems with commensurate or incommensurate properties, it becomes possible to tailor nanotribological characteristics to specific applications.

Regarding the dissipation processes, yet, the measurement of energies dissipated in experimental settings is not a straightforward task. Experimental methods are constrained by the resolution of the apparatus and the methodology employed for measurements. Taking dynamic AFM as an illustrative example, one important factor influencing resolution pertains to the size of oscillation amplitudes. Utilizing tuning fork sensors with amplitudes in the range of tens of picometers becomes imperative, enabling the study of intramolecular characteristics and enhancing the resolution of the experimental setup. As we navigate this multifaceted landscape, it becomes evident that the relentless pursuit of understanding the internal molecular degrees of freedom in nanocontacts is not just an academic endeavor, but a portal to unlocking transformative applications in nanoscience and nanotechnology.

References

1. C.H. Scholz, Earthquakes and friction laws. Nature **391**, 37–42 (1998)
2. N. Chan, S.G. Balakrishna, A. Klemenz, M. Moseler, P. Egberts, R. Bennewitz, Contraast in nanoscale friction between rotational domains of graphene on Pt(111). Carbon **113**, 132–138 (2017)
3. A. Socoliuc, E. Gnecco, S. Maier, O. Pfeiffer, A. Baratoff, R. Bennewitz, E. Meyer, Atomic-scale control of friction by actuation of nanometer-sized contacts. Science **313**, 207–210 (2006)
4. A. Fall, B. Weber, M. Pakpour, N. Lenoir, N. Shahidzadeh, J. Fiscina, C. Wagner, D. Bonn, Sliding friction on wet and dry sand. Phys. Rev. Lett. **112**, 175502 (2014)
5. I.M. Hutchings, Leonardo da Vinci's studies of friction. Wear **360–361**, 51–66 (2016)
6. K. Bashandeh, V. Tsigkis, P. Lan, A.A. Polycarpu, Extreme environment tribological study of advanced bearing polymers for space applications. Tribol. Int. **153**, 106634 (2021)
7. X. Fan, Q. Xue, L. Wang, Carbon-based solid–liquid lubricating coatings for space applications: a review. Friction **3**, 191–207 (2015)
8. S. Erbas-Cakmak, D.A. Leigh, C.T. McTernan, A.L. Nussbaumer, Artificial molecular machines. Chem. Rev. **115**, 9655–10406 (2015)
9. S. Kassem, T. van Leeuwen, A.S. Lubbe, M.R. Wilson, B.L. Feringa, D.A. Leigh, Artificial molecular motors. Chem. Soc. Rev. **46**, 2592–2621 (2017)
10. P. Fan, J. Gao, H. Mao, Y. Geng, Y. Yan, Y. Wang, S. Goel, X. Luo, Scanning probe lithography: state-of-the-art and future perspectives. Micromachines **13**, 228 (2022)
11. E. Gnecco et al., *Fundamentals of Friction An Wear on the Nanoscale* (Springer, New York, 2015)
12. N. Manini, G. Mistura, G. Paolicelli, E. Tosatti, A. Vanossi, Current trends in the physics of nanoscale friction. Adv. Phys. X **2**(3), 569–590 (2017)
13. A. Vanossi, N. Manini, M. Urbakh, S. Zapperi, E. Tosatti, *Colloquium*: modeling friction: from nanoscale to mesoscale. Rev. Mod. Phys. **85**, 529–548 (2013)
14. L. Gigli, N. Manini, A. Benassi, E. Tosatti, A. Vanossi, R. Guerra, Graphene nanoribbons on gold: understanding superlubricity and edge effects. 2D Mater. **4**, 045003 (2017)
15. S. Kawai, A. Benassi, E. Gnecco et al., Superlubricity of graphene nanoribbons on gold surfaces. Science **351**(6276), 957–961 (2016)
16. A. Klemenz, L. Pastewka, S.G. Balakrishna, A. Caron, R. Bennewitz, M. Moseler, Atomic scale mechanisms of friction reduction and wear protection by graphene. Nano Lett. **14**, 7145–7152 (2014)

17. A. Kolberg, C. Wenzel, K. Hackenstrass, R. Schwarzl et al., Opposing temperature dependence of the stretching response of single PEG and PNiPAM polymers. J. Am. Chem. Soc. **141**, 11603–11613 (2019)
18. Q. Li, Y. Dong, D. Perez, A. Martini, R.W. Carpick, Speed dependence of atomic stick-slip friction in optimally matched experiments and molecular dynamics simulations. Phys. Rev. Lett. **106**, 126101 (2011)
19. R. Pawlak, J.G. Vilhena, A. Hinaut et al., Conformations and cryo-force spectroscopy of spray-deposited single-strand DNA on gold. Nat. Commun. **10**, 685 (2019)
20. J.G. Vilhena, R. Pawlak, P. D'Astolfo et al., Flexible superlubricity unveiled in sidewinding motion of individual polymeric chains; Phys. Rev. Lett. **128**, 216102 (2022)
21. M. Calvaresi, A.-S. Duwez, D.A. Leigh, D. Sluysmans, Y. Song, F. Zerbetto, L. Zhang, Mechanical tightening of a synthetic molecular knot. Chemicals **9**, 65–75 (2023)
22. D.M. Eigler, E.K. Schweizer, Positioning single atoms with a scanning tunnelling microscope. Nature **344**, 524–526 (1990)
23. D.M. Eigler, C.P. Lutz, E.E. Rudge, An atomic switch realized with the scanning tunnelling microscope. Nature **352**, 600–603 (1991)
24. L. Bartels, G. Meyer, K.-H. Rieder, Basic steps of lateral manipulation of single atoms and diatomic clusters with a scanning tunneling microscope tip. Phys. Rev. Lett. **79**(4), 697–700 (1997)
25. G. Dujardin, A. Mayne, O. Robert, F. Rose, C. Joachim, H. Tang, Vertical manipulation of individual atoms by a direct STM tip-surface contact on Ge(111). Phys. Rev. Lett. **80**(14), 3085–3088 (1998)
26. T.A. Jung, R.R. Schlittler, J.K. Gimzewski, H. Tang, C. Joachim, Controlled room-temperature positioning of individual molecules: molecular flexure and motion. Science **271**, 181–184 (1996)
27. B.C. Stipe, M.A. Rezaei, W. Ho, Inducing and viewing the rotational motion of a single molecule. Science **279**, 1907–1909 (1998)
28. J.R. Hahn, W. Ho, Oxidation of a single carbon monoxide molecule manipulated and induced with a scanning tunneling microscope. Phys. Rev. Lett. **87**(16), 166102 (2001)
29. H.J. Lee, W. Ho, Single-bond formation and characterization with a scanning tunneling microscope. Science **286**, 1719–1722 (1999)
30. N. Fournier, C. Wagner, C. Weiss, R. Temirov, F.S. Tautz, Force-controlled lifting of molecular wires. Phys. Rev. B **84**, 035435 (2011)
31. S. Kawai, M. Koch, E. Gnecco et al., Quantifying the atomic-level mechanics of single long physisorbed molecular chains. Proceed. Natl. Acad. Sci. USA **111**, 3968–3972 (2014)
32. G. Langewisch, J. Falter, H. Fuchs, A. Schirmeisen, Forces during the controlled displacement of organic molecules. Phys. Rev. Lett. **110**, 036101 (2013)
33. R. Pawlak, W. Ouyang et al., Single-molecule tribology: force microscopy manipulation of a porphyrin derivative on a copper surface. ACS Nano **10**, 713–722 (2015)
34. M. Ternes, C.P. Lutz, C.F. Hirjibehedin, F.J. Giessibil, A.J. Heinrich, The force needed to move an atom on a surface. Science **319**, 1066–1069 (2008)
35. C. Wagner, N. Fournier, V.G. Ruiz, C. Li, K. Mullen, M. Rohlfing, A. Tkatchenko, R. Temirov, F.S. Tautz, Non-additivity of molecule-surface van der Waals potentials from force measurements. Nat. Commun. **5**, 5568 (2014)
36. J.G. Vilhena, E. Gnecco, R. Pawlak, F. Moreno-Herrero, E. Meyer, Stick-slip motion of ssDNA over graphene. J. Phys. Chem. B **122**, 840–846 (2017)
37. S. Lu, W. Cai, S. Zhang, S. Cui, Single-chain mechanics at low temperatures: a study of three kinds of polymers from 153 to 300 K. Macromolecules **56**, 3204–3212 (2023)
38. T. Kudernac, N. Ruangsupapichat, M. Parschau et al., Electrically driven directional motion of a four-wheeled molecule on a metal surface. Nature **479**, 208–211 (2011)
39. F. Moresco, G. Meyer, K.-H. Rieder, H. Tang, A. Gourdon, C. Joachim, Recording intramolecular mechanics during the manipulation of a large molecule. Phys. Rev. Lett. **87**, 088302 (2001)

40. P. Stacko, J.C.M. Kistemaker, T. van Leeuwen, M.-C. Chang, E. Otten, B. Feringa, Locked synchronous rotor motion in a molecular motor. Science **356**, 964–968 (2017)
41. B.L. Feringa, The art of building small: from molecular switches to motors (Nobel Lecture). Angew. Chem. Int. **56**, 11060–11078 (2017)
42. H.-P. Jacquot de Rouville, G. Vives, G. Rapenne, A family of electron-triggered molecular motors based on aromatic building blocks. Pure Appl. Chem. **80**, 659–667 (2008)
43. R. Pawlak, S. Fremy, S. Kawai, T. Glatzel et al., Directed rotations of single porphyrin molecules controlled by localized force spectroscopy. ACS Nano **6**, 6318–6324 (2012)
44. R. Pawlak, T. Meier, N. Renaud, M. Kisiel, A. Hinaut, T. Glatzel et al., Design and characterization of an electrically powered single molecule on gold. ACS Nano **11**, 9930–9940 (2017)
45. L. Gross, F. Mohn, N. Moll, P. Loljeroth, G. Meyer, The chemical structure of a molecule resolved by atomic force microscopy. Science **325**, 1110–1114 (2009)
46. M. Rief, M. Gautel, F. Oesterhelt, J.M. Fernandez, H.E. Gaub, Reversible unfolding of individual titin immunoglobulin domains by AFM. Science **276**, 1109–1112 (1997)
47. R. Pawlak, S. Kawai, T. Meier, T. Glatzel, A. Baratoff, E. Meyer, Single-molecule manipulation experiments to explore friction and adhesion. J. Phys. D Appl. Phys. **50**, 113003 (2017)
48. R. Pawlak, J.G. Vilhena, P. D'Astolfo et al., Sequential bending and twisting around C-C single bonds by mechanical lifting of a pre-adsorbed polymer. Nano Lett. **20**, 652–657 (2020)
49. B. Alberts, A. Johnson, J. Lewis, M. Raff, K. Roberts, *Molecular Biology of the Cell*, 4th edn. (Garland Science, New York, 2002)
50. M.P. Allen, D.J. Tildesley, *Computer Simulation of Liquids*, 2nd edn (Oxford University Press; Oxford, 2017)
51. D. Frenkel, B. Smit, *Understanding Molecular Simulation* (Academic Press, Great Britain, 2002)
52. L. Verlet, Computer "experiments" on classical fluids. I. Thermodynamical properties of Lennard-Jones molecules. Phys. Rev. **159**, 98–103 (1967)
53. R. Santamaria, A. Alvarez de la Paz, L. Roskop, L. Adamowicz, Statistical contact model for confined molecules. J. Stat. Phys. **164**, 1000–1025 (2016)
54. T. Schneider, E. Stoll, Molecular-dynamics study of a three-dimensional one-component model for distortive phase transitions. Phys. Rev. B **17**, 1302 (1978)
55. H.C. Andersen, Molecular dynamics simulations at constant pressure and/or temperature. J. Chem. Phys. **72**, 2384 (1980)
56. W.G. Hoover, Canonical dynamics: equilibrium phase-space distributions. Phys. Rev. A **31**, 1695 (1985)
57. S. Nosé, A unified formulation of the constant temperature molecular dynamics methods. J. Chem. Phys. **81**, 511–519 (1984)
58. R. Santamaria, J. Soullard, Revisiting particle dynamics in the NPT ensemble under the extended Lagrangian approach. Mol. Simul. **49**, 855–866 (2023)
59. R. Santamaria, *Molecular Dynamics* (Springer, New York, 2023)
60. H.M. Berman, J. Westbrook, Z. Feng, G. Gilliland, T.N. Bhat, H. Weissig, I.N. Shindyalov, P.E. Bourne, The protein data bank. Nucl. Acids Res. **28**, 235–242 (2000)
61. D. Case et al., *AMBER 14* (University of California, San Francisco, CA, 2014)
62. A.I. Jewett, D. Stelter, J. Lambert et al., Moltemplate: a tool for coarse-grained modeling of complex biological matter and soft condensed matter physics. J. Mol. Biol. **433**, 166841 (2021)
63. W.D. Cornell, P. Cieplak, C.I. Bayly et al., A second generation force field for the simulation of proteins, nucleic acids, and organic molecules. J. Am. Chem. Soc. **117**, 5179–5197 (1995)
64. W.L. Jorgensen, J. Tirado-Rives, The OPLS [optimized potentials for liquid simulations] potential functions for proteins, energy minimizations for crystals of cyclic peptides and crambin. J. Am. Chem. Soc. **110**, 1657–1666 (1988)
65. W.L. Jorgensen, D.S. Maxwell, J. Tirado-Rives, Development and testing of the OPLS all-atom force field on conformational energetics and properties of organic liquids. J. Am. Chem. Soc. **118**, 11225–11236 (1996)

66. I. Cacelli, G. Prampolini, Parametrization and validation of intramolecular force fields derived from DFT calculations. J. Chem. Theory Comput. **3**(5), 1803–1817 (2007)
67. V. Diez-Cabanes, G. Prampolini, A. Francés-Monerris, A. Monari, M. Pastore, Iron's wake: the performance of quantum mechanical-derived versus general-purpose force fields tested on a luminescent iron complex. Molecules **25**, 3084 (2020)
68. G. Prampolini, F. Ingrosso, J. Cerezo, A. Iagatti, P. Foggi, M. Pastore, Short- and long-range solvation effects on the transient UV–Vis absorption spectra of a Ru(II)–polypyridine complex disentangled by nonequilibrium molecular dynamics. J. Phys. Chem. Lett. **10**, 2885–2891 (2019)
69. I. Cacelli, A. Ferretti, G. Prampolini, Perturbative multireference configuration interaction (CI-MRPT2) calculations in a focused dynamical approach: a computational study of Solvatochromism in pyrimidine. J. Phys. Chem. A **119**, 5250–5259 (2015)
70. J. Cerezo, D. Aranda, F.J. Avila-Ferrer, G. Prampolini, F. Santoro, Adiabatic-molecular dynamics generalized vertical hessian approach: a mixed quantum classical method to compute electronic spectra of flexible molecules in the condensed phase. J. Chem. Theory Comput. **16**, 1215–1231 (2020)
71. N. De Mitri, S. Monti, G. Prampolini, V. Barone, Absorption and emission spectra of a flexible dye in solution: a computational time-dependent approach. J. Chem. Theory Comput. **9**, 4507–4516 (2013)
72. J. Cerezo, G. Prampolini, I. Cacelli, Developing accurate intramolecular force fields for conjugated systems through explicit coupling terms; Theor. Chem. Acc. **137**, 80 (2018)
73. F. Iori, R. Di Felice, E. Molinari, S. Corni, GolP: an atomistic force-field to describe the interaction of proteins with Au(111) surfaces in water. J. Comput. Chem. **30**(9), 1465–1476 (2009)
74. W. Ouyang, D. Mandelli, M. Urbakh, O. Hod, Nanoserpents: graphene nanoribbon motion on two-dimensional hexagonal materials. Nano Lett. **18**, 6009–6016 (2018)
75. W. Ouyang, R. Sofer, X. Gao et al., Anisotropic interlayer force field for transition metal dichalcogenides: the case of molybdenum disulfide. J. Chem. Theory Comput. **17**, 7237–7245 (2021)
76. W. Ouyang, O. Hod, R. Guerra, Registry-dependent potential for interfaces of gold with graphitic systems. J. Chem. Theory Comput. **17**, 7215–7223 (2021)
77. C. Brunken, M. Reiher, Self-parametrizing system-focused atomistic models. J. Chem. Theory Comput. **16**, 1646–1665 (2020)
78. I. Cacelli, A. Cimoli, P.R. Livotto, G. Prampolini, An automated approach for the parameterization of accurate intermolecular force-fields: pyridine as a case study. J. Comput. Chem. **33**, 1055 (2012)
79. M. Campetella, N. De Mitri, G. Prampolini, Automated parameterization of quantum-mechanically derived force-fields including explicit sigma holes: a pathway to energetic and structural features of halogen bonds in gas and condensed phase. J. Chem. Phys. **153**, 044106 (2020)
80. S.A. Grimme, General quantum mechanically derived force field (QMDFF) for molecules and condensed phase simulations. J. Chem. Theory Comput. **10**, 4497–4514 (2014)
81. J.T. Horton, A.E.A. Allen, L.S. Dodda, D.J. Cole, QUBEKit: automating the derivation of force field parameters from quantum mechanics. J. Chem. Inf. Model. **59**, 1366–1381 (2019)
82. J.G. McDaniel, J.R. Schmidt, First-principles many-body force fields from the gas phase to liquid: a 'universal' approach. J. Phys. Chem. B **118**, 8042–8053 (2014)
83. D. Semrouni, C.J. Cramer, L. Gagliardi, AMOEBA force field parameterization of the azabenzenes. Theor. Chem. Acc. **134**, 1590 (2015)
84. L. Vanduyfhuys, S. Vandenbrande, T. Verstraelen, R. Schmid, M. Waroquier, V. Van Speybroeck, QuickFF: a program for a quick and easy derivation of force fields for metal-organic frameworks from ab initio input. J. Comput. Chem. **36**, 1015–1027 (2015)
85. J.G. Vilhena, L. Greff da Silveira, P.R. Livotto, I. Cacelli, G. Prampolini, Automated parameterization of quantum mechanically derived force fields for soft materials and complex fluids: development and validation. J. Chem. Theory Comput. **17**, 4449–4464 (2021)

86. B. Waldher, J. Kuta, S. Chen, N. Henson, A. Clark, ForceFit: a code to fit classical force fields to quantum mechanical potential energy surfaces. J. Comput. Chem. **31**, 2307–2316 (2010)

Nanoisland Manipulation Experiments at Oxidized, Contaminated and Nanorough Interfaces: Structural Superlubricity and Directional Locking

Wai H. Oo, Alper Özoğul, Franciszek Krok, Enrico Gnecco, and Mehmet Z. Baykara

Abstract This chapter reports on atomic force microscopy based nano-manipulation experiments performed on noble metal nanoislands (gold and platinum), which were previously shown to exhibit structurally superlubric sliding under ambient conditions on highly oriented pyrolytic graphite (HOPG). Experiments performed on oxidized platinum nanoislands on HOPG demonstrate an increase in interfacial shear stress when compared with non-oxidized islands, but not a breakdown of structural superlubricity. In addition, an effect reminiscent of contact aging is observed on a sample system that comprises gold nanoislands on HOPG, which interestingly is suppressed in the presence of environmental contamination. Nanomanipulation of gold islands is also performed on molybdenum disulfide (MoS_2). Here, the high degree of commensurability at the interface does not result in superlubricity, but rather in a specific "directional locking" effect. The effect was observed not only on freshly cleaved flat surfaces but also on bilayers grown on a nanorough silicon wafer, where atomic-scale resolution of the complex cross-section of the system could be achieved using HAADF-STEM.

W. H. Oo · M. Z. Baykara (✉)
Department of Mechanical Engineering, University of California Merced, Merced, CA 95343, USA
e-mail: mehmet.baykara@ucmerced.edu

A. Özoğul
Institute for Materials Science and Max Bergmann Center for Biomaterials, TU Dresden, 01069 Dresden, Germany

F. Krok · E. Gnecco
Marian Smoluchowski Institute of Physics, Jagiellonian University, 30-348, Krakow, Poland

E. Gnecco and E. Meyer (eds.), *Fundamentals of Friction and Wear on the Nanoscale*, NanoScience and Technology, https://doi.org/10.1007/978-3-031-63065-1_8

1 Introduction

In classical mechanics, friction arises as a force that resists the relative motion of two contacting surfaces. From the slipping of tectonic plates to the melodic chords of a violin, friction is present in myriad forms all around nature. While it is essential and desirable in some processes, friction is a major source of wasted energy in many mechanical systems, taking up about one fifth of the annual energy consumption on earth [1]. From a technological standpoint, friction should be reduced and controlled in various systems and processes, to increase the efficiency and lifetime of mechanical systems, and to reduce the strain on our limited and non-renewable energy resources.

The conventional method to reduce friction (and wear) at interfaces involves the use of liquid lubricants [2]. While liquid lubricants are being used successfully in various applications, certain operational conditions (including but not limited to high temperatures and exotic environments such as vacuum) necessitate the use of solid lubricants [3], which typically comprise layered, lamellar solids such as highly oriented pyrolytic graphite (HOPG) and molybdenum disulfide (MoS_2) [4].

While most state-of-the-art research on liquid and solid lubricants is of an applied nature aimed at, e.g. increasing their lifetime and operational parameter space, research groups around the world are also conducting fundamental research in friction and lubrication, both theoretically and experimentally [5]. One of the main premises of such work is that enabling the formation of a fundamental understanding of frictional processes on nanoscopic and even atomic length scales would allow controlling and mitigating friction in practically relevant scenarios. Complementarily, fundamental work in friction also aims at devising novel ways for friction mitigation, i.e. lubrication at interfaces.

An exciting concept that arose from fundamental friction research is *structural superlubricity* [6]. First conceptualized theoretically [7–9], structural superlubricity involves the idea that friction at an interface formed by two atomically flat yet structurally incommensurate surfaces should practically vanish, due to a systematic, pairwise cancellation of lateral forces experienced by the majority of the atoms at the two sides of the interface due to symmetry arguments. The concept also can be visualized by a one-dimensional model, whereby the potential energy barrier encountered by each atom of a slider on top of a periodic, incommensurate substrate drops as the slider size increases, ultimately resulting in the slider "floating" over the substrate, as schematically described in Fig. 1 [10].

Spurred on by the promise of "nearly perfect lubrication without lubricants", experimental research in structural superlubricity flourished. In particular, Dienwiebel et al. [11] experimentally validated that friction experienced at the interface formed by two graphite surfaces all but vanished at all rotational registries except for well-defined configurations corresponding to commensurate structures. Subsequently, various atomic force microscopy (AFM) based nano-manipulation experiments by the group of André Schirmeisen showed that gold and antimony nanoislands exhibited ultra-low friction forces while sliding on flat HOPG and MoS_2 surfaces under clean, ultrahigh vacuum (UHV) conditions, whereby interfacial shear stresses

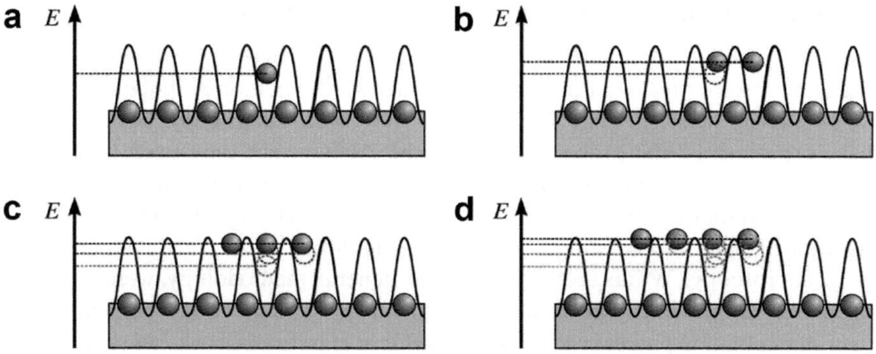

Fig. 1 Schematic illustration describing the concept of structural superlubricity, whereby the potential energy barrier experienced by slider atoms gradually drops with increasing slider size, from (**a**–**d**). Figure adapted from [10]

dropped with increasing contact size, a key signature of structural superlubricity [12–14]. More recently, AFM based nano-manipulation experiments were again employed to show that structurally superlubric sliding can also be achieved under uncontrolled ambient conditions for nanoislands of gold and platinum on HOPG, with measured shear stresses being essentially on the same order as those recorded under UHV conditions [15, 16]. While this observation strongly challenged the theoretical requirement of molecular cleanliness at the interface for structural superlubricity, it also raised questions as to how resilient the ultra-low friction state would be to chemical changes at the interface, e.g. those induced by oxidation and contaminant molecules.

Motivated as above, in this chapter we report the results of AFM based nano-manipulation experiments performed on oxidized platinum nanoislands, whereby an increase in shear stresses but not a complete breakdown of the superlubric state is observed. We also employ an improved experimental methodology to investigate the superlubric sliding of gold nanoislands on HOPG more robustly, revealing the presence of an effect reminiscent of contact aging, which is suppressed under environmental contamination. Gold islands have been also manipulated on MoS_2 surfaces. In this case, the high degree of commensurability between the two contacting surfaces results in preferential directions of motion for the islands. Quite interestingly, this effect was observed not only on flat MoS_2 surfaces obtained by the cleaving (i.e. peeling) of a bulk mineral crystal, but also on a MoS_2 bilayer standing on a nanorough silicon wafer.

2 Nano-Manipulation via Atomic Force Microscopy

The experiments presented in this chapter involve the nano-manipulation of noble metal islands on HOPG and MoS_2 via atomic force microscopy. Specifically, in these experiments, the probe (i.e. tip) of the AFM is not only used for imaging (utilizing the common tapping and contact modes) but also to manipulate, i.e. laterally move islands on the substrate, while recording the interfacial friction force directly as deduced from the twisting of the AFM cantilever, as in conventional friction force microscopy. While the details of the general methodology were discussed extensively in previously published work [15–17], it should be highlighted that the nano-manipulation process can be performed in at least two separate ways (Fig. 2). In particular, the majority of AFM-based nano-manipulation experiments are performed in the "tip at side" mode, whereby the AFM probe first lands on the substrate, in close proximity to the nanoisland to be manipulated. The probe is then approached to the island and starts manipulating (i.e. pushing) it from the side after establishing contact. The additional twisting experienced by the probe during island pushing is directly related to the interfacial friction force. While practical, this method, when applied on superlubric material systems, suffers from reproducibility issues as well as data collection rates. This is simply because the islands slide away from the AFM-imaged area soon after contact with the probe is established, owing to nearly vanishing frictional resistance. As a result, a particular island can be tested typically only once, with very limited control over parameters such as the direction and speed of sliding. On the other hand, the "tip on top" manipulation mode [18] offers much more reliable data collection as the AFM probe is now placed *on* the island and is used to slide the island in any direction with great control over sliding speed. In this mode, the same island can be manipulated hundreds of times, significantly improving both the quantity and the statistical relevance of the recorded friction data. While the more conventional tip at side mode was used to record the friction data on oxidized platinum islands discussed in Sect. 3, the utilization of the "tip on top" mode allowed for uncovering the contact aging phenomenon that is reported on gold nanoislands in Sect. 4. The discussion would not be complete without mentioning that nanoislands can be also manipulated in dynamic AFM modes, i.e. by gently "hammering" them with a tip oscillating in a controlled way while scanning [19]. Although this method does not allow a direct quantification of friction forces, it can be more reliable when the lateral size of the islands and the tip apex are comparable, as in the case of the gold islands manipulated on MoS_2 presented in Sect. 5.

3 Structural Superlubricity at Oxidized Interfaces

Considering that a significant number of practical surfaces experience chemical degradation in the form of oxidation over time, the question of how robust structural superlubricity would be in the face of interfacial oxidation gains importance.

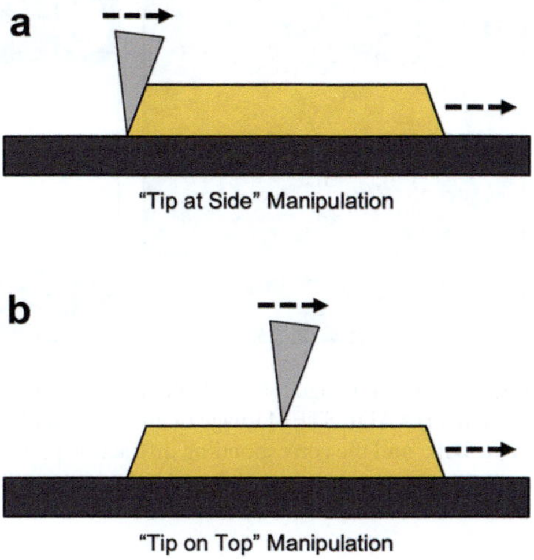

Fig. 2 Schematic illustrations of the **a** "tip at side" and **b** "tip on top" manipulation modes

In order to obtain the first clues toward answering this question, and to additionally probe if friction at structurally superlubric interfaces can be chemically tuned, AFM based nano-manipulation experiments were performed on oxidized platinum nanoislands.

E-beam evaporation followed by post-deposition annealing was employed to grow platinum nanoislands on cleaved HOPG substrates. The oxidation of the nanoislands was achieved by oxygen plasma exposure in a reactive ion etching (RIE) system. The details of sample synthesis can be found in Refs. [16, 20]. Figure 3a shows the general morphology of the sample system after oxidation as imaged by scanning electron microscopy (SEM), whereby the HOPG substrate is covered by flat, "pancake-like" platinum nanoislands of various size (from a few tens of nm to a few hundreds of nm, laterally). While the islands mainly exhibit straight edges that indicate crystalline character, high angle annular dark field scanning transmission electron microscopy (HAADF-STEM) was employed to confirm crystallinity, yielding images showing atomic planes arranged on top of each other in (111) configuration, parallel to the top surface of the HOPG substrate (Fig. 3b). The corresponding Fourier Transform (Fig. 3b, inset) provides further validation for the crystalline structure of the nanoislands.

In order to confirm that the platinum nanoislands were indeed oxidized, X-ray photoelectron spectroscopy (XPS) experiments were performed, a representative result of which is presented in Fig. 4, focusing on the Pt4f region. The doublet peak (red) that accompanies the metallic Pt4f doublet peaks (blue) corresponds to $PtO4f_{(7/2)}$ and confirms that the nanoislands are indeed oxidized on and near the top surface [21, 22]. To confirm that the bottom surfaces of the nanoislands (which are in contact with the HOPG substrate) are also oxidized, high-resolution, cross-sectional energy-dispersive X-ray spectroscopy (EDX) maps were recorded. The representative results in Fig. 5 indeed point to an increase in oxygen intensity on the bottom surface of the nanoisland, albeit weaker than the top island. While the

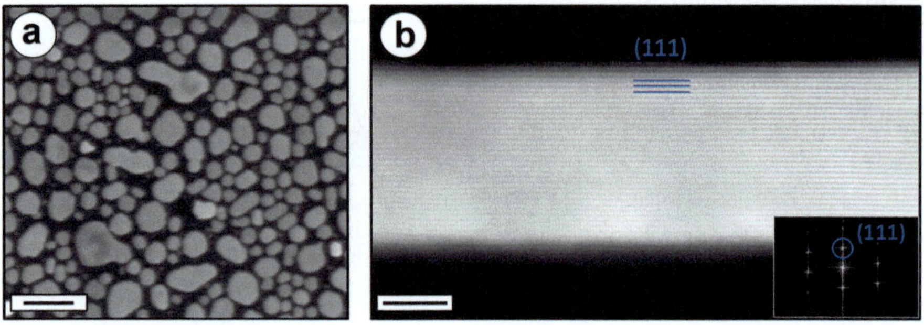

Fig. 3 **a** SEM image of oxidized platinum nanoislands on HOPG. **b** High-resolution, cross-sectional HAADF STEM image of an oxidized platinum nanoisland showing pristine atomic planes in the bulk, and the corresponding diffraction pattern (inset). Scale bars correspond to 200 and 3 nm, respectively. Figure adapted from [20]

relatively weak oxygen peak at the bottom surface can be explained by the fact it is protected from plasma exposure owing to its contact with the HOPG substrate, the results show that oxidation, at least partially, also took place at the sliding interface.

Following the confirmation of (partial) oxidation at the bottom surface of the platinum nanoislands, nano-manipulation experiments were performed by the "tip at side" method described in Sect. 2. In particular, 21 oxidized platinum nanoislands were investigated. The results of the experiments, in terms of interfacial friction force as a function of contact area (as inferred simply from topographical AFM imaging with the assumption of atomically flat interfaces [23]), are presented in Fig. 6a, together with results obtained on non-oxidized platinum, as well as gold

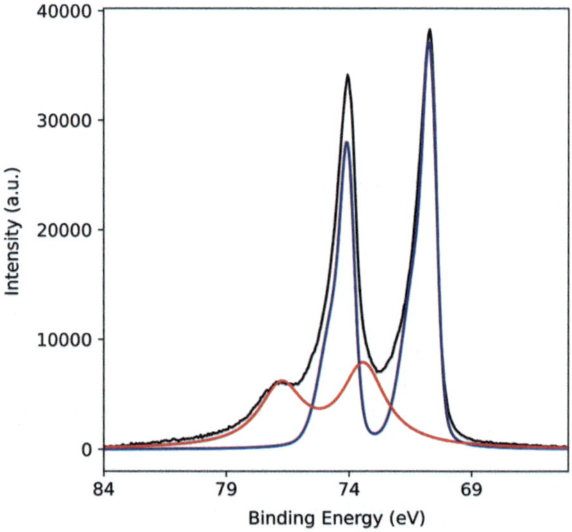

Fig. 4 Representative XPS spectra recorded on oxidized platinum nanoislands. The experimentally measured spectrum (background subtracted) is shown in black, whereas the reference spectrum corresponding to the metallic Pt4f orbital is shown in blue, and the reference spectrum corresponding to $PtO4f_{(7/2)}$ is shown in red. Figure adapted from [20]

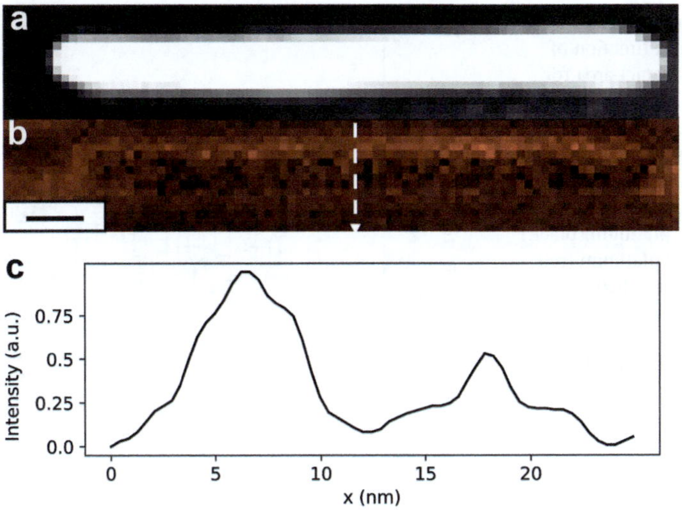

Fig. 5 **a** Cross-sectional HAADF STEM image of an oxidized platinum nanoparticle. **b** The corresponding EDX map for the oxygen Kα peak, where brightness indicates a higher oxygen net count (background subtracted). Scale bar corresponds to 15 nm. **c** Line profile of oxygen net intensity taken along the white, dashed arrow in (**a**), from the top to the bottom of the nanoparticle. Two peaks corresponding to increased oxygen content on the top (~5000 oxygen net counts) and the bottom surfaces (~2500 oxygen net counts) can be observed. Figure adapted from [20]

nanoislands. While it is clear that the great majority of oxidized platinum nanoislands exhibit higher friction forces than non-oxidized platinum and gold nanoislands (mean values of 9.25, 1.41, and 0.33 nN, respectively), it should be remarked that friction force is a sub-linear function of contact area in structural superlubricity [24]. As such, a much better metric for comparison between the different types of nanoislands would be shear stress, i.e. friction force normalized with respect to contact area. The comparison of shear stresses is thus presented in Fig. 6b, from which two main conclusions can be drawn: (i) all three types of nanoislands undergo structurally superlubric sliding in accordance with theory, as shown by an overall decrease in shear stress with contact area, and (ii) oxidized platinum nanoislands exhibit significantly higher shear stresses when compared with non-oxidized platinum and gold nanoislands (mean values of 0.22, 0.13, and 0.02 MPa, respectively).

In conclusion, it was shown that (partial) oxidation at metal/graphite interfaces does lead to a significant increase in shear stresses, without loss of structural superlubricity. An increase in chemical interaction forces between the islands and the substrate due to oxidation can be a prime candidate for the physical mechanism behind the shear stress increase [25]. Regardless, the experimental results presented here hint at the somewhat surprising resilience of the structurally superlubric state against oxidation at the metal/graphite interface.

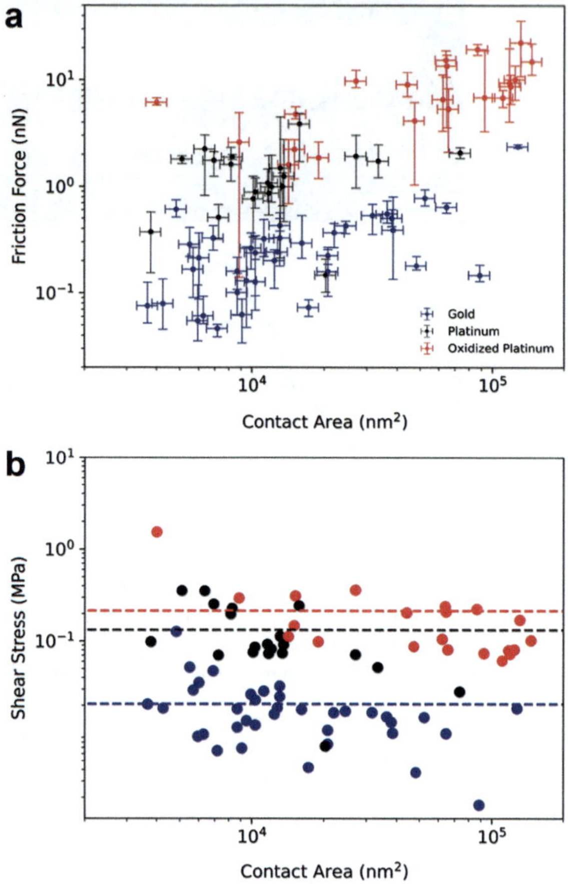

Fig. 6 **a** Friction force measured as a function of interfacial contact area for gold (blue), platinum (black) and oxidized platinum (red) nanoislands. **b** Shear stress measured as a function of interfacial contact area for gold (blue), platinum (black) and oxidized platinum (red) nanoislands. Dashed horizontal lines indicate mean values for each island type. Figure adapted from [20]

4 Environmental Contamination and Contact Aging in Structural Superlubricity

As described earlier in this chapter, gold nanoislands were previously shown to exhibit structurally superlubric sliding on HOPG substrates under uncontrolled ambient conditions [15]. While these early results led to renewed hope for the realization of structurally superlubric mechanical systems operating under practical conditions, they suffered from the main issue of data acquisition rate associated with the "tip at side" manipulation approach. Motivated in this fashion, measurements were repeated on this material system with the "tip on top" manipulation mode described in Sect. 2, allowing for the manipulation of the same island hundreds of times as opposed to just once or twice which was the case for earlier work [15]. Employing such an improved measurement technique led to the observation of intriguing features. In

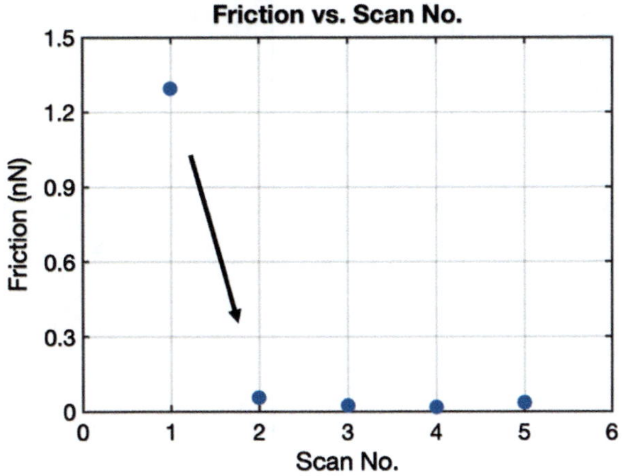

Fig. 7 Mean friction forces extracted from consecutive manipulation experiments (i.e. scans) performed on a single gold island under ambient conditions. Each data point corresponds to the mean friction force extracted from the corresponding scan. As highlighted by the downward arrow, the initial scan yields a significantly higher friction force when compared with subsequent scans, which is reminiscent of "contact aging"

particular, a "contact aging" effect was encountered, whereby the initial manipulation experiment (i.e. scan) performed on a gold nanoisland (with mean friction force extracted from the probe dragging the island back and forth a total of 256 times, over lateral distances of several tens of nanometers) almost always yielded significantly higher friction forces than subsequent manipulation experiments performed on the same island within a manner of minutes (Fig. 7). This effect, which was observed for 10 out of 13 (i.e. 77%) gold islands on a freshly synthesized sample (measured within a few days after synthesis), was interestingly suppressed when the same sample was instead measured a few weeks after synthesis, with only 6 out of 17 (i.e. 35%) islands exhibiting aging. At the same time, phase images taken during tapping-mode imaging on the "old" sample revealed contamination in the form of dispersed *puddles* that are up to 10 Å in height (Fig. 8), which were absent in phase images recorded on the "fresh" sample. While the chemical origin of the contaminant layer cannot be discerned from AFM imaging, it is highly likely that it consists of airborne hydrocarbons adsorbed on the sample surface, similar to what was reported for van der Waals materials recently [26]. While these experimental observations hint at the non-negligible role of environmental contamination on structurally superlubric sliding under ambient conditions, elucidating the details of the involved physical mechanisms would certainly require support from theoretical modeling.

5 Directional Locking Due to Commensurability

Gold islands of about a few tens of nm in width and a few nm in height were also deposited and manipulated on MoS_2 bilayer and multilayer surfaces. Figure 9a shows a representative SEM image of several islands, as distinguished individually on bilayer MoS_2. Due to the crystallographic symmetry of the substrate, the islands

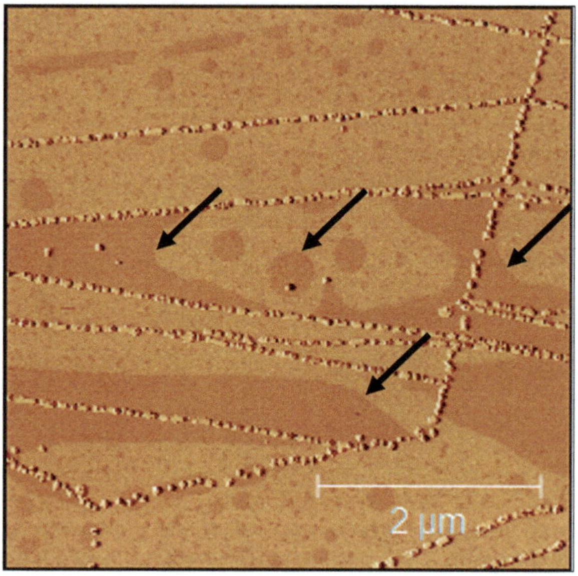

Fig. 8 Phase image acquired on an HOPG substrate decorated with gold nanoislands several weeks after synthesis, showing contamination in the form of dispersed "puddles", some of which are highlighted by arrows

present a triangular shape in all cases. Their size could be easily tuned by the amount of Au deposited and the temperature of the substrate during the deposition process.

In the case of the multilayer MoS_2, atomically resolved STM imaging showed a (111) atomic plane arrangement of the top surface of the islands [28]. The same arrangement is expected for the Au atoms at the contact interface. The level of commensurability between Au(111) and MoS_2 (lattice mismatch = 11%) was low enough to allow manipulation with an oscillating AFM tip. The Au islands were found

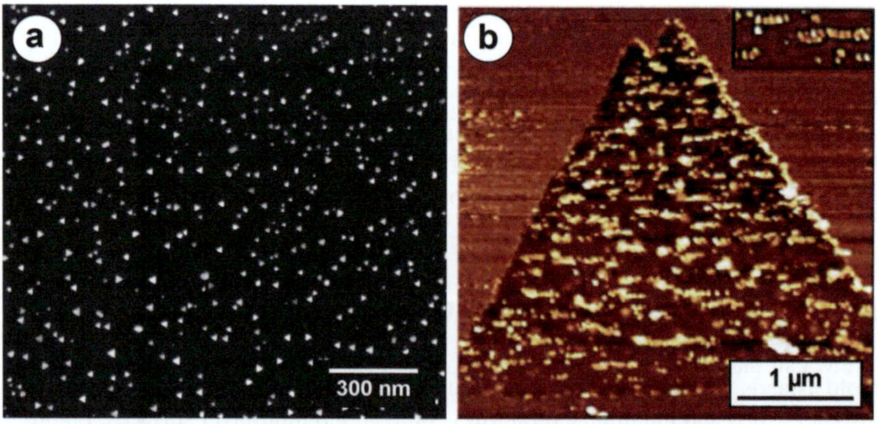

Fig. 9 **a** HR-SEM morphology of triangular Au nanoislands assembled on a double-layer MoS_2/ SiO_2 surface after MBE-deposition of 2 ML of Au at 400 °C. **b** Overview of the AFM morphology of Au islands manipulated by the AFM tip in tapping mode. Figure adapted from [27]

to move only along preferential directions on both bilayer and multilayer MoS_2. The directional locking was confirmed by accompanying MD simulations, which also allowed to identify the preferential directions of motion with the zigzag alignment of the Mo and S atoms on the MoS_2 surface [28]. A similar effect was also observed in the manipulation of Au islands deposited on the bilayer [27], which could be ultimately exploited for the realization of low-friction "nanohighways" for metallic islands and even for nano-objects carried on them.

Figure 9b shows the assemblies of Au islands on several-μm-wide triangular flakes, which were observed after the manipulation took place. The islands arranged themselves in long stripes along the fast-scanning direction of the AFM, which, in this specific case, iwass also close to the zigzag direction of MoS_2 (corresponding to the orientation of the MoS_2 flake edges [29]). High-resolution SEM imaging of the resulting structure (not shown) revealed that the triangular shape of the Au islands did not change during the manipulation. Furthermore, all islands forming a stripe did not merge but rather remained well-separated by narrow nanogaps. This unfortunate side effect (preventing possible applications of Au stripes, e.g. as nanoelectrodes) remained also after annealing the samples up to 550 °C. To understand if the effect was caused by electrostatic repulsion among the islands or by the nano-roughness of the SiO_2 substrate possibly reproduced by the MoS_2 bilayer, a lamella of the Au/MoS_2/SiO_2 system was precisely cut along the zigzag direction of MoS_2, and its cross-section was imaged by HAADF-STEM [30]. In this way not only the MoS_2 bilayer was found to follow the corrugation of the SiO_2 substrate, but its atomic structure could be also resolved (Fig. 10). It was observed that (i) the measured distance between the S–S atoms forming the outer layer of the modulated MoS_2 layer could differ (locally) by up to 15% from the S–S distance of unstressed MoS_2 [30] and (ii) the density of lattice sites occupied by Au atoms in the plane constituting the nanoisland/substrate was lower than in the Au bulk. While the tensile stress suggested by the first observation was confirmed by independent Raman measurements, MD simulations would be helpful to address possible relations between these results and the high mobility of the Au islands on the MoS_2 bilayer.

6 Summary and Outlook

In conclusion, the examples provided in this Chapter demonstrate the potential of AFM-based nanomanipulation in investigating the nanotribological and superlubric properties of metal nanoislands on solid surfaces of layered materials. Combined with complementary techniques such as HAAD-STEM, EDX and Raman spectroscopy, AFM can play a pivotal role in understanding the influence of oxidation, contamination and surface roughness on interfacial friction on the nanoscale. In the specific case of gold nanoislands on a MoS_2 bilayer, HAAD-STEM has even allowed to recognize a lower density of atoms in the bottom of the metal nanoislands and to quantify the increased distance between the atoms in the top layer of the substrate.

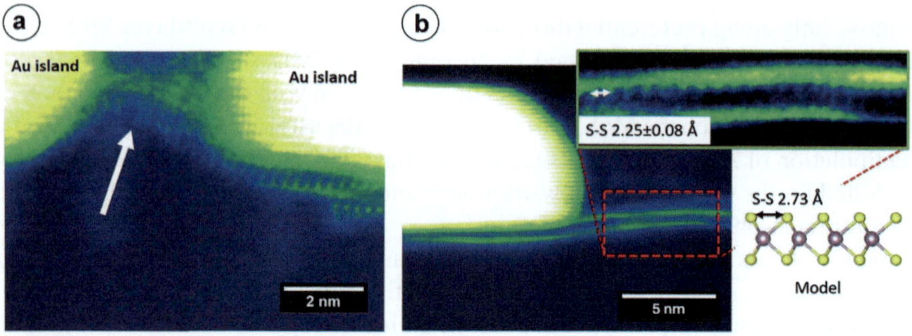

Fig. 10 **a** HAADF-STEM cross-section of Au islands grown on MoS_2/SiO_2 substrate. **b** Au/MoS_2 interface showing the bilayer structures of MoS_2. The magnified section of the MoS_2 layer shows the atomic arrangement with the intralayer distances

Extensions to other systems would represent an important step forward in our current understanding of tribology at the nanoscale.

References

1. K. Holmberg, A. Erdemir, Influence of tribology on global energy consumption, costs and emissions. Friction **5**, 263–284 (2017)
2. T. Mang, K. Bobzin, T. Bartels, *Industrial Tribology: Tribosystems, Friction, Wear and Surface Engineering, Lubrication* (Wiley-VCH, Weinheim, 2011)
3. C. Donnet, A. Erdemir, Historical developments and new trends in tribological and solid lubricant coatings. Surf. Coat. Technol. **180**, 76–84 (2004)
4. M.R. Vazirisereshk, A. Martini, D.A. Strubbe, M.Z. Baykara, Solid lubrication with MoS_2: a review. Lubricants **7**, 57 (2019)
5. A. Vanossi et al., Recent highlights in nanoscale and mesoscale friction. Beilstein J. Nanotechnol. **9**, 1995–2014 (2018)
6. M.Z. Baykara, M.R. Vazirisereshk, A. Martini, Emerging superlubricity: a review of the state of the art and perspectives on future research. Appl. Phys. Rev. **5**, 041102 (2018)
7. M. Hirano, K. Shinjo, Atomistic locking and friction. Phys. Rev. B **41**, 11837–11851 (1990)
8. J.B. Sokoloff, Theory of energy dissipation in sliding crystal surfaces. Phys. Rev. B **42**, 760–765 (1990)
9. M. Hirano, K. Shinjo, Superlubricity and friction anisotropy. Wear **168**, 121–125 (1993)
10. H. Holscher, A. Schirmeisen, U.D. Schwarz, Principles of atomic friction: from sticking atoms to superlubric sliding. Philos. Trans. Roy. Soc. A Math. Phys. Eng. Sci. **366**, 1383–1404 (2008)
11. M. Dienwiebel et al., Superlubricity of graphite. Phys. Rev. Lett. **92**, 126101 (2004)
12. D. Dietzel, J. Brndiar, I. Stich, A. Schirmeisen, Limitations of structural superlubricity: chemical bonds versus contact size. ACS Nano **11**, 7642–7647 (2017)
13. D. Dietzel, M. Feldmann, U.D. Schwarz, H. Fuchs, A. Schirmeisen, Scaling laws of structural lubricity. Phys. Rev. Lett. **111**, 235502 (2013)
14. E. Cihan, D. Dietzel, B.R. Jany, A. Schirmeisen, Effect of amorphous-crystalline phase transition on superlubric sliding. Phys. Rev. Lett. **130**, 126205 (2023)
15. E. Cihan, S. Ipek, E. Durgun, M.Z. Baykara, Structural lubricity under ambient conditions. Nat. Commun. **7**, 12055 (2016)

16. A. Özoğul, S. Ipek, E. Durgun, M.Z. Baykara, Structural superlubricity of platinum on graphite under ambient conditions: the effects of chemistry and geometry. Appl. Phys. Lett. **111**, 211602 (2017)
17. D. Dietzel et al., Interfacial friction obtained by lateral manipulation of nanoparticles using atomic force microscopy techniques. J. Appl. Phys. **102**, 084306 (2007)
18. D. Dietzel, M. Feldmann, H. Fuchs, U.D. Schwarz, A. Schirmeisen, Transition from static to kinetic friction of metallic nanoparticles. Appl. Phys. Lett. **95**, 053104 (2009)
19. D. Dietzel, U.D. Schwarz, A. Schirmeisen, Nanotribological studies using nanoparticle manipulation: principles and application to structural lubricity. Friction **2**, 114 (2014)
20. A. Özoğul, B.R. Jany, F. Krok, E. Gnecco, M.Z. Baykara, Influence of interfacial oxidation on friction in structural superlubricity. Tribol. Lett. **69**, 105 (2021)
21. K.S. Kim, N. Winograd, R.E. Davis, Electron spectroscopy of platinum-oxygen surfaces and application to electrochemical studies. J. Am. Chem. Soc. **93**, 6296–6297 (1971)
22. Z.Y. Li, P. Beck, D.A.A. Ohlberg, D.R. Stewart, R.S. Williams, Surface properties of platinum thin films as a function of plasma treatment conditions. Surf. Sci. **529**, 410–418 (2003)
23. C. Ritter, M. Heyde, B. Stegemann, K. Rademann, U.D. Schwarz, Contact-area dependence of frictional forces: Moving adsorbed antimony nanoparticles. Phys. Rev. B **71**, 085405 (2005)
24. A.S. de Wijn, (In)commensurability, scaling, and multiplicity of friction in nanocrystals and application to gold nanocrystals on graphite. Phys. Rev. B **86**, 085429 (2012)
25. Q.P. He, D.C. Joy, D.J. Keffer, Impact of oxidation on nanoparticle adhesion to carbon substrates. RSC Adv. **3**, 15792–15804 (2013)
26. A. Pálinkás et al., The composition and structure of the ubiquitous hydrocarbon contamination on van der Waals materials. Nat. Commun. **13**, 6770 (2022)
27. F. Trillitzsch, A. Janas, A. Özogul, C. Neumann, A. George, B.R. Jany, F. Krok, A. Turchanin, E. Gnecco, Scanning-probe-induced assembling of gold striations on mono- and bi-layered MoS_2 on SiO_2. MRS Adv. **5**, 2201 (2020)
28. F. Trillitzsch, R. Guerra, A. Janas, N. Manini, F. Krok, E. Gnecco, Directional and angular locking in the driven motion of Au islands on MoS_2. Phys. Rev. B **98**, 165417 (2018)
29. A.M. van der Zande, P.Y. Huang, D.A. Chenet, T.C. Berkelbach, Y.M. You, G.H. Lee, T.F. Heinz, D.R. Reichman, D.A. Muller, J.C. Hone, Grains and grain boundaries in highly crystalline monolayer molybdenum disulphide. Nat. Mater. **12**, 554 (2013)
30. E. Gnecco, A. Janas, B.R. Jany, A. George, A. Turchanin, G. Cempura, A. Kruk, M. Tripathi, F. Lee, A.B. Dalton, F. Krok, Atomic-scale characterization of contact interfaces between thermally self-assembled Au islands and few-layer MoS_2 surfaces on SiO_2. Appl. Surf. Sci. **616**, 156483 (2023)

Tribological Aspects of in Situ Manipulation of Nanostructures Inside Scanning Electron Microscope

Boris Polyakov, Leonid Dorogin, Sven Oras, Rynno Lõhmus, and Sergei Vlassov

Abstract This chapter is dedicated to manipulation of nanostructures inside a scanning electron microscope (SEM) employed for real-time tribological measurements. Different approaches to force registration and calculation of static and kinetic friction are described. Application of the considered methodologies to manipulation of nanostructures with different materials and geometries are demonstrated. Visually guided assembly of nanoscale devices is demonstrated. Advantages and limitations of the technique in comparison to traditional AFM-based manipulation techniques are discussed.

1 Introduction

Macroscopic laws of friction had been thoroughly studied during last centuries. Recently significant progress has been made in research of friction phenomena at the atomic scale. However, it is important to improve our understanding of tribological behavior of nanoscale and mesoscale structures and systems, in order to provide both fundamental and practical knowledge for emerging numerous applications in nanotechnology. These structures are typically too complex to be modelled from

B. Polyakov (✉)
Institute of Solid State Physics, University of Latvia, Ķengaraga Iela 8, Riga, Latvia
e-mail: boris.polyakov@cfi.lu.lv

L. Dorogin
Department of Molecules and Materials, MESA+ Institute for Nanotechnology, University of Twente, Enschede 7500 AE, The Netherlands

S. Oras
Tartu Ülikooli Tehnoloogiainstituut, Nooruse 1, Tartu, Tartumaa, Estonia

R. Lõhmus · S. Vlassov
Tartu Ülikooli Füüsika Instituut, W.Ostwaldi Tn 1, Tartu, Tartumaa, Estonia

E. Gnecco and E. Meyer (eds.), *Fundamentals of Friction and Wear on the Nanoscale*, NanoScience and Technology, https://doi.org/10.1007/978-3-031-63065-1_9

atomic/molecular point of view, and at the same time too small, so that the macroscopic models often break down. Therefore, strong demands arise for advanced experimental and modelling techniques aimed at providing more information on characteristics of investigated nanoobjects (nanoparticles, nanorods, nanowires, nanotubes, etc.).

Tribological characterization of nanostructures often implies controllable displacement of investigated objects. In contrast to macroscopic objects, where a large fraction of applied energy is spent to overcome inertia of a body to put it to motion, at the nanoscale most of the energy is dissipated on adhesion and static friction forces. Taking into account the contact area dependence of the nanoscale friction [1], it is extremely important that the investigated nanostructures had well-defined geometry and atomically smooth surfaces.

Typical objects in nanotribological studies include various types of crystalline nanoparticles (NPs) or nanowires (NWs).

NP material can be chosen in a wide range from noble metals to metal oxides, sulphides, halides, etc. By choosing a proper synthesis method, the NP shape and size can be well controlled. For example, gold NPs can be produced in tetrahedral, decahedral, icosahedral shapes and size from several to a few hundred nanometres. Using thermal or laser assisted annealing the faceted NPs can be transformed into spherical ones [2]. Moreover, complex core–shell structures [3, 4] with core consisting of inorganic nanoparticles surrounded by inorganic or organic materials as a shell can be investigated.

NWs of metal (e.g. gold, silver, etc.) or covalent materials (e.g. silicon, indium arsenide, zinc oxide, etc.) can be fabricated in monocrystalline form having well-defined shape [5–8], making them good candidates for nanotribological experiments. Being nanoscale in one dimension, NWs are convenient objects for bridging nano and microscale. Moreover, the profile of NW during the manipulation experiments can be used as a tool for measuring friction force, significantly expanding assortment of experimental techniques [9].

The most common instrument used for the manipulation of individual nanostructures is an atomic force microscope (AFM). AFM enables a high level of control over the applied force and precise manipulation. Besides its obvious benefits, AFM has several significant limitations. In a single scan, AFM can be used either for visualization or manipulation. Therefore, only initial and final positions of the object can be determined, while manipulation process is performed blindly without visual feedback. AFM manipulations are also time-consuming since it is necessary to take image of manipulated NP before and after displacement. It also makes finding of the object of interest challenging, since scan rate of AFM is significantly lower than that of SEM, especially for larger areas. AFM maximal lateral scan range remains in the order of 100×100 μm and depending on AFM model single image acquiring time can reach few hours.

It is possible to integrate AFM with SEM and perform in situ manipulations and frictional measurements for nanostructures. There are several commercial SEM-compatible AFM instruments [10, 11]. In such approach prominent AFM features are complemented with advantages of SEM. Independent visualization and manipulation functions provide real-time observation of the behaviour of the objects during the manipulations and simultaneous recording of the interaction force. Visual guidance is especially important for manipulation of 1D nanostructures like NWs or nanotubes (NTs). It is possible to observe the profile evolution of a NW during its initial displacement and find degree of elastic deformation of the NW translated over substrate surface. In addition, SEM enables to identify rolling, sliding, or rotational motions of nano objects during manipulations. Moreover, SEM provides vacuum environment, excluding the humidity effect on tribological measurements. In combined AFM-SEM systems imaging can be realized both with scanning electron beam or AFM probe. However, for inside-SEM tribological studies AFM is mainly used for manipulations and force measurements. In this case the feedback is not essential and the whole set-up can be significantly simplified. The last concept together with application examples will be introduced in more details in the following sections.

2 Instrumentation

2.1 Nanomanipulators

A typical nanomanipulation system consists of a sharp needle (tip) and a positioning device that enables the displacement of the tip. Precisely controlled movement of the tip is the essence of nanoscale experiments. Nanomanipulation system should provide both coarse positioning (mm range) and precise (nm range) motion of the probe relative to substrate. Some commercially available nanomanipulators, such as e.g. 3D Manipulators by SmarAct GmbH [12], combine both functions in one device using inertial piezo slider principle in which slow elongation of piezo is used to reposition the mass followed by abrupt contraction resulting in sliding without affecting the current position of the mass. This procedure can be repeated many times allowing to achieve relatively long travel distances (up to several centimeters). However, separation of coarse and fine motions is preferable because of stability and precision. The coarse movements can be realized either manually using precise screws, or can be automated by means of piezo-, stepper- or servo motors. For fine movements piezo actuators (scanners), providing down to subangstrom motion resolution, are commonly utilized. Among them, a tubular scanner design combining *X–Y–Z* motion in a single piezotube has emerged as a simple and widely used technique. The tube scanners can be relatively easily installed inside any electron microscopes and can serve as a basis of a simple home-made manipulation system [13]. A serious drawback of the piezotube scanners is a crosstalk and spherical coordinates of *X–Y* movement. Should be noted that the Z movement range is approximately magnitude smaller compared to lateral scan. A special design of piezoscanner mechanics should be used to enable exclusion of all degrees of freedom but one. One example of advantageous solutions is a flexure guidance system actuated by a piezoelement. A flexure guidance system is a mechanical system used in precision engineering and positioning applications. It provides guidance and support for the movement of an object along a predetermined path while minimizing friction and backlash. This system relies on flexure elements, which are compliant structures made of materials like metals or ceramics, to achieve high-precision and repeatable motion. Modular principle of stacking and combining piezoscanners enables building systems according to specific user requirements using commercially available components like e.g. attocube scanners [14]. If independent *X*, *Y* and *Z* scanners are used, then the linear motion in Cartesian coordinates can be realized. When maximum precision and stability is required, decoupling of X and Y from Z scanners could be beneficial. Instead of a single stack, XY and Z could be separately fixed to the same base for reducing mechanical/vibrational closed loop.

There are several general drawbacks inherent to all piezoelectric scanners, including nonlinear response to applied electric field, hysteresis, creep, and depolarization with time [15]. Various hardware and software solutions can be used to minimize these effects such as closed loop scanner equipped with position sensors (electrical, optical, etc.) or alternatively special scanner calibration, which compensate its hysteresis [16]. In order to minimize thermal drift, parts of the manipulation system should have similar thermal expansion coefficients. Titanium is a popular choice for metal parts as its thermal expansion coefficient is among the lowest for metals and is similar to the one of piezoceramic.

General requirements for nanomanipulation systems are rigid, compact, and stable design. Additional requirements arise for manipulation inside an electron microscope. No magnetic components should be used in close proximity to the electron beam to avoid distortions of the image. Both samples and probes should have sufficient electrical conductivity to exclude charging effect. Components of nanomanipulation system should be vacuum compatible, surfaces should be of high processing quality, and design should exclude closed volumes to minimize degassing and accelerate vacuum pumping.

2.2 Force Measurements

Utilization of nanomanipulation methods for tribological studies requires force detection capability. One way is to use conventional AFM cantilevers with an external deflection detection systems [16–19] like e.g. optical readout, where laser beam is reflected by a cantilever and angular deflection of the cantilever can be read out by a position sensing photodetector. However, in addition to technical challenges related to introducing optical system into SEM chamber, laser irradiation will generate heat that is not easily dissipated in vacuum conditions and will contribute to thermal noise and drift. Alternative concept is based on self-actuating and self-sensing force probes where electronic excitation and readout are used [20, 21], like piezoresistive cantilevers or electromechanical resonators. The self-sensing probes are more convenient for applications inside SEM due to compact design, fully digital-electronic control, and simple integration into vacuum chamber.

Force Sensors Based on Quartz Tuning Fork

Quartz tuning fork (QTF) is a very common type of electromechanical resonator, originally designed to serve as a frequency standard in various devices. The QTF can be used as force sensors since oscillation parameters (amplitude, frequency, and phase) are correlated with the applied force. Due to the piezoelectric properties of a quartz crystal, oscillations in QTF sensors can be excited and registered electrically, making them small, robust, and simple to operate compared to the optical force detection methods. Moreover, QTF force sensors have several advantages: high mechanical Q-factor, large spring constant, adjustable oscillation amplitude, and high sensitivity [22]. For obtaining numerical force values, QTF-based sensor should be calibrated [23, 24].

To make a force sensor suitable for measurements and manipulations inside SEM, sharp probe should be attached to one prong of QTF and electrically connected with one of its electrodes [25]. General recommendation for the probes used for in situ nanomanipulations is sufficient electrical conductivity to dissipate an electric charge, which will unavoidably arise due to the electron beam and will disturb an image in case of insufficient conductivity.

The tip can be fabricated on QTF directly using Focused Ion Beam (FIB) technique [26], or the tip can be glued to QTF sensor [27]. Tips can be made from different materials. For example, tungsten tips can be easily prepared on-site by electrochemical etching [28]. However, sharp metallic tips are relatively easily deformed plastically and therefore are suitable only for very fine manipulations where the forces in order of a few nN or smaller are involved and contact with hard substrate is avoided. Alternatively, conventional AFM cantilevers can be glued to QTF using optical microscope and micromanipulators [25]. The tip of AFM cantilever is usually hidden below the cantilever and is not seen on the SEM image, therefore both cantilever and the substrate should be tilted to make the tip visible from the top during manipulations [25]. More convenient is to use cantilevers with the tip located at the very end of the cantilever and tilted outward [29]. Such configuration makes the tip visible from the top (Fig. 1).

Depending on how the tip is glued to QTF, either normal or shear operational modes are possible. In normal mode the tip oscillates perpendicular to the substrate surface (Fig. 2a), while in shear mode the tip oscillates parallel to the surface (Fig. 2b). Conventional AFM microscopes typically operate in normal mode, where cantilever bends in the vertical plane. However, in contact mode measurements twisting component can additionally provide frictional data. AFM modes were originally developed for imaging and indentation of the surfaces by using tip-sample interaction data via segmented photodiode. The interpretation of force interaction of the tip with manipulated object is complicated (not impossible though) since the directions of the tip movements and tip oscillations do not match. In shear mode the directions of the tip movement and oscillation coincide, which is favorable for the displacement of the

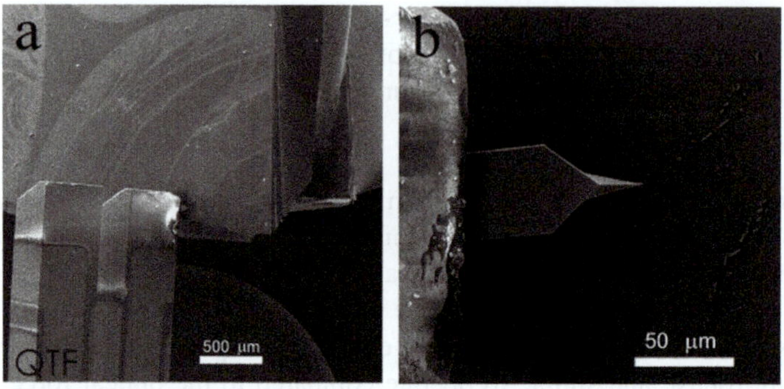

Fig. 1 SEM image of QTF force sensor with manually glued AFM cantilever having a tip visible from the top (**a**). Zoomed image (**b**)

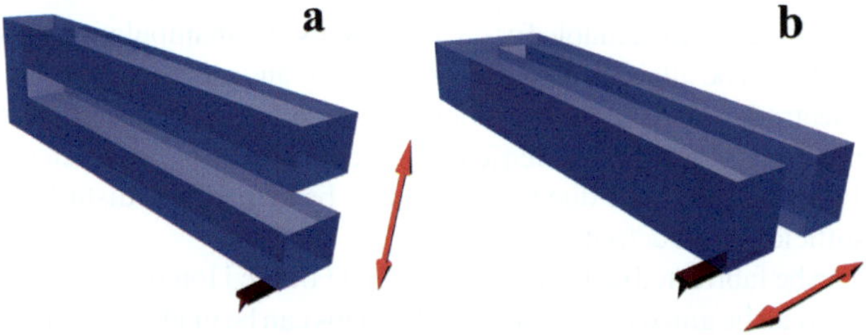

Fig. 2 Schematics of QTF force sensor for normal (**a**) and shear regime (**b**)

object and interpretation of the interaction force. However, maintaining the constant height above the substrate and avoiding jump-into-contact is more challenging.

Akyama-Probes

Since the early 2000s, self-sensing and self-actuating probes utilizing QTF design in combination with micromachined cantilevers became commercially available. The original probe was invented by Terunobu Akiyama and therefore called "Akiyama-probe" [30]. Akiyama-probe features a symmetrical arrangement of a U-shaped silicon cantilever attached to both prongs of a QTF (Fig. 3). The tuning fork serves as a high Q-factor mechanical oscillatory force sensor. The ends of the QTF prongs are moving in-plane and have opposite phases. This motion causes a small vibration at the glued ends of the cantilever, which always contains a Z-component (i.e., the axis of the tip) due to the twisting motion of the prongs (i.e., the cantilever disturbs the symmetry of the TF). The cantilever amplifies the vibration and out-of-plane motion of the tip is obtained. The direction of the Akiyama-probe tip oscillation is similar to conventional tapping mode AFM cantilevers. The tip oscillates at QTF resonance frequency.

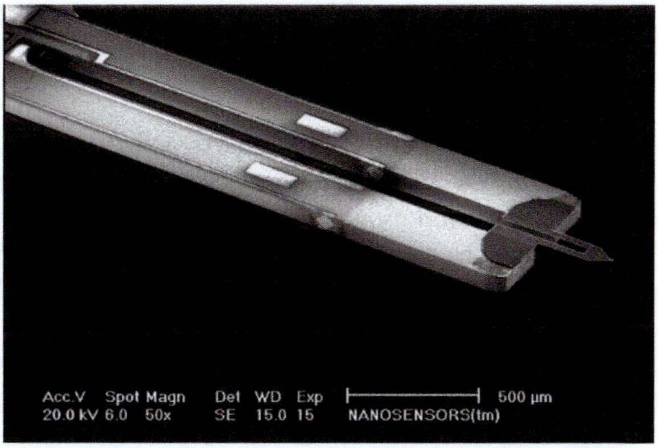

Fig. 3 SEM image of Akyama probe. Copyright by NanoWorld AG

Piezoresistive Cantilevers

Working principle of piezoresistive cantilevers is based on a property of piezoresistive materials to change their bulk resistivity when mechanical stress is applied. Silicon is a well-known piezoresistive material and suitable for fabrication of cantilevers. There are different designs of piezoresistive cantilever.

In general, a silicon resistor made from doped silicon is placed on a U-shaped or single slab cantilever and electrically isolated from it. The resistor is connected to an external DC-biased Wheatstone bridge, which directly measures the deflection by measuring the cantilever's resistance. Elastic deformation in piezoresistor is proportional to voltages difference $V_1 - V_2 = -V\Delta R/4R$. The resistor can be placed near to the most highly stressed region close to the base of the cantilever. The resistor should be made as thin as possible to ensure maximal stress and consequently high piezoresistive response.

Alternatively, whole cantilever can be fabricated as a piezoresistor (Fig. 4). In this case cantilever will produce zero net piezoresistive response due to mutual compensation of stress on the top and bottom side of the cantilever and zero total contribution to the change in resistance [31, 32].

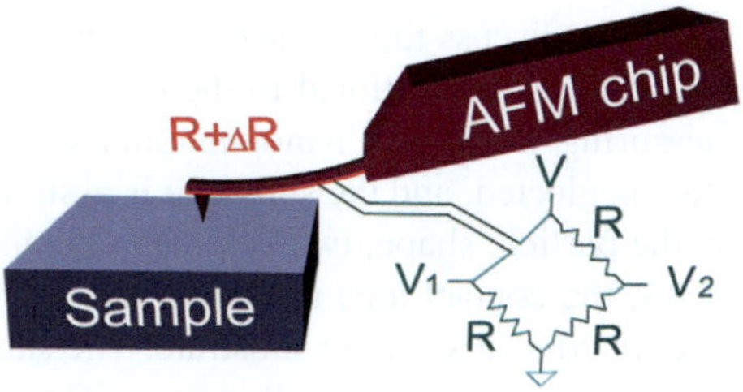

Fig. 4 Detection of piezoresistive cantilever deflection using Wheatstone bridge

3 Manipulation of Nanostructures with Direct Force Registration

Frictional properties of NPs were extensively studied in AFM-based manipulation experiments either in contact or dynamic (oscillating) modes. Following a common AFM manipulation protocol, NPs are first visualized in dynamic mode using a very gentle force. Then several manipulation strategies can be applied. According to one method a tip is positioned behind the NP to be manipulated and the particle is pushed in contact mode with disabled feedback whilst lateral force signal is recorded. Then image is taken again to examine the result of manipulation [33]. Another contact method, so-called "tip-on-top", was introduced by Dietzel et al. [34]. In this method, the AFM tip is first positioned on top of the NP approximately at its center and normal force is applied. The tip is then moved aside and then the NP follows the tip motion. The measured torsional signal (lateral force) is directly proportional to the interfacial friction between the particle and the substrate. NP should be flattened to be manipulated by "tip-on-top" method. In oscillating mode NPs can be moved during the scanning process with enabled feedback when the amplitude of the tip oscillations is increased above a certain threshold value [35]. Force applied on particle is proportional to the tip oscillation amplitude. Tip oscillates perpendicular to the substrate plane, and lateral component of the applied force pushing a NP parallel to the substrate plane depends on NP diameter and tip radius. Increasing the scan rate above a certain value rather than increasing the oscillation amplitude also leads to displacement of NPs [36].

If experiments are performed inside SEM using one of the force-sensors described earlier in this chapter, simultaneous force registration and visualization of particle movement is possible providing additional information and higher degree of control over the manipulation process. In this section we will discuss such an approach illustrated by chosen studies performed on nanostructures of different geometries.

3.1 Contact Area

Let's first discuss the contact area problem. It is known that the friction force at the nanoscale is proportional to the contact area (in the absence of wear) [37]. When measuring friction in nanomanipulation experiments, roughness of a substrate is often neglected, and the substrate is assumed to be atomically flat. Then, depending on the particle shape, two cases can be distinguished. If particle have well-defined facets, the contact area is considered to be simply equal to the surface area of the facet in contact with the substrate. The same is valid for surface area of the base of flattened particles (also called "nanoislands" [34]). Static friction force can then be calculated from simple relation $F_{friction} = \tau S_r$, where τ is the interfacial shear stress/strength and S_r is the contact area [38]. The shear strength is defined as an ultimate shear stress τ before the object starts to move or slips and can be estimated using the relation $\tau_{theo} = G^*/Z$ between the theoretical shear strength and the combined shear modulus, $G* = [(2 - \nu_1)/G_1 + (2 - \nu_2)/G_2]^{-1}$, where $G_{1,2} = E_{1,2}/2(1 + \nu_{1,2})$ [39]. Z is an empirical material-dependent coefficient ranging from 5 to 30 [40].

The situation is different for a spherical particle. A Spherical particle (nanoball) is a special case of a particle having geometrical point contact with the substrate. However, for real systems the contact area is always finite and typically can be estimated on the basis of continuum elasticity models for deformable spheres such as JKR [41] or DMT-M model [42]. According to Tabor [43], the choice of the most suitable model is determined by the parameter:

$$\eta = \left(\frac{16R\gamma^2}{9K^2 z_0^3} \right)^{1/3} \tag{1}$$

where R is the radius of the sphere, γ is the work of adhesion, and z_0 is the equilibrium spacing for the Lennard–Jones potential of the surfaces. K is the combined elastic modulus of the sphere and substrate, defined as $K = 4/3\left[\left(1 - \nu_1^2\right)/E_1 + \left(1 - \nu_2^2\right)/E_2\right]^{-1}$ in which $\nu_{1,2}$ and $E_{1,2}$ are the Poisson ratios and Young's moduli of the substrate and sphere, respectively. For small η, the DMT-M theory is more appropriate. According to the DMT-M model, the contact area $A_{\text{DMT-M}}$ is expressed:

$$A_{\text{DMT}-\text{M}} = \pi \left(\frac{2\pi\gamma}{K} \right)^{2/3} R^{4/3} \tag{2}$$

In the next sections two works on real-time manipulation of faceted and rounded NPs are described.

3.2 *Manipulation of Polyhedron-Like Nanoparticles*

Vlassov et al. [25] applied in situ nanomanipulation technique to study static friction of chemically synthesized Au NPs (nominal diameter 150 nm) on oxidized silicon wafers inside SEM. Manipulation set-up consisted of a nanopositioner equipped with a self-made QTF-based force sensor.

Au particles used in the experiments were, in general, not spherical but had facets of different geometries (Fig. 5a). Therefore, the contact area between the particle and the surface also varied from particle to particle. Contact areas and static friction values were estimated analytically considering common geometries of the particles used in experiment [6] (Table 1). It should be noted that real particles often exhibit truncated edges and apexes. Moreover, NPs were annealed at 500 °C to remove surfactant. Thermal treatment may lead to rounding of particles [25]. Therefore, the contact areas and static frictions should be generally lower than the maximal expected values calculated for perfect geometries and perfectly smooth surfaces (Table 1).

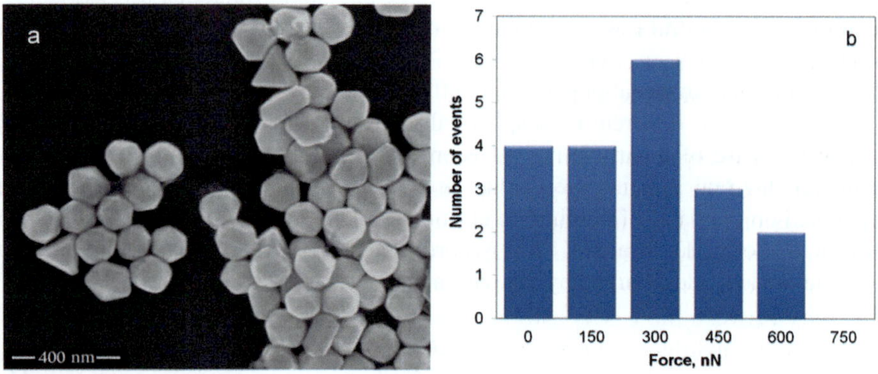

Fig. 5 SEM micrograph of Au NPs (**a**); Distribution of the static friction force for Au NPs (**b**)

Table 1 Estimated static friction forces for 150 nm Au particles of different geometries

Shape	Contact area, nm^2	Static friction, nN
Tetrahedral	9743	2768
Decahedral	3652	1038
Icosahedral	2693	765

In the manipulation experiment the particles were firstly displaced ("kicked") by an abrupt tip motion to a fresh site in order to reduce the initial adhesion. Initial displacement was followed by controlled manipulation of the particle by pushing it with oscillating tip in the smooth motion of the manipulator (scan mode) with simultaneous force recording. During manipulation, the tip moved parallel to the surface along a straight line without feedback loop.

A significant force was required to overcome the static friction and displace an Au NP. After the displacement, NPs moved easily at a very low applied force. This finding is in agreement with other studies that have demonstrated that the kinetic friction vanishes for clean surfaces in vacuum [44]. Nonstop motion of NPs during the manipulation was an essential condition for staying in the vanishing friction regime; rested NPs had a tendency to adhere strongly to the substrate. Static friction was measured for 19 NPs and ranged from 50 to 750 nN (Fig. 5b).

It should be noted that for some particles the oscillation amplitude dropped to zero during manipulation, which corresponded to a force higher than 1500–2500 nN (depending on the particular QTF force sensor). Variations in the experimental values of static friction were explained by the different contact areas between a particle and a substrate due to different shapes of the NPs.

3.3 Manipulation of Nanoballs and Nanodumbbells

Similar manipulation technique was used for manipulation of silver nanoballs and dumbbells produced by pulsed laser-induced partial melting of 100 nm diameter silver NWs deposited on an oxidized silicon wafer [45, 46]. Due to concentration of electric field on a NW ends, silver melts under laser irradiation and governed by the surface tension, rounded structures are formed on both ends of a NW. Shorter NWs were completely melted into "nanoballs" (Fig. 6).

Friction of nanoballs and nanodumbbells was measured with a QTF-based force sensor operated in shear mode. Nanoballs were first displaced by abrupt motion of the tip and then static friction was measured in one-line slow scan.

The chosen frames and typical manipulation force curve are shown in Fig. 7. The force needed to overcome the static friction ($F_{st} = 162$ nN) and displace the nanoball (region b, c in Fig. 7) is significantly higher than the force needed to maintain the motion (region c–e in Fig. 7). Kinetic friction force in this particular case spontaneously decreased from $F_{kin} = 26$–7 nN (Fig. 7), which can be attributed to switching between sliding and rolling types of motion. Such behavior was not typical and was observed only in few cases.

Assuming the following parameters for silicon and silver: $E_1 = 71.7$ GPa, $\nu_1 = 0.17$, $E_2 = 82.5$ GPa, $\nu_2 = 0.36$, $\gamma = 50$ mJ/m^2 [36, 47], $R = 200$ nm and $z_0 = 0.3$ nm, we obtained $\eta \approx 0.22$. For such η, the DMT-M theory is more appropriate. According to the DMT-M model, the contact area for 400 nm spherical Ag NPs is $A_{\text{DMT-M}} \approx 117$ nm^2. Then for Z in the range from 5 to 30 [40] calculated ultimate static friction $F_{\text{friction}} = \tau S_r$ ranging from 34 to 205 nN. Experimentally measured static friction force having value 65 nN agrees well with the calculated range of friction force.

The same manipulation technique was then applied to Ag nanodumbells. Nanodumbbells of 1–2 μm in length can be roughly considered as two spheroid NPs connected by a NW. It should be noted, that rounded ends of dumbbells often had considerable deviations from ideal spherical shape related to contact angle of liquid metal droplet with substrate. Therefore, DMT-M model for a sphere was not valid

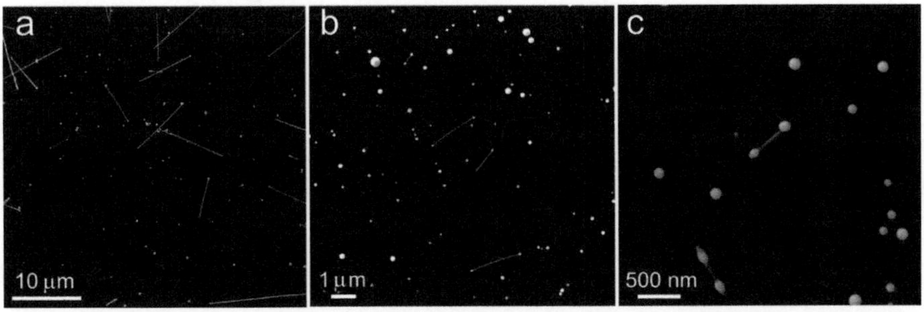

Fig. 6 Nanoballs and dumbbels nanostructures produced by laser processing of Ag NWs (**a**) and Au NWs (**b**, **c**)

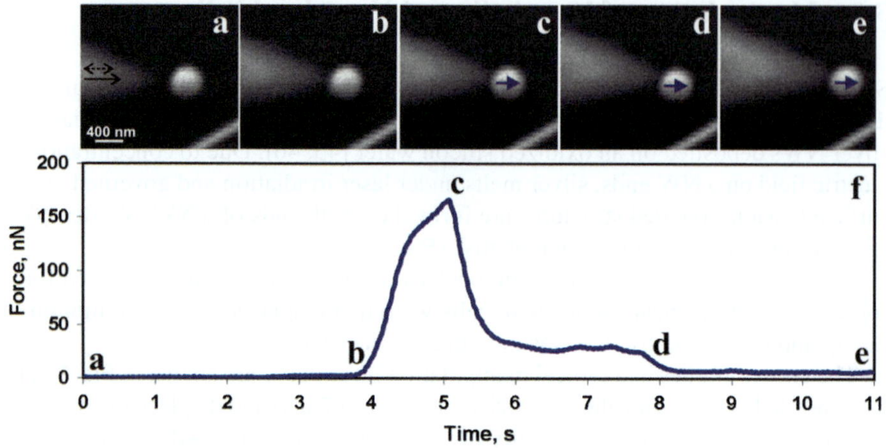

Fig. 7 Manipulation of a silver nanoball. The tip approaches the nanoball (**a**, **b**). Nanoball is displaced (**c**) and continuously moved by the tip (**c**–**e**). Corresponding tip-nanoball force interaction data recorded by QTF sensor (**f**). Solid black arrow indicates tip motion, while dashed line indicates tip oscillation direction. Direction of NP movement indicated by short arrows

anymore. Nevertheless, dumbbells still ensured a relatively small contact area with the substrate and therefore reduced adhesion and static friction compared to that of an untreated pentagonal Ag NWs. Displacement of Ag NWs on a flat silicon substrate is hardly possible without severe damage and plastic deformation of NWs. Moreover, for dumbbell-type structures both rolling and sliding motions are possible, which can be distinguished visually when manipulated inside SEM. It makes them attractive objects for nanotribological studies by manipulations. Still, in some cases static friction of the dumbbell was high enough to keep one end fixed during manipulation which led to plastic deformation of the structure.

A typical dumbbell manipulation experiment alongside a corresponding force graph is shown in Fig. 8. Significant force ($F_{st} = 1050$ nN) was required to overcome static friction and displace the dumbbell. In the demonstrated case, the dumbbell first rolled over (Fig. 8a, b), and then rotated around one end (Fig. 8c, d). In some cases, one end completely lost contact with the substrate and dumbbell rotated around adhered end out of substrate plane. Frequently, rolling of the dumbbell up to approx. 90° around one end was observed.

As was demonstrated above, SEM-assisted visual guiding is especially useful for manipulation of elongated objects by probe-equipped force sensor. This leads us to the next section dedicated to manipulation of individual NWs.

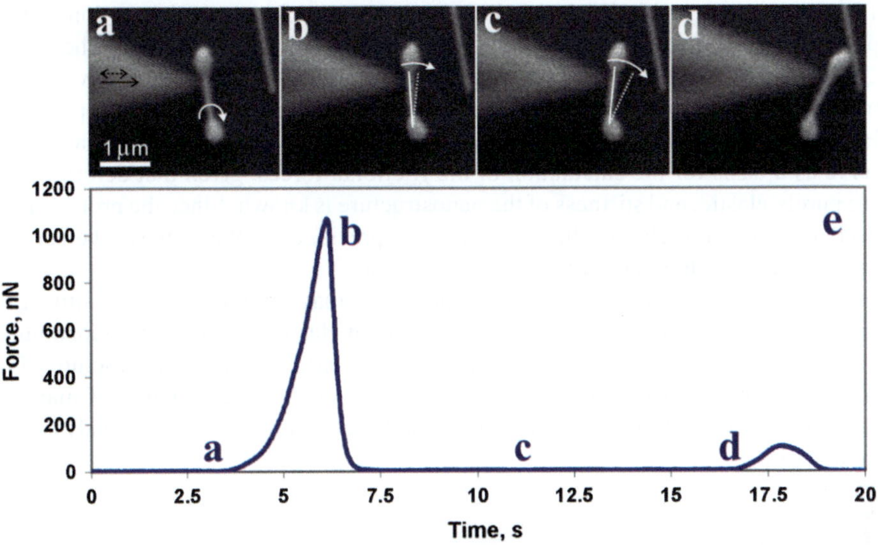

Fig. 8 Manipulation of a silver nanodumbbell. The dumbbell rolls over approx. 90° (**a**, **b**), then rotates around one of its "balls" (bottom ball, **b**–**d**), and finally runs into NW (**d**). Arrows indicate the motion type. Corresponding tip-dumbbell force interaction data recorded by a QTF sensor (**e**)

4 Manipulation of 1D Nanostructures Without Direct Force Measurement

There are a number of experimental methods and theoretical models for investigation and description of mechanical properties of one-dimensional (1D) nanostructures such as NWs and nanotubes (NTs). Nevertheless, only a few works have focused on measuring the static or kinetic friction of NWs (or NTs) on a flat substrate. Most of the friction measurements on NTs and NWs were performed using a conventional AFM at ambient conditions (or even under optical microscope in case of larger structures [48]). Hsu et al. [49] manipulated BN NW on a silicon substrate and simultaneously measured the lateral force. Falvo et al. [50] studied sliding and the rotation of multiwall carbon NTs on HOPG using lateral force AFM microscopy. It is necessary to emphasize that visual guidance for NW manipulation experiments is even more critical than for NPs manipulation. Only a short or very stiff NW behaves as a rigid rod during manipulation (slides or rotates as a rigid object without bending). High aspect ratio 1D nanostructures can be bent significantly or even broken prior displacement during manipulation if static or kinetic friction force exceeds bending strength of the structure [51].

In situ manipulations inside SEM can reveal additional information such as evolution of the bending profile of 1D nanostructure in contrast to AFM-based manipulations in which only the initial and final positions of the manipulated object are obtained. On the other hand, direct measurement of the forces needed to displace the

1D nanostructure inside SEM is challenging. Absence of the optical detection system prevents straightforward detection of lateral displacement of the probe. The use of QTF-based force sensor is inefficient as significantly higher forces are involved in the manipulation of 1D nanostructures compared to manipulation of smaller particles. However, the evolution of the 1D nanostructure profile during the manipulation can serve as a basis for the calculation of the frictional forces, given that deformations are purely elastic, and stiffness of the nanostructure is known. Since the probe is now used only for manipulation, there is no need for precise control over the probe-surface interaction regarding the forces and contact areas.

In the following section we will demonstrate that combination of in situ SEM visual observation and mathematical modelling of elastic bending during manipulation can be a powerful tool for tribological investigations of 1D nanostructures. For simplicity, we will focus on manipulation of NWs only. Interpretation of manipulation results for NTs is more challenging and requires more complex analysis and modelling.

4.1 Elastic Beam Theory Employed for Tribomechanical Studies of Nanowires

Theoretical methods for investigation of mechanical and tribological properties of NWs commonly employ the framework of classical physics and range from simplest Newton's mass point models [52] to continuum models and finite element method (FEM) calculations [53]. Due to the relatively large sizes of the systems (which may account in millions of atoms), application of molecular dynamics is limited to NW with dimensions of several nm. Therefore, the continuum approach is attractive for modelling of NWs because of extensively developed applicable analytical and numerical methods with different levels of accuracy.

One of the most typical deformations of NW is elastic bending. Using the assumption of pure elastic bending, it is possible to investigate many important properties of NW exposed to the friction and other forces occurring during manipulations. The advantage of this assumption is its relative simplicity, which we will demonstrate throughout the current section.

The most known model based on the assumption of pure elastic bending is Timoshenko elastic beam theory (EBT) [54]. It combines the equilibrium equations of a continuum body with the laws of linear elasticity approximated for one-dimensional structure. The bending profile of a NW comes as a result of the interplay between elastic, friction, and other external forces distributed along the NW with certain densities. The EBT can assist in measuring NW-substrate friction.

The present section will include description of several methods of theoretical approach to NW-on-substrate friction during and after manipulations alongside with some experimental examples [55–57]. These methods enable to gather information on NW-substrate distributed static and kinetic friction as functions along NW's length, ultimate static friction, as well as flexural strength at the breaking point. Additionally, the case of a NW half-suspended on a trench will be considered that enables measuring Young's modulus for individual NWs and can be used as an alternative way to measure the ultimate static friction.

Elastic Forces

Let us consider a prismatic-shaped NW of length L and diameter D, bent under external in-plane forces (Fig. 9). For the implementation of the EBT we will need to introduce variables of the force and momentum of elastic stresses in a cross-section of the NW. They represent the force and momentum between the elastically deformed sides of the NW at an imaginary cut and will be designated as **F** and **M**, respectively. Their components can be written as integrals over the cross-section area S at any given point l of NW axis [54, 58]:

$$F_i = \int_S \sigma_{i\gamma} n_\gamma dS$$

$$\mathrm{M_i} = \int \varepsilon_{i\alpha\beta} \mathrm{r}_\alpha \sigma_{\beta\gamma} \mathrm{dn}_\gamma \tag{3}$$

$$M_i = \int_S e_{i\alpha\beta} r_\alpha \sigma_{\beta\gamma} n_\gamma dS \tag{4}$$

where $\sigma_{\alpha\beta}$ are the components of stress tensor, n_γ are the components of the normal vector to the element of cross-section area dS, r_α are the components of the radius vector from the axial point l and $e_{\alpha\beta\gamma}$ represents the unit anti-symmetric tensor. Both the elastic force **F** and the momentum **M** are considered as functions of the coordinate l along the NW axis (Fig. 9).

Here and below, we assume that the NW is undergoing a pure bending deformation. This can be a good approximation for the given problem provided the external forces are close to the normal direction of the NW axis and lie on the same plane as

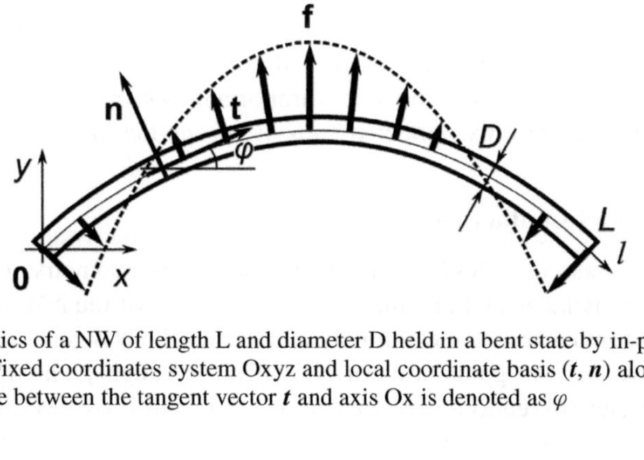

Fig. 9 Schematics of a NW of length L and diameter D held in a bent state by in-plane distributed lateral force ***f***. Fixed coordinates system Oxyz and local coordinate basis (***t***, ***n***) along the NW axis l are used. Angle between the tangent vector ***t*** and axis Ox is denoted as φ

the NW itself. It assures the negligibly small tensile, compressive, and twisting deformations. The case of the pure bending of prismatic shaped NW yields the following equation for momentum in the framework of linear isotropic elasticity [58]:

$$\mathrm{M} = \mathrm{EIt} \times \frac{\mathrm{dt}}{\mathrm{dl}}\mathbf{M} = EI\mathbf{t} \times \frac{d\mathbf{t}}{dl}, \tag{5}$$

where E is the Young's modulus of the NW, I is the area moment of inertia of the NW, and $\mathbf{t}$ is the tangent vector of the NW axis. Equation (5) describes the momentum of elastic forces inside a NW bent purely with the given curvature. It can be projected to the Oz axis and rewritten as

$$\mathrm{M} = \mathrm{EI}\frac{\mathrm{d}\varphi}{\mathrm{dl}} = \mathrm{EI}\kappa M = EI\frac{d\varphi}{dl} = EI\kappa, \tag{6}$$

where $\varphi(l)$ is the tangent angle function over the length of the NW and $\kappa(l) = 1/R(l)\kappa(\mathrm{l}) = \frac{\mathrm{d}\varphi}{\mathrm{dl}} = \frac{1}{\mathrm{R}} = d\varphi/dl$ is the curvature function related to the local radius of curvature $R(l)$R. Physical meaning of Eqs. (5) and (6) for pure bending deformation is of similar kind as Hooke's law for an elastic spring. In fact, this is the "Hooke's law" written for the specific case, where κ plays a role of "displacement", M stands for "elastic force" of restoration, and EI is a "spring constant" (which is sometimes referred to as *flexural rigidity*).

By following this analogy, we can obtain the expressions for elastic energy per unit length u^{st}, and the total elastic energy U^{st} of the bent NW can be calculated, respectively, as

$$u^{\mathrm{st}} = \frac{EI}{2}\kappa^2 \mathrm{M_i} = \int \varepsilon_{\mathrm{i}\alpha\beta}\mathrm{r}_\alpha\sigma_{\beta\gamma}\mathrm{dn}_\gamma \tag{7}$$

$$U^{\mathrm{st}} = \int_0^L \frac{EI}{2}\kappa^2 dl \tag{8}$$

It must be noted that single crystalline solids possess anisotropic elastic properties and therefore should be described with multiple elastic moduli. However, due to simplicity, the isotropic model is often used. Although this approximation of isotropic body is acceptable in the most cases, it is important to keep in mind that isotropic elastic moduli describe only "effective" elastic properties in the specific axes of deformation.

Cross-Section and Contact Area

The elastic response of a NW is directly related to its geometry (cross-section), particularly the diameter D. In its turn, the cross-section of the NW determines the NW-substrate contact area that essentially determines the NW-substrate friction.

Typical cross-section geometries of NWs can be roughly classified as polygonal and round cross-sections. For example, hexagonal cross-section yields $I =$

$5\sqrt{3}D^4/256$. Contact regions for polygonal shapes are easily defined. Determination of contact region for circular cross-section is a non-trivial problem that requires employment of contact mechanics.

It should be noted that from the expression for *I*, which scales as power 4 of the NW lateral dimensions, it is evident that an accurate measurement of NW cross-section geometry is crucial for correct application of EBT. However, accurate measurement even of the diameter is not a trivial task, not to mention the measurements of the cross-section distortions. In literature, for simplicity reasons a round cross-section is often assumed instead of the real polygonal one. Moreover, possible distortions of the NW cross-section are commonly not considered. This should be kept in mind when interpreting the manipulation results for NWs.

Friction and Other External Forces

External forces acting on a NW may have different origins: friction, adhesion, actuator load etc. In the case of pure bending of a NW on a flat substrate the lateral forces will play the major role, whereas normal forces can be neglected.

The forces can be distributed or concentrated. A distributed force can be defined as a continuous function $\mathbf{f}(l)$ along the NW. Then $\mathbf{f}(l)dl$ corresponds to the force exerted on an infinitesimal element dl in the point l of the NW. Friction forces are a typical example of the distributed external forces, which may vary from point to point. The concentrated force $\mathbf{f}_0$ can be expressed in the same manner via Dirac's delta-function as $\mathbf{f}(l) = \mathbf{f}_0\delta(l\text{-}l_0)$. Concentrated forces in NWs commonly occur as a result of a load from an external actuator (e.g. manipulator's tip).

Boundary Conditions

Boundary conditions for the elastic momentum and force in the NW depend on the specific situation. For a free end of the NW, the conditions $\mathbf{F} = 0$ and $\mathbf{M} = 0$ must be satisfied. In a typical case if the end of NW is pushed by an actuator, then the boundary value of $\mathbf{F}$ is non-zero and is defined, while $\mathbf{M} = 0$ as there is no tip-induced momentum.

Equilibrium of the Nanowire

Let us now consider a NW in mechanical equilibrium due to the interplay of intrinsic elastic force in the NW and the external forces of various origins. The full system of equilibrium equations for the NW involves conditions for the elastic force $\mathbf{F}$ and momentum $\mathbf{M}$ [58]:

$$\frac{d\mathbf{F}}{dl} = -\mathbf{f}, \tag{9}$$

$$\frac{d\mathbf{M}}{dl} = -\mathbf{t} \times \mathbf{F} \tag{10}$$

where $\mathbf{f}$ is the distributed external force per unit length. Equation (9) can be interpreted as the condition of zero total force $d\mathbf{F} + \mathbf{f}dl = 0$ acting on a prismatic element dl of the

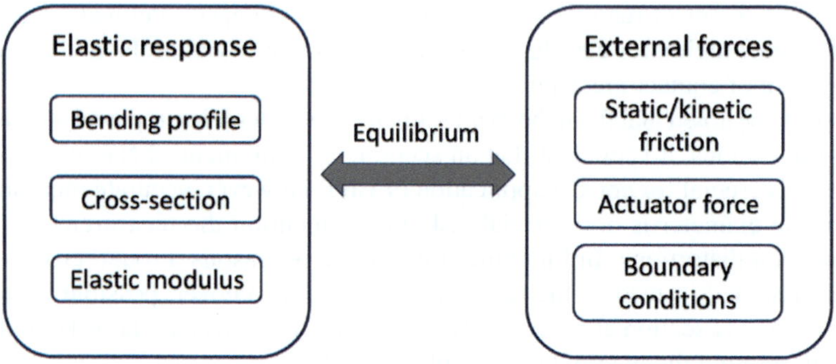

Fig. 10 Logical scheme describing the intrinsic and external factors for elastically bent NW to be in mechanical equilibrium

NW. The other equation reflects the condition of zero total momentum, respectively. **M** always has the same direction *Oz* (perpendicular to the bending plane) and the right-hand side of the Eq. (10) has magnitude of the normal component of the force $F_n = \mathbf{F} \cdot \mathbf{n}$. Therefore, it can be rewritten as $dM + F_n dl = 0$, which had to be demonstrated.

General view on the flexural mechanics of NW can be summarized in a form of diagram (Fig. 10), disclosing the internal factors of elastic response and external forces of various nature being in mutual equilibrium. In other words, the bending profile of resting, being displaced, or uniformly moving NW, can be calculated from condition of balance of NW's elastic response (if Young's modulus and cross-section are known) and external forces (friction forces and applied actuator force). Respectively, an inverse problem—determination of friction force—can be solved, if bending profile and elastic response of NW are known. It should be noted that even the kinetic friction can be analyzed with the equilibrium equations providing the NW is moving uniformly.

4.2 *Nanowire Loaded at One End*

In this subsubsection we will discuss tribological aspects of the configuration when part of the NW is fixed on the substrate by adhesion/static friction, while the other part is either suspended or can slide along the substrate.

Model

Let us consider a specific situation, when initially straight NW is loaded at the end and a part of the NW is fixed to the substrate by static friction (see Fig. 11). The mobile part of the NW is subject to kinetic friction and the concentrated load force. The distributed force **f** consists of the kinetic friction force $\mathbf{q}^{\text{kin}}$ per unit length distributed uniformly along the moving part of the NW:

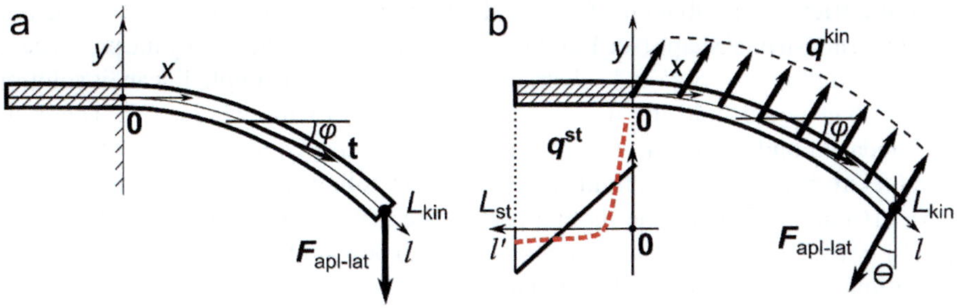

Fig. 11 Schematics of a NW loaded by concentrated force $F_{\text{apl-lat}}$ at the end in point L_{kin}. NW is half-suspended and friction-free at the protruding part (**a**). NW is affected by distributed kinetic friction force q^{kin} on the interval (0, L_{kin}) along l axis (**b**). The left end is fixed by strong static friction force q^{st} on the interval (0, L_{st}) along l' axis. The linear-fit static friction q^{st} (solid line) can provide a lower-bound estimate for actual friction force (dashed line). The angle between the tangent vector and axis Ox is denoted as φ and $F_{\text{apl-lat}}$ is directed by angle θ to Oy axis

$$\mathbf{f} = \mathbf{q}^{\text{kin}}. \tag{11}$$

Zero elastic force and momentum at the NW's free end dictate the boundary conditions:

$$\mathbf{F}|_{l=L_{\text{kin}}} = \mathbf{F}_{\text{apl-lat}} \tag{12}$$

$$\mathbf{M}|_{l=L_{\text{kin}}} = 0. \tag{13}$$

And elastic force **F** according to Eq. (9) yields together with Eq. (11):

$$\mathbf{F} = -\mathbf{q}^{\text{kin}} \cdot (l - L_{\text{kin}}) + \mathbf{F}_{\text{apl-lat}}. \tag{14}$$

Finally, the equation for $\varphi(l)$ describing the profile of NW with boundary conditions:

$$EI\frac{d^2\phi}{dl^2} = \{q^{\text{kin}}(l - L_{\text{kin}}) - F_{\text{apl-lat}}\} \cdot \cos(\phi + \theta) \tag{15}$$

$$\varphi|_{l=0} = 0 \tag{16}$$

$$\left.\frac{d\varphi}{dl}\right|_{l=L_{\text{kin}}} = 0, \tag{17}$$

where Eq. (16) defines orientation of the NW with respect to the coordinate system, Eq. (17) follows directly from Eqs. (13) and (12) is automatically satisfied.

Static friction distribution q_y^{st} along the motionless ("static") part L_{st} of the real NW has a non-trivial shape (see Fig. 11b) which can be complicated by the presence of surface roughness of NW and substrate. Let us compose a simple linear-fit solution for q_y^{st} that would hold the mechanical equilibrium of the whole NW. There are 2 boundary conditions that govern that: elastic force $F_y|_{l=0}$ and momentum $M|_{l=0}$ at the start point of the motionless region must follow from the solution of Eqs. (15), (16) and (17) from the mobile part of the NW. Consider static friction force $q_y^{st}(l') = a \cdot l' + b$, where a and b are constants to be found with the named boundary conditions, which can be explicitly put as:

$$F_y\big|_{l=0} = \int_0^{L_{st}} q_y^{st}(l')dl' \tag{18}$$

$$M\,|_{l=0} = \int_0^{L_{st}} q_y^{st}(l')l'dl' \tag{19}$$

This allows to find such parameters as maximal static friction q_{max}^{st} and averaged static friction q_{avg}^{st} after $q_y^{st}(l')$ is revealed. The results obtained from this consideration can be used as a lower bound estimate of static friction as discussed below.

In the particular case, when the kinetic friction is absent or negligible $\mathbf{q}^{kin} = 0$, we have situation of half-suspended NW loaded at the protruding end. This case is sometimes referred to as "clamped" or "cantilevered beam" in textbooks.

Young's Modulus Measurements

As it was previously demonstrated, Young's modulus is an essential parameter in the framework of EBT. Typically, bulk values or values found in literature are used in tribological studies of NWs. However, considering high scattering of reported mechanical properties of nanostructures [59] and the fact that they may depend on a quality of material and synthesis conditions, it is more accurate to measure Young's modulus for a particular set of NWs or even for each single NW separately.

Polyakov et al. [55] and Vlassov et al. [59] measured elastic properties and ultimate strength of ZnO and CuO NWs half-suspended on the edge of a trench in a patterned silicon wafer as shown in Fig. 12. The cantilevered beam bending technique [60] was then applied to determine Young's modulus and bending strength. The fixed part of the NW stayed motionless due to high adhesion. The suspended part of the NW was pushed by the tip of the QTF-based force sensor. Sensor signal and the grabbed SEM images were recorded simultaneously during the experiment. To find the value of Young's modulus, the SEM image of the NW profile was numerically fitted to the curve given by Eq. (15). A typical force curve and the corresponding SEM images are presented in Fig. 12. Integral static friction force in adhered part was usually higher than applied force and no displacement of NW was observed besides few cases for very short adhered part. Often, internal stress inside NW generated by applied force exceeded ultimate strength of NW and led to fracture of NW (Fig. 13). Measurements carried out on 9 CuO NWs and 14 ZnO NWs resulted in an average

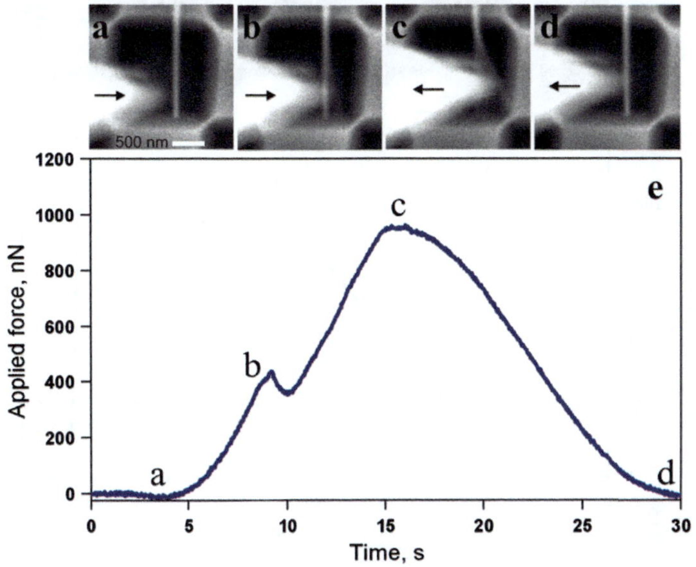

Fig. 12 Young's modulus measurement of a CuO NW. Arrows indicate direction of the tip movement. Sequence of SEM images of CuO NW bending (**a**–**d**) and the corresponding force curve (**e**)

Young's moduli of 204 and 58 GPa respecively. The average bending strengths of CuO and ZnO NWs were found to be 8.2 and 4.4 GPa.

Static Friction Measurements

If elastic properties of NW are known the NW itself can serve as a force sensor for measurements of friction forces given that manipulation is visually guided.

In the method introduced by Polyakov et al. [56] the NW-substrate static friction can be calculated by fitting of an experimentally found bending profile to the theoretical profile of a NW lying on a flat substrate and pushed from one end, while the second end is fixed by static friction. The adhered end of the NW stays motionless as long as the applied external force $\mathbf{F}_{\text{apl-lat}}$ does not exceed the total static friction force. When the elastic force overcomes the total static friction force, the whole NW is displaced. The NW bending profile (or the profile in the "most bent state") just before complete displacement of the NW is then used for fitting to find the maximum static friction and average static friction, which determine the total static friction force.

The described method was applied to study the static friction of ZnO NWs on oxidized silicon wafer. Figure 14 represents a typical manipulation experiment. The NW "most bent state" profile of the whole NW was calculated according to Eq. (15) and laid over the SEM image (Fig. 14c). From the elastic deformation of the NW the generated elastic force F and momentum M were calculated, enabling to find counteracting friction forces in the adhered part.

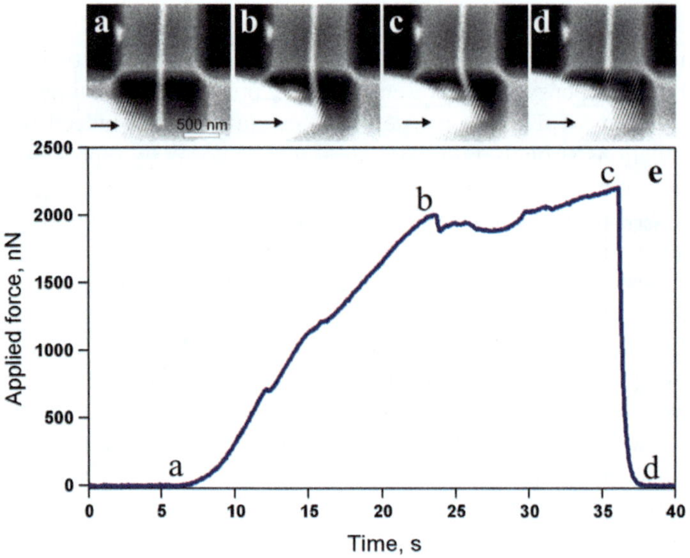

Fig. 13 Bending strength measurement of a CuO NW. Arrows indicate the direction of tip movement. Sequence of SEM images of bent and broken NW (**a**–**d**) and the corresponding force curve (**e**)

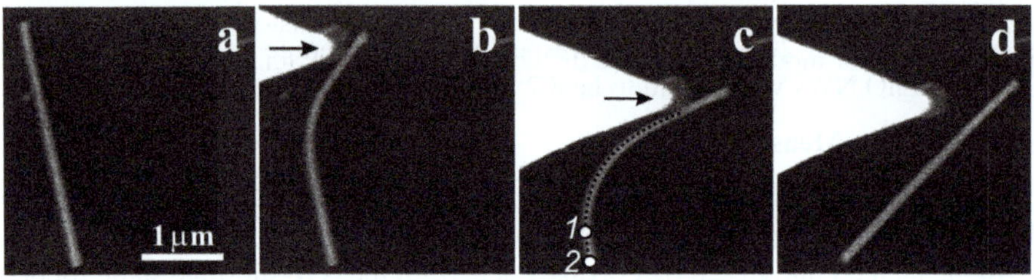

Fig. 14 Series of SEM images of the ZnO NW bent by AFM tip from one end while other end is adhered to the substrate by static force. Intact NW (**a**); partially displaced NW (**b**); "most bent state" of the NW with fitting profile lied over the NW image (**c**); completely displaced NW (**d**)

The median value of the averaged static friction was $q^{st}_{avg} \sim 5$ nN/nm for 39 measured ZnO NWs. The corresponding value of averaged static shear stress was $\sigma^{st}_{avg} \sim 67$ MPa. Those values can be referred to as the lower bound estimation of static friction (see Fig. 11b). Many NWs were bent up to 90° before displacement, but some NWs broke during bending. Fractured NWs did not necessarily break in the maximal tensile strain region but in any region with a structural defect. The median value of tensile stress in successfully displaced NWs was $\sigma_{\text{tens-displ}} \sim 2.6$ GPa (for a set of 32 NWs), while in broken NWs, $\sigma_{\text{tens-fract}} \sim 3.3$ GPa (for a set of 21 NWs).

More complex tribomechanical characterization was reported by Vlassov et al. [51]. In their work, special trenches were etched in Si wafer to serve as a substrate

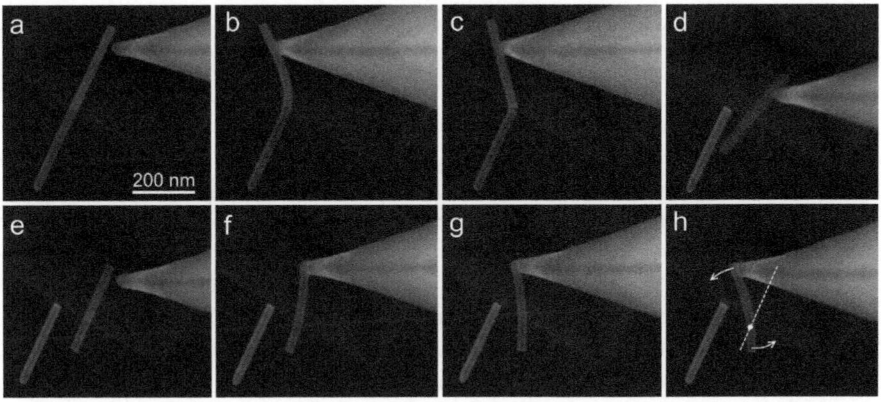

Fig. 15 Chosen frames from complex manipulation experiment consisted in breaking the suspended part of ZnO NW and positioning it to a fresh location for tribological test [51].

for testing. ZnO NWs were transferred onto the Si substrate so that some NWs were half-suspended over the trenches. Then, Young's modulus of the NWs, required for the subsequent friction calculation, was separately measured in AFM three-point bending tests. Next, bending tests were performed inside SEM using AFM probe without direct force detection. For the majority of NWs, the displacement from initial locations by bending was impossible due to the contact ageing. To perform tribological tests and assure fresh contacts, the 'pick-and-place' approach was applied by breaking the suspended part and transferring it to a new location followed by immediate friction measurements in a bending test (Fig. 15). For freshly transferred parts, friction was high enough to detect bending of NW prior and even during the displacement (Fig. 15) allowing to calculate friction forces based on the bending curvature.

Let's now shortly discuss frictional aspects of the pick-and-place manipulation approach. As can be seen from Fig. 15c, the broken part of the NW does not have contact with the substrate and is barely touching the adhered part of the NW. As a result, the broken part stays attached to the probe by the adhesion. Next, when detached part of the NW is placed half-suspended to the new location on the substrate, the contact area between freshly adhered part significantly exceeds the small contact area between the NW and the AFM probe. Therefore, NW is released from the AFM probe and stays on the substrate. Note that partially suspended NW configuration simplifies the positioning of the tip and helps to avoid the tip damage by avoiding the tip-sample interaction.

Force calculations were performed by simulating every experiment individually using Finite Element Method (FEM). Geometrical parameters and stiffness were taken from experimental data. Each NW was modelled as a prismatic elastic beam with a hexagonal cross-section. Part of the NW was rigidly fixed to simulate the contact between the NW and the substrate in a static condition when NW is loaded from one end, but the adhered part is still fixed allowing to calculate interfacial shear stresses. The maximal interfacial shear strength for most bent state (corresponding to curvature of NW profile right before the displacement of the adhered part in experiment) can serve as a measure of static friction. Shear stress values obtained with FEM model were in order of 1 GPa, which is significantly higher than values obtained with analytical linear-fit model described earlier in this subsection.

Disagreement between models reflects the complexity of the static friction problem. It is important to understand that both models describe idealistic situations with perfectly smooth flat surfaces and infinitely rigid contact with contact area defined purely geometrically. In real conditions, both substrate and NW have surface defects and surface impurities, making real contact area less uniform and smaller (the effect of substrate roughness on static friction is described in next section). Moreover, contact regions can undergo elastic deformation causing redistribution of interfacial stresses. In contrast to the smooth linear distribution $q_y^{st}(l\prime)$ shown in Fig. 11b and assumed in the analytical model, in FEM simulations stresses are concentrated mostly near the interface between adhered and free part of the NW (Fig. 16). It is obvious that for the relatively long adhered part, FEM model will give much higher local shear stress values compared to the analytical model. For longer adhered parts the analytical model can serve as a lower bound estimation, because it will strongly underestimate maximal interfacial shear strength. At the same time, FEM model can considerably overestimate interfacial stress values due to abrupt transition between adhered and free part of the NW.

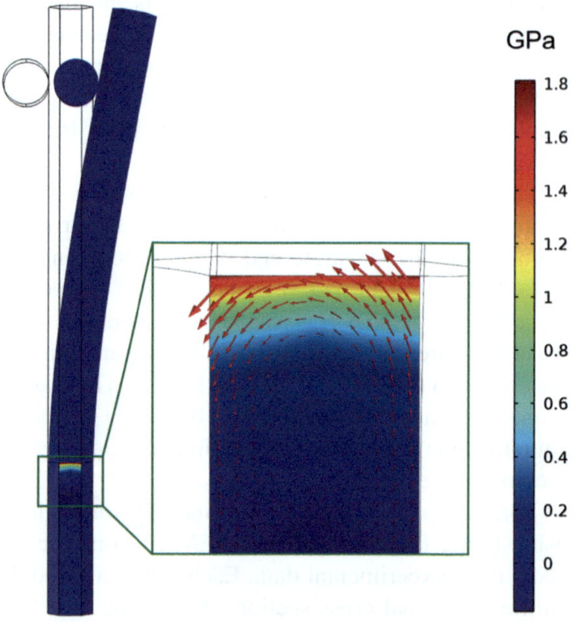

Fig. 16 FEM model of interfacial stress distribution in elastically bent ZnO NW contacting the Si substrate. The deflection corresponds to the most bent state before displacement in the real test [51]. © IOP Publishing. Reproduced with permission. All rights reserved

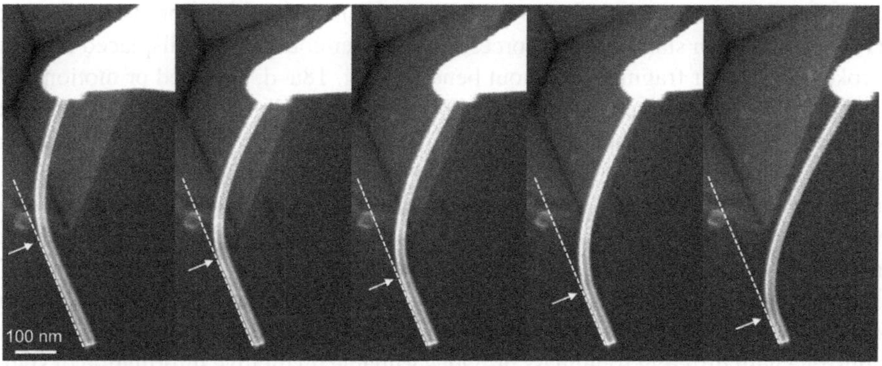

Fig. 17 Stepwise displacement of adhered ZnO NW. Arrows indicate the point that separates the moving part of the NW from the part rigidly adhered to the substrate [51].

High values obtained with FEM simulations agree well with experimental observations. First, curvatures of bent NWs in strength and static friction tests are comparable. It indicates that breaking the NW and breaking the NW/substrate contact are two competitive processes and should have comparable stress values. Second, static friction does not directly correlate with the length of the adhered part [51] as was shown in experiment where a stepwise displacement of the NW with relatively long initial adhered part was observed in bending test (Fig. 17). When the interfacial shear strength was exceeded, some fraction of the adhered part was abruptly displaced, while NW was still held on the substrate by the remaining part. The scenario was repeated five times until also the last adhered section was displaced. Every displacement occurred at a similar curvature of the NW. According to the FEM simulation, maximal interfacial shear strength values in each step were 5.7/6.6/6.9/5.7/5.1 GPa.

Effect of a Substrate Roughness on Static Friction

It is known, that for typical surfaces, macroscopic dry friction is usually nearly independent of the surface roughness, unless the surfaces are either very rough or very smooth [61]. However, in nanoscopic systems, roughness of the contacting surfaces is directly associated to the actual contact area and the average distance between them, determining magnitude of van der Waals (vdW) force, which plays the same role as external loading force or gravity at the macroscale. The effect of nanoscale roughness on adhesion for MEMS has been addressed in many works [62, 63].

Polyakov et al. [64] applied previously described "NW loaded at one end" approach for investigation of the effect of surface roughness on the CuO NW-substrate static friction. Four substrates of roughness R_{RMS} ranging from 0.7 to 16.1 nm were prepared by chemical etching of amorphous silicon. CuO NWs were manipulated according to the scheme shown at Fig. 14.

Four distinct scenarios of the NW behavior during manipulation were observed (Fig. 18). At high static friction forces, NWs either could not be displaced at all or broke into shorter fragments without bending (Fig. 18a–d, "crashed or motionless" in Fig. 18q). At lower static friction, NWs were gradually displaced from one end of the NW with the other end fixed by the static friction (Fig. 18i–l, "bent and displaced" in Fig. 18q). If the static friction force exceeded the ultimate strength of the NW, the NW broke (Fig. 18e–h, "bent and broken" in Fig. 18q). If static friction force was smaller than the NW interfacial shear strength, the NW was displaced as a rigid object (Fig. 18a–d, "zero friction"). Only in the "bent and displaced" scenario the "most bent state" of the NW could be registered and static friction can be analyzed quantitatively. Nevertheless, visual observation of NW behavior on substrate with different roughness provides valuable qualitative information of static friction. Frequency of occurrence of named scenarios for substrates with different roughness is summarized on the graph in Fig. 18q.

From Fig. 18 it is evident that nanoscale roughness has drastic impact on static friction for nanostructures. It can be explained by considering that the smaller the roughness of the surfaces, the "closer" the two surfaces are. It leads to higher vdW force, and consequently, higher adhesion and static friction. Numerical values of the static friction forces calculated for the cases when the NW was bent and displaced using its "most bent state" profile gives an approximately order of magnitude difference between smooth and most rough substrates. Data on averaged static friction q_{avg}^{st} for all substrates are summarized in Table 2. It is necessary to note that used statistics does not account the cases of "zero static friction" (friction below the sensitivity of the method) and motionless NWs (friction beyond the applicability of the method), which means that actual difference should be even higher. Even for the cases where calculation of friction was possible, the difference of friction force between smoothest and most rough substrate is one order of magnitude.

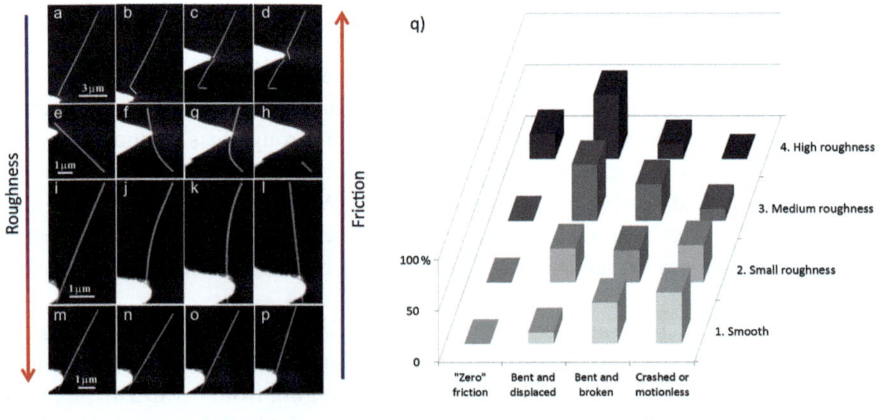

Fig. 18 Behavior of CuO NWs during manipulation on substrates of different roughness. Chosen frames from manipulation videos (**a**–**p**). Statistics for occurrence of different scenarios depending on the surface roughness (**a**)

Table 2 Effect of substrate roughness on NW-substrate static friction

Substrate Nr	R_{RMS}, nm	Aver. static friction, nN/nm
1. Smooth	0.7	>5.8
2. Small roughness	2.1	3.9
3. Medium roughness	5.4	1.4
4. High roughness	16.1	<0.6

4.3 Nanowire Pushed in the Midpoint: Kinetic Friction

At the moment when the nanomanipulator's tip starts to push the NW in the middle point, the static friction force q_{st} is distributed over a portion of the NW of unknown length, which may be smaller than the entire NW length L (Fig. 19a). Only a fraction l' of the NW where the applied force exceeds the ultimate value of the distributed static friction force (corresponding to the maximum shear strength) is displaced by the tip (Fig. 19b). Static friction acts on the intact part of the NW of the length $L - l'$, while kinetic friction affects the displaced portion l' of the NW. For a partially displaced NW, the effective friction force is the combination of the static q_{st} and kinetic q_{kin} friction forces. It is difficult to measure the length of the moving part of the NW and to analyze the contribution of the static and kinetic friction forces during the initial stage of NW displacement. Upon further displacement of the NW (uniform translation), we consider the friction force as purely kinetic and uniformly distributed along the length of the NW (Fig. 19c).

Model

If a NW is being uniformly dragged at its midpoint without rolling and all parts of the NW have the same constant velocity, the equilibrium equations Eqs. (13) and (14) are still applicable due to the principle of relativity.

In this case, the profile of the deformed NW is determined by the balance of the external driving tip force, the kinetic friction between the NW and the substrate, and the intrinsic elastic forces of the NW. The distributed driving force $\mathbf{F}_{\text{apl-lat}}$ can be modelled via the delta function, and the kinetic friction $\mathbf{q}^{\mathrm{kin}}$ maintains a constant vector opposite to the direction of motion and $\mathbf{F}_{\text{apl-lat}}$ (Fig. 11):

$$\mathbf{f} = \mathbf{q}^{\mathrm{kin}} + \mathbf{F}_{\mathrm{apl-lat}} \cdot \delta\left(l - \frac{L}{2}\right), \tag{20}$$

where $\delta(x)$ is Dirac's delta-function.

The condition of zero total force yields $\mathbf{F}_{\text{apl-lat}} = -\mathbf{q}^{\mathrm{kin}}L$.

Zero elastic force and momentum at the free ends of the NW dictate the boundary conditions:

$$\mathbf{F}|_{l=0} = \mathbf{F}|_{l=L} = 0\mathrm{M}|_{\mathrm{l=0}} = \mathrm{M}|_{\mathrm{l=L}} = 0 \tag{21}$$

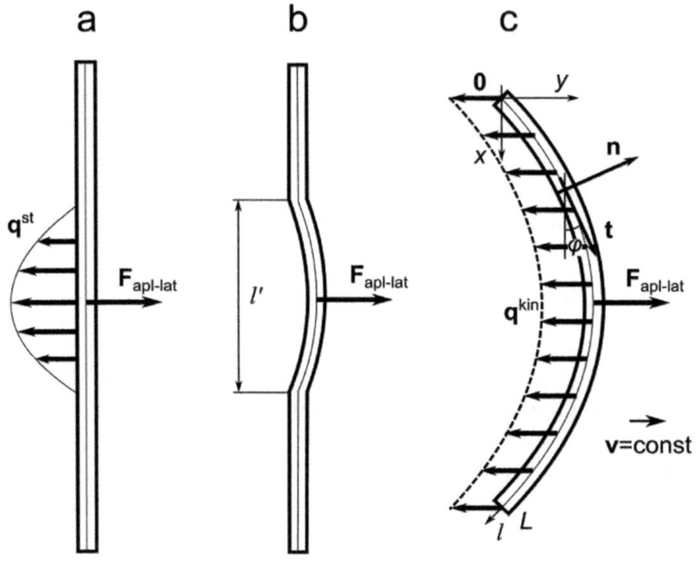

Fig. 19 NW pushed at the midpoint by concentrated force $\boldsymbol{F}_{\text{apl-lat}}$. NW does not move until the applied force exceeds maximal static friction or shear strength (**a**). Partially displaced NW (**b**). Entirely displaced NW. Kinetic friction force q_{kin} equally distributed over NW length (**c**). Fixed coordinate system Oxyz and local coordinate basis (***t***, ***n***) along the NW axis *l*. Angle between the tangent vector ***t*** and axis Ox is denoted as φ

$$\mathbf{M}|_{l=0} = \mathbf{M}|_{l=L} = 0. \tag{22}$$

The differential equation of "kinetic" equilibrium of the NW on the interval (*0*, *L*) directly follows from Eqs. (5), (9) and (10):

$$\mathrm{IE}\frac{\mathrm{d}^2\varphi}{\mathrm{dl}^2} = -\mathrm{q}^{\mathrm{kin}}\left[1\sin\varphi + \mathrm{LH}\left(1-\frac{\mathrm{L}}{2}\right)\right] IE\frac{d^2\varphi}{dl^2} = -q^{\mathrm{kin}}\left[l - LH\left(l-\frac{L}{2}\right)\right]\cos\varphi, \tag{23}$$

where $H(x)$ is the Heaviside step function. The Eq. (23) can be solved numerically to obtain the profile of NW. It is easy to check that the solution of Eq. (23) together with the initial condition $\varphi'(0) = 0$ fully complies with the free boundary conditions Eqs. (21) and (22).

Kinetic Friction Measurements

Polyakov et al. [57] utilized the described model to calculate distributed kinetic friction of ZnO NWs on a flat surfaces (HOPG or silicon wafer) inside SEM. The NW was translated along the substrate surface by pushing it at its midpoint with AFM tip (Fig. 20). NW profile during translation was used to determine the kinetic friction force distributed along the NW by fitting it to theoretical profile calculated

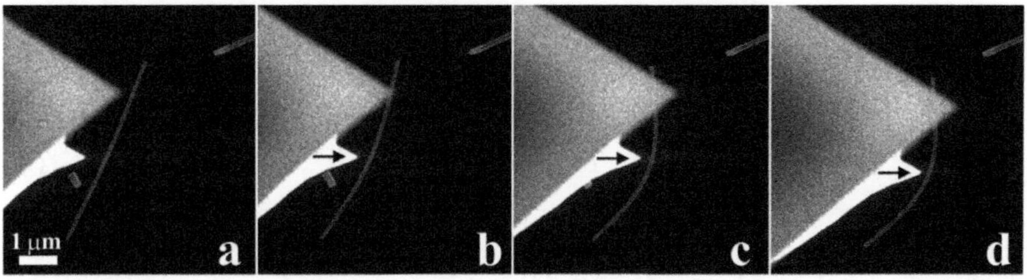

Fig. 20 Sequence of SEM images of the ZnO NW dragged by AFM tip on HOPG substrate. The arrow indicates the direction of tip movement (**a**); partially displaced NW (**b**); completely displaced NW (**c**); constant characteristic profile during NW motion (**d**)

using Eq. (23). The measurements were performed for a set of 8 NWs on silicon wafer and for 6 NWs on HOPG. Interfacial shear stress σ_{kin} were found to be 3.2 and 2.75 MPa respectively. These values are in average about twenty times lower than static interfacial stress ($\sigma_{\text{avg}}^{\text{st}} \sim 67$ MPa) for ZnO NWs on silicon wafer [56]. It agrees well with known fact that for nanoscale objects static friction is commonly significantly higher than kinetic friction [34] (see Fig. 7).

4.4 Redistributed Static Friction of a Bent Nanowire Relaxed After Manipulation on a Flat Substrate

Static friction can hold an elastically bent NW on a flat surface in a deformed state after the removal of external load. The first attempt to use the profile of the elastically bent NW for calculation of static friction was made by Bordag et al. [65]. They assumed the most bent part of an arc-shaped InAs NW to be circular and used the bending curvature as a basis for the static friction calculation. Then Strus et al. [66] introduced a method, where an AFM image of bent CNT was used for determining the static friction and elastic stresses in the framework of the EBT. Stan [67] used parabolas to fit through a defined region of the NW centerline for the analysis with Strus' equilibrium expressions.

Below we will consider elaborated method proposed by Dorogin et al. [9, 68] for modelling of the distributed static friction force in the frameworks of the EBT. Method benefits from fully satisfied boundary conditions and complete equilibrium equations, which enables to apply it for analysis of complex shaped NWs.

Model

The equations of equilibrium for an elastically bent NW held in a bent state by distributed static friction force $\mathbf{f}(l) = \mathbf{q}^{\text{st}}(l)\ \text{q}^{\text{st}}(\text{l})$ give (see Fig. 9):

$$\frac{dF_t}{dl} - \kappa F_n = -q_t^{\text{st}}\text{EI}\frac{\text{d}\kappa}{\text{dl}} = -\text{F}_\text{n} \tag{24}$$

$$\frac{dF_n}{dl} + \kappa F_t = -q_n^{st} \tag{25}$$

$$EI\frac{d\kappa}{dl} = -F_n \tag{26}$$

where F_t and F_n are the projections of elastic force **F**, q_t^{st} and q_n^{st} are the projections of $\mathbf{q}^{st}$ to the local coordinates (**t**, **n**)(t, n).

The complete set of boundary conditions Eqs. (21) and (22) applied to the system of Eqs. (24), (25) and (26) yields:

$$\kappa\prime(0) = \kappa\prime(\mathrm{L}) = 0 \tag{27}$$

$$\kappa|_{l=0} = \kappa|_{l=L} = 0$$

$$\left.\frac{d\kappa}{dl}\right|_{l=0} = \left.\frac{d\kappa}{dl}\right|_{l=L} = 0 \tag{28}$$

We will extend the system of 3 equilibrium Eqs. (24), (25) and (26) for 4 unknown functions (F_t, F_n, q_t^{st} and q_n^{st}) with the condition of absence of the tangential component of the friction $q_t^{st} = 0$. Thus, the system of equations has become complete and can be easily solved for F_t and q_n^{st}:

$$F_t = -EI\int_0^l \kappa\frac{d\kappa}{dl}dl = -EI\frac{\kappa^2}{2} \tag{29}$$

$$q_n^{st} = EI\left(\frac{d^2\kappa}{dl^2} + \frac{\kappa^3}{2}\right), \tag{30}$$

with the initial condition $F_t|_{l=0} = 0$. $\mathrm{F_t}|_{\mathrm{l}=0} = 0$ The absence of a tangential friction component does not lead to the vanishing of F_t which is directly linked with F_n according to the Eq. (24) and is necessary for the exact NW equilibrium.

It is important to note that the assumption $q_t^{st} = 0$ was dictated by the consideration that the direction of $\mathbf{q}^{st}$ should be close to the direction at which the NW tends to unbend. This "unbending" direction correspondingly lies close to normal to the NW's line. Formally it means that the integral contribution of q_t^{st} along the length of the NW is much smaller than that of q_n^{st}. Moreover, the tangential component of the force does not contribute to the bending in the framework of the current model (as can be seen from Eq. (10)) and is considered to be a small effect of higher order.

Redistributed Static Friction

By using the above-described model, the equilibrium of NW elastically bent on a flat surface can be employed for calculation of the static force distribution $\mathbf{q}^{st}(l)$ as shown in Eq. (30), which in turn is determined by the curvature $\kappa(l)$. Therefore, extraction of a differentiable curvature function is needed, which can be realized in the following procedure: (1) NW centerline vectorization; and (2) interpolation of tangent angle $\varphi(l)$ and curvature $\kappa(l)$.

The first stage of centerline vectorization belongs merely to digital image processing and will not be discussed here. The output of the vectorization stage is a set of n discrete points (x_i, y_i), representing the NW centerline.

In the second stage, differential analysis of the centerline is needed to convert the Cartesian coordinates (x_i, y_i) to natural coordinates (l_i, φ_i), which are more suitable for the subsequent calculations. The discrete mesh for subsequent interpolation of $\varphi(l)$ is produced with the following expressions:

$$\phi_i = \tan^{-1}\left(\frac{y_{i+1} - y_i}{x_{i+1} - x_i}\right), \quad 1 \le i < n - 1 \tag{31}$$

$$l_i = l_{i-1} + \sqrt{(x_i - x_{i-1})^2 + (y_i - y_{i-1})^2}, \quad 2 \le i < n \tag{32}$$

where $l_1 = 0$.

Correct choice of the interpolation function is important due to the errors, which may arise during the identification and differentiation of the centerline curve. Regular polynomial interpolation would likely break the boundary conditions Eqs. (27) and (28). The interpolation of $\varphi(l)$ can be performed through the curvature $\kappa(l)$ with the linear combination of the selected polynomial functions as follows:

$$\frac{d\varphi}{dl} = \kappa(l) = \sum_{i=0}^{N} A_i l^{i+2}(L - l)^2,$$

$$\frac{d\varphi}{dl} = \kappa(l) = \sum_{i=0}^{N} A_i l^{i+2}(L - l)^2$$
$$= \sum_{i=0}^{N} B_i l^2(L - l)^{i+2} \tag{33}$$

where A_i are coefficients to be determined by the numerical interpolation. It is apparent that $\varphi(l)$ defined according to Eq. (33) complies with the boundary conditions Eqs. (27) and (28).

The NW-substrate interaction can be characterized directly by the friction distribution and indirectly by the elastic energy stored in the bent NW. The stronger the NW-substrate friction interaction, the higher strain energy can be conserved in the elastically bent NW. It is notable that energy distribution can be calculated with higher accuracy compared to the friction force distribution due to its dependence on the NW curvature $\kappa(l)$ and not on the curvature's derivatives.

Dorogin et al. [9] applied the described method to ZnO NW manipulated to arc shape on a silicon wafer and calculated the static friction force q_n^{st} together with distributed elastic energy u^{st} along the wire. NW was pushed near its midpoint

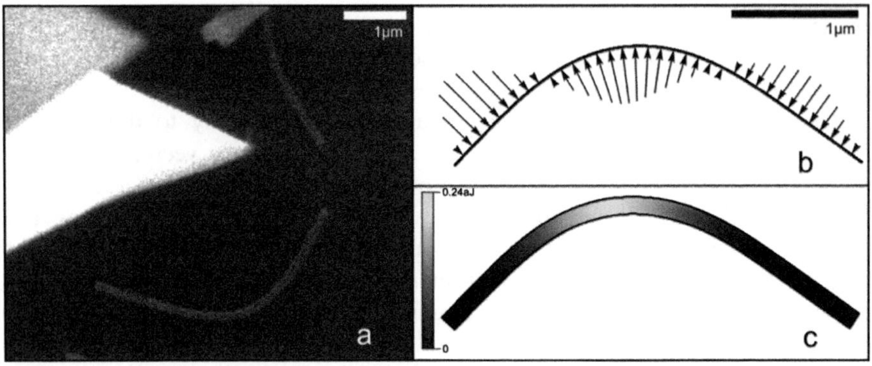

Fig. 21 SEM image of ZnO NW in a bent state self-balanced by static friction (**a**). Calculated distributed static friction along the NW (**b**). Elastic energy distribution along the NW (**c**)

perpendicular to the NW axis by an AFM tip inside SEM. Over several microns of travel, the initially straight profile of the NW transformed into the characteristic arc shape. After the removal of the tip, the NW relaxed and came to equilibrium non-straight shape determined by the interplay of elastic and friction forces (Fig. 21a).

From Fig. 21b it is clearly seen that static friction distribution has non-trivial wavy profile. While u^{st} has only one extremum near the NW center, the distributed friction q_n^{st} has three extrema of different signs. This evidences a non-local character of the NW-substrate friction. Moreover, the model enables certain cases when $|q_n^{st}|$ has absolute maxima near the NW ends, rather than in the middle where the absolute maximum strain is located. Consequently, it is impossible to judge where the maximum friction is applied without detailed analysis in the framework of the presented distributed NW friction model.

5 Assembly of Nanoscale Mechanisms

In addition to benefits described in previous sections, nanomanipulations inside SEM can be used for visually guided assembly and testing of prototype devices (or their key elements) that operate at sub-micrometer scales. Such activities involve displacement and positioning of individual nanostructures, giving rise to nanotribology-related problems and challenges, especially when assembled device should have movable parts. As was shown in previous sections, high surface energy of individual nanostructures causes strong interactions between them and/or with the substrate, resulting in contact strengths comparable to or even exceeding the strength of contacting materials preventing sliding or intermitted contact between parts. Therefore, when developing the nanoscale mechanisms, it is not enough to just downscale the existing macroscopic designs. Alternative approaches are required to tune the strength of nanoscale contacts.

As an example, we will discuss a study dedicated to the assembly and testing of low-friction nanojoint prototype [69]. Authors irradiated Ag and Au NWs by pulsed laser to form rounded end bulbs either at one end of NW, resulting in so-called nano-pin configuration, or at both ends (nanodumbbells configuration). Irradiated nanostructures were deposited onto Si substrate that had chemically etched inverted pyramids (i-pyramids) of different sizes. Then, individual nanostructures were placed vertically into the i-pyramids by manipulation with the rounded end bulb facing downwards (Fig. 22). In such configuration i-pyramid confines in-plane mobility of nanostructure while having small contact area with substrate (wall of i-pyramid). Rotational motion is still available allowing tilting the nanostructure in any direction while the end bulb remains inside the pyramid forming a joint-type system.

The most challenging part of the whole study was pick-and-place manipulation of nanostructures. Displacement of longer nanostructures was hardly possible without severe plastic deformation (Fig. 23). Therefore, only shorter structures that had minimal or no contact (in case of short nanodumbbells) of the non-melted part with the substrate, were used for pick-and-place manipulations.

Assembly of each ‘nanojoint’ started by contacting a selected nanostructure with a Si tip in such a way that contact area (and hence adhesion) of the nanostructure

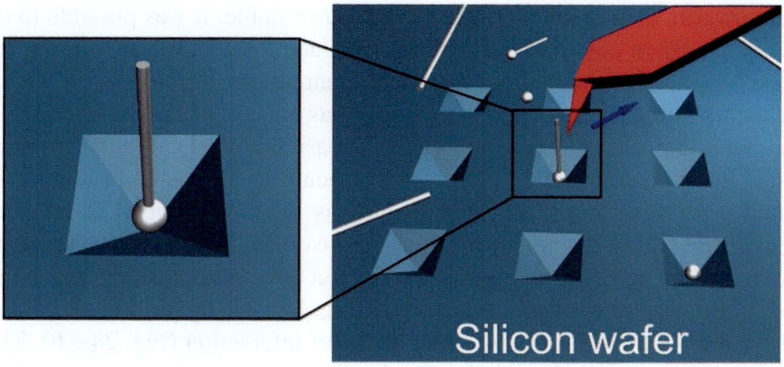

Fig. 22 Schematic of a Ag nanopin deposited in the i-pyramid with the help of the nanomanipulator

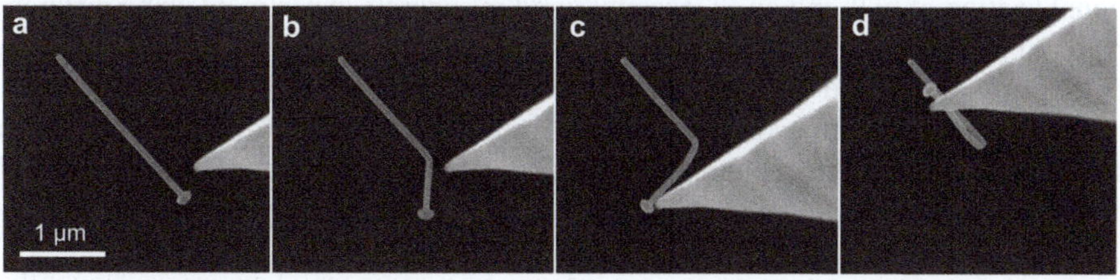

Fig. 23 Manipulation of the Ag nano-pin on a flat Si surface. The friction between the nano-pin and Si surface is too high to allow the displacement without severe deformation of the nano-pin [69].

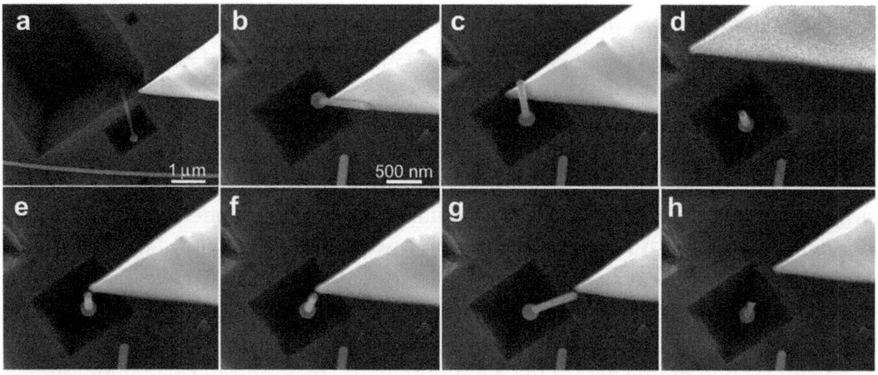

Fig. 24 Chosen frames from assembly (**a–d**) and manipulation (**e–h**) of nanojoint prototype inside SEM [69].

with the tip is higher than with the substrate, allowing to lift the nanostructure off the substrate. Next, the picked structure was transferred and placed into to the i-pyramid by pushing its ball-end down and at the same time moving the tip sideways by series of abrupt motions. In some cases, the object was released without deformation and took an upward position that appeared to be rather stable. It was possible to tilt the object repeatedly in any direction without displacing the ball-end. Chosen frames from one typical assembly and testing experiment are show in Fig. 24.

The manipulation system used in the study was not equipped with a force sensing feature excluding direct quantitative force measurement. Qualitative estimations suggested that rotational frictional forces between the ball-end and sidewalls of the i-pyramid were very low. Such conclusion was based on several facts. First, no visible deformation of the non-melted part of the object was observed during tilting in contrast to severe deformation of similar structures on flat Si surface (Fig. 24). Second, tilting was possible not only by pushing, but also by pulling the protruding end of the nanojoint with the tip using only vdW interaction (Fig. 24e–h). Third, a certain voltage was applied to the tip, and the distance at which nanojoint jumps into the contact with the tip due driven by electrostatic attraction was measured. With the help of FEM simulations, total force was estimated as approx. 1 nN. Finally, rotational friction of ball-end on Si was measured in separate experiment using different manipulation setup (Fig. 8c), and it was found that the force needed to rotate such object around ball-end is in the order of 1 nN.

Assembly and testing of the nanojoint prototype described in this section is a good illustration of the capabilities of nanomanipulation techniques, as well as of the non-trivial friction-related problems that arise alongside the nanomanipulation.

6 Specific Problems of Manipulations Inside SEM

There are several issues to keep in mind when planning experiments inside SEM. First, typical scanning rate of the electron beam is limited to a few Hz, therefore only relatively slow processes can be visualized. Another limitation is associated with SEM resolution. Resolution and signal intensity are reversely proportional to scanning speed. Visible dimensions of the nanostructures at magnification close to the resolution limit are sensitive to brightness/contrast settings of a microscope. It can be problematic to identify the exact shape of the NW cross-section, for example, to distinguish between square and rectangular cross-sections or estimate degree of hexagon or pentagon distortions. Precise determination of NWs diameter and cross-section, however, is essential for correct measurement of the contact area and frictional forces in the framework of EBT.

Vacuum conditions inside the SEM chamber significantly limit the dissipation of any heat that may be generated either locally or by parts of the manipulation system. This may lead to thermal drifts.

Another thing that should be considered is the effect of electron beam on a substrate surface and nanostructures. Nanostructures are more sensitive to the heating than bulk structures, so that even moderate heat-treatment can cause morphological changes in nanostructures far below melting temperature [70]. Focused e-beam is capable of introducing significant energy and increase the local temperature in nanostructures. Additional effects can be substrate surface activation or electrostatic charging, which also can influence nanotribological experiments [71]. Degradation of some materials under electron beam should is another issue. For example, electron beam can induce severe defects in carbon-based materials like graphene, graphite and carbon nanotubes, or self-assembled monolayers of organic molecules, which are frequently used for tribological experiments [72].

Serious problem is e-beam assisted carbon deposition [73]. High energy electrons can decompose hydrocarbons molecules, which are present in the vacuum chamber and amorphous carbon can be deposited on the areas exposed to electron beam for prolonged periods. The amount of deposited carbon depends on the vacuum level, e-beam energy, the presence of hydrocarbon inside the chamber and exposition time.

Therefore, special care should be taken when interpreting the results obtained by in situ SEM manipulations and comparing them with experiments obtained with AFM, especially in ambient conditions.

7 Outlook

It was demonstrated that visual guidance during the nanomanipulations inside SEM is extremely helpful and provides essential information on behavior of manipulated nanostructures. Method is well suitable for manipulation of nanoparticles and especially fruitful for elongated objects. In-SEM technique is not limited to in-plane manipulations, but also capable of performing experiments in three dimensions. Moreover, experiments are less time-consuming in comparison to AFM manipulations due to separation of visualization and manipulation functions. Besides the apparent advantages, there are certain peculiarities and limitations that should be considered when planning and interpreting experiments inside SEM.

References

1. Y. Mo, K.T. Turner, I. Szlufarska, Friction laws at the nanoscale. Nature **457**(7233), 1116–1119 (2009). https://doi.org/10.1038/nature07748
2. L. Liu, P. Peng, A. Hu, G. Zou, W.W. Duley, Y.N. Zhou, Highly localized heat generation by femtosecond laser induced plasmon excitation in ag nanowires. Appl. Phys. Lett. **102**(7), 073107 (2013). https://doi.org/10.1063/1.4790189
3. L. Wang, H.-Y. Park, S.I.-I. Lim, M.J. Schadt, D. Mott, J. Luo, X. Wang, C.-J. Zhong, Core@shell nanomaterials: gold-coated magnetic oxide nanoparticles. J. Mater. Chem. **18**(23), 2629–2635 (2008). https://doi.org/10.1039/B719096D
4. L.M. Dorogin, S. Vlassov, A.L. Kolesnikova, I. Kink, R. Lõhmus, A.E. Romanov, Crystal mismatched layers in pentagonal nanorods and nanoparticles. Phys. Status Solidi B **247**(2), 288–298 (2010). https://doi.org/10.1002/pssb.200945385
5. D. Hidayat, A. Purwanto, W.-N. Wang, K. Okuyama, Preparation of size-controlled tungsten oxide nanoparticles and evaluation of their adsorption performance. Mater. Res. Bull. **45**(2), 165–173 (2010). https://doi.org/10.1016/j.materresbull.2009.09.025
6. D. Seo, C.I. Yoo, I.S. Chung, S.M. Park, S. Ryu, H. Song, Shape adjustment between multiply twinned and single-crystalline polyhedral gold nanocrystals: decahedra, icosahedra, and truncated tetrahedra. J. Phys. Chem. C **112**(7), 2469–2475 (2008). https://doi.org/10.1021/jp7109498
7. C.-L. Chiang, M.-B. Hsu, L.-B. Lai, Control of nucleation and growth of gold nanoparticles in AOT/Span80/isooctane mixed reverse micelles. J. Solid State Chem. **177**(11), 3891–3895 (2004). https://doi.org/10.1016/j.jssc.2004.07.003
8. C. Cao, S. Park, S.J. Sim, Seedless synthesis of octahedral gold nanoparticles in condensed surfactant phase. J. Colloid Interf. Sci. **322**(1), 152–157 (2008). https://doi.org/10.1016/j.jcis.2008.03.031
9. L.M. Dorogin, B. Polyakov, A. Petruhins, S. Vlassov, S. Rünno, I. Kink, A.E. Romanov, Modeling of kinetic and static friction between an elastically bent nanowire and a flat surface. J. Mater. Res. **27**(3), 580–585 (2012). https://doi.org/10.1557/jmr.2011.339
10. Semilab: scanning electron microcope integrated atomic force microscopy (SEM AFM), https://semilab.com/jp/category/products/sem-afm. Accessed 10 March 2023
11. Nanonics AFM SEM/FIB integrated nanoscale characterization. Nanonics imaging LTD, AFM SEM/FIB integrated nanoscale characterization. https://www.nanonics.co.il/solutions/afm-sem-fib. Accessed 10 March 2023
12. SmarAct nanomanipulators. https://www.smaract.com/en/. Accessed 10 March 2023
13. D. Erts, A. Lõhmus, R. Lõhmus, H. Olin, A.V. Pokropivny, L. Ryen, K. Svensson, Force interactions and adhesion of gold contacts using a combined atomic force microscope and transmission electron microscope. Appl. Surf. Sci. **188**(3), 460–466 (2002). https://doi.org/10.1016/S0169-4332(01)00933-3
14. Attocube. Attocube systems AG. https://www.attocube.com/en. Accessed 10 March 2023
15. PI Ceramic: Piezo technology, actuators and components. https://www.piceramic.com/en/. Accessed 10 March 2023
16. R. Wiesendanger, *Scanning Probe Microscopy and Spectroscopy: Methods and Applications* (Cambridge University Press, Cambridge, 1994)

17. M. Troyon, H.N. Lei, Z. Wang, G. Shang, A scanning force microscope combined with a scanning electron microscope for multidimensional data analysis. Microsc. Microanal. Microstruct. **8**(6), 393–402 (1997). https://doi.org/10.1051/mmm:1997130
18. K. Fukushima, D. Saya, H. Kawakatsu, Development of a versatile atomic force microscope within a scanning electron microscope. Jpn. J. Appl. Phys. **39**(6S), 3747 (2000). https://doi.org/10.1143/JJAP.39.3747
19. I. Joachimsthaler, R. Heiderhoff, L.J. Balk, A universal scanning-probe-microscope-based hybrid system. Meas. Sci. Technol. **14**(1), 87 (2002). https://doi.org/10.1088/0957-0233/14/1/313
20. U. Mick, V. Eichhorn, T. Wortmann, C. Diederichs, S. Fatikow, Combined nanorobotic AFM/SEM system as novel toolbox for automated hybrid analysis and manipulation of nanoscale objects, in *Proceedings of the 2010 IEEE International Conference on Robotics and Automation* (2010), pp. 4088–4093. https://doi.org/10.1109/ROBOT.2010.5509414
21. U. Stahl, C.W. Yuan, A.L. De Lozanne, M. Tortonese, Atomic force microscope using piezoresistive cantilevers and combined with a scanning electron microscope. Appl. Phys. Lett. **65**(22), 2878–2880 (1994). https://doi.org/10.1063/1.113030
22. M. Barbic, L. Eliason, J. Ranshaw, Femto-Newton force sensitivity quartz tuning fork sensor. Sens. Actuat. A Phys. **136**(2), 564–566 (2007). https://doi.org/10.1016/j.sna.2007.01.001
23. S. Fain, K.A. Barry, M.G. Bush, B. Pittenger, R.N. Louie, Measuring average tip-sample forces in intermittent-contact (tapping) force microscopy in air. Appl. Phys. Lett. **76**, 930–932 (2000). https://doi.org/10.1063/1.125633
24. C. Su, L. Huang, K. Kjoller, Direct measurement of tapping force with a cantilever deflection force sensor. Ultramicroscopy **100**(3), 233–239 (2004). https://doi.org/10.1016/j.ultramic.2003.11.007
25. S. Vlassov, B. Polyakov, L.M. Dorogin, A. Lõhmus, A.E. Romanov, I. Kink, E. Gnecco, R. Lõhmus, Real-time manipulation of gold nanoparticles inside a scanning electron microscope. Solid State Commun. **151**(9), 688–692 (2011). https://doi.org/10.1016/j.ssc.2011.02.020
26. H. Hida, M. Shikida, K. Fukuzawa, S. Murakami, K. Sato, K. Asaumi, Y. Iriye, K. Sato, Fabrication of a quartz tuning-fork probe with a sharp tip for AFM systems. Sens. Actuat. A Phys. **148**(1), 311–318 (2008). https://doi.org/10.1016/j.sna.2008.08.021
27. V.T. Tung, S.A. Chizhik, T.X. Hoai, N.T. Tinh, V.V. Chikunov, V.T. Tung, S.A. Chizhik, T.X. Hoai, N.T. Tinh, V.V. Chikunov, Tuning fork scanning probe microscopes - applications for the nano-analysis of the material surface and local physico-mechanical properties, in *Scanning Probe Microscopy: Physical Property Characterization at Nanoscale* (IntechOpen, 2012). https://doi.org/10.5772/36187
28. S. Kerfriden, A.H. Nahlé, S.A. Campbell, F.C. Walsh, J.R. Smith, Short communication the electrochemical etching of tungsten STM tips. Electrochim. Acta **43**(12), 1939–1944 (1998). https://doi.org/10.1016/S0013-4686(97)00316-2
29. AdvancedTEC^{TM} Cont. http://www.nanosensors.com/. Accessed 10 March 2023
30. Akiyama-Probe (A-Probe) guide, NANOSENSORS, NanoWorld AG. https://www.akiyamaprobe.com/. Accessed 10 March 2023
31. J.L. Arlett, J.R. Maloney, B. Gudlewski, M. Muluneh, M.L. Roukes, Self-sensing micro- and nanocantilevers with attonewton-scale force resolution. Nano Lett. **6**(5), 1000–1006 (2006). https://doi.org/10.1021/nl060275y
32. M. Li, H.X. Tang, M.L. Roukes, Ultra-sensitive NEMS-based cantilevers for sensing, scanned probe and very high-frequency applications. Nat. Nanotech **2**(2), 114–120 (2007). https://doi.org/10.1038/nnano.2006.208
33. C. Baur, A. Bugacov, B.E. Koel, A. Madhukar, N. Montoya, T.R. Ramachandran, A.A.G. Requicha, R. Resch, P. Will, Nanoparticle manipulation by mechanical pushing: underlying phenomena and real-time monitoring. Nanotechnology **9**(4), 360 (1998). https://doi.org/10.1088/0957-4484/9/4/011
34. D. Dietzel, M. Feldmann, H. Fuchs, U.D. Schwarz, A. Schirmeisen, Transition from static to kinetic friction of metallic nanoparticles. Appl. Phys. Lett. **95**(5), 053104 (2009). https://doi.org/10.1063/1.3193551

35. K. Mougin, E. Gnecco, A. Rao, M.T. Cuberes, S. Jayaraman, E.W. McFarland, H. Haidara, E. Meyer, Manipulation of gold nanoparticles: influence of surface chemistry, temperature, and environment (vacuum versus ambient atmosphere). Langmuir **24**(4), 1577–1581 (2008). https://doi.org/10.1021/la702921v
36. S. Kim, D.C. Ratchford, X. Li, Atomic force microscope nanomanipulation with simultaneous visual guidance. ACS Nano **3**(10), 2989–2994 (2009). https://doi.org/10.1021/nn900606s
37. A.M. Homola, J.N. Israelachvili, M.L. Gee, P.M. McGuiggan, Measurements of and relation between the adhesion and friction of two surfaces separated by molecularly thin liquid films. J. Tribol. **111**(4), 675–682 (1989). https://doi.org/10.1115/1.3261994
38. R.W. Carpick, M. Salmeron, Scratching the surface: fundamental investigations of tribology with atomic force microscopy. Chem. Rev. **97**(4), 1163–1194 (1997). https://doi.org/10.1021/cr960068q
39. A.H. Cottrell, *Dislocations and Plastic Flow in Crystals*, 1st edn. (Oxford University Press, Oxford, 1953)
40. J.P. Hirth, J. Lothe, *Theory of Dislocations; McGraw-Hill Series in Materials Science and Engineering* (McGraw-Hill, New York, 1968)
41. K.L. Johnson, K. Kendall, A.D. Roberts, Surface energy and the contact of elastic solids. Proc. R. Soc. Lond. A **324**(1558), 301–313 (1971). https://doi.org/10.1098/rspa.1971.0141
42. B.V. Derjaguin, V.M. Muller, Yu.P. Toporov, Effect of contact deformations on the adhesion of particles. J. Colloid Interf. Sci. **53**(2), 314–326 (1975). https://doi.org/10.1016/0021-9797(75)90018-1
43. D. Tabor, Surface forces and surface interactions. J. Colloid Interf. Sci. **58**(1), 2–13 (1977). https://doi.org/10.1016/0021-9797(77)90366-6
44. D. Dietzel, C. Ritter, T. Mönninghoff, H. Fuchs, A. Schirmeisen, U.D. Schwarz, Frictional duality observed during nanoparticle sliding. Phys. Rev. Lett. **101**(12), 125505 (2008). https://doi.org/10.1103/PhysRevLett.101.125505
45. B. Polyakov, S. Vlassov, L. Dorogin, J. Butikova, K. Smits, M. Antsov, S. Oras, R. Zabels, R. Lõhmus, Metal nanodumbbells for nanomanipulations and tribological experiments. Phys. Scr. **90**(9), 094007 (2015). https://doi.org/10.1088/0031-8949/90/9/094007
46. B. Polyakov, S. Vlassov, L.M. Dorogin, N. Novoselska, J. Butikova, M. Antsov, S. Oras, R. Lõhmus, I. Kink, Some aspects of formation and tribological properties of silver nanodumbbells. Nanoscale Res. Lett. **9**(1), 186 (2014). https://doi.org/10.1186/1556-276X-9-186
47. D.R. Smith, F.R. Fickett, Low-temperature properties of silver: PMC. J. Res. Natl. Instit. Stand. Technol. **100**(2), 119–171 (1995)
48. M.P. Manoharan, A.V. Desai, G. Neely, M.A. Haque, Synthesis and elastic characterization of zinc oxide nanowires. J. Nanomater. **2008**, 1–7 (2008). https://doi.org/10.1155/2008/849745
49. J.-H. Hsu, S.-H. Chang, Surface adhesion between hexagonal boron nitride nanotubes and silicon based on lateral force microscopy. Appl. Surf. Sci. **256**(6), 1769–1773 (2010). https://doi.org/10.1016/j.apsusc.2009.10.001
50. M.R. Falvo, J. Steele, R.M. Taylor, R. Superfine, Evidence of commensurate contact and rolling motion: AFM manipulation studies of carbon nanotubes on HOPG. Tribol. Lett. **9**(1), 73–76 (2000). https://doi.org/10.1023/A:1018808511550
51. S. Vlassov, B. Polyakov, S. Oras, M. Vahtrus, M. Antsov, A. Šutka, K. Smits, L.M. Dorogin, R. Lõhmus, Complex tribomechanical characterization of ZnO nanowires: nanomanipulations supported by FEM simulations. Nanotechnology **27**(33), 335701 (2016). https://doi.org/10.1088/0957-4484/27/33/335701
52. R. Mohan, Y. Liang, Chapter 10 tensile and flexural deformation of nickel nanowires via molecular dynamics simulations, in *Cutting Edge Nanotechnology* (2010). https://doi.org/10.5772/201
53. M.A. Schubert, S. Senz, M. Alexe, D. Hesse, U. Gösele, Finite element method calculations of ZnO nanowires for nanogenerators. Appl. Phys. Lett. **92**(12), 122904 (2008). https://doi.org/10.1063/1.2903114

54. S. Timoshenko, J.N. Goodier, *Theory of Elasticity*, 2nd edn. (McGraw-Hill Book Company Inc., New York, 1951)
55. B. Polyakov, L.M. Dorogin, S. Vlassov, I. Kink, A. Lohmus, A.E. Romanov, R. Lõhmus, Real-time measurements of sliding friction and elastic properties of ZnO nanowires inside a scanning electron microscope. Solid State Commun. **151**(18), 1244–1247 (2011). https://doi.org/10.1016/j.ssc.2011.05.045
56. B. Polyakov, L.M. Dorogin, S. Vlassov, I. Kink, A.E. Romanov, R. Lõhmus, Simultaneous measurement of static and kinetic friction of ZnO nanowires in situ with a scanning electron microscope. Micron **43**(11), 1140–1146 (2012). https://doi.org/10.1016/j.micron.2012.01.009
57. B. Polyakov, L.M. Dorogin, A. Lõhmus, A.E. Romanov, R. Lõhmus, In situ measurement of the kinetic friction of ZnO nanowires inside a scanning electron microscope. Appl. Surf. Sci. **258**(7), 3227–3231 (2012)
58. L.D. Landau, E.M. Lifshitz, *Theory of Elasticity*, 3rd edn. (Butterworth-Heinemann, Oxford, 1984)
59. S. Vlassov, D. Bocharov, B. Polyakov, M. Vahtrus, A. Šutka, S. Oras, V. Zadin, A. Kyritsakis, Critical review on experimental and theoretical studies of elastic properties of Wurtzite-structured ZnO nanowires. Nanotechnol. Rev. **12**(1), 505 (2023). https://doi.org/10.1515/ntrev-2022-0505
60. E.C.C.M. Silva, L. Tong, S. Yip, K.J. Van Vliet, Size effects on the stiffness of silica nanowires. Small **2**(2), 239–243 (2006). https://doi.org/10.1002/smll.200500311
61. B.N.J. Persson, Sliding friction, in *NanoScience and Technology* by eds. K. Von Klitzing, R. Wiesendanger (Springer, Berlin, 1998). https://doi.org/10.1007/978-3-662-03646-4
62. N.S. Tambe, B. Bhushan, Scale dependence of micro/nano-friction and adhesion of MEMS/NEMS materials, coatings and lubricants. Nanotechnology **15**(11), 1561 (2004). https://doi.org/10.1088/0957-4484/15/11/033
63. F.W. DelRio, M.P. de Boer, J.A. Knapp, E. David Reedy, P.J. Clews, M.L. Dunn, The role of van Der Waals forces in adhesion of micromachined surfaces. Nat. Mater. **4**(8), 629–634 (2005). https://doi.org/10.1038/nmat1431
64. B. Polyakov, S. Vlassov, L.M. Dorogin, P. Kulis, I. Kink, R. Lohmus, The effect of substrate roughness on the static friction of CuO nanowires. Surf. Sci. **606**(17), 1393–1399 (2012). https://doi.org/10.1016/j.susc.2012.04.029
65. M. Bordag, A. Ribayrol, G. Conache, L. Fröberg, S. Gray, L. Samuelson, L. Montelius, H. Pettersson, Shear stress measurements on InAs nanowires by AFM manipulation. Small **3**(8), 1398–1401 (2007). https://doi.org/10.1002/smll.200700052
66. M.C. Strus, R.R. Lahiji, P. Ares, V. López, A. Raman, R. Reifenberger, Strain energy and lateral friction force distributions of carbon nanotubes manipulated into shapes by atomic force microscopy. Nanotechnology **20**(38), 385709 (2009). https://doi.org/10.1088/0957-4484/20/38/385709
67. G. Stan, S. Krylyuk, A.V. Davydov, R.F. Cook, Bending manipulation and measurements of fracture strength of silicon and oxidized silicon nanowires by atomic force microscopy. J. Mater. Res. **27**(3), 562–570 (2012). https://doi.org/10.1557/jmr.2011.354
68. L.M. Dorogin, S. Vlassov, B. Polyakov, M. Antsov, R. Lõhmus, I. Kink, A.E. Romanov, Real-time manipulation of ZnO nanowires on a flat surface employed for tribological measurements: experimental methods and modeling. Phys. Status Solidi B **250**(2), 305–317 (2013). https://doi.org/10.1002/pssb.201248445
69. S. Vlassov, S. Oras, M. Antsov, J. Butikova, R. Lõhmus, B. Polyakov, Low-friction nanojoint prototype. Nanotechnology **29**(19), 195707 (2018). https://doi.org/10.1088/1361-6528/aab163
70. S. Vigonski, V. Jansson, S. Vlassov, B. Polyakov, E. Baibuz, S. Oras, A. Aabloo, F. Djurabekova, V. Zadin, Au nanowire junction breakup through surface atom diffusion. Nanotechnology **29**(1), 015704 (2018). https://doi.org/10.1088/1361-6528/aa9a1b
71. G. Conache, A. Ribayrol, L.E. Fröberg, M.T. Borgström, L. Samuelson, L. Montelius, H. Pettersson, S.M. Gray, Bias-controlled friction of InAs nanowires on a silicon nitride layer studied by atomic force microscopy. Phys. Rev. B **82**(3), 035403 (2010). https://doi.org/10.1103/PhysRevB.82.035403

72. D. Teweldebrhan, A.A. Balandin, Modification of graphene properties due to electron-beam irradiation. Appl. Phys. Lett. **94**(1), 013101 (2009). https://doi.org/10.1063/1.3062851
73. S. Zaitsev, O. Shtempluck, E. Buks, Effects of electron beam induced carbon deposition on the mechanical properties of a micromechanical oscillator. Sens. Actuat. A Phys. **179**, 237–241 (2012). https://doi.org/10.1016/j.sna.2012.02.039

Probing Superlubricity of Layered Van der Waals Contact Junctions by Colloidal Probe Atomic Force Microscopy

Renato Buzio, Andrea Gerbi, Cristina Bernini, Luca Repetto, Andrea Silva, and Andrea Vanossi

Abstract This chapter reviews Colloidal Probe—Atomic Force Microscopy (CP-AFM) friction experiments probing superlubricity in layered Van der Waals (mesoscopic) contact junctions. Specifically, we cover all the aspects of these experiments, from the preparation of the probes to the modeling required to achieve insight into the physical mechanisms underlying the observed phenomenology. The main fabrication methods of the graphene-coated colloidal probes are comprehensively described. We also report on key features that accompany the emergence of superlubricity for graphene-based homo-junctions, as well as for hetero-junctions of graphene with common transition metal dichalcogenides. Finally, we discuss atomic friction data in light of the single-asperity thermally-activated Prandtl-Tomlinson (PT) model. The PT model is shown to effectively rationalize the results of CP-AFM experiments involving different colloidal probes, material substrates and sliding regimes. In particular, nearly-dissipationless sliding occurs only for the lower corrugated interface of graphene homo-junctions, as a result of the load-controlled transition from dissipative stick–slip to continuous superlubric sliding taking place in the low-loads limit.

R. Buzio (✉) · A. Gerbi · C. Bernini
CNR-SPIN Institute for Superconductivity, Innovative Materials and Devices, National Research Council CNR, C.so F.M. Perrone 24, 16152 Genova, Italy
e-mail: renato.buzio@spin.cnr.it

L. Repetto
Physics Department, University of Genova, Via Dodecaneso 33, 16146 Genova, Italy

A. Silva · A. Vanossi
CNR-IOM Materials Foundry Institute, National Research Council CNR, c/o SISSA, Via Bonomea 265, 34136 Trieste, Italy

International School for Advanced Studies (SISSA), Via Bonomea 265, 34136 Trieste, Italy

E. Gnecco and E. Meyer (eds.), *Fundamentals of Friction and Wear on the Nanoscale*, NanoScience and Technology, https://doi.org/10.1007/978-3-031-63065-1_10

1 Probing Superlubric Features at the Mesoscale

Superlubricity is an umbrella term for situations in which sliding friction vanishes or very nearly vanishes [1]. It is also used for physical systems displaying a very weak dependence of sliding friction on normal load, so that the (differential) friction coefficient becomes ultrasmall (i.e. ≤ 0.01). To achieve superlubricity, dry or liquid lubrication can be implemented. A central theme of the current tribology research regards the investigation of dry superlubricity mechanisms enabled by two-dimensional (2D) materials. This certainly originates from the long-standing interest in the lubricating properties of two well-known lamellar solids, namely graphite and MoS_2 (e.g. see [2, 3] and references therein), yet it has seen an upsurge in the last two decades thanks to new achievements in the synthesis and characterization of high-quality layered materials samples, prepared as monoatomic sheets or few-layers-thick specimens displaying unique physical properties [4, 5].

In general, dry superlubricity assisted by 2D materials can involve the manifestation and interplay of different complex mechanisms, for example related to tribochemical reactions and third-body effects. This is the case when the targeted 2D materials show enhanced reactivity against ambient environment or the couterbodies' surfaces under the chosen testing conditions, so that severe mechanical or chemical degradations take place [6–8]. In turn, graphitic systems such as bulk graphite and graphene flakes are often considered ideal candidates to explore dry superlubricity mechanisms within simplified and more tractable scenarios [9–13]. In fact the graphite-mediated (or graphene-mediated) contact phenomena usually involve only weak van der Waals forces, and dry superlubricity may occur thanks to the formation of easy share planes across atomically smooth contact interfaces [14].

Specifically for bulk graphite, both macro- and nano-scale experiments showed that superlubricity appears after the formation of a transfer layer (TL), derived by tribo-induced transfer of material from a graphite substrate to the sliding countersurface [9, 10]. This TL mostly consists of nanosized graphitic flakes, that are thought to arrange in misfit configurations assuring interfacial crystalline incommensurability and (differential) friction coefficients as small as 0.001. This superlow friction condition, named structural lubricity, is characterized by the suppression of the energy barriers to sliding motion, due to the cancellation of the atomistic lateral forces in an incommensurate crystalline interface [15] (Fig. 1a). Evidence for structural lubricity was obtained by Dienwiebel et al. [9] using AFM, in single-asperity graphite nanocontacts. Authors considered a sharp tungsten nanoprobe sliding on Highly Oriented Pyrolytic Graphite (HOPG); they observed continuous nearly-dissipationless sliding when the TL—formed as a result of the tribo-transfer of an individual nanoflake from graphite to the sliding AFM tip—assumed an incommensurate misfit configuration with respect to the underlying crystalline substrate. Structural lubricity was later recognized in other nanoscale graphitic homo-junctions, e.g. based on carbon nanotubes, nanoribbons and fullerene-based nanobearings [15].

More recently, the capability to prepare 2D flakes-wrapped AFM nanoprobes has enabled the investigation of interfacial forces and mechanical dissipation in

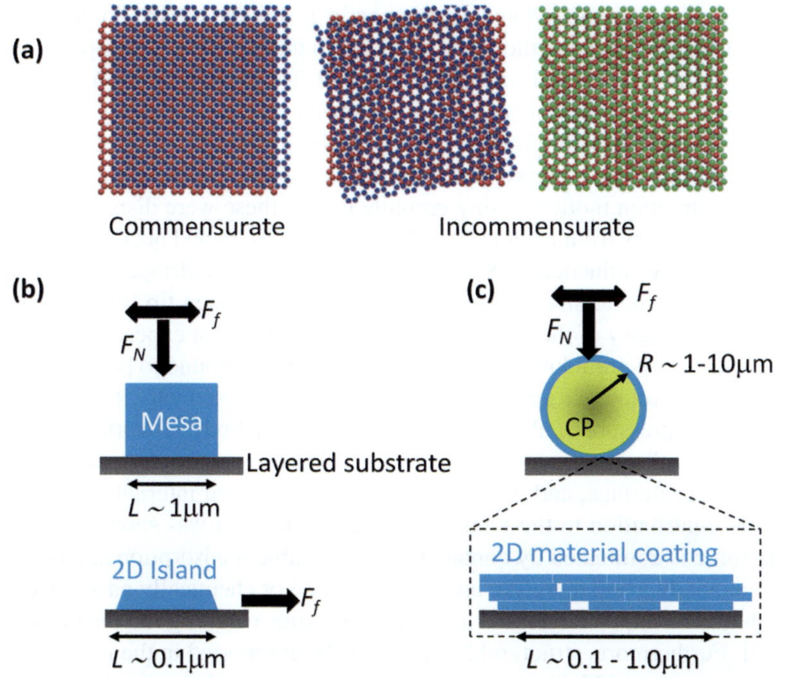

Fig. 1 **a** Mating identical surfaces with the same lattice spacing (red and blue) may give rise to both commensurate interfaces when aligned (left panel) and incommensurate ones when orientationally misaligned at a misfit generic angle (central panel). Structural incommensurability may also emerge when considering crystalline lattice-mismatched surfaces (right panel). Reproduced from [15] under the CC BY 4.0 license (https://creativecommons.org/licenses/by/4.0/). **b** Prototypical experiments on single-asperity, layered homo-/hetero-junctions: a microscopic graphitic mesa (top, under the action of the normal force F_N) and an atomic island (bottom) are driven along the surface of an atomically-flat layered bulk substrate, meanwhile the friction force F_f is measured. **c** Similar experiments but for a multi-asperity layered homo-/hetero-junction: a micrometric colloidal probe CP—coated by a random assembly of 2D nanoflakes—is sled over the surface of an atomically-smooth layered bulk substrate

layered van der Waals hetero-junctions [11, 16]. The practical realization of crystalline heterogeneous contacts is often claimed to be a necessary condition to obtain the best incommensurate sliding and—possibly—structural lubricity between mismatched lattices. Quite unexpectedly, however, AFM experiments with graphene-coated probes sled against common transition metal dichalcogenides (TMDS), such as MoS_2 and WS_2, have attested an increase of the friction force compared to the graphitic homo-junctions despite the natural lattice incommensurability of such hetero-junctions [11]. At present, this is an open issue that needs to be fully elucidated.

Another relevant issue concerns the capability to scale-up structural lubricity from nanoscale contact junctions to mesoscopic ones. Experimentally, the realization of a sub-micrometric (or even micrometric) single-asperity junction free from atomic-defects and contaminants, represents a formidable task. Strategies implemented to this end regard the AFM-assisted manipulation of nanoparticles [17–19], 2D atomic islands and nanostructures [20–24] and graphitic mesas [25–27] over atomically-flat substrates (Fig. 1b). Nowadays, the sliding dynamics of the mesas is certainly providing major support to the realization of structural lubricity at the microscale. Early reports attesting this capability concerned the observation of the self-retraction motion of tiny graphite flakes: these were displaced to various suspended positions from islands of HOPG, and they retracted back onto the islands under no external influences [28]. This effect was later confirmed by shearing and cleaving a micrometer-sized graphite mesa using a tungsten tip [29]. In particular the sixfold symmetry of the self-retraction provided the first experimental evidence for microscale structural lubricity. The phenomenon was found to persist in ambient air, with sliding velocities up to 25m/s [30], and it was practically load-independent up to a normal pressure of about 1.7 MPa [31]. A large body of work was dedicated to the understanding of the role played by interfacial adsorbates and graphite defects at the contact interface, and it revealed that typically such internal structures have negligible contribution to the total friction. In contrast, it was shown that the total friction force is dominated by contact edges, as ambient adsorption and contamination are inevitable, and therefore under coordinated or chemically passivated carbon atoms may exist at the edges and act as pinning sites resulting in considerable friction [27]. Furthermore, structural lubricity has been reported in the manipulation of graphite mesa over different crystalline surfaces, such as hexagonal-Boron Nitride (hBN) and mica [25, 26], as well as over non-Van der Waals layered materials with sub-nanometer roughness [32] and nanostructured surfaces [33]. The investigation of these prototypical systems has shown that in many situations ultralow (differential) friction coefficients (≤ 0.001) are achieved, even when the tribo-couples are assembled and/or manipulated under ambient environment [34].

In principle, Colloidal Probe (CP)—AFM experiments represent an alternative, powerful tool to study the scale up of structural lubricity in mesoscopic tribo-pairs. Originally introduced by Ducker et al. [35] for normal-force spectroscopy purposes, CP-AFM is usually implemented by gluing at the end of a tipless cantilever a sphere (made of silica, glass, TiO_2, , ZnS, metals, polystyrene and common polymers, PDMS…) of micrometric radius (~1–50 μm) by means of an epoxy resin. The advantage of this technique lies in the exact knowledge of the tip geometry. In fact, thanks to the use of complementary high-resolution microscopy techniques like scanning electron microscopy and reverse AFM imaging, both the radius R of the sphere and its surface morphology can be monitored in the course of the experiments (e.g. [36–38]). Hence, one can predict with greater accuracy the magnitude of the normal forces acting between the CP and a flat counter surface under different working conditions [39–41]. Importantly, contact mechanics models for a sphere-on-flat contact geometry with $R \sim 2$–50 μm typically place the size of the nominal contact spot in the mesoscopic range, i.e. ~100–500 nm, which is particularly useful for the scale-up of contact phenomena from the nanoscale towards the larger scales [42, 43]. The CP-AFM technique is actively used to characterize different properties of lamellar solids

[44–47], including their friction response (e.g. [48–51]), as well as to investigate the manifestation of superlubricity for different tribosystems [52–56].

Some CP-AFM studies have recently appeared (see Sect. 3), in which the surface of the CPs was effectively coated by thin 2D layers using different means. These approaches make it possible to explore sliding friction and superlubricity in mesoscopic layered homo- and hetero-junctions (Fig. 1c), which is indeed the main topic of the present Chapter. In fact, CP-AFM experiments not only allow to identify the general friction phenomenology of layered Van der Waals junctions—in fact indicating a novel potential scheme to achieve superlubricity—but they also give direct access to the elementary (atomic-scale) mechanisms that control energy dissipation. Remarkably, the thermally-activated single-asperity Prandtl–Tomlinson (PT) model [57, 58], whose details we will briefly outline hereafter, is found to comprehensively describe friction experiments involving different colloidal probes, material substrates, and sliding regimes. This ultimately bridges mesoscopic CP-AFM tribo-pairs to a well-established theoretical framework of single-asperity nanotribology.

The PT model is indeed one of the most adopted and influential 'minimalistic' models so far suggested for addressing the phenomenology of single-asperity friction. Specifically, it describes effectively friction force microscopy experiments, where tribological forces are measured by dragging a probing tip along a surface. Qualitative conclusions drawn within this framework provide guidance to understand friction at the nanoscale, often retaining their validity in more advanced models and molecular dynamics simulations. In its standard formulation, the PT model assumes that a point mass (mimicking, e.g., the AFM tip) is dragged over a periodic (e.g., sinusoidal) potential representing the interaction between the tip and the underneath crystalline substrate. The probing single-asperity contact is pulled by a spring extending between the tip position and the position of the microscopic support stage, externally driven with constant velocity relative to the substrate. Thus, the total potential energy $V(x, t)$ characterizing the PT tribological system consists of two parts: (i) the tip-substrate interaction, with corrugation amplitude E_0 and lattice periodicity a and (ii) the elastic interaction with effective spring constant k between the tip (at position x) and the external support moving at constant speed v:

$$V(x, t) = -\frac{E_0}{2}\cos\left(\frac{2\pi x}{a}\right) + \frac{1}{2}k(vt - x)^2$$

It should be noted that in a typical AFM-like experiment, the effective 'spring constant', mimicked here by k in the PT model, is not only due to the external torsional stiffness of the cantilever but also comprises the contribution from the lateral stiffness of the mechanical contact. Since in the model there is no attempt to explicitly describe how energy is dissipated into the substrate, a viscous damping coefficient γ is introduced in the equation of motion. This damping is a phenomenological representation of the resultant effect of all the real physical dissipation mechanisms, such as vibrational and electronic excitations, which are too complex and computationally demanding to be made fully explicit. The instantaneous lateral force measured in tribological experiments reads $F_L = k(vt - x)$, and the kinetic friction F_f is evaluated as sufficiently long time average over a dynamical steady-state regime of F_L.

2 Probing Layered Van der Waals Homo- and Hetero-Junctions by CP-AFM: General Phenomenology

In the last decade there has been increasing interest for CP-AFM experiments addressing the settling of superlubricity—under ambient conditions—for prototypical homo-/hetero-junctions based on layered Van der Waals materials. To prepare such contact junctions, colloidal beads have been typically coated by Few-Layers-Graphene (FLG) flakes or graphite, then they have been sled against bulk layered material substrates like HOPG, h-BN or TMDs (MoS_2, WS_2, ...). Normal loads from a few hundreds of nanoNewtons up to a few microNewtons have been used. To obtain the graphene-coated microparticles, four distinct methodologies have been developed: (i) direct graphene growth [59], (ii) triboinduced graphene transfer [12, 43, 60], (iii) all-dry graphene transfer [45, 49], or (iv) wet graphene transfer [61]. Besides these approaches, some studies have taken into consideration the emergence of superlubricity for metal-coated CP sled against graphite [62]. Despite unavoidable differences in the lateral size, morphology, composition and cleanliness of the contact interfaces thus prepared, well-recognizable features clearly emerged from such experiments, that are briefly summarized hereafter:

- **Ultralow friction coefficients**: friction force versus normal load (F_f versus F_N) characteristics, measured on micrometric and sub-micrometric length scales, typically show a linear trend, yielding ultralow values ($\ll 0.01$) of the (differential) friction coefficient μ;
- **Small contact adhesion**: normal load F_N versus probe-substrate distance D (F_N versus D) characteristics attest a small contact adhesion F_A, varying from a few tens to (at most) a few hundreds of nanoNewtons;
- **Robustness of the friction response against experimental conditions:** friction force F_f is almost insensitive to variations of the sliding velocity (from below 1 μm/s up to 100 μm/s), sliding direction (of the long-axis of the AFM cantilever relative to the substrate) and ambient Relative Humidity (RH);

As a matter of fact, Liu et al. [59] showed that the friction coefficient between SiO_2 microspheres (nominal radius $R = 4$ μm) and HOPG decreased under ambient conditions ($RH = 18\% \pm 3\%$) from $\mu = 0.046$ to $\mu = 0.003$ upon direct growth of a nm-thick Multi-Layer-Graphene (MLG) coating on the CPs (Fig. 2a). Such a tribo-pair displayed a remarkably stable friction force $F_f \sim$ 10nN for about 7000 s under $F_N = 1.45$ μN, that corresponds to $\sim$ 1 GPa contact pressure. A small friction coefficient $\mu = 0.0025$ was also achieved for the MLG/HBN interface. Notably, the typical adhesion F_A at the MLG/HOPG tribo-pair amounted to only $\sim$ 70 nN. Li et al. [12] in turn demonstrated that the friction coefficient between pristine SiO_2 microspheres (nominal radius $R = 11.5\ \mu m$) and HOPG decreased under ambient conditions ($RH = 40 - 70\%$) from $\mu = 0.012$ to only $\mu = 0.0003$ upon the tribo-transfer of nanosized Few-Layer-Graphene (FLG) flakes at the bead surface (Fig. 2b). The friction force F_f was in such a case largely below 1 nN, and it was stable over sliding distances as long as 40 mm under a continuous normal load of 185 nN. Again, the typical adhesion force at the FLG/HOPG tribo-pair amounted to about $F_A \sim 70$ nN. Li et al. [45] studied the friction response of graphene-coated CPs prepared by dry transfer of micrometric MLG flakes onto silica beads (nominal radius $R = 11.5\ \mu m$). The F_f versus F_N characteristics measured under ambient conditions ($RH = 40 - 60\%$) on HOPG, h-BN, MoS_2 and WS_2 were linear up to a maximum normal load of $\sim$ 300nN, with friction coefficients in the range 0.0005–0.003 depending on the specific tribo-pair. Friction force was stable over sliding distances of 80mm under the normal load 180 nN. The adhesion force was as small as $F_A \sim 20 - 30$nN. Similar results were reported by other studies, and they are summarized for simplicity in the general F_f (measured at $F_N = 0$nN) versus F_A plot (Fig. 2c). This plot has the advantage that allows to compare different AFM experiments, since the friction force F_f (at $F_N = 0$nN) represents a well-defined quantity that can be readily estimated from experimental F_f versus F_N curves, and it is affected only by the (small) adhesion force F_A.

One can observe that the range of the spanned friction forces is quite broad (0.1–10 nN), with the hetero-junctions systematically showing a higher friction than the graphene-based homo-junctions. Quite interestingly, the graph shows that all the CP-AFM experiments fall into a rather restricted adhesion region, indeed not far from the outcome of AFM measurements conducted with conventional nanosized probes. In this respect it is also instructive to consider that the measured adhesion F_A is one order of magnitude smaller than theoretically expected for a micrometric colloidal bead in contact with an HOPG substrate under standard ambient conditions. In fact assuming a bead with nominal curvature radius $R = 12.6\ \mu m$ (as in [60]), i.e. slightly larger that $R = 11.5\ \mu m$ used in [12, 45], and a sphere-on-flat contact geometry, one can estimate a Van der Waals (VdW) attraction of about:

$$F_{\mathrm{VdW}}(\delta_0) = \frac{A_H R}{6(\delta_0 + D_0)^2} \sim 200\, nN$$

where $A_H = 1.35 \times 10^{-9}$J is the Hamaker constant for the silica-HOPG interface [63] and $D_0 = 0.17$nm is the interatomic spacing at the contact spot [64]. The silica interfacial roughness here limits the distance of closest approach between the two contacting surfaces to a separation $\delta_0 = 1$nm. On the other side, using the theoretical model by Farshchi-Tabrizi et al. [64], the capillary force of the expected water meniscus condensed around the bead-HOPG contact amounts to more than ~1 μN for $RH > 40\%$. Therefore one concludes that the total adhesion of a smooth micrometric bead with $R = 12.6\ \mu m$ in contact with HOPG is expected to be dominated by the capillary force, so that it should overcome the adhesion F_A measured in experiments by more than a factor 10. In turn, calculations of the capillary adhesion for

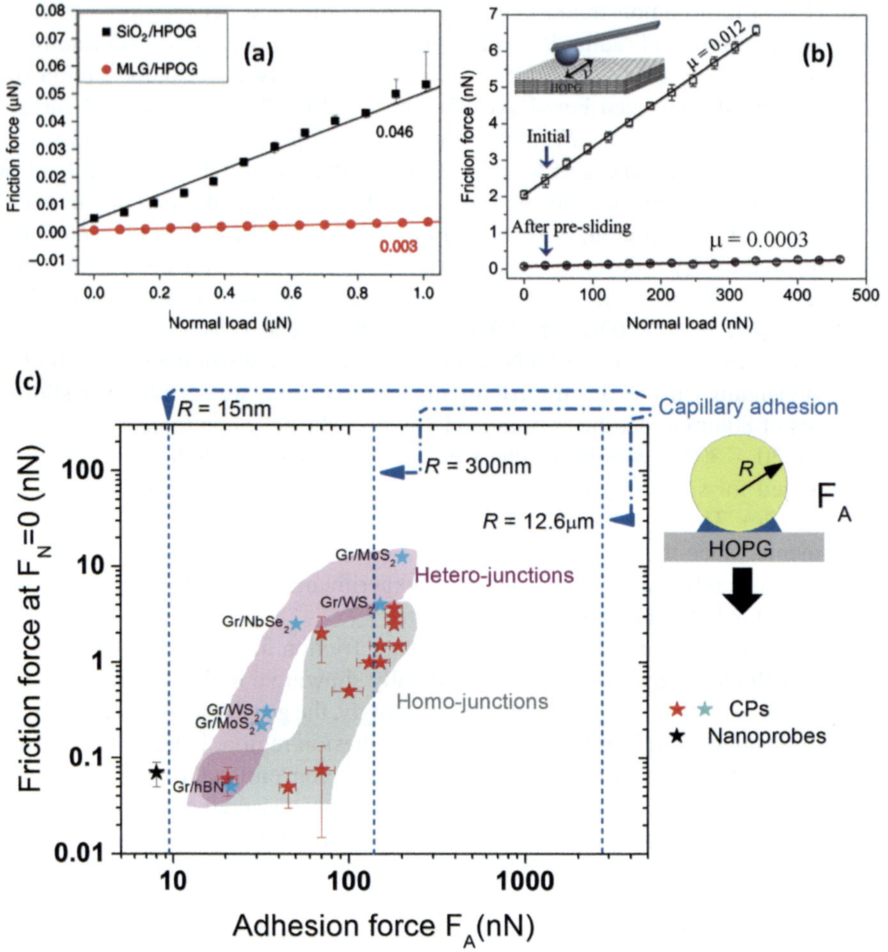

Fig. 2 **a** Friction force as a function of the applied normal load for a SiO_2 microsphere sliding against HOPG (SiO_2/HOPG) and a MLG-coated SiO_2 microsphere against HOPG (MLG/HOPG), respectively. Adapted from [59] under the CC BY 4.0 license (https://creativecommons.org/licenses/by/4.0/). **b** Friction force versus normal load for a SiO_2 microsphere sled on HOPG, respectively in the pristine ('initial') state and after tribo-transfer of nanosized FLG flakes onto the microsphere contact region (named 'pre-sliding'). Adapted from [12] under the CC BY 4.0 license (https://creativecommons.org/licenses/by/4.0/). **c** Plot F_f (measured at $F_N = 0$nN) versus F_A obtained from AFM experiments conducted with CPs [12, 45, 59, 60, 62] and standard nanoprobes [11]. Here, the abbreviation Gr stands for the graphene-coated probe

much smaller probes, namely with $R = 15$nm and $R = 300$nm respectively (e.g. see [60, 64]), give lower and upper bounds to the capillary adhesion F_A measured in CPs experiments. Hence, Fig. 2c strongly suggests that CP-AFM experiments mainly probe superlubricity of nanoscopic (rather than mesoscopic) contact spots.

In this respect, both Liu et al. [59] and Li et al. [12, 45] speculated that superlubricity relies on the surface roughness of the microspheres, that incorporates elevated nanoasperities covered by randomly oriented graphene nanopatches. Such a multi-asperity interfacial geometry involves the cumulative effect of many incommensurate nanocontacts and is expected to be robust against variations of load, humidity or orientational effects. On one hand this scenario holds great promises, as it suggests a conceptual route to extend structural lubricity from nanoscopic contacts to macroscopic (rough) ones. On the other hand, it demands a deeper understanding of the general mechanisms that might assist the formation of the incommensurate graphene nanopatches. This is particularly relevant for the situations in which the patches do not originate from the direct graphene growth (as in [59]) but from a more common tribo-induced TL (e.g. [12, 45]), so that the following key questions need to be properly addressed:

- How does the tribo-induced TL roughness depend on common properties (size, composition) of the CPs as well as on the environmental humidity?
- How does the TL morphology impact contact mechanics?
- Is the ultralow coefficient of friction accompanied by atomic-scale nearly-dissipationless sliding, as prescribed by structural lubricity? More in general, which are the atomic-scale dissipation mechanisms for such interfaces?
- What is the role of the normally applied load and of the sliding velocity?
- Does superlubricity manifest when CPs are lubricated by FLG flakes obtained from technologically-relevant, high-throughput scalable production routes?

The next sections contain experimental evidences and theoretical arguments that take us a step forward in understanding sliding friction phenomena with the graphene-coated CPs.

3 The Surface Morphology of CPs: Evidences for a Few Superlubric Contact Nanoasperities

The morphology of the CPs can be routinely inspected in the course of friction experiments using complementary high-resolution microscopy techniques. Investigations typically focus on the morphological, chemical and structural variations taking place at the CPs' surface close to the nominal contact area. In general, when a smooth CP is placed in contact with an atomically-smooth substrate like HOPG, the nominal contact area can be estimated by Hertzian contact mechanics to be at the mesoscale, roughly in between 10^3nm^2 ($\sim$ 100 nm contact spot for a bead radius $R \sim 2$ μm) and 10^6 nm^2 ($\sim$ 500nm contact spot for $R \sim 25$ μm) (e.g. see [43]). As discussed below, however, such contact area provides only an upper limit to the real contact area. Due to the ubiquitous roughness of the micrometric beads, it is very likely that each CP interacts with an elastically-stiff and atomically-smooth substrate only at a few dominant nanoasperities.

3.1 CPs Coated by Direct Graphene Growth

Liu et al. [59] characterized the surface morphology of representative graphene-coated CPs prepared by chemical vapor deposition (CVD) of MLG onto silica microspheres. They claimed that after CVD, the silica surface was typically covered with a continuous and homogeneous MLG film, as attested by Raman spectroscopy and cross-sectional High Resolution Transmission Electron Microscopy (HRTEM). Yet, the graphene film was nanorough due to the original root mean square (rms) roughness $\sigma \sim 23$nm of the bare microspheres. This evidence, together with the linear (Amonton's like) F_f versus F_N characteristics, led the authors to claim that the contact between the graphene-coated beads and HOPG is in the multi-asperity regime. More specifically, by exploitation of contact mechanics numerical simulations carried out in the elastic limit, Liu et al. [59] found that only a few nanoasperities of the graphene-coated CPs were in direct contact with HOPG under a normally applied load as high as 1.4 μN.

3.2 CPs Coated by Tribo-Induced Graphene Transfer

Li et al. [12] investigated the surface morphology of graphene-coated CPs prepared by tribo-induced transfer of FLG flakes from a graphite substrate to the CPs. To this end, silica CPs were repeatedly sled against HOPG until a remarkable friction drop was observed. It is well known that during the sliding motion, graphite nanoflakes are transferred from graphite to the sliding probe [10]. The topographies of the CPs were investigated by AFM as well as by Field-Emission Scanning Electron Microscopy (FESEM) respectively before and after the interaction with HOPG (called 'pre-sliding'). AFM morphologies attested the presence of nanoasperities over the pristine silica surface with an overall surface roughness around 1 nm (Fig. 3a). After 'pre-sliding', a similar surface roughness was detected by AFM, but FESEM additionally revealed the presence of small FLG flakes dispersed within a $\sim$ 500 nm contact zone (Fig. 3b). The chemical composition detected by X-ray photoelectron spectroscopy was in accordance with the expectations for pristine graphene. Analysis by HRTEM of the structure of the tribo-transferred nanoflakes (Fig. 3c) provided an approximate thickness of $\sim$ 2 nm and a composite structure with multiple layers mixed with amorphous structures. The spacing of the multiple layers was about 0.38 nm, a little larger than that of pristine graphite, which may be attributed to the introduction of oxygen-containing functional groups during the tribo-exfoliated process. The authors concluded that the real contact area between the graphene-coated CPs and HOPG was indeed very small owing to the graphene-coated nanoasperities within the nominal contact zone (multi-asperity regime).

Buzio et al. [43, 60] explored in more detail the build-up of the tribo-induced graphene-based TLs. On one side, they went beyond the case of the ideally smooth silica CPs addressed by Li et al. [12], by considering a representative set of nominally rigid CPs having different curvature radius (R ~ 2–25 μm) and pristine surface

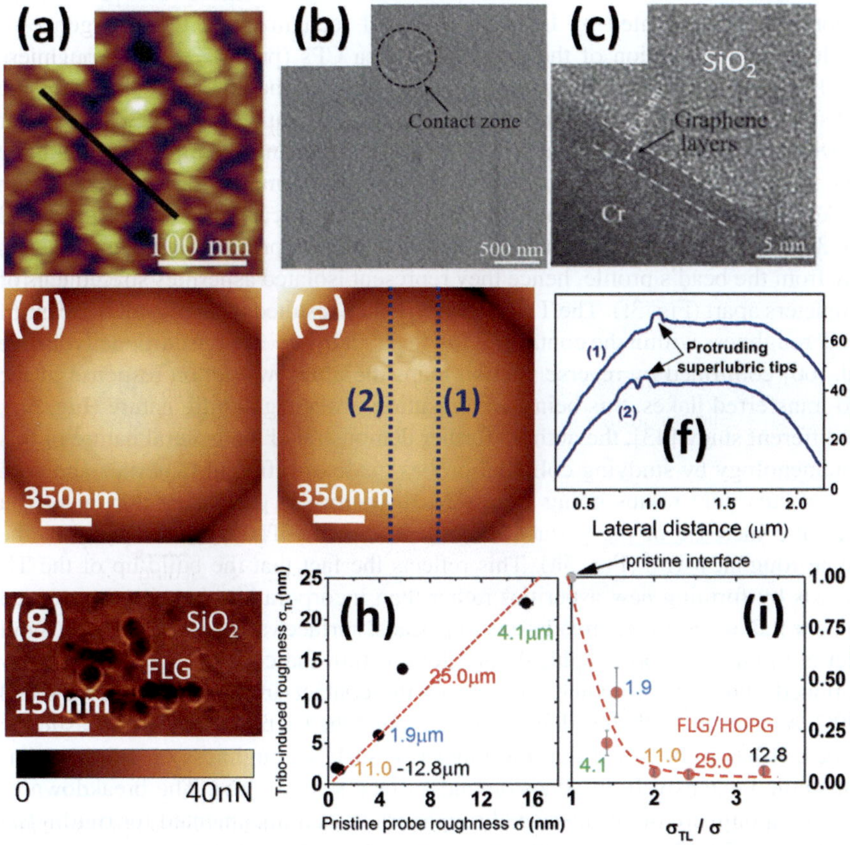

Fig. 3 **a** AFM topography on the top central region of a pristine SiO_2 probe, **b** FESEM image of the same probe after TL formation within the contact zone, **c** HRTEM image of the cross-section of the top center surface region in (**b**): reprinted from [12] under the CC BY 4.0 license (https://creativecommons.org/licenses/by/4.0/). **d** AFM topography of a pristine silica bead (radius $R = 12.8$ μm), **e** morphology of the same bead after the formation of the TL, **f** cross sections along the dashed lines in (**e**); **g** reverse AFM friction map of the TL in (**e**): reprinted from [60] copyright σ_{TL} versus σ RR © 2021 Elsevier Ltd. with permission from Elsevier. **h** Plot of and **i** plot of normalized adhesion versus normalized roughness, obtained using CPs with different micrometric curvature radius(the value of is indicated in color text; see the main text for the symbols meaning); reprinted with permission from [43], copyright © 2022 American Chemical Society

roughness ($\sigma \sim 0.5 - 15$nm). On the other hand, they extensively characterized the TL morphology by means of reverse AFM and friction force microscopy imaging.

To assist the triboinduced material transfer from the sliding HOPG substrate to the pristine beads, CPs were intentionally driven with the sliding direction perpendicular to the local orientation of the surface steps. This choice in fact promotes material transfer from HOPG to the AFM probe [49]. As scanning progressed, a permanent drop of the friction force to a few nanoNewton took place, this being a typical signature for the formation of the TL. Adhesion force F_A was also observed to decrease below $\sim$ 200 nN. The overall phenomenology resembles the 'pre-sliding' phenomenon documented by Li et al. [12] and recognized as the emergence of superlubricity. Inspection of the smoothest silica CPs (pristine surface roughness $\sigma \sim 0.5$nm) by reverse AFM imaging (Fig. 3d) showed that in the pristine state, the bead's surface was free from bumps; cross sections were satisfyingly fitted by circular fits, with R close to the nominal one (12.6 μm). AFM analysis of a representative probe after repeatedly sliding on HOPG revealed the formation of a TL, consisting of flakes with lateral size $\sim$ 50 nm clustered within an apical region of the silica bead (Fig. 3e). Cross sectional heights showed that the flakes protrude a few nanometers away from the bead's profile, hence they represent isolated asperities spaced tens of nanometers apart (Fig. 3f). The TL formation hence resulted in an overall increase of the CP roughness within the contact region to about $\sigma_{TL} \sim 2$nm. Additionally, Buzio et al. [60] confirmed by reverse friction maps the ultralow friction response of the tribo-transferred flakes, this being an indication of their graphitic nature (Fig. 3g). In a different study [43], the authors further demonstrated the general nature of this phenomenology by studying colloidal probes made of different materials and with nominal curvature radius going from ~2 μm up to ~25 μm. They documented a systematic increase of the surface roughness with the TL (σ_{TL}) compared to the pristine roughness (σ) (Fig. 3h). This reflects the fact that the build-up of the TL proceeds by forming new asperities rather than incorporating the tribo-transferred graphene flakes within the already-existing beads' surface valleys, so that their overall effect is to enhance topographical variations within the contact region. As briefly mentioned above, TL roughness decreases the contact area between the CPs and HOPG. As a result, both the adhesive force F_A and the friction force were drastically reduced. It has to be noted that the impact of surface roughness on adhesion has been studied extensively in adhesion and surface science [65]. The breakdown of adhesion upon increase of interfacial roughness is well documented for rigid interfaces [38, 66–71]. It is noteworthy that a plot of the normalized adhesion (i.e. the ratio $F_{A,TL}/F_A$ between adhesion measured with/without the TL) versus normalized roughness (σ_{TL}/σ) (Fig. 3i) indicates the existence of a "minimum adhesion plateau" for the TL/HOPG interface (for $\sigma_{TL} > 2\sigma$), in analogy with similar trends found for other systems [67, 68, 70]. This plateau corresponds to the adhesion experienced at a graphitic nanosized contact spot, and it gives reason for the small adhesion force experienced by such tribointerfaces.

Buzio et al. [43] also addressed the impact of the TL nanoasperities on the CP-HOPG contact mechanics. They speculated that the existence of distinct lubric asperities over the beads' surface does not always imply a multiasperity contact regime.

To clarify this issue, they observed that the TL formation was accompanied also by emergence of multiple-tip effects in AFM imaging [43, 60]. Topographies, in such a case, displayed repetitive patterns revealing the arrangement of the most protruding asperities affecting contact mechanics. It is however well known that in presence of multiple-tips, normal and shear forces are very often supported by one asperity only i.e., the most prominent one. Buzio et al. [43] further supported this picture by exploiting scaling arguments derived from large scale molecular dynamics simulations [70, 72], that suggest that the TL/HOPG interface can evolve toward a single-asperity nanocontact in the small-loads limit (e.g., $F_N \sim 10$nN). Accordingly, neither the nominal curvature radius R nor the pristine surface roughness σ of the CPs are expected to significantly impact the TL/HOPG contact mechanics and superlubricity very likely reflects the friction response of an individual nanocontact or at most of a few nanocontacts. This conclusion also agrees with the friction phenomenology reported by Liu et al. [59] for CPs coated by direct graphene growth.

3.3 CPs Coated by Graphene Wet-Transfer Methods

According to background literature, coating AFM probes by means of liquid-processed 2D nanomaterials provides an effective strategy to test the fundamental contact properties of novel homo- and hetero-junctions (e.g. see [73, 74]), as well as to explore the occurrence of superlubricity in sliding friction experiments. Compared to mechanically cleaved or grown samples, liquid dispersions of 2D flakes are in general more suitable for low-cost mass production. Additionally, they are virtually able to readily and conformally coat interfaces with arbitrary geometry. As a result, liquid dispersions are gaining attention to reduce friction in practical applications [75]. In the field of AFM experiments, nanoprobes coated by liquid-processed graphene flakes have shown both smaller adhesion and friction together with improved lifetime [76], albeit in-depth investigations of their contact properties and ultimate friction performances were not attempted. State-of-the-art solution-processed SLG/FLG flakes can display ultralow friction response as bulk graphite or cleaved graphene [77], nonetheless the output of the preparation, in terms of achieved lubrication, is affected by a high degree of uncertainty arising from their broad distribution in thickness and size. These factors—together with the variability associated with their random stacking over the probes' surface—ultimately result in a high degree of inhomogeneity for the wet-transferred graphene coatings.

Recently, Buzio et al. [61] used a high-quality aqueous dispersion of Single-Layer-Graphene (SLG) and FLG flakes to obtain lubricious wet-transferred coatings over silica CPs via a modified drop-casting method. A combination of FESEM and reverse AFM imaging was then exploited to identify the (wet-transferred) graphene asperities that bear contact forces. Importantly, the use of a surfactant-free dispersion provided lubricious wet-transferred coatings without the need to include a high-temperature annealing step [77]. Micrographs showed that the graphene coating was inhomogeneous at the sub-micrometric scale, the silica surface being only partially coated

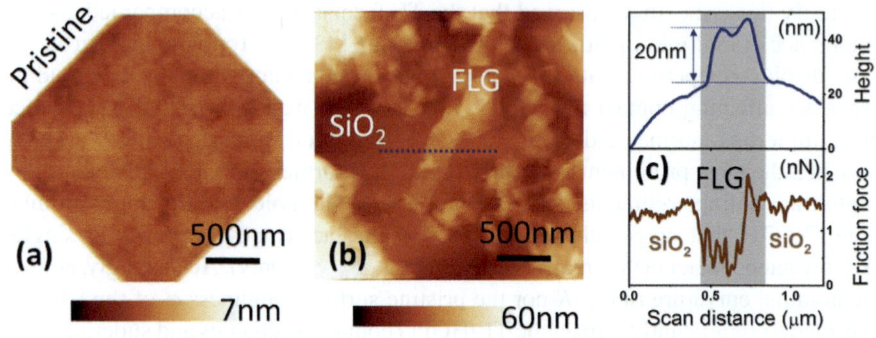

Fig. 4 **a** Reverse AFM topography of the top central region of a pristine SiO_2 CP (after subtraction of the spherical shape) and **b** AFM topography of the same probe after deposition of FLG flakes from a water-based liquid dispersion; **c** cross-section height and friction force along the dotted line in (**b**), showing the lubricious behavior of the deposited flakes: adapted and reprinted from [61] under the CC BY 4.0 license (https://creativecommons.org/licenses/by/4.0/)

by tens-of-nanometers tall protrusions formed by randomly stacked and/or crumpled flakes (Fig. 4a, b). The agglomerated flakes clearly maintained a graphite-like lubricious behavior after the wet-transfer process (Fig. 4c). In view of the inhomogeneous graphene coverage, the study showed that graphene-mediated adhesion and friction varied from probe to probe. To better explore the impact of the wet-transferred graphene on the junctions' contact mechanics, the surface morphology of an ideally-smooth CP was systematically characterized by AFM in the course of successive graphene deposition runs, together with the evolution of the adhesive force F_A against HOPG. The authors demonstrated that an adhesion breakdown (from an initial value $F_A = 3\ \mu N$ to about 500 nN) took place as soon as the newly deposited FLG flakes formed the topographically highest contact asperity nearby the expected contact zone. Accordingly, Buzio et al. [61] concluded that the main effect of the deposited flakes is to add a lubricious roughness to the CPs, that in turn drives the decrease of the contact area—from meso to nanoscale—and ultimately bears the contact forces. This description qualitatively agrees with the phenomenology depicted in Fig. 2c. The authors provided a rough estimate of $\sim 2 \times 10^2 nm^2$ for the contact area at the topographically highest contact spot with graphite. Furthermore, they showed through sliding friction experiments that the differential friction coefficients was ultralow ($\mu \leq 0.003$) against graphite as well as against representative TMDs substrates (MoS_2 and WS_2).

3.4 CPs Coated by All-Dry Graphene Transfer

A number of studies have reported on the preparation of graphene-coated CPs by all-dry transfer of individual graphene flakes using micromanipulation techniques [45,

78, 79]. In practice, a micromanipulation stage operating under an optical microscope, and equipped with long-working-distance objectives, is employed to manipulate large (tens-of-micrometers-sized) graphene flakes and place them in contact with the surface of a CP, e.g. using a polymer-assisted stamping approach [80]. The graphene flakes are typically prepared by micromechanical exfoliation of HOPG, they have thickness of several nanometers and they are firmly attached to the surface of the CP using a glue (either epoxy resin or UV curable adhesives have been used for the purpose). Complementary high-resolution microscopies and spectroscopies (e.g. HRTEM, FESEM, Raman spectroscopy) can be implemented to characterize the quality of the transferred flakes.

Li et al. [45] have used graphene-coated CPs prepared by all-dry transfer method to study superlubricity in graphene-based homo- and hetero-junctions (the latter with hBN, MoS_2, WS_2). They attested the achievement of (differential) friction coefficients in the range $\mu \sim 0.0005 - 0.003$, for loads up to 280 nN and sliding velocities from 500 nm/s up to 16 μm/s. Experiments were performed under ambient conditions ($RH = 40 - 60\%$), yet the adhesive forces between the graphene-coated CPs and each substrate were as small as 20–30 nN. Indeed, a careful inspection of the surface of representative CPs by reverse AFM imaging indicated that the transferred flakes were not atomically flat, but a roughness of a few nanometers was found. Li et al. [45] concluded that the graphene-coated CPs formed a multi-asperity contact with an atomically-flat countersurface. Yu et al. [79] recently reported about friction measurements conducted with graphene-coated CPs against HOPG and hBN. Ultralow values of the friction coefficient $\mu \sim 0.005 - 0.009$ and an adhesive force $F_A = 20-26$nN were measured under ambient conditions, which are consistent with results by Li et al. [45] despite substantial differences in the transferred flakes thickness and lateral size, as well as in the details of the transfer method. Yu et al. [79] did not provide detailed information on the surface roughness of the transfer flakes nearby the contact zone. It is noteworthy that the contact properties of the graphene-based CPs prepared by dry-transfer methods do agree with the phenomenology resumed in Fig. 2c.

4 Probing the Elementary Dissipation Mechanisms by CP-AFM

4.1 Stick–Slip Dynamics in Graphene-Based Homo- and Hetero-Junctions

It is well established within AFM nanotribology that structural lubricity is characterized, at the atomic scale, by the suppression of stick–slip dissipative motion in favor of nearly continuous sliding [9, 81]. This is the result of the ideal suppression of all the energy barriers to sliding motion due to contact interface incommensurability, which mutually cancels out substrate lateral forces. The direct observation by CP-AFM of atomic-scale lateral force traces hence represents an invaluable route to unravel the mechanisms that underpin superlubricity in mesoscopic contact junctions. In principle, the phenomenology of atomic-scale friction for the graphene-coated CPs is expected to agree with that reported for other graphitic nanocontacts [9, 82], in view of the small size of the protruding nanoasperities (Figs. 3f and 4c) and the small contact forces that usually come into play ($F_N < 1$ μN, $F_A \sim 10$–500 nN, $F_f \sim 1$ nN; see Fig. 2).

Atomic-friction data for, respectively, graphene-wrapped AFM nanoprobes and graphene-coated CPs sliding on HOPG, have been reported in recent studies [11, 49, 79], albeit no systematic investigation of their load and velocity dependence were attempted. Buzio et al. [43, 60] originally implemented atomic-scale friction force spectroscopy for CPs made of different materials (SiO_2, borosilicate glass, titanium) and coated by tribo-induced FLG flakes. Experiments first of all considered the case of graphitic homo-junctions, prepared by sliding the graphene-coated CPs onto freshly-cleaved HOPG. Within a broad range of normal loads, 100nN $< F_N <$ 700nN, atomic-scale friction maps acquired with different CPs invariably showed stick–slip motion with the hexagonal symmetry of the underneath graphite lattice (Fig. 5a).

Despite the measurement of extremely low (differential) friction coefficients $\mu \sim 3 \times 10^{-4}$ (Fig. 5b), the ubiquitous presence of stick–slip motion indeed signaled the appearance of an interlocking mechanism between the tribo-induced graphitic nanoasperities of each CP and the underneath surface. This indicates that structural lubricity was likely hampered—within that specific range of normal loads—by different contact pinning effects. Quite interestingly, F_f versus F_N curves were affected by some variability, reflecting changes in the fine details of the sliding interface during the experiments (Fig. 5b). On one hand there were curves with a weak dependence on F_N and finite friction down the pull-off point (that were named "type I" friction curves). On the other side, there were entirely different characteristics showing vanishing friction at the lowest load values, and a monotonic increase at some point (named "type II" curves). Notably, each CP usually showed characteristics of both types regardless of the bead material, curvature radius and pristine surface roughness. Such a friction duality, with two distinct regimes of finite and vanishing friction, was already observed in nanoparticles manipulation experiments conducted on HOPG [83]. Also the AFM experiments addressing superlubricity in nanoscale graphene homo-junctions [9] recognized two entirely different types of friction behaviors, named respectively, 'high-friction state' and 'low friction state'. The most relevant feature for superlubric sliding of CPs is certainly related to the "type II" friction curves, that showed an ultralow friction force ~ 0.1 nN persisting up to compressive loads ~10–100 nN. These friction curves perfectly match the results previously discussed in the framework of incommensurate contacts [12]. Buzio et al. [43, 60] inspected the AFM friction loops and observed that the mechanical dissipation greatly evolved with the normal load F_N, from nearly-frictionless continuous sliding to regular stick–slip, and to highly-dissipative (multi-slip) stick–slip (Fig. 5c). This behavior was found to hold for probes having different curvature radii, up to $R \sim 25$ μm, and under different ambient humidity conditions (relative

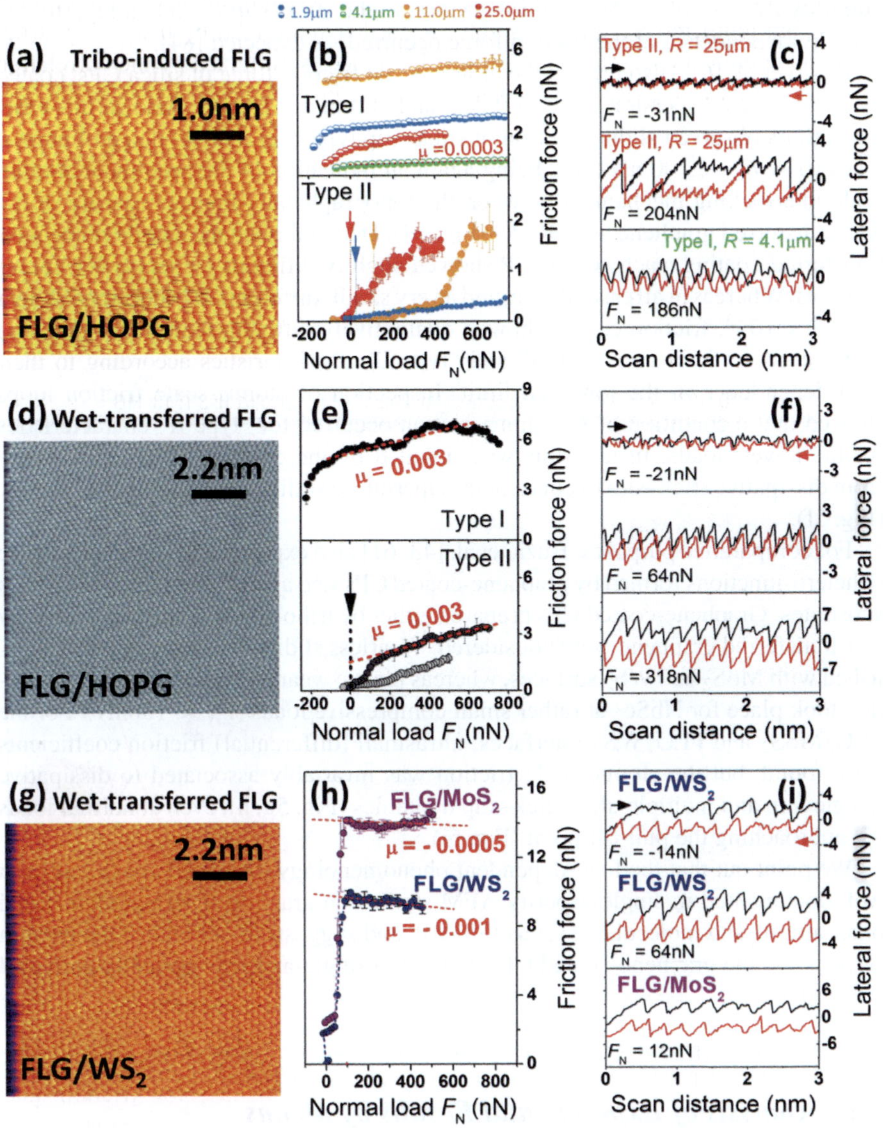

Fig. 5 **a** Atomic-scale friction map recorded on HOPG by means of large CP ($R = 25$ μm) coated by a tribo-induced TL, **b** representative friction versus load curves acquired with graphene-coated CPs having different radius R and **c** representative load-dependent friction loops; reprinted with permission from [43], copyright $WS_2MoS_2WS_2$ © 2022 American Chemical Society. **d** Atomic-scale friction map recorded on HOPG by means of a CP coated by wet-transferred FLG flakes, **e** typical friction versus load curves obtained with three graphene-coated CPs, **f** load-dependent friction loops, **g** atomic-scale friction map recorded on by means of the graphene-coated CP in (**d**–**f**), **h** representative friction versus load curves acquired on and: adapted and reprinted from [61] under the CC BY 4.0 license (https://creativecommons.org/licenses/by/4.0/). **i** Load-dependent friction loops for layered hetero-juntions (R. Buzio, A. Gerbi, C. Bernini, L. Repetto, A. Silva and A. Vanossi, *unpublished data*)

humidity $RH \sim 30\% - 60\%$). Furthermore, in the stick–slip regime, a logarithmic velocity-dependence of the friction force occurred, as expected [84].

Buzio et al. [61] also probed the atomic-scale friction force of silica CPs, coated by an aqueous dispersion of FLG flakes and sliding on HOPG. Also in this case, the authors observed sliding motion of stick–slip type over a broad range of normal loads ($0 < F_N < 700$ nN), this being again an indication of the prominent atomic-scale interlocking mechanism between the topographically highest asperity of the wet-transferred graphene and HOPG (Fig. 5d). Depending on the details of the wet-transferred coating, each coated CP showed slightly different F_f versus F_N curves (Fig. 5e). Whereas in all cases they found a very small (differential) friction coefficient $\mu \sim 3 \times 10^{-3}$, friction curves showed qualitatively different features. Again, one might classify them into "type I" and "type II" characteristics according to their load dependence in the low-load limit. Inspection of atomic-scale friction loops showed that a condition of vanishing friction occurred for "type II" characteristics for the lowest loads. In fact, this was signaled by the change of the lateral force from dissipative stick–slip to continuous superlubric sliding on reducing the load F_N (Fig. 5f).

For comparative purposes, Buzio et al. [43, 61] also explored the friction response of hetero-junctions formed by graphene-coated CPs sled against single-crystal TMDs substrates. Graphene-coated CPs prepared either by tribo-transferred graphene or by wet-transferred graphene were considered. Wearless sliding was successfully established with MoS_2 and WS_2 surfaces, whereas plastic wear and interfacial reorganization took place for $NbSe_2$ at rather small compressive loads ($F_N < 150$nN). For the FLG/MoS_2 and FLG/WS_2 interfaces, ultrasmall (differential) friction coefficients were found, but the atomic-scale friction was invariably associated to dissipative (single-slip and multiple-slip) stick–slip dynamics (Fig. 5g, h), even at normal loads F_N approaching the pull-off point (Fig. 5i).

We point out that the load-dependent phenomenology discussed above has been well documented in single-asperity AFM studies on graphite, TMDs, [85–87] and other atomic and molecular crystals [88–90], and suggests to rationalize the friction response of the graphene-coated CPs within the single-asperity thermally-activated PT model.

4.2 *Analysis of Experimental Results by Means of the Thermally-Activated PT Model*

In order to address and interpret friction force microscopy data, where resistive lateral forces are typically measured by dragging a proximal probe along a surface, we need to detail some further ingredients of the previously introduced PT approach. The PT model predicts two distinct sliding regimes for the tip motion depending on the dimensionless parameter $\eta = 2\pi^2 E_0/(ka^2)$, which represents the ratio between the tip-substrate potential energy and the effective elastic energy of the driving spring.

When $\eta \leq 1$, the total potential energy has a single minimum at any time and the proximal probe moves smoothly over the potential with vanishing dissipation. On the contrary, two or more minima appear in the energy landscape for $\eta > 1$, so that the tip dynamics becomes intermittent (i.e. of stick–slip type) and dissipation occurs. The value $\eta = 1$ marks the transition from smooth sliding to slips by one substrate-lattice spacing. Stick–slip motion corresponds to jumps of the tip between successive minima of the energy landscape due to mechanical instabilities induced by the effective system elasticity (comprising both the intrinsic contact stiffness and the external driving spring). The possibility of slips of higher multiplicity (multiple-slip regime [86]) occurs for larger values of η, as clearly exemplified, by $T = 0$ numerical simulations, in Fig. 6a. It should be noted that this is a necessary but not sufficient condition to observe multiple slips, since the frictional dynamics also depends on the damping coefficient γ [91]. Specifically, by favoring inertial effects, an underdamped regime (small γ) promotes more easily multiple slips compared to an overdamped dynamics (large γ), as illustrated for different values of the damping in Fig. 6b.

It should be noted that, in real systems, at finite temperature, hysteresis and dissipation must always disappear in the zero-speed limit of adiabatic sliding, where stick–slip instabilities are preempted by thermal fluctuations (thermolubricity). A

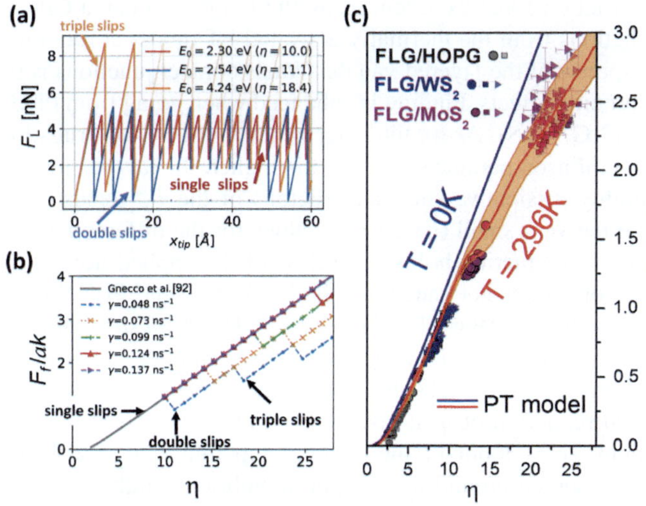

Fig. 6 **a** Force traces as a function of tip position for the three values of E_0 (thus η) reported in the legend. The labels mark the traces corresponding to single, double, and triple slips. The traces were obtained at $\gamma = 0.048\text{ns}^{-1}$. **b** Normalized friction force F_f/ak as a function of η, for the values of the damping γ reported in the legend. The solid gray line shows the analytical result from [92]. **c** Comparison of experimental normalized-friction-force versus normal-load (F_f/ak versus F_N) data with predictions from the PT model, for three graphene-coated CPs (indicated by circles, squares and triangles respectively). The PT confidence level is shown in orange; reprinted from [61] under the CC BY 4.0 license (https://creativecommons.org/licenses/by/4.0/)

useful way to summarize the rich sliding phenomenology of the PT model lies in the inspection of the graph of Fig. 6b, displaying the normalized friction force F_f/ak as a function of the Tomlinson parameter η. Here one can observe that as the system transitions from overdamped (large γ) to underdamped (small γ), dynamics evolves with η from single stick–slip to multiple stick–slip, following specific branches that impart a characteristic downward flexion to the plot (compared to the overdamped linear limit case [92]).

Buzio et al. [43, 60, 61] established a tight comparison between predictions of the PT model and CP-AFM experiments using graphene-coated CPs. To this end, the load-dependent interfacial parameters E_0, k and η have been systematically extracted from F_f versus F_N curves measured by means of different CPs [93]. It has been shown that the potential corrugation E_0 depends weakly on F_N and typically assumes the highest values for the FLG/MoS_2 and FLG/WS_2 hetero-interfaces. For the FLG/HOPG homo-junctions, E_0 achieved values as small as $\sim$ 0.5 eV in the low-loads limit $F_N < 100$ nN. Since the lateral contact stiffness k in turn varied less from one interface to the other, the Tomlinson parameter $\eta \propto E_0/k$ acquired different values for the FLG/HOPG, FLG/MoS_2, and FLG/WS_2 junctions that mainly reflected the splitting of the interfacial corrugation E_0. By renormalization of the friction force by ak, F_f versus F_N curves were thus mapped into the adimensional F_f^* versus η curves, with $F_f^* \equiv F_f/ak$ (Fig. 6c). This graph demonstrates that the friction force experienced by the graphene-coated CPs agrees very well with predictions from the thermally-activated PT model at room temperature. The friction response of the layered van der Waals contact junctions is thus greatly influenced by pinning effects. Pinning is enhanced when going from FLG/HOPG to FLG/WS_2 and FLG/MoS_2. As for the latter two, the effective corrugation potential E_0 (hence the Tomlinson parameter $\eta \propto E_0$) is higher, the layered hetero-junctions fall into the highly dissipative stick–slip regions of the F_f^* versus η plot. This is the case despite the very small (measured) values for the differential coefficient of friction μ. Also, this situation holds regardless of the applied normal load F_N, so that the friction force does not vanish for the FLG/WS_2 and FLG/MoS_2 junctions even in the low loads limit (Fig. 5h). For the FLG/HOPG system, however, continuous superlubric sliding is possible since η assumes smaller values ($2.0 < \eta < 5$) and the contact moves from stick–slip to continuous superlubric sliding thanks to a load-controlled reduction of η. It has to be pointed out that at the sliding speed of the experiments ($v \sim 30$nm/s), the system temperature ($T = 296$K) makes this transition to be observed around $\eta \sim 2$ (thermolubricity) rather than at $\eta = 1$ as predicted by the athermal PT model. On the other hand, the hetero-junctions enter the multi-slip sliding regime at $\eta < 10$ [86]. This change is accompanied by a downward flexion of both the theoretical curve and the experimental data in Fig. 6c. In this underdamped, multi-slip regime, the details of the contact become progressively more important as η increases, resulting in a sizable spread of the experimental data and increased confidence level of the theoretical curve. The origin of the double slips is thus rooted in the underdamped nature of the contact (see Fig. 6b).

Buzio et al. [43, 60, 61] thus revealed that, despite differences from probe to probe originated by the variability of the graphene coating (either tribo-induced or wet-transferred), the thermally-activated PT model provides a fruitful and comprehensive framework to rationalize mechanical dissipation and superlubricity in layered van der Waals contact junctions as probed by CP-AFM.

5 Outlook

The CP-AFM technique is a powerful and flexible methodology to address superlubricity phenomena in 2D-material-based homo- and hetero-junctions. To date, CP-AFM friction experiments have mostly involved CPs coated by FLG flakes, namely tribo-induced ones, or deposited by micromechanical manipulation or from the liquid phase respectively. Most CP-AFM studies have consistently confirmed the achievement of superlubricity when sliding over selected atomically-smooth crystalline substrates (HOPG, hBN, MoS_2, WS_2…). This is a low-adhesion condition characterized by ultrasmall values of the (differential) friction coefficient, $\mu \ll 0.01$. Superlubric sliding was shown to be robust against variations of the CPs composition and geometry (curvature radius, pristine surface roughness), ambient humidity and sliding parameters (normal load $F_N < 1\ \mu$N, sliding velocity $v = 1$ nm/s–10 μm/s). Nowadays, a deeper insight into the underlying mechanisms has been achieved thanks to the systematic investigation of the CPs morphology and of the atomic-scale lateral force dynamics. In fact:

- High-resolution microscopy techniques reveal that the graphene-coated CPs surface is dominated by ubiquitous, lubricious nanoroughness. Both experimental evidences and theoretical arguments—invoking large-scale contact mechanics simulations and modeling of adhesive forces—suggest the contact mechanics to be nearby the crossover from single-asperity to multiasperities. Asperities decrease the CP-substrate true contact area from mesoscale to nanoscale, and therefore lead to low adhesion.
- Mechanical dissipation is controlled by the specific energy landscape experienced by a few contact nanoasperities. Experiments show well-known signatures of nanofriction dynamics, such as (i) the appearance of friction duality with two distinct regimes of finite and vanishing friction, and (ii) a load-driven transition from continuous sliding to stick–slip at the atomic scale.
- A quantitative analysis of atomic-scale friction loops reveals the excellent agreement between the friction dynamics of the graphene-coated CPs and the thermally-activated single-asperity PT model. A truly vanishing friction state is found only in the low-loads limit ($F_N < 100$nN) and for the graphitic homo-junctions (having indeed the smallest values of the interfacial potential corrugation).

Accordingly, one can ascribe the origin of the vanishing friction state to interfacial lattice mismatch (structural lubricity) at the nanosized contact interface typically formed between the graphene-coated CPs and a graphite substrate. The most common stick–slip dynamics in turn arises from pinning effect, either related to nanoscale interfacial defects or ambient contaminants.

As the current tribology research progresses towards the investigation of novel classes of 2D nanomaterials, CP-AFM offers a unique platform to probe the fundamental energy dissipation mechanisms and deepens our general understanding of those materials properties that assure stable superlubric sliding under technologically relevant working conditions. The biggest challenges in such CP-AFM studies certainly come from the needs to (i) improve control over the coverage, adhesion and surface roughness of the CPs 2D coatings, and to (ii) implement spectro-microscopy techniques offering a clearer perspective on the origin of the pinning sources (e.g. ambient or fabrication-related contaminants, sliding induced defects etc.).

Acknowledgements The Authors gratefully acknowledge financial support from the Italian Ministry of University and Research (MIUR) under the PRIN2017 program, project 20178PZCB5 "UTFROM—Understanding and tuning friction through nanostructure manipulation". A.V. also acknowledges support from the ERC Advanced Grant ULTRADISS, contract 86344023.

References

1. A. Erdemir, J.M. Martin, J. Luo, Introduction, in *Superlubricity*, 2nd edn, eds. by A. Erdemir, J.-M. Martin, J. Luo (Elsevier, 2021), pp. xvii–xix. https://doi.org/10.1016/B978-0-444-64313-1.00031-4
2. J. Spreadborough, The frictional behaviour of graphite. Wear **5**, 18–30 (1962). https://doi.org/10.1016/0043-1648(62)90177-1
3. J.M. Martin, C. Donnet, T. Le Mogne, T. Epicier, Superlubricity of molybdenum disulphide. Phys. Rev. B **48**, 10583–10586 (1993). https://doi.org/10.1103/PhysRevB.48.10583
4. K.S. Novoselov, A. Mishchenko, A. Carvalho, A.H.C. Neto, 2D materials and van der Waals heterostructures. Science **353**, 9439 (2016). https://doi.org/10.1126/science.aac9439
5. A.K. Geim, I.V. Grigorieva, Van der Waals heterostructures. Nature **499**, 419–425 (2013). https://doi.org/10.1038/nature12385
6. D. Marchetto, P. Restuccia, A. Ballestrazzi, M.C. Righi, A. Rota, S. Valeri, Surface passivation by graphene in the lubrication of iron: a comparison with bronze. Carbon **116**, 375–380 (2017). https://doi.org/10.1016/j.carbon.2017.02.011
7. P.G. Grützmacher, S. Suarez, A. Tolosa, C. Gachot, G. Song, B. Wang, V. Presser, F. Mücklich, B. Anasori, A. Rosenkranz, Superior wear-resistance of Ti_3C_2Tx multilayer coatings. ACS Nano **15**, 8216–8224 (2021). https://doi.org/10.1021/acsnano.1c01555
8. G. Tang, Z. Wu, F. Su, H. Wang, X. Xu, Q. Li, G. Ma, P.K. Chu, Macroscale superlubricity on engineering steel in the presence of black phosphorus. Nano Lett. **21**, 5308–5315 (2021). https://doi.org/10.1021/acs.nanolett.1c01437
9. M. Dienwiebel, G.S. Verhoeven, N. Pradeep, J.W.M. Frenken, J.A. Heimberg, H.W. Zandbergen, Superlubricity of graphite. Phys. Rev. Lett. **92**, 126101 (2004). https://doi.org/10.1103/PhysRevLett.92.126101
10. A.P. Merkle, L.D. Marks, Friction in full view. Appl. Phys. Lett. **90**, 2007–2009 (2007). https://doi.org/10.1063/1.2456192
11. Y. Liu, A. Song, Z. Xu, R. Zong, J. Zhang, W. Yang, R. Wang, Y. Hu, J. Luo, T. Ma, Interlayer friction and superlubricity in single-crystalline contact enabled by two-dimensional flake-wrapped atomic force microscope tips. ACS Nano **12**, 7638–7646 (2018). https://doi.org/10.1021/acsnano.7b09083

12. J. Li, T. Gao, J. Luo, Superlubricity of graphite induced by multiple transferred graphene nanoflakes. Adv. Sci. **5**, 1700616 (2018). https://doi.org/10.1002/advs.201700616
13. T.D. Sha, H. Pang, L. Fang, H.X. Liu, X.C. Chen, D.M. Liu, J.B. Luo, Superlubricity between a silicon tip and graphite enabled by the nanolithography-assisted nanoflakes tribo-transfer. Nanotechnology. **31**, 70 (2020). https://doi.org/10.1088/1361-6528/ab70cd
14. O. Hod, E. Meyer, Q. Zheng, M. Urbakh, Structural superlubricity and ultralow friction across the length scales. Nature **563**, 485–492 (2018). https://doi.org/10.1038/s41586-018-0704-z
15. A. Vanossi, C. Bechinger, M. Urbakh, Structural lubricity in soft and hard matter systems. Nat. Commun. **11**, 1–11 (2020). https://doi.org/10.1038/s41467-020-18429-1
16. J. Tian, X. Yin, J. Li, W. Qi, P. Huang, X. Chen, J. Luo, Tribo-induced interfacial material transfer of an atomic force microscopy probe assisting superlubricity in a WS2/graphene heterojunction. ACS Appl. Mater. Interf. **12**, 4031–4040 (2020). https://doi.org/10.1021/acsami.9b14378
17. D. Dietzel, U.D. Schwarz, A. Schirmeisen, Nanotribological studies using nanoparticle manipulation: principles and application to structural lubricity. Friction. **2**, 114–139 (2014). https://doi.org/10.1007/s40544-014-0054-2
18. E. Cihan, S. İpek, E. Durgun, M.Z. Baykara, Structural lubricity under ambient conditions. Nat. Commun. **7**, 12055 (2016). https://doi.org/10.1038/ncomms12055
19. A. Özoğul, S. Ipek, E. Durgun, M.Z. Baykara, Structural superlubricity of platinum on graphite under ambient conditions: the effects of chemistry and geometry. Appl. Phys. Lett. **111**, 8529 (2017). https://doi.org/10.1063/1.5008529
20. P.E. Sheehan, C.M. Lieber, Friction between van der Waals solids during lattice directed sliding. Nano Lett. **17**, 4116–4121 (2017). https://doi.org/10.1021/acs.nanolett.7b00871
21. M. Liao, P. Nicolini, L. Du, J. Yuan, S. Wang, H. Yu, J. Tang, P. Cheng, K. Watanabe, T. Taniguchi, L. Gu, V.E.P. Claerbout, A. Silva, D. Kramer, T. Polcar, R. Yang, D. Shi, G. Zhang, Ultra-low friction and edge-pinning effect in large-lattice-mismatch van der Waals heterostructures. Nat. Mater. **21**, 47–53 (2022). https://doi.org/10.1038/s41563-021-01058-4
22. S. Kawai, A. Benassi, E. Gnecco, H. Sode, R. Pawlak, X. Feng, K. Mullen, D. Passerone, C.A. Pignedoli, P. Ruffieux, R. Fasel, E. Meyer, Superlubricity of graphene nanoribbons on gold surfaces. Science **351**, 957–961 (2016). https://doi.org/10.1126/science.aad3569
23. X. Feng, S. Kwon, J.Y. Park, M. Salmeron, Superlubric sliding of graphene nanoflakes on graphene. ACS Nano **7**, 1718–1724 (2013). https://doi.org/10.1021/nn305722d
24. M. Liao, A. Silva, L. Du, P. Nicolini, V.E.P. Claerbout, D. Kramer, R. Yang, D. Shi, T. Polcar, G. Zhang, Twisting dynamics of large lattice-mismatch van der Waals heterostructures. ACS Appl. Mater. Interf. **15**, 19616–19623 (2023). https://doi.org/10.1021/acsami.3c00558
25. Y. Song, D. Mandelli, O. Hod, M. Urbakh, M. Ma, Q. Zheng, Robust microscale superlubricity in graphite/hexagonal boron nitride layered heterojunctions. Nat. Mater. **17**, 894–899 (2018). https://doi.org/10.1038/s41563-018-0144-z
26. B. Liu, J. Wang, S. Zhao, C. Qu, Y. Liu, L. Ma, Z. Zhang, K. Liu, Q. Zheng, M. Ma, Negative friction coefficient in microscale graphite/mica layered heterojunctions. Sci. Adv. **6**, 1–9 (2020). https://doi.org/10.1126/sciadv.aaz6787
27. C. Qu, K. Wang, J. Wang, Y. Gongyang, R.W. Carpick, M. Urbakh, Q. Zheng, Origin of friction in superlubric graphite contacts. Phys. Rev. Lett. **125**, 126102 (2020). https://doi.org/10.1103/PhysRevLett.125.126102
28. Q. Zheng, B. Jiang, S. Liu, Y. Weng, L. Lu, Q. Xue, J. Zhu, Q. Jiang, S. Wang, L. Peng, Self-retracting motion of graphite microflakes. Phys. Rev. Lett. **100**, 1–4 (2008). https://doi.org/10.1103/PhysRevLett.100.067205
29. Z. Liu, J. Yang, F. Grey, J.Z. Liu, Y. Liu, Y. Wang, Y. Yang, Y. Cheng, Q. Zheng, Observation of microscale superlubricity in graphite. Phys. Rev. Lett. **108**, 5503 (2012). https://doi.org/10.1103/PhysRevLett.108.205503
30. J. Yang, Z. Liu, F. Grey, Z. Xu, X. Li, Y. Liu, M. Urbakh, Y. Cheng, Q. Zheng, Observation of high-speed microscale superlubricity in graphite. Phys. Rev. Lett. **110**, 1–5 (2013). https://doi.org/10.1103/PhysRevLett.110.255504

31. C.C. Vu, S. Zhang, M. Urbakh, Q. Li, Q.C. He, Q. Zheng, Observation of normal-force-independent superlubricity in mesoscopic graphite contacts. Phys. Rev. B **94**, 1–6 (2016). https://doi.org/10.1103/PhysRevB.94.081405
32. D. Peng, J. Wang, H. Jiang, S. Zhao, Z. Wu, K. Tian, M. Ma, Q. Zheng, 100 km wear-free sliding achieved by microscale superlubric graphite/DLC heterojunctions under ambient conditions. Natl. Sci. Rev. (2022). https://doi.org/10.1093/nsr/nwab109
33. X. Huang, T. Li, J. Wang, K. Xia, Z. Tan, D. Peng, X. Xiang, B. Liu, M. Ma, Q. Zheng, Robust microscale structural superlubricity between graphite and nanostructured surface. Nat. Commun. **14**, 2931 (2023). https://doi.org/10.1038/s41467-023-38680-6
34. M. Ma, Q. Zheng, Chapter 6: structural superlubricity in large-scale heterogeneous layered material junctions, in *Superlubricity*, eds. by A. Erdemir, J.M. Martin, J. Luo (Elsevier, Amsterdam, 2021), pp. 85–111. https://doi.org/10.1016/B978-0-444-64313-1.00006-5
35. W.A. Ducker, T.J. Senden, R.M. Pashley, Measurement of forces in liquids using a force microscope. Langmuir **8**, 1831–1836 (1992). https://doi.org/10.1021/la00043a024
36. S.M. Notley, S. Biggs, V.S.J. Craig, Calibration of colloid probe cantilevers using the dynamic viscous response of a confined liquid. Rev. Sci. Instrum. **74**, 4026–4032 (2003). https://doi.org/10.1063/1.1597950
37. R. Buzio, A. Bosca, S. Krol, D. Marchetto, S. Valeri, U. Valbusa, Deformation and adhesion of elastomer poly(dimethylsiloxane) colloidal AFM probes. Langmuir **23**, 9293–9302 (2007). https://doi.org/10.1021/la700941c
38. R. Buzio, U. Valbusa, Interfacial stiffness and adhesion of randomly rough contacts probed by elastomer colloidal AFM probes. J. Phys. Condens. Matter **20**, 354014 (2008). https://doi.org/10.1088/0953-8984/20/35/354014
39. B. Cappella, G. Dietler, Force-distance curves by atomic force microscopy. Surf. Sci. Rep. **34**, 1–104 (1999). https://doi.org/10.1016/S0167-5729(99)00003-5
40. H.J. Butt, B. Cappella, M. Kappl, Force measurements with the atomic force microscope: technique, interpretation and applications. Surf. Sci. Rep. **59**, 1–152 (2005). https://doi.org/10.1016/j.surfrep.2005.08.003
41. H.-J. Butt, M. Kappl, Normal capillary forces. Adv. Colloid Interf. Sci. **146**, 48–60 (2009). https://doi.org/10.1016/j.cis.2008.10.002
42. R. Buzio, Exploring mesoscale contact mechanics by atomic force microscopy, in *Scanning Probe Microscopic Nanosciences Nanotechnology*, ed. by B. Bhushan (Springer, Berlin, Heidelberg, 2012), pp. 55–75. https://doi.org/10.1007/978-3-642-25414-7_3
43. R. Buzio, A. Gerbi, C. Bernini, L. Repetto, A. Vanossi, Sliding friction and superlubricity of colloidal AFM probes coated by tribo-induced graphitic transfer layers. Langmuir **38**, 12570–12580 (2022). https://doi.org/10.1021/acs.langmuir.2c02030
44. P. Ganjeh-Anzabi, H. Jahandideh, S.A. Kedzior, M. Trifkovic, Precise quantification of nanoparticle surface free energy via colloidal probe atomic force microscopy. J. Colloid Interf. Sci. **641**, 404–413 (2023). https://doi.org/10.1016/j.jcis.2023.03.061
45. J. Li, J. Li, L. Jiang, J. Luo, Fabrication of a graphene layer probe to measure force interactions in layered heterojunctions. Nanoscale **12**, 5435–5443 (2020). https://doi.org/10.1039/C9NR09528D
46. N.T. Garabedian, H.S. Khare, R.W. Carpick, D.L. Burris, AFM at the macroscale: methods to fabricate and calibrate probes for millinewton force measurements. Tribol. Lett. **67**, 1134 (2019). https://doi.org/10.1007/s11249-019-1134-2
47. Y.H. Ding, P. Zhang, H.M. Ren, Q. Zhuo, Z.M. Yang, X. Jiang, Y. Jiang, Surface adhesion properties of graphene and graphene oxide studied by colloid-probe atomic force microscopy. Appl. Surf. Sci. **258**, 1077–1081 (2011). https://doi.org/10.1016/j.apsusc.2011.09.005
48. P. Shi, Y. Lu, J. Sun, C. Tang, Y. Wang, L. Jiang, L. Qian, L. Chen, Towards a deeper understanding of superlubricity on graphite governed by interfacial adhesion. Carbon **199**, 479–485 (2022). https://doi.org/10.1016/j.carbon.2022.08.035
49. K. Yu, Y. Peng, H. Lang, S. Ding, Y. Huang, Material transfer mechanism for fabrication of superlubricity interface by reciprocating rubbing on graphite under high contact stress. Carbon **188**, 420–430 (2022). https://doi.org/10.1016/j.carbon.2021.12.041

50. T. Van Tran, K.H. Chung, Tribological characteristics of single-layer h-BN measured by colloidal probe atomic force microscopy. Coatings **10**, 530 (2020). https://doi.org/10.3390/COATINGS10060530
51. T. Gao, J. Li, S. Yi, J. Luo, Potential-dependent friction on a graphitic surface in ionic solution. J. Phys. Chem. C **124**, 23745–23751 (2020). https://doi.org/10.1021/acs.jpcc.0c07358
52. H. Cai, C. Zhang, F. Li, M. Liu, T. Zhang, H. Chu, Z. Liu, Study on the microcosmic superlubricity mechanism of PVPA affected by metal cations. Friction **11**, 1150–1164 (2023). https://doi.org/10.1007/s40544-022-0632-7
53. S. Li, Y. Li, P. Bai, Y. Meng, Y. Tian, Potential-dependent interfacial frictional behavior between charged microspheres and gold in aqueous solutions. J. Phys. Chem. C **126**, 4555–4562 (2022). https://doi.org/10.1021/acs.jpcc.1c08651
54. S. Feng, Y. Liu, J. Li, S. Wen, Superlubricity achieved with zwitterionic brushes in diverse conditions induced by shear actions. Macromolecules **54**, 5719–5727 (2021). https://doi.org/10.1021/acs.macromol.1c00430
55. J. Li, Z. Dou, Y. Liu, J. Luo, J. Xiao, Improvement of load bearing capacity of nanoscale superlow friction by synthesized fluorinated surfactant micelles. ACS Appl. Nano Mater. **1**, 953–959 (2018). https://doi.org/10.1021/acsanm.7b00367
56. J. Li, C. Zhang, P. Cheng, X. Chen, W. Wang, J. Luo, AFM studies on liquid superlubricity between silica surfaces achieved with surfactant micelles. Langmuir **32**, 5593–5599 (2016). https://doi.org/10.1021/acs.langmuir.6b01237
57. L. Prandtl, Ein Gedankenmodell zur kinetischen theorie der festen Körper. ZAMM J. Appl. Math. Mech. Zeitschrift Für Angew. Math. Und Mech. **8**, 85–106 (1928). https://doi.org/10.1002/zamm.19280080202
58. G.A. Tomlinson, CVI: a molecular theory of friction, London, Edinburgh. Dublin. Philos. Mag. J. Sci. **7**, 905–939 (1929). https://doi.org/10.1080/14786440608564819
59. S. Liu, H.-P. Wang, Q. Xu, T. Ma, G. Yu, C. Zhang, D. Geng, Z. Yu, S. Zhang, W. Wang, Y. Hu, H. Wang, J. Luo, Robust microscale superlubricity under high contact pressure enabled by graphene-coated microsphere. Nat. Commun. **8**, 14029 (2017). https://doi.org/10.1038/ncomms14029
60. R. Buzio, A. Gerbi, C. Bernini, L. Repetto, A. Vanossi, Graphite superlubricity enabled by triboinduced nanocontacts. Carbon **184**, 875–890 (2021). https://doi.org/10.1016/j.carbon.2021.08.071
61. R. Buzio, A. Gerbi, C. Bernini, L. Repetto, A. Silva, A. Vanossi, Dissipation mechanisms and superlubricity in solid lubrication by wet-transferred solution-processed graphene flakes: implications for micro electromechanical devices. ACS Appl. Nano Mater. **6**, 11443–11454 (2023). https://doi.org/10.1021/acsanm.3c01477
62. J. Li, J. Li, X. Chen, Y. Liu, J. Luo, Microscale superlubricity at multiple gold–graphite heterointerfaces under ambient conditions. Carbon **161**, 827–833 (2020). https://doi.org/10.1016/j.carbon.2020.01.070
63. B.A. Krajina, L.S. Kocherlakota, R.M. Overney, Direct determination of the local Hamaker constant of inorganic surfaces based on scanning force microscopy. J. Chem. Phys. **141**, 799 (2014). https://doi.org/10.1063/1.4898799
64. M. Farshchi-Tabrizi, M. Kappl, Y.J. Cheng, J. Gutmann, H.J. Butt, On the adhesion between fine particles and nanocontacts: An atomic force microscope study. Langmuir **22**, 2171–2184 (2006). https://doi.org/10.1021/la052760z
65. B.N.J. Persson, O. Albohr, U. Tartaglino, A.I. Volokitin, E. Tosatti, On the nature of surface roughness with application to contact mechanics, sealing, rubber friction and adhesion. J. Phys. Condens. Matter **17**, R1–R62 (2005). https://doi.org/10.1088/0953-8984/17/1/R01
66. R. Buzio, E. Gnecco, C. Boragno, U. Valbusa, Friction force microscopy investigation of nanostructured carbon films. Carbon **40**, 883–890 (2002). https://doi.org/10.1016/S0008-6223(01)00208-1
67. A. Kumar, T. Staedler, X. Jiang, Role of relative size of asperities and adhering particles on the adhesion force. J. Colloid Interf. Sci. **409**, 211–218 (2013). https://doi.org/10.1016/j.jcis.2013.07.039

68. S.N. Ramakrishna, L.Y. Clasohm, A. Rao, N.D. Spencer, Controlling adhesion force by means of nanoscale surface roughness. Langmuir **27**, 9972–9978 (2011). https://doi.org/10.1021/la201727t
69. S. Yang, H. Zhang, S.M. Hsu, Correction of random surface roughness on colloidal probes in measuring adhesion. Langmuir **23**, 1195–1202 (2007). https://doi.org/10.1021/la0622828
70. J.M. Monti, P.M. McGuiggan, M.O. Robbins, Effect of roughness and elasticity on interactions between charged colloidal spheres. Langmuir **35**, 15948–15959 (2019). https://doi.org/10.1021/acs.langmuir.9b02161
71. S. Peillon, A. Autricque, M. Redolfi, C. Stancu, F. Gensdarmes, C. Grisolia, O. Pluchery, Adhesion of tungsten particles on rough tungsten surfaces using atomic force microscopy. J. Aerosol. Sci. **137**, 5431 (2019). https://doi.org/10.1016/j.jaerosci.2019.105431
72. L. Pastewka, M.O. Robbins, Contact area of rough spheres: Large scale simulations and simple scaling laws. Appl. Phys. Lett. **108**, 802 (2016). https://doi.org/10.1063/1.4950802
73. R.A. Soler-Crespo, W. Gao, L. Mao, H.T. Nguyen, M.R. Roenbeck, J.T. Paci, J. Huang, S.T. Nguyen, H.D. Espinosa, The role of water in mediating interfacial adhesion and shear strength in graphene oxide. ACS Nano **12**, 6089–6099 (2018). https://doi.org/10.1021/acsnano.8b02373
74. Y. Li, S. Huang, C. Wei, D. Zhou, B. Li, C. Wu, V.N. Mochalin, Adhesion between MXenes and other 2D materials. ACS Appl. Mater. Interf. **13**, 4682–4691 (2021). https://doi.org/10.1021/acsami.0c18624
75. M. Marian, D. Berman, A. Rota, R.L. Jackson, A. Rosenkranz, Layered 2D nanomaterials to tailor friction and wear in machine elements—a review. Adv. Mater. Interf. **9**, 1–19 (2022). https://doi.org/10.1002/admi.202101622
76. F. Hui, S. Chen, X. Liang, B. Yuan, X. Jing, Y. Shi, M. Lanza, Graphene coated nanoprobes: a review. Crystals **7**, 269 (2017). https://doi.org/10.3390/cryst7090269
77. R. Buzio, A. Gerbi, S. Uttiya, C. Bernini, A.E. Castillo, F. Palazon, A.S. Siri, V. Pellegrini, L. Pellegrino, F. Bonaccorso, Ultralow friction of ink-jet printed graphene flakes. Nanoscale **9**, 7612–7624 (2017). https://doi.org/10.1039/C7NR00625J
78. M. Ishikawa, M. Ichikawa, H. Okamoto, N. Itamura, N. Sasaki, K. Miura, Atomic-scale peeling of graphene. Appl. Phys. Exp. **5**, 3–6 (2012). https://doi.org/10.1143/APEX.5.065102
79. K. Yu, P. Xu, Y. Peng, Y. Huang, H. Lang, S. Ding, Ultra-low friction and stiffness dependence of interlayer friction in graphite flakes under various rotation angles. Mater. Today Adv. **18**, 100380 (2023). https://doi.org/10.1016/j.mtadv.2023.100380
80. A. Castellanos-Gomez, M. Buscema, R. Molenaar, V. Singh, L. Janssen, H.S.J. Van Der Zant, G.A. Steele, Deterministic transfer of two-dimensional materials by all-dry viscoelastic stamping. 2D Mater. **1**, 1–20 (2014). https://doi.org/10.1088/2053-1583/1/1/011002
81. M.Z. Baykara, M.R. Vazirisereshk, A. Martini, Emerging superlubricity: a review of the state of the art and perspectives on future research. Appl. Phys. Rev. **5**, 1445 (2018). https://doi.org/10.1063/1.5051445
82. E. Meyer, E. Gnecco, Superlubricity on the nanometer scale. Friction. **2**, 106–113 (2014). https://doi.org/10.1007/s40544-014-0052-4
83. D. Dietzel, C. Ritter, T. Mönninghoff, H. Fuchs, A. Schirmeisen, U.D. Schwarz, Frictional duality observed during nanoparticle sliding. Phys. Rev. Lett. **101**, 125505 (2008). https://doi.org/10.1103/PhysRevLett.101.125505
84. E. Gnecco, E. Riedo, R. Bennewitz, E. Meyer, H. Brune, Thermally activated phenomena in nanoscopic sliding friction. Tribotest **12**, 169–174 (2006). https://doi.org/10.1002/tt.14
85. S. Fujisawa, K. Yokoyama, Y. Sugawara, S. Morita, Analysis of experimental load dependence of two-dimensional atomic-scale friction. Phys. Rev. B Condens. Matter Mater. Phys. **58**, 4909–4916 (1998). https://doi.org/10.1103/PhysRevB.58.4909
86. S.N. Medyanik, W.K. Liu, I.-H. Sung, R.W. Carpick, Predictions and observations of multiple slip modes in atomic-scale friction. Phys. Rev. Lett. **97**, 136106 (2006). https://doi.org/10.1103/PhysRevLett.97.136106
87. C. Lee, Q. Li, W. Kalb, X.Z. Liu, H. Berger, R.W. Carpick, J. Hone, Frictional characteristics of atomically thin sheets. Science **328**, 76–80 (2010). https://doi.org/10.1126/science.1184167

88. A. Socoliuc, R. Bennewitz, E. Gnecco, E. Meyer, Transition from stick-slip to continuous sliding in atomic friction: entering a new regime of ultralow friction. Phys. Rev. Lett. **92**, 134301–134311 (2004). https://doi.org/10.1103/PhysRevLett.92.134301
89. P. Steiner, R. Roth, E. Gnecco, A. Baratoff, S. Maier, T. Glatzel, E. Meyer, Two-dimensional simulation of superlubricity on NaCl and highly oriented pyrolytic graphite. Phys. Rev. B **79**, 45414 (2009). https://doi.org/10.1103/PhysRevB.79.045414
90. R. Buzio, A. Gerbi, M. Barra, F. Chiarella, E. Gnecco, A. Cassinese, Subnanometer resolution and enhanced friction contrast at the surface of perylene diimide PDI8-CN2 thin films in ambient conditions. Langmuir **34**, 3207–3214 (2018). https://doi.org/10.1021/acs.langmuir.7b04149
91. A. Vanossi, N. Manini, M. Urbakh, S. Zapperi, E. Tosatti, Colloquium: modeling friction: from nanoscale to mesoscale. Rev. Mod. Phys. **85**, 529–552 (2013). https://doi.org/10.1103/RevModPhys.85.529
92. E. Gnecco, R. Roth, A. Baratoff, Analytical expressions for the kinetic friction in the Prandtl-Tomlinson model. Phys. Rev. B **86**, 35443 (2012). https://doi.org/10.1103/PhysRevB.86.035443
93. K.B. Jinesh, S.Y. Krylov, H. Valk, M. Dienwiebel, J.W.M. Frenken, Thermolubricity in atomic-scale friction. Phys. Rev. B **78**, 155440 (2008). https://doi.org/10.1103/PhysRevB.78.155440

Frictional Characteristics of Graphene on Textured Surfaces

Manoj Tripathi, Sathvik Ajay Iyengar, Md. Hasan-ur Rahman, Venkataramana Gadhamshetty, Pulickel Madhava Ajayan, and Alan B. Dalton

Abstract The significance of graphene and other 2D materials as solid-state lubricants has been known for a decade. These atomically thin sheets contain sensitive surface atoms that are responsible for unique frictional characteristics that mainly depend on the nature of the underlying substrates. Thus, interfacial interactions between atomically thin sheets and engineered surfaces enhances interfacial adhesion force and induces strain in the sheets are useful for tuning the friction behaviour. This degree of regulation offers phenomenal advantages in nanoscale electromechanical systems and nanoscale robotics, where a moving mechanical system is needed. The present chapter discusses the importance of straining 2D materials for friction force regulation through engineered surfaces. Several sophisticated methodologies for preparing textured surfaces are highlighted, and modern characterisation techniques, including machine learning tools, which are useful for analysing strain and mechanics in 2D materials, are discussed. Using graphene as a case study, several results from nanoscale friction force microscopy on engineered surface are presented.

1 Introduction

Friction forces are inevitable in any mechanical system with movable components operating in relative motion and are often responsible for mechanical failure, energy loss and performance inefficiencies. Several strategies have been adopted to mitigate the adverse effects of friction forces by using solid-state and liquid lubricants at the interface [1]. Traditionally, liquid lubricants (such as organic oils and ionic liquids)

M. Tripathi (✉) · A. B. Dalton
Department of Physics and Astronomy, University of Sussex, Brighton BN1 9RH, UK
e-mail: M.Tripathi@sussex.ac.uk

S. A. Iyengar · P. M. Ajayan
Department of Materials Science and Nanoengineering, Rice University, Houston, USA

Md. Hasan-ur Rahman · V. Gadhamshetty
Department Civil and Environmental Engineering, South Dakota School of Mines and Technology, Rapid City, SD 57701, USA

E. Gnecco and E. Meyer (eds.), *Fundamentals of Friction and Wear on the Nanoscale*, NanoScience and Technology, https://doi.org/10.1007/978-3-031-63065-1_11

have been used to control friction forces; nevertheless, their undesirable effects on the environment have led to concerted efforts toward limited application. Additionally, liquid-state lubricant inefficiency under confined conditions has been observed due to viscosity modifications that hinder tribological performance [2]. On the other hand, solid-state lubricants obtained mainly from layered materials, such as graphite, transition metal dichalcogenide (e.g., MoS_2 and WS_2), and hexagonal boron nitride (h-BN), are favorable for engineering applications [3]. The isolation of the atomic layer of graphite as graphene and the discovery of other 2D materials paved the way for the thinnest solid-state lubricants and revolutionised several fields, including flexible electronics and mechanical systems (N/MEMS) [4, 5].

Several studies have revealed that only one or a few atomic layers (<10) of lamellar materials deposited on a surface lead to a friction coefficient (COF) on the order of 0.01, often referred to as "superlubricity", a condition of extreme slipperiness usually predicted by hypothetical geometry. Nevertheless, the in-plane elasticity of 2D materials, such as graphene, and their interaction with the substrate, which acts as a support, influence frictional characteristics. weak van der Waals interactions with the substrate lead to higher friction values in graphene under the "puckering" effect [6]. In contrast, stronger interfacial interactions minimise the frictional response; for instance, under the influence of ionic and covalent interactions, the puckered structure at the interface between the nanoscopic slider and graphene reduces the friction coefficient [5, 6]. Thus, one can regulate the frictional characteristics of graphene through interactions with pretreated surfaces that induce several effects, such as pinning, flattening and straining of graphene lattices, which could also be applicable in other similar 2D materials.

In the present chapter, we have highlighted the effect of surface treatments through the development of nanoscale to microscale engineered structures. These textured architectures induce strain in the lattice of graphene during conformation and alter the electrical, mechanical, and magnetic properties [7] and, consequently, frictional characteristics [8]. We have discussed the techniques generally implemented for the measurement of strain through probing techniques (scanning tunnelling microscopy, friction force microscopy), optical detection (Raman spectroscopy), and electronic imaging (scanning transmission electron microscopy) with the supported by machine learning tools. Several methods for developing engineered structures at different metric scale lengths that apply to friction tuning strategies have been discussed. Strain-induced physical effects in graphene are used to develop next-generation nano/microelectromechanical systems, sensors and energy-efficient technologies.

2 Preparation of Textured Surfaces

2.1 Top-Down Nanofabrication

In nanotechnology, the ability to manipulate materials at the atomic and molecular scale has unlocked unprecedented possibilities for studying exotic physical properties. A prominent avenue in this area of exploration involves advanced nanoscale texturing techniques, particularly e-beam lithography (EBL) followed by reactive ion etching (RIE), and cutting-edge methods such as focused ion beam (FIB) milling. Top-down nanofabrication techniques such as e-beam lithography (EBL) and reactive ion etching (RIE) play crucial roles in creating high-quality textured substrates for investigating exotic frictional properties in 2D materials such as graphene [9].

EBL, with its exceptional precision and adaptability to various materials, allows for the fabrication of intricate nanostructures, making it a fundamental tool for subnanometer precision needs [10]. Its key features contributing to widespread adoption include exceptional precision, adaptability to various materials, and scalability. The unparalleled resolution of EBL, stemming from the small de Broglie wavelength of electrons, enables the fabrication of intricate nanostructures. The ability to control the position and intensity of the electron beam ensures faithful pattern reproduction, establishing the EBL as a fundamental tool for subnanometer precision needs. EBLs demonstrate versatility across a range of materials, including semiconductors, metals, and insulators, making it essential to study diverse 2D materials with tailored fabrication approaches. These techniques are essential for investigating the frictional properties of 2D materials such as graphene and other 2D materials, where factors such as surface topography and material thickness significantly influence frictional behavior [9, 11, 12]. The ability to control the position and intensity of the electron beam in EBL ensures faithful pattern reproduction, addresses scalability challenges and allows for efficient patterning of larger areas, which is crucial for academic research and industrial applications [10].

Grayscale EBL enhances the lithographic process by enabling the fabrication of three-dimensional structures with smoothly varying heights, which is valuable in studying frictional properties at the nanoscale [12]. Grayscale EBLs prove advantageous in designing surfaces with intricate topographical features, allowing researchers to create graded slopes, ramps, and other three-dimensional structures by controlling the electron dose across the substrate. This feature is particularly valuable in the context of studying frictional properties, where surface topography plays a critical role in material interactions. The ability to smoothly transition between different exposure levels in grayscale EBL is crucial for applications where gradual changes in surface properties are desired. This can be especially relevant when exploring the bending-to-slipping transition in 2D materials, as it allows for the creation of surfaces with controlled variations in topography, influencing frictional behavior at the nanoscale.

Following the precision of EBL, the next crucial step is reactive ion etching (RIE). In RIE, the patterned resist acts as a mask, while a plasma containing reactive ions selectively etches the exposed substrate material. A specialised variant within RIE, known as deep reactive ion etching (DRIE), has gained prominence. RIE, particularly deep reactive ion etching (DRIE), is advantageous for creating high-aspect-ratio structures with sidewall profiles approaching vertical, making it valuable for fabricating intricate nanostructures in 2D materials [13]. Its ability to achieve deep, anisotropic etching makes DRIE particularly advantageous for applications where structural depth is a critical parameter. In the study of exotic frictional properties, DRIE offers the potential to create three-dimensional textures with enhanced complexity, contributing to a more comprehensive understanding of how surface morphology influences frictional behavior at the nanoscale. FIB milling, another state-of-the-art nanoscale texturing technique, employs a focused beam of ions to remove material [14]. While FIB excels in three-dimensional sculpting, EBL and RIE are often preferred for 2D patterning due to their higher resolution and throughput. The choice between these techniques depends on the specific requirements of the study and the desired substrate characteristics. FIB provides a unique perspective in scenarios requiring a broader view of surface interactions. FIB milling, while excelling in three-dimensional sculpting, is often preferred for creating larger surface structures and exploring frictional phenomena at a coarser scale, positioning it as a complementary tool to EBL and RIE [15]. In addition to EBL, (D)RIE, and FIB, other modern examples consist of combining top-down nanofabrication methods, such as nanoimprinting, and bottom-up nanofabrication methods, such as self-assembly (as described by Kong et al. [12] by fabricating a controlled array of Au nanoparticles) demonstrates the interdisciplinary nature of nanofabrication techniques.

2.2 *Substrate Considerations*

The study of interactions between 2D materials and substrates is crucial for understanding and optimising the properties of these materials. When working with insulating substrates such as fused silica, it is important to consider the substrate's electronic band structure to achieve high resolution. Frictional interactions observed under such conditions are typically more agnostic of the substrate. Spin coating of a conductive polymer onto a substrate before the EBL process helps dissipate charge buildup during electron exposure, preventing pattern distortions [16]. On the other hand, when 2D materials are transferred onto textured magnetic metal surfaces, the interplay between the magnetic properties of the metal substrate and the electronic properties of the 2D materials can lead to intriguing phenomena. This interplay can induce alterations in the electronic transport properties of 2D materials, influencing spin-dependent scattering, spin–orbit coupling, and conductance [17].

The influence of textured substrates on overlaying 2D materials is substantial when modulating the properties of the materials. The combination of a superhydrophobic state, nano- and microfeatures of the hierarchical surface, and the appropriate composition of the surface textured layer provides outstanding resistance of superhydrophobic coatings to external chemical and mechanical impacts [18]. Additionally, the growth or deposition of two dimensions enables the creation of ultrathin flexible films suitable for controlled wrinkling, leading to the creation of textured surface topographies for directing cell alignment and morphology [19].

Moreover, the growth of 2D materials on textured substrates, such as molten glass, has advantages for low-cost, transparent substrates for photonic applications [20]. The textured substrate also plays a crucial role in the epitaxial growth behavior of 2D materials, where the strong interaction between the edge of the 2D material and the pristine substrate facet determines the alignment of the 2D material [21]. Furthermore, the concept of texturing can be extended to other 2D material coatings, which have potential applications in various fields, such as antifouling substrates, stretchable electronics, barriers, and advanced electrode architectures for energy storage, catalysis, and sensing [22]. The surface texture profiles of existing concrete substrates have been shown to have an important effect on the adhesion strength between existing concrete and ultrahigh-performance fibre concrete overlays, as demonstrated by the short- and long-term results of splitting cylinder tensile and slant shear tests [23]. Additionally, the electrical resistivity values of films grown on textured substrates increased, which is indicative of surface defects [24]. Textured substrates also contribute to light trapping benefits and reduce reflection losses at the front surface of solar cells through a process known as the double-bounce effect [25]. Therefore, textured substrates profoundly impact the properties and behaviour of overlaying 2D materials, influencing adhesion strength, electrical resistivity, light trapping, and various potential applications.

2.3 Property-Informed Texturing Approaches

The investigation of exotic frictional properties in 2D materials, particularly graphene, has revealed the intriguing influence of textured surfaces on the transition from bending to slipping at short distances. The hexagonal lattice structure of graphene dynamically evolves when encountering specific topographical features, such as microscale ridges or nanopatterned surfaces, leading to a shift in frictional behavior [26]. This transition has significant implications for applications such as nanoscale lubrication, emphasising the importance of understanding how specific textures impact the shape and morphology of graphene layers. These insights pave the way for tailored materials, such as graphene-coated surfaces with precisely engineered textures, enabling advanced friction control in diverse technological applications. This nuanced exploration illuminates the intricate forces at the nanoscale and highlights the transformative potential of understanding and manipulating frictional properties in 2D materials (Fig. 1).

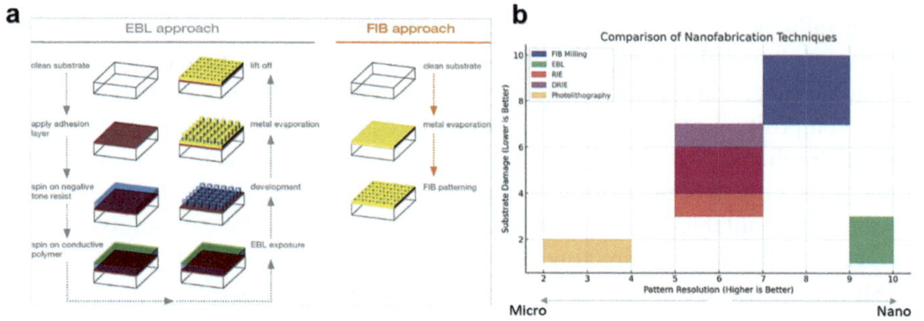

Fig. 1 Nanofabrication techniques for textured surfaces. **a** Process workflow for the EBL and FIB approaches. **b** Comparison of substrate damage (or etch) versus pattern resolution across various techniques [27]. *Note* while low damage and high resolution are generally desirable, there are several use cases in which higher beam damage/etch and lower resolution have great utility. Reprinted from Fabrication [28]

Delving into the diverse landscape of nanoscale texturing techniques reveals a rich variety of possibilities across various types of 2D materials—semimetallic (graphene), semiconducting (MoS_2 [n-type], WSe_2 [p-type]), and insulating (h-BN) materials. Crafting unique methodologies for each material type allows for the exploration of diverse properties and applications. In semimetallic graphene, for example, nanoscale texturing can engineer strain-induced pseudomagnetic fields, influencing charge carrier dynamics and band structures, thereby offering implications for creating graphene-based devices with tunable electronic properties. The semiconducting materials such as MoS_2 (n-type) and WSe_2 (p-type), surface texturing has become a tool for generating quantum dots or nanostructured interfaces, influencing charge transport and optoelectronic properties. This is pivotal for developing efficient semiconductor devices, including transistors and photodetectors. In insulating materials such as hexagonal boron nitride (h-BN), nanoscale texturing can be harnessed to engineer phononic properties, opening avenues for designing materials with tailored thermal conductivity. This versatility extends thermal management and insulation applications, demonstrating the capability of nanoscale manipulation across a spectrum of 2D materials.

By expanding the scope, the impact of nanoscale texturing techniques on diverse 2D materials can be further enhanced by considering different phases, such as 2H, 1T', etc. In semimetallic graphene, varying the surface texture can influence the strain-induced pseudomagnetic fields and alter the stacking configurations, leading to distinct phases with unique electronic properties. Similarly, for semiconducting MoS_2 and WSe_2, surface texturing has become a versatile tool for tailoring quantum dots or nanostructured interfaces, which, when combined with specific phases, adds an additional layer of complexity to charge transport and optoelectronic properties. Furthermore, harnessing the benefits of heterostructuring, both laterally and vertically, offers unprecedented control over material interactions. This approach becomes particularly useful when exploring twist-controlled heterostructures, revealing intricate moiré patterns that can dramatically influence the electronic and optical properties of the combined materials. The ability to engineer moiré superlattices through precise twist control opens up new avenues for designing materials with emergent properties, demonstrating the profound impact of nanoscale manipulation and heterostructuring in the rich landscape of 2D materials.

3 Mechanics of Strain Engineering in Graphene

Atomically thin sheets of graphite and other 2D materials have an incredible capacity for deformation and can withstand enormous elastic strain (up to 25% for graphene) before fracture [29]. These sheets on a substrate are susceptible to stress distribution with localised compression and tension through interfacial bonding, folding, wrinkling, and contact conditions that modulate the electronic properties and phonons [30]. Thus, an in-depth relationship between the surface structures of graphene leads to different functionalities useful for scientific and engineering applications. Engineered surfaces with nano- to microscale dimples, grooves, pillars, and other geometries are useful for regulating friction and adhesion forces [31]. One of the advantages of using engineered structures is that 2D materials are kept in a constant local strained state (i.e., static strain), which is challenging for prestrained flexible polymeric surfaces such as polydimethyl siloxane and polystyrene [32]. The strain induced in 2D materials through the engineered surface is closely related to the textured design morphology (aspect ratio). Thus, changing the substrate nanostructure patterns can alter the magnitude of strain and strain-induced properties [8, 33]. Nevertheless, the interfacial interaction between 2D materials and textured surfaces is difficult to interpret owing to the involvement of several surface forces and possibly nonlinear elastic behaviour of 2D materials. Li and Zhang presented a theoretical framework for determining the substrate-regulated graphene morphology, especially over grooved surfaces. In their investigation, they delineated the graphene equilibrium morphology by considering the interplay between three types of free energies: (1) the graphene-substrate interaction energy, (2) the graphene strain energy, and (3) the substrate strain energy (mostly applicable to flexible substrates such as polymers or elastomers) (Fig. 2) [34]. Under the influence of substrate interaction energy, all types of bonding, either weak physical forces (e.g., van der Waals) or strong chemical bonds (e.g., covalent), may coexist in graphene over the substrate structure. Thus, the contribution of the chemical bond between graphene and the substrate (for epitaxially grown graphene over metallic surfaces) is additive to the weak van der Waals interactions. The strain energy in graphene increases during conformation, leading to out-of-plane bending and in-plane stretching. The substrate strain energy depends on its stiffness and the external mechanical load applied during the deposition of graphene. In the absence of a mechanical load for the deposition of graphene over a thick, rigid structure, negligible strain energy is induced in the substrate. Nevertheless, soft surfaces result in large deformations, where the strain energy of

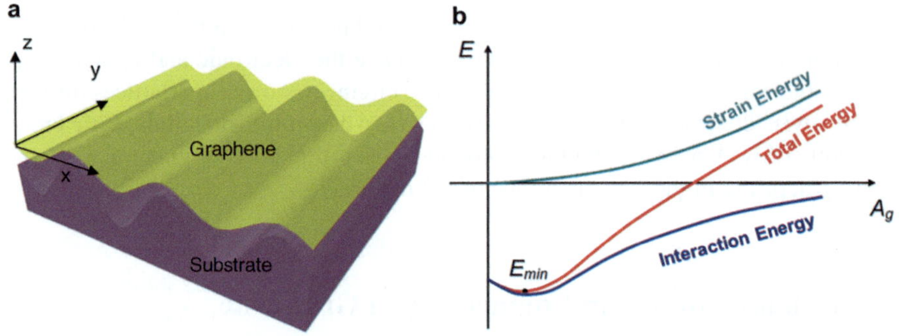

Fig. 2 **a** Schematic of graphene over a periodic groove surface. **b** Schematic of the energetics for determining the substrate-regulated graphene morphology. A_g is the graphene corrugation amplitude, and E_{min} is the total free energy minimisation. Reprinted from Li and Zhang [34]

the substrate becomes comparable to that of graphene and the graphene-substrate interaction energy.

Bunch and Dunn [35] noted the consequences of decreasing the lateral size of 2D materials from the microscale to the nanoscale that influence the surface forces toward the substrate. At different aspect ratios of textured surfaces, different configurations of graphene adhere to micromechanical structures. For instance, the adhesion of graphene to a stepped surface is influenced by morphological parameters (step height, vicinal angle, terrace width, and orientation) [36]. It is observed that 2D materials that are separated by small distances in textured-induced conformations are sensitive to the operant range of surface forces. The structural stiffness decreases with decreasing size, and a nonlinear relation persists between surface forces and the structural stiffness of the materials [35]. An interesting phenomenon has been demonstrated theoretically [37] and experimentally [8, 38] where a sudden morphological transition of graphene from a conformation to a suspended state, "snap-through", occurs due to a balance between the surface adhesion energy and the conformed strain energy. Using linear elasticity theory, one can model the relationship between the conformation of thin films and adhesion to substrates [37].

4 Techniques for Measuring Strain in 2D Materials

The most common experimental techniques used to measure localised strain in graphene and other 2D materials are scanning probe microscopy (SPM), Raman spectroscopy, and transmission electron microscopy (TEM). Scanning tunnelling microscopy (STM) is a sophisticated probing technique used to resolve the nanoscale landscape of conductive substrates and achieve atomic-scale resolution. Here, the current (~ nA) is a tunnel between the conductive probe and the substrates vertically separated by an optimised distance (Fig. 3a STM [39]). The STM morphology map revealed that the physical conformation of the graphene on the substrate led to smoother outcomes. Ishigami and coworkers observed a nearly 60% smoother surface for mechanically deposited graphene over a silica substrate. This reduction in the roughness (rms ~ 0.19 nm) of the graphene surface is due to the energy cost of the graphene closely following the sharp, rough (rms = 0.31 nm) features of the silica substrate. This degree of extrinsic corrugation in graphene by the silica substrate is measured at the maximum limit of strain of approximately 1% [40]. STM mapping is limited to conductive substrates, and its noncontact mode of operation cannot induce any strain on the testing surfaces. This limitation is covered by atomic force microscopy (AFM), where high-resolution imaging provides the surface morphology of 2D materials and induces pressure in the z-axis through direct contact between the tip-apex and 2D materials. The induced strain (ε) under pressure can be expressed as

$$\varepsilon = F/(AE) \tag{1}$$

where F and A are the applied force and cross-sectional area of the tip apex, respectively, and E is Young's modulus of the 2D material. The suspended membrane deformed significantly under higher applied forces (microwongtons) (Fig. 3b); however, when supported by a rigid substrate (such as Si), the maximum applied force

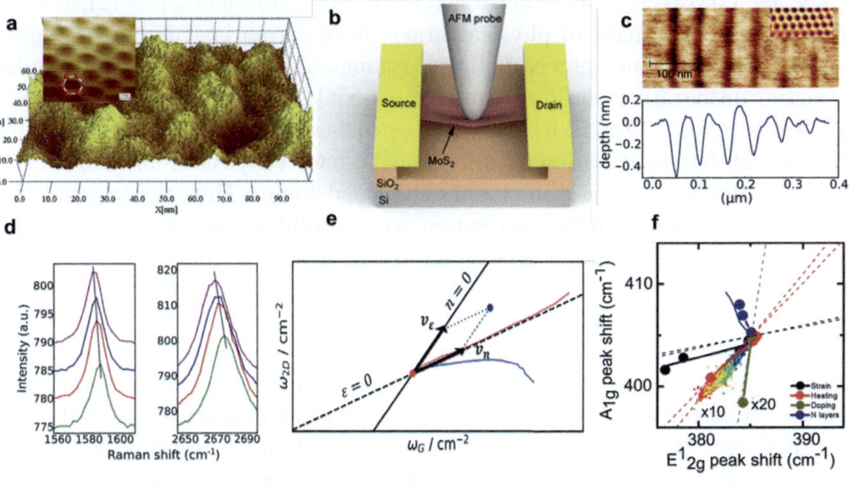

Fig. 3 **a** Nanoscale morphology of graphene and its atomic-scale resolution (inset) determined by scanning tunneling microscopy (STM). **b** Schematic illustration showing the physical deformation of a suspended MoS_2 connected to electrode (source and drain). **c** AFM-induced lithographic patterns of graphene over silica substrates at different penetration depths; the inset shows the atomic-scale resolution of the lattice structure. **d** Shifting of Raman modes (G and 2D) in graphene under different strains induced by lithographic texturing. Correlation between the phonon modes (Gω vs 2Dω) deconvoluting strain (ε) and carrier concentration in **e** graphene and **f** MoS_2. Reprinted from Chaste et al. [33], Manzeli et al. [41], Manzeli et al. [42], Li et al. [43]

reached 25 nN [32, 41]. Using the apex of a diamond-like carbon-coated probe, one can generate a lithographic pattern with an optimised depth of approximately 5 nm in a graphene-over silicon (covered with an oxide layer) wafer. It has been observed that the pinning of graphene onto textured surfaces generated during scratching is achieved due to the greater interfacial adhesion energy between graphene and silica compared to its elastic energy [42]. It is assumed that the adhesion of graphene to the patterned surface is improved during indentation (i.e., pushing toward the surface) compared to that on a flat surface where graphene is partially suspended. The estimated strain in graphene for conforming around the lithographic geometry is approximately 0.15%.

In contrast to atomic force microscopy, Raman spectroscopic mapping comprises light-matter interactions to obtain Raman scattering, typically weak signals. It is a nondestructive technique if operated at controlled laser power [44]. Raman spectroscopy has been used as a key diagnostic tool for monitoring the strain in graphene and other 2D materials through its phononic response and energy band structures (Fig. 3d–e). In graphene, Raman spectroscopy can monitor the strain by analysing the frequency shift in G (~ 1580 cm^{-1}) and 2D (~ 2700 cm^{-1}) peak positions by a factor determined through respective Grüneisen parameter ($\Delta\omega_G$, $\Delta\omega_{2D}$) and $\Delta\omega$ $E^1{}_{2g}$ and $\Delta\omega$ A_{1g} in MoS_2 for $E^1{}_{2g}$ (~ 384 cm^{-1}) and A_{1g} (~ 404 cm^{-1}) [45, 46]. Under the influence of uniaxial strain, these parameters vary $\partial\omega 2D/\partial\varepsilon \approx -83$ cm^{-1}/%, $\partial\omega G^+/\partial\varepsilon \approx -36$ cm^{-1}/%, and $\partial\omega G^-/\partial\varepsilon \approx -18$ cm^{-1}/%.

Lee and coworkers first demonstrated the use of a correlation plot (2Dω vs Gω) to distinguish the effects of physical strains from carrier concentrations (doping) (Fig. 3e). The correlation plot is applicable for uniaxial and biaxial strain in graphene depending on the ratio of the Raman modes; uniaxial 2Dω/Gω $\approx 2.2 \pm 0.2$ and biaxial variation between 2.45 and 2.8 [47]. The slope of the charge doping axis is 2Dω/Gω $\approx$ 0.7. The coordinates of the doping and strain axis intersection are considered intrinsic characteristics of graphene with minimal strain and doping. Thus, the deviation of the experimental results from the intersection axis is useful for interpreting the localised strain and doping. The nature of the strain is tensile if the Raman mode shifts below the intersection axis. Moreover, the vector projects the upwards strain axis across the intersection, which is considered a compressive strain in the graphene [47]. Zhang and coworkers observed the localised distribution of areal strain (hydrostatic strain) in graphene over nanospheres with different diameters ranging from 20–200 nm. The graphene deposited over nanospheres has a higher tensile strain than that deposited over the flat surface, revealing the stretching of graphene lattices over the nanospheres (Fig. 4a). Andrea and coworkers, as shown in Fig. 4b, observed a similar trend (with distinct values) of graphene transferred over textured Si surfaces of different pitch lengths, varying from 40 to 250 nm [8]. Generally, graphene on a flat surface has relatively less tension, which is increased over textured surfaces during the conformation of the textured geometry. Similarly, textured (pillar geometry)-induced geometrical deformation in MoS_2 follows the relation $\Delta A_{1g} = 0.26\ \Delta E^1{}_{2g}$), leading to a strain of approximately 2%.

Electron microscopy is another powerful platform for providing in-depth information for measuring the mechanical properties of 2D materials by enabling the

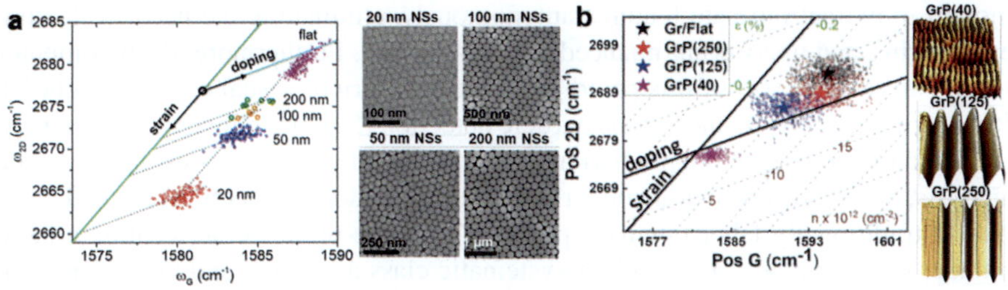

Fig. 4 **a** Correlation of graphene Raman modes over nanospheres of different diameters varying from 20 to 200 nm, and **b** over textured Si surfaces with different pitch lengths: 40 nm, 125 nm and 250 nm. Reprinted from Zhang et al. [48], Mescola et al. [8]

investigation of their conformation and strain at atomic resolution. Recently, cross-sectional imaging of a few graphene layers was carried out by scanning transmission electron microscopy (STEM) to measure the bending rigidity of the layers. In this nontrivial measurement, sample preparation plays a crucial role, where a few layers of graphene are draped over the steps of h-BN. This led to the development of a vertical heterostructure where the graphene layer/s was bent around the atomically sharp edges of hBN. The bending profile of few-layer/s graphene was taken to calculate the stiffness following geometrical parameters obtained from STEM: bending radius (R), thickness (N), step height (H) and bending angle (θ). These geometrical parameters are used in the following relation under the assumption of negligible in-plane strain for the incommensurate interface.

$$B = R\Gamma(\mathrm{H} - 2\mathrm{R}(1 - \cos\theta))/\sin^2\theta \quad (2)$$

where B is the bending stiffness and Γ is the graphene/h-BN interfacial adhesion energy ~ 0.126 J/m^2. The bending stiffness of bilayer graphene is measured between 2.6 and 5.8 eV, and for 10 layers, it is approximately 18 eV. Additionally, thicker graphene sheets have a greater radius of curvature and higher bending angle, which are directly related to the step height of the subsurface [49].

4.1 *Machine Learning for Measuring Strain in 2D Materials*

In the realm of 2D material science, machine learning (ML) in conjunction with imaging techniques has proven to be a pivotal tool for efficiently extracting insights from intricate datasets. One of the emerging techniques that incorporates the ML tool is the deep learning neural network, which plays an instrumental role in predicting design, synthesis, and characterisation, particularly addressing challenges posed by traditional methods in thin films and 2D materials for high-throughput screening [50, 51]. Diving into the world of deep learning, its exceptional ability to extract insights from unprocessed image inputs transforms image analysis. Deep learning empowers models to extract intricate features autonomously, eliminating the need for feature engineering and uncovering nuanced patterns in data. Furthermore, deep computer vision, an integral part of this paradigm, excels in various visual recognition tasks. Its exceptional performance spans particle detection [52], brain image segmentation [53], and material structure prediction [54]. Specifically, in atomic-resolution image analysis, deep learning neural networks revolutionise defect detection [55] and automate a traditionally labor-intensive process. They are proficient at locating a large number of atomic defects, enabling systematic class averaging and minimising the need for manual intervention or image processing methods.

The key challenges associated with structural defect and strain characterisations of 2D materials are subpicometer precision limitations and signal-to-noise ratio constraints. Achieving subpicometer precision in tracking individual atom strains remains a challenge and is primarily constrained by the precision limits and detection limits of atom-by-atom electron microscopy [56]. The high radiation doses required for precise atom positioning led to ionisation and knock-on damage, compromising defect structures [57]. Current techniques are effective in bulk materials but face limitations in 2D materials due to damage thresholds, resulting in a tradeoff between spatial resolution and precision in strain measurements.

Deep learning techniques combined with aberration-corrected STEM offer unprecedented benefits in visualising strain fields induced by single-atom defects in 2D materials [58]. These methods provide subpicometer precision, enabling direct observation of strain fields, even around beam-sensitive materials. For instance, a fully convolutional network (FCN) was deployed to identify four types of point defects in a monolayer 2D transition metal dichalcogenide ($WSe_{2-2x}Te_{2x}$) (Fig. 5).

These include double vacancies (DVs), single vacancies (SVs), single Te substitutions (SeTe), and double Te substitutions (2Te). The classification of the defects involved the following steps: (1) First, sequential STEM images were acquired by averaging 10 frames with short dwell times to minimise distortions from sample drift, resulting in high-precision strain measurements. The study analysed images of 9 regions covering approximately 130,000 atoms in a total area of 4000 nm^2. (2) Next, a deep learning model employing fully convolutional networks (FCNs) with the deep residual U-Net (ResUNet) architecture was trained to generate 2D maps of defect positions for data analysis. (3) The model was trained with simulated data, incorporating postprocessing steps such as adding Gaussian noise, probe jittering, image shear, and varying spatial sampling in the final training dataset. (4) Subsequently, class-averaged images were generated from the FCN outputs, and isolated defects with a distance $\geq$ 6.6 Å were selected. The original images were sectioned into defect-centered windows (Fig. 5a–d), grouped by type, and aligned to obtain high signal-to-noise class-averaged images. Images were categorised by defect type, forming stacks with 180 2Te, 312 SeTe, 576 SV, and 18 DV individual defects (Fig. 5e–h). (5) Next, the precision was measured by comparing single images versus class-averaged images. Class averaging, assisted by an FCN, achieves subpicometer precision in measuring atomic spacings and local strains. Single images exhibit 6–8 pm precision comparable to that of reported measures for 2D materials [59, 60], while class-averaged images demonstrate a 20-fold improvement (0.3 pm). This

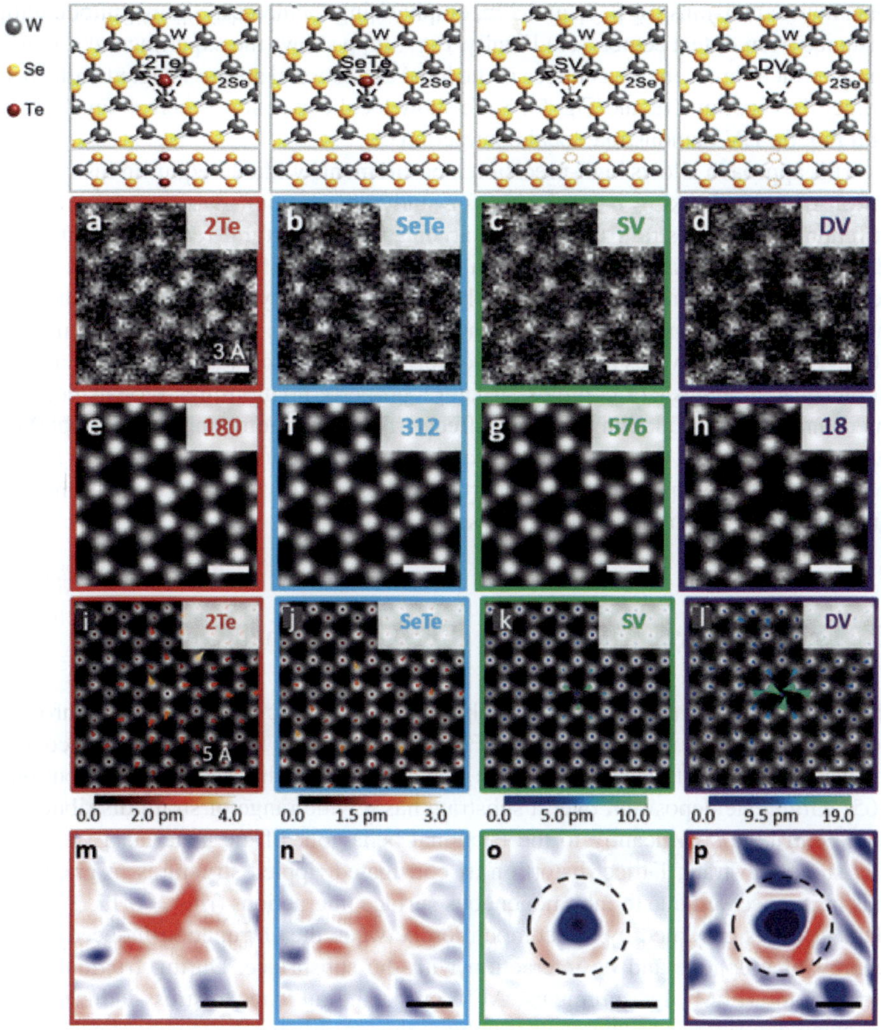

Fig. 5 Four types of defects in $WSe_{2-2x}Te_{2x}$: 2Te, SeTe, SV, and DV. **a–d** Representative single images showing defects identified by the fully convolutional network (FCN); **e–h** class-averaged images, **i–l** visualisation of the two-dimensional displacement vector field overlaid on class-averaged images highlighting chalcogen site defects, **m–p** visual representation of experimental dilation maps depicting local projected area changes associated with defects. Reproduced from Chia-Hao Lee et al., Nano Letters, 2020. Copyright 2020, American Chemical Society [58])

method will be instrumental in the electron microscopy assessment of minor strains around atomic defects in 2D materials while minimising electron beam damage. (6) Furthermore, displacement and strain fields were measured for each defect type, as illustrated in Fig. 5i–l), with magnified displacement vectors overlaid on class-averaged defect images. The vectors depict local lattice contractions for SV and DV but show expansions for SeTe and 2Te, emphasising the decrease in magnitude with distance from defect centers. (7) Finally, Fig. 5m–p shows the experimental dilation maps, which represent the changes in the local 2D material area associated with each defect. In Fig. 5m 2Te and Fig. 5n SeTe show local expansion, whereas Fig. 5o SV and Fig. 5p DV demonstrate local contraction.

This approach allows high-precision measurements without requiring instrumentation changes, making it invaluable for studying radiation-sensitive crystals and revealing complex strain phenomena arising from structural rearrangements and defect interactions in 2D materials. In addition to identifying point defects, deep learning has been used to successfully identify other types of defects, such as nanopores [61], dopants [62], graphene nanoholes [63] and carbon contaminants [64]. Thus, ML provides a powerful tool for 2D material characterisation, enabling sophisticated analysis of complex structure–property relationships such as doping, mechanical, and oxidation straining effects. The continuous improvement of ML algorithms, coupled with rich datasets and reduced computational costs, will accelerate progress in exploring new properties and extracting meaningful relationships from atomic-resolution electron microscopy data.

5 Friction Force Microscopy (FFS) of Strained Graphene

The frictional characteristics of graphene at the nanoscale are coupled through interfacial interactions between the AFM probe apex and the graphene surface and between the supporting surface and the mechanical and adhesion force contributions [65]. Graphene deposition over a substrate has a wide range of strain distributions that contribute significantly to the frictional behaviour of the material. Even on a flat silica substrate, lattice distortion as corrugated rippled structures in graphene has been observed due to inhomogeneous surface interactions (rms $= 0.12 \pm 0.01$). This effect is governed by the competition between the corrugation-induced strain energy of graphene and its interaction with the subsurface. During friction force microscopy (FFM) measurements, AFM probe sliding leads to stress-induced out-of-plane deformation, a "buckled structure" responsible for the frictional behaviour of graphene that has been observed in other 2D materials [66]. One of the typical distorted atomically resolved stick–slip motions is illustrated in Fig. 6a, where it is evident that surface-induced corrugation influences localised graphene lattices and, consequently, the friction map. On the other hand, the suspended graphene depicts a uniform distribution of stick–slip motion due to the undistorted lattice structure (Fig. 6b).

With increasing thickness of graphene, the friction force decreases compared to that of a single layer (1LG) due bending rigidity in the structure, which mitigates the out-of-plane deformation and electron–phonon coupling in epitaxially grown silicon carbide [67]. These rippled structures and their associated domains were not observed in the micrographic image of the topography by AFM showed some distinguishable frictional domains for a fixed sliding direction of the AFM probe. The anisotropy

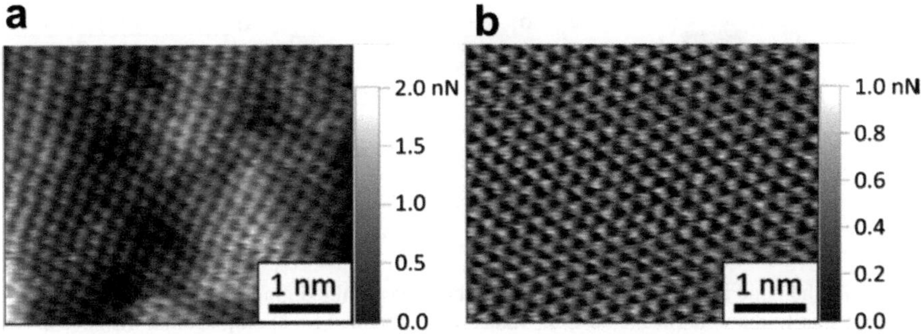

Fig. 6 **a** Stick–slip map of the atomic layer of graphene over a silica substrate indicating that a distorted lattice structure influences the overall frictional behaviour. **b** Stick–slip map of suspended graphene showing the substrate-free frictional map and undistorted lattice. Reprinted from Deng et al. [65]

in the friction force map arises depending on the sliding direction of the probe with respect to the ripple alignment; motion in the rippled axis direction causes lower friction values because the deformation resistance is greater perpendicular to the strain.

The increase in the ripple dimensions in the atomic layer of graphene occurs at an approximate height of 1–5 nm, width = 7–18 nm and length varies from nm to microns, leading to wrinkles, commonly observed with compressive and bending stresses in graphene. Like for the ripple structure, the orientation of the wrinkles significantly influences the frictional characteristics of strained graphene. The coefficient of friction (COF) of the perpendicularly aligned wrinkle is nearly 194% greater than the values measured for parallelly aligned wrinkles [68]. This is due to the generation of a greater contact area between the probe apex and the deformed wrinkle oriented perpendicular to the fast scan axis. The parallel orientation of the wrinkles, especially at the top region of the wrinkled architecture, is relatively rigid, leading to less deformation; consequently, a lower friction force is achieved (Fig. 7a, b). The variation in the COF at different orientations of the fixed wrinkle dimension against the sliding AFM probe follows the periodicity of 180° and agrees with the fitting mode 1 ASin|θ|, where ± 90° (Fig. 7c). These variations have not been observed over the basal plane of graphene (i.e., periodicity of 60°) or at the interface (graphene edge) between the atomic sheet and the substrate [66, 68]. Nevertheless, in the latter case, the graphene edge was found to be the weakest attribute in the graphene layer against the sliding tip apex, which wears quickly compared to wrinkles [69].

The deposition of graphene over a flat surface leads to the uncontrolled distribution of ripples and wrinkles, revealing friction force anisotropy. The friction and anisotropy characteristics of a strained structure can be tuned by depositing graphene over the textured surface during peculiar ridge formation with a groove separation distance of 40 nm and depth of 10 nm. The periodicity between the grooves defines the strain distribution in the conformed graphene along the crest and trough regions. The coverage of graphene not only decreases the friction force by tenfold compared

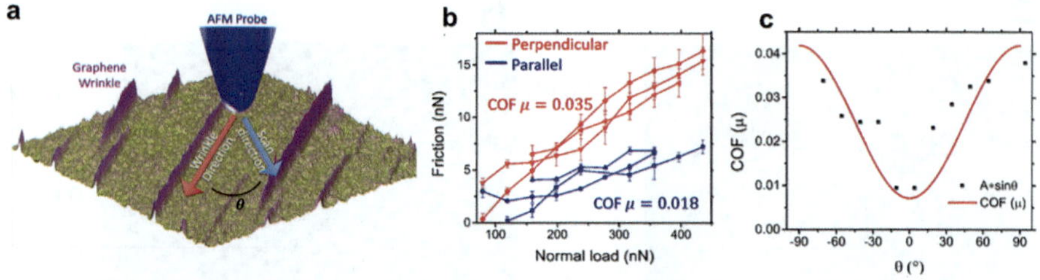

Fig. 7 **a** Schematic illustration of the AFM probe over wrinkled graphene scanning parallel and perpendicular to the wrinkle axis. **b** COF values were obtained for sliding against wrinkles in two parallel and perpendicular directions. **c** COF values are measured at different orientations of wrinkle axes; the lowest value is measured for the tip apex sliding parallel to the wrinkle axis. Reprinted from Long et al. [68]

to that of the bare Si-textured surface but is also regulated over the localised sites of the crest and trough regions for scanning either parallel or orthogonal to the groove axis (Fig. 8). The trough region is responsible for the higher friction force values than the crest for scanning perpendicular to the textured surface. Nevertheless, the friction is suppressed during the scan in the parallel direction. These anisotropies were not observed in the friction force values of the bare textured regions, indicating that the straining of the graphene occurred differently on the ridge-shaped textured surfaces [8]. The COF was measured for a flat surface, and the textured regions revealed that the Gr/flat surface had intermediate values between the orthogonal (higher) and parallel grooved surfaces (Fig. 9a, b). The texture-induced anisotropic friction response in graphene can be explained by anisotropic stretching/compression of the C–C bonds at different regions of the textured surfaces. Molecular dynamics (MD) simulations revealed that the net tensile strain resulted from the combined effect of substrate adhesion and adjacent suspended graphene (Fig. 9c–f).

The nanoscopic textured pitch periodicity (such as 40 nm) stretches the suspended graphene at the centre. It minimises the out-of-plane deformation during the event of a sliding tip. Nevertheless, the stretching of graphene through clamping can also be regulated by increasing the separation between grooves. The suspension for large spacing grooves is relatively relaxed in the central region, and the uniaxial strain dominates near the groove axis. The resultsResults distinct from those of nanoscopic textured surfaces have been obtained, where the friction force recorded is nearly three times greater while scanning parallel to the groove axis than when scanning orthogonal to the groove axis under the similar load conditions (Fig. 10) [70]. The indentation depth (elastic) increases while sliding in the parallel direction, which leads to greater deformation. This load-induced deformation is limited to the clamp region. In-depth simulation analysis revealed that sliding orthogonal to the groove axis stiffens the graphene and decreases the force required to drag it, while graphene deforms easily to slide parallel to the groove axis, and a larger force is required to drag the probe apex.

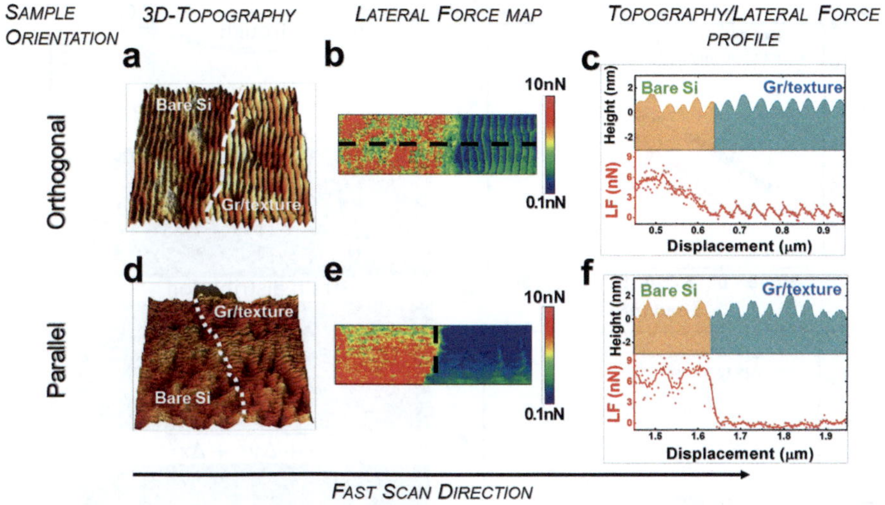

Fig. 8 Effect of scan direction on the friction force for sliding through the probe apex parallel and perpendicular to the groove surface at a periodicity of 40 nm. **a** Topography of partially covered graphene over the textured surface separated by a white dashed line. **b, c** Friction map and associated friction profile of bare and graphene-covered textured surfaces for sliding perpendicular to the groove axis. The surface is rotated to 90° to measure the **d** topography, **e** friction map, and **f** friction force profile. Reprinted from Mescola et al. [8]

The textural design for inducing in-plane biaxial strain in graphene can be carried out by deposition over prefabricated circular holes in a silica substrate. The in-plane strain is induced through the pressure difference between the optimised pressurised chamber and the ambient conditions that lead to controlled in-plane tension in the suspended graphene. In the absence of external pressure, the central region of the suspended graphene becomes flat with a slight sinking near the wall of the circular periphery of the substrate due to the adhesion from textured walls. The friction forces for graphene over the supported surface are the highest, and the lowest values are obtained in the region where graphene is suspended with in-plane strain (0.37%) (Fig. 11). This frictional behaviour influenced the COF, which systematically decreased from supported graphene to tension-induced strained graphene. Figure 11c reveals that the COF measured for highly strained graphene is sixfold lower than that of supported graphene and reaches the superlubricity regime [71]. These observations were reproducible regardless of the probe type and geometry. The atomic-scale simulation revealed a decrease in the asymmetry of the stick–slip phases achieved under higher-strain conditions. This weakness in the stick-and-slip asymmetry indicates the minimisation of energy corrugation, leading to stick–slip transformation to the continuous sliding state [72]. Thus, applying in-plane strain to graphene influences the out-of-plane flexibility and affects the contact quality [73].

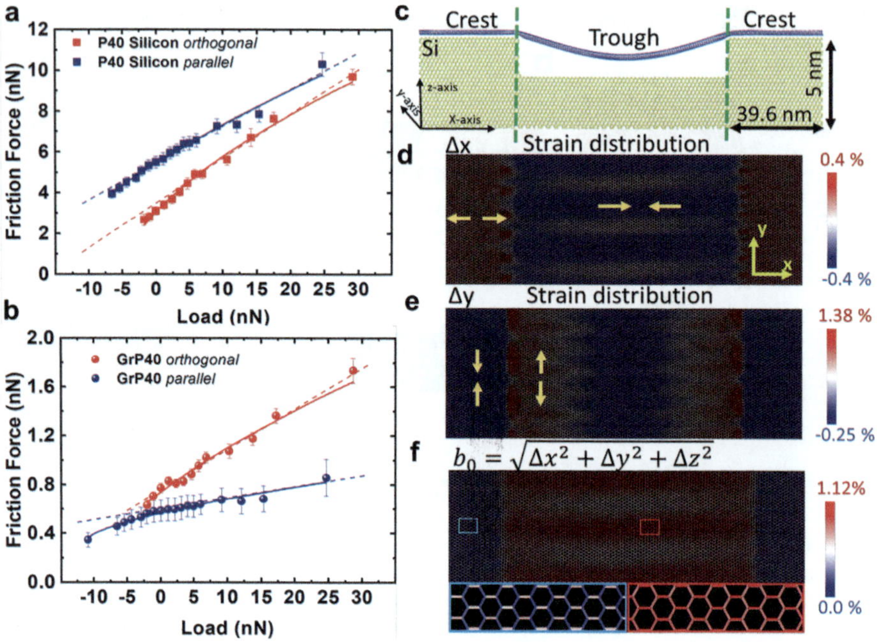

Fig. 9 Friction force as a function of applied load over a textured surface with a periodicity of 40 nm for **a** bare and **b** graphene-covered surfaces. **c** MD simulation setup of the suspended graphene over the textured region showing the crest and trough regions. **d–f** The strain distribution is based on the bond strain variation along the x-axis (Δx), **e** y-axis (Δy), and **f** total bond length (b_o). The asymmetry in the b_o between different regions and along different axes is readily apparent, as shown in the enlarged inset image. Reprinted from Mescola et al. [8]

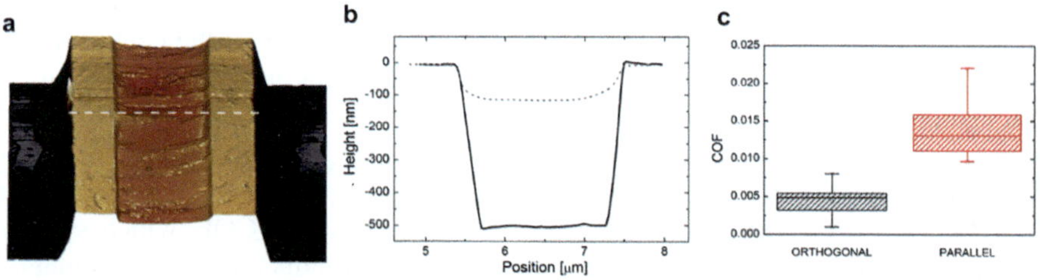

Fig. 10 **a** 3D topography of suspended graphene over a groove surface with spacings between the grooves of nearly 3 microns and a depth of 500 nm. **b** The line profile over the groove surface between two crest regions shows the clamping of graphene at the crest and its suspension at the centre. **c** The COF values measured over the suspended graphene surface moving orthogonal and parallel to the groove axis. Reprinted from Mescola et al. [70]

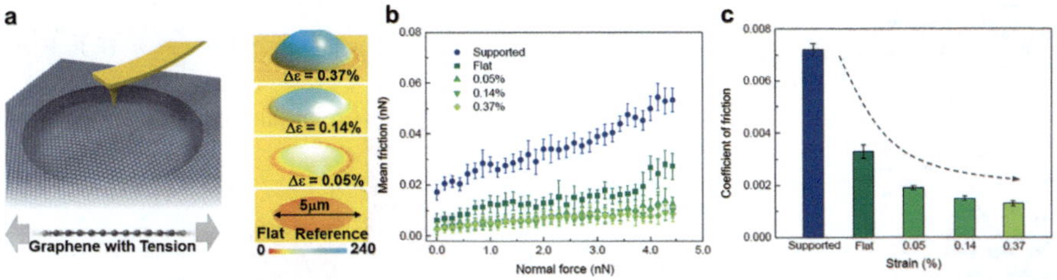

Fig. 11 **a** Schematic showing the suspended graphene over a circular hole. The tension at the centre of the suspended graphene was probed by the AFM tip apex. The inset shows the bulging of the suspended graphene corresponding to changes in the strain (Δε). **b** Trend of the mean friction force as a function of the normal force (applied), where tension-induced graphene reveals lower friction values. **c** Histogram of the friction coefficient of graphene over the supported surface (compressive strain) to that of the suspended graphene under biaxial tensile strain. Reprinted from Zhang et al. [71]

The frictional characteristics of graphene can also be modulated through deposition over another 2D material with different lattice constants and surface energies. This structural configuration leads to the development of a vertical heterostructure with intriguing interfacial properties. Cho and coworkers [74] investigated the frictional response of graphene deposited over different substrates, including h-BN and silica. The topography revealed the flattening of graphene over the atomically smooth h-BN compared to that of silica. The interfacial interaction between h-BN and graphene reduces friction-induced energy dissipation threefold, as observed through an atomic-scale stick–slip profile (Fig. 12). With increasing displacement (up to 2 nm), a consistent friction force is measured over the graphene/h-BN heterostructure, and in contrast, there is a systematic increase in the amplitude of the stick–slip profile with distance for graphene over silica, indicating the puckering effect. Thus, there is an out-of-plane deformation in graphene over silica under the influence of shear force from the apex of the AFM tip. It is anticipated that graphene could have a larger intimate contact area over the atomically smooth h-BN surface than on the corrugated silica substrate, which can induce strain on the deposited graphene. There is a strong correlation between the interfacial adhesion of graphene and the underlying substrates, and ultraflat smooth graphene that adheres well to its subsurface is a desirable condition for reducing the friction force.

6 Summary

Nanoscale textured designs carried out by sophisticated techniques such as EBL, FIB and RIE are wonderful platforms for precision engineering to develop 3D nanostructures; nevertheless, comprehensive knowledge of these materials is needed. Atomically thin sheets of graphene and other 2D materials are susceptible to conforming over textured surfaces. The degree of conformation can be determined through several

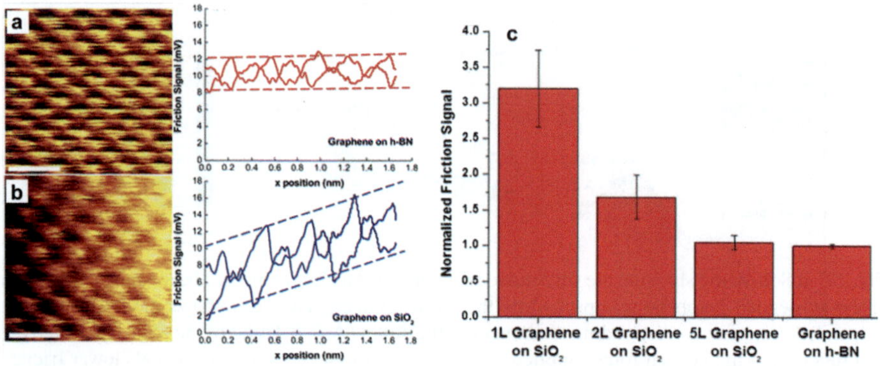

Fig. 12 Atomic-scale friction map of graphene over a, **a** h-BN substrate and its associated stick–slip, and **b** graphene over silica along with the stick–slip profile at relatively higher altitudes, indicating greater friction dissipation. Histograms summarising the normalised friction force during scanning of the AFM probe over different substrates reveal that graphene over the h-BN surface has a threefold reduction in friction force compared with that over the silica surface. Reprinted from Cho et al. [74]

imaging and probing techniques that are useful for monitoring strain in 2D materials. The interfacial interaction and geometry of the textured design govern the extent to which the conformation is responsible for inducing strain in 2D materials. Under the influence of tension, stretched graphene is reluctant to deform out-of-plane, leading to less friction energy dissipation; however, compression in 2D materials is equally important for influencing the surface chemistry and frictional response. Thus, by controlling the strain distribution in 2D materials, one can achieve a regulated friction force, which has been a complex task in traditional techniques. In the era of modern science and compact nanoscale machinery, solid-state lubricants are required for complex structures, where a fundamental understanding of strained structures is pivotal.

References

1. S. Zhang, T. Ma, A. Erdemir, Q. Li, Tribology of two-dimensional materials: from mechanisms to modulating strategies. Mater. Today **26**, 67–86 (2019)
2. J. Klein, E. Kumacheva, Confinement-induced phase transitions in simple liquids. Science **269**(5225), 816–819 (1995)
3. T.W. Scharf, S.V. Prasad, Solid lubricants: a review. J. Mater. Sci. **48**(2), 511–531 (2013)
4. Q.H. Wang, K. Kalantar-Zadeh, A. Kis, J.N. Coleman, M.S. Strano, Electronics and optoelectronics of two-dimensional transition metal dichalcogenides. Nat. Nanotechnol. **7**(11), 699–712 (2012)
5. G. Paolicelli, M. Tripathi, V. Corradini, A. Candini, S. Valeri, Nanoscale frictional behavior of graphene on SiO_2 and Ni(111) substrates. Nanotechnology **26**(5), 055703 (2015)
6. C. Lee, Q. Li, W. Kalb, X.-Z. Liu, H. Berger, R.W. Carpick, J. Hone, Frictional characteristics of atomically thin sheets. Science **328**(5974), 76–80 (2010)

7. A.A. Bukharaev, A.K. Zvezdin, A.P. Pyatakov, Y.K. Fetisov, Straintronics: a new trend in micro-and nanoelectronics and materials science. Phys. Usp. **61**(12), 1175 (2018)
8. A. Mescola, G. Paolicelli, S.P. Ogilvie, R. Guarino, J.G. McHugh, A. Rota, E. Iacob, E. Gnecco, S. Valeri, N.M. Pugno, Graphene confers ultralow friction on nanogear cogs. Small **17**(47), 2104487 (2021)
9. L. Liu, M. Zhou, L. Jin, L. Li, Y. Mo, G. Su, X. Li, H. Zhu, Y. Tian, Recent advances in friction and lubrication of graphene and other 2D materials: mechanisms and applications. Friction **7**, 199–216 (2019)
10. X. Zheng, L. Gao, Q. Yao, Q. Li, M. Zhang, X. Xie, S. Qiao, G. Wang, T. Ma, Z. Di, Robust ultra-low-friction state of graphene via moiré superlattice confinement. Nat. Commun. **7**(1), 13204 (2016)
11. J. Yang, B. Chen, H. Song, H. Tang, C. Li, Synthesis, characterization, and tribological properties of two-dimensional Ti_3C_2. Cryst. Res. Technol. **49**(11), 926–932 (2014)
12. S. Kwon, K.E. Lee, H. Lee, S.J. Koh, J.-H. Ko, Y.-H. Kim, S.O. Kim, J.Y. Park, The effect of thickness and chemical reduction of graphene oxide on nanoscale friction. J. Phys. Chem. B **122**(2), 543–547 (2018)
13. H. Li, J. Wang, S. Gao, Q. Chen, L. Peng, K. Liu, X. Wei, Superlubricity between MoS_2 monolayers. Adv. Mater. **29**(27), 1701474 (2017)
14. A. Rota, M. Tripathi, G. Gazzadi, S. Valeri, Origin of hydrophobicity in FIB-nanostructured Si surfaces. Langmuir **29**(17), 5286–5293 (2013)
15. P. Serles, M. Hamidinejad, P.G. Demingos, L. Ma, N. Barri, H. Taylor, C.V. Singh, C.B. Park, T. Filleter, Friction of Ti_3C_2T x MXenes. Nano Lett. **22**(8), 3356–3363 (2022)
16. X.-X. Zhang, S. Jiang, J. Lee, C. Lee, K.F. Mak, J. Shan, Spin dynamics slowdown near the antiferromagnetic critical point in atomically thin $FePS_3$. Nano Lett. **21**(12), 5045–5052 (2021)
17. R. Friedrich, V. Caciuc, N. Atodiresei, S. Blügel, Exchange interactions of magnetic surfaces below two-dimensional materials. Phys. Rev. B **93**(22), 220406 (2016)
18. L.B. Boinovich, E.B. Modin, A.R. Sayfutdinova, K.A. Emelyanenko, A.L. Vasiliev, A.M. Emelyanenko, Combination of functional nanoengineering and nanosecond laser texturing for design of superhydrophobic aluminum alloy with exceptional mechanical and chemical properties. ACS Nano **11**(10), 10113–10123 (2017)
19. Z. Wang, D. Tonderys, S.E. Leggett, E.K. Williams, M.T. Kiani, R.S. Steinberg, Y. Qiu, I.Y. Wong, R.H. Hurt, Wrinkled, wavelength-tunable graphene-based surface topographies for directing cell alignment and morphology. Carbon **97**, 14–24 (2016)
20. J. Chen, X. Zhao, S.J. Tan, H. Xu, B. Wu, B. Liu, D. Fu, W. Fu, D. Geng, Y. Liu, Chemical vapor deposition of large-size monolayer $MoSe_2$ crystals on molten glass. J. Am. Chem. Soc. **139**(3), 1073–1076 (2017)
21. J. Dong, L. Zhang, X. Dai, F. Ding, The epitaxy of 2D materials growth. Nat. Commun. **11**(1), 5862 (2020)
22. P.Y. Chen, J. Sodhi, Y. Qiu, T.M. Valentin, R.S. Steinberg, Z. Wang, R.H. Hurt, I.Y. Wong, Multiscale graphene topographies programmed by sequential mechanical deformation. Adv. Mater. **28**(18), 3564–3571 (2016)
23. B.A. Tayeh, B.H. Abu Bakar, M.A. Megat Johari, M.M. Ratnam, Existing concrete textures: their effect on adhesion with fibre concrete overlay. Proc. Instit. Civ. Eng.-Struct. Build. **167**(6), 355–368 (2014)
24. J. Rajendran, S.P. Ramalingam, M. Piraviperumal, Properties of indium tin oxide films grown on microtextured glass substrates. ACS Appl. Electron. Mater. **3**(12), 5459–5467 (2021)
25. P.I. Widenborg, A.G. Aberle, Polycrystalline silicon thin-film solar cells on AIT-textured glass superstrates. Adv. Opto Electron. (2007)
26. M.R. Vazirisereshk, H. Ye, Z. Ye, A. Otero-De-La-Roza, M.-Q. Zhao, Z. Gao, A.C. Johnson, E.R. Johnson, R.W. Carpick, A. Martini, Origin of nanoscale friction contrast between supported graphene, MoS_2, and a graphene/MoS_2 heterostructure. Nano Lett. **19**(8), 5496–5505 (2019)
27. New Whitepaper FIB-SEM—Raith—Nanofabrication systems for EBL, FIB and LBL. (n.d.). Raith—Nanofabrication Systems for EBL, FIB and LBL. https://raith.com/new-whitepaper-fib-sem/

28. R. Fabrication, Overcoming the limitations of nanofabrication—new white paper available. https://raith.com/new-whitepaper-fib-sem/
29. G. Cocco, E. Cadelano, L. Colombo, Gap opening in graphene by shear strain. Phys. Rev. B **81**(24), 241412 (2010)
30. H. Feng, Z. Cheng, D. Long, T. Yang, Z. Lu, Q. He, How do substrates affect the friction on graphene at the nanoscale? Lubricants **11**(11), 465 (2023)
31. Y. Zhang, X. Wang, H. Li, B. Wang, Adhesive behavior of micro/nano-textured surfaces. Appl. Surf. Sci. **329**, 174–183 (2015)
32. S. Yang, Y. Chen, C. Jiang, Strain engineering of two-dimensional materials: methods, properties, and applications. InfoMat **3**(4), 397–420 (2021)
33. J. Chaste, A. Missaoui, S. Huang, H. Henck, Z. Ben Aziza, L. Ferlazzo, C. Naylor, A. Balan, A.T.C. Johnson, Jr., R. Braive, A. Ouerghi, Intrinsic properties of suspended MoS_2 on SiO_2/Si pillar arrays for nanomechanics and optics. ACS Nano **12**(4), 3235–3242 (2018)
34. T. Li, Z. Zhang, Substrate-regulated morphology of graphene. J. Phys. D Appl. Phys. **43**(7), 075303 (2010)
35. J.S. Bunch, M.L. Dunn, Adhesion mechanics of graphene membranes. Solid State Commun. **152**(15), 1359–1364 (2012)
36. Y. He, W. Yu, G. Ouyang, Effect of stepped substrates on the interfacial adhesion properties of graphene membranes. Phys. Chem. Chem. Phys. **16**(23), 11390–11397 (2014)
37. S. Scharfenberg, D. Rocklin, C. Chialvo, R.L. Weaver, P.M. Goldbart, N. Mason, Probing the mechanical properties of graphene using a corrugated elastic substrate. Appl. Phys. Lett. **98**(9) (2011)
38. F. Lee, M. Tripathi, P. Lynch, A.B. Dalton, Configurational effects on strain and doping at graphene-silver nanowire interfaces. Appl. Sci. **10**(15), 5157 (2020)
39. E. Stolyarova, K.T. Rim, S. Ryu, J. Maultzsch, P. Kim, L.E. Brus, T.F. Heinz, M.S. Hybertsen, G.W. Flynn, High-resolution scanning tunneling microscopy imaging of mesoscopic graphene sheets on an insulating surface. Proc. Natl. Acad. Sci. **104**(22), 9209–9212 (2007)
40. M. Ishigami, J.-H. Chen, W.G. Cullen, M.S. Fuhrer, E.D. Williams, Atomic structure of graphene on SiO_2. Nano Lett. **7**(6), 1643–1648 (2007)
41. S. Manzeli, A. Allain, A. Ghadimi, A. Kis, Piezoresistivity and strain-induced band gap tuning in atomically thin MoS_2. Nano Lett. **15**(8), 5330–5335 (2015)
42. P. Nemes-Incze, G. Kukucska, J. Koltai, J. Kürti, C. Hwang, L. Tapasztó, L.P. Biró, Preparing local strain patterns in graphene by atomic force microscope based indentation. Sci. Rep. **7**(1), 3035 (2017)
43. X. Li, Y. Wei, G. Lu, Z. Mei, G. Zhang, L. Liang, Q. Li, S. Fan, Y. Zhang, Gate-tunable contact-induced Fermi-level shift in semimetal. Proc. Natl. Acad. Sci. **119**(17), e2119016119 (2022)
44. M. Tripathi, A. King, G. Fratta, M. Meloni, M. Large, J.P. Salvage, N.M. Pugno, A.B. Dalton, Laser-based texturing of graphene to locally tune electrical potential and surface chemistry. ACS Omega **3**(12), 17000–17009 (2018)
45. T. Mohiuddin, A. Lombardo, R. Nair, A. Bonetti, G. Savini, R. Jalil, N. Bonini, D. Basko, C. Galiotis, N. Marzari, Uniaxial strain in graphene by Raman spectroscopy: G peak splitting, Grüneisen parameters, and sample orientation. Phys. Rev. B **79**(20), 205433 (2009)
46. F. Lee, M. Tripathi, R.S. Salas, S.P. Ogilvie, A.A. Graf, I. Jurewicz, A.B. Dalton, Localised strain and doping of 2D materials. Nanoscale (2023)
47. J.E. Lee, G. Ahn, J. Shim, Y.S. Lee, S. Ryu, Optical separation of mechanical strain from charge doping in graphene. Nat. Commun. **3**(1), 1024 (2012)
48. Y. Zhang, M. Heiranian, B. Janicek, Z. Budrikis, S. Zapperi, P.Y. Huang, H.T. Johnson, N.R. Aluru, J.W. Lyding, N. Mason, Strain modulation of graphene by nanoscale substrate curvatures: a molecular view. Nano Lett. **18**(3), 2098–2104 (2018)
49. E. Han, J. Yu, E. Annevelink, J. Son, D.A. Kang, K. Watanabe, T. Taniguchi, E. Ertekin, P.Y. Huang, A.M. van der Zande, Ultrasoft slip-mediated bending in few-layer graphene. Nat. Mater. **19**(3), 305–309 (2020)

50. K.T. Butler, D.W. Davies, H. Cartwright, O. Isayev, A. Walsh, Machine learning for molecular and materials science. Nature **559**(7715), 547–555 (2018)
51. M.H.U. Rahman, M. Tripathi, A. Dalton, M. Subramaniam, S.N.L. Talluri, B.K. Jasthi, V. Gadhamshetty. Mach. Learn. 2D Mater. Sci. 163 (2023)
52. Y. Zhu, Q. Ouyang, Y. Mao, A deep convolutional neural network approach to single-particle recognition in cryo-electron microscopy. BMC Bioinf. **18**, 1–10 (2017)
53. T.M. Quan, D.G.C. Hildebrand, W.-K. Jeong, Fusionnet: a deep fully residual convolutional neural network for image segmentation in connectomics. Front. Comput. Sci. **3**, 613981 (2021)
54. H. Kirschner, R. Hillebrand, Neural networks for HREM image analysis. Inf. Sci. **129**(1–4), 31–44 (2000)
55. A. Maksov, O. Dyck, K. Wang, K. Xiao, D.B. Geohegan, B.G. Sumpter, R.K. Vasudevan, S. Jesse, S.V. Kalinin, M. Ziatdinov, Deep learning analysis of defect and phase evolution during electron beam-induced transformations in WS2. NPJ Comput. Mater. **5**(1), 12 (2019)
56. P.Y. Huang, S. Kurasch, J.S. Alden, A. Shekhawat, A.A. Alemi, P.L. McEuen, J.P. Sethna, U. Kaiser, D.A. Muller, Imaging atomic rearrangements in two-dimensional silica glass: watching silica's dance. Science **342**(6155) 224–227 (2013)
57. K. Elibol, T. Susi, G. Argentero, M. Reza Ahmadpour Monazam, T.J. Pennycook, J.C. Meyer, J. Kotakoski, Atomic structure of intrinsic and electron-irradiation-induced defects in $MoTe_2$. Chem. Mater. **30**(4), 1230–1238 (2018)
58. C.-H. Lee, A. Khan, D. Luo, T.P. Santos, C. Shi, B.E. Janicek, S. Kang, W. Zhu, N.A. Sobh, A. Schleife, Deep learning enabled strain mapping of single-atom defects in two-dimensional transition metal dichalcogenides with sub-picometer precision. Nano Lett. **20**(5), 3369–3377 (2020)
59. S. Wang, G.-D. Lee, S. Lee, E. Yoon, J.H. Warner, Detailed atomic reconstruction of extended line defects in monolayer MoS_2. ACS Nano **10**(5), 5419–5430 (2016)
60. A. Azizi, Y. Wang, G. Stone, A.L. Elias, Z. Lin, M. Terrones, V.H. Crespi, N. Alem, Defect coupling and sub-angstrom structural distortions in W1–x Mo x S2 monolayers. Nano Lett. **17**(5), 2802–2808 (2017)
61. C. Wang, Q. Zou, Z. Cheng, J. Chen, C. Luo, F. Liang, C. Cai, H. Bi, X. Lian, X. Ji, Tailoring atomic 1T phase CrTe2 for in situ fabrication. Nanotechnology **33**(8), 085302 (2021)
62. M. Ziatdinov, O. Dyck, A. Maksov, X. Li, X. Sang, K. Xiao, R.R. Unocic, R. Vasudevan, S. Jesse, S.V. Kalinin, Deep learning of atomically resolved scanning transmission electron microscopy images: chemical identification and tracking local transformations. ACS Nano **11**(12), 12742–12752 (2017)
63. M. Ziatdinov, O. Dyck, S. Jesse, S.V. Kalinin, Atomic mechanisms for the Si atom dynamics in graphene: chemical transformations at the edge and in the bulk. Adv. Func. Mater. **29**(52), 1904480 (2019)
64. K. Lee, J. Park, S. Choi, Y. Lee, S. Lee, J. Jung, J.-Y. Lee, F. Ullah, Z. Tahir, Y.S. Kim, Stem image analysis based on deep learning: identification of vacancy defects and polymorphs of MoS_2. Nano Lett. **22**(12), 4677–4685 (2022)
65. Z. Deng, N.N. Klimov, S.D. Solares, T. Li, H. Xu, R.J. Cannara, Nanoscale interfacial friction and adhesion on supported versus suspended monolayer and multilayer graphene. Langmuir **29**(1), 235–243 (2013)
66. J.S. Choi, J.-S. Kim, I.-S. Byun, D.H. Lee, M.J. Lee, B.H. Park, C. Lee, D. Yoon, H. Cheong, K.H. Lee, Friction anisotropy–driven domain imaging on exfoliated monolayer graphene. Science **333**(6042), 607–610 (2011)
67. T. Filleter, J.L. McChesney, A. Bostwick, E. Rotenberg, K.V. Emtsev, T. Seyller, K. Horn, R. Bennewitz, Friction and dissipation in epitaxial graphene films. Phys. Rev. Lett. **102**(8), 086102 (2009)
68. F. Long, P. Yasaei, W. Yao, A. Salehi-Khojin, R. Shahbazian-Yassar, Anisotropic friction of wrinkled graphene grown by chemical vapor deposition. ACS Appl. Mater. Interfaces. **9**(24), 20922–20927 (2017)
69. M. Tripathi, F. Awaja, R.A. Bizao, S. Signetti, E. Iacob, G. Paolicelli, S. Valeri, A. Dalton, N.M. Pugno, Friction and adhesion of different structural defects of graphene. ACS Appl. Mater. Interfaces. **10**(51), 44614–44623 (2018)

70. A. Mescola, A. Silva, A. Khosravi, A. Vanossi, E. Tosatti, S. Valeri, G. Paolicelli, Anisotropic rheology and friction of suspended graphene. Phys. Rev. Mater. **7**(5), 054007 (2023)
71. S. Zhang, Y. Hou, S. Li, L. Liu, Z. Zhang, X.-Q. Feng, Q. Li, Tuning friction to a superlubric state via in-plane straining. Proc. Natl. Acad. Sci. **116**(49), 24452–24456 (2019)
72. A. Socoliuc, R. Bennewitz, E. Gnecco, E. Meyer, Transition from stick-slip to continuous sliding in atomic friction: entering a new regime of ultralow friction. Phys. Rev. Lett. **92**(13), 134301 (2004)
73. S. Li, Q. Li, R.W. Carpick, P. Gumbsch, X.Z. Liu, X. Ding, J. Sun, J. Li, The evolving quality of frictional contact with graphene. Nature **539**(7630), 541–545 (2016)
74. D.-H. Cho, L. Wang, J.-S. Kim, G.-H. Lee, E.S. Kim, S. Lee, S.Y. Lee, J. Hone, C. Lee, Effect of surface morphology on friction of graphene on various substrates. Nanoscale **5**(7), 3063–3069 (2013)

Early-Stage Wear of Polymer Surfaces and Layered Materials Scraped by a Nanotip

Enrico Gnecco, Hesam Khaksar, and Juan J. Mazo

Abstract In this chapter we describe two representative applications of atomic force microscopy to investigating early-stage wear on polymeric surfaces and layered materials on the nanoscale. Ripples and exfoliated flakes or chips are the most typical surface structures obtained in each case. On polymers the ripple formation can be interpreted within the Prandtl-Tomlinson mechanism for atomic-scale friction with the evolving surface profile defining the energy landscape sensed by the scanning probe. If the scratching is repeated, nanoplastic particles are extruded from the crests of the ripples and displaced from the places where they were formed. Layered materials are exfoliated all along the scanned track, with the wear products either folded or bent depending on the thickness of the worn material. On multilayer surfaces, but not on monolayers, stick–slip is also observed in the friction force signal acquired while scratching.

1 Introduction

In contrast with the progress achieved in our understanding of sliding friction on the nanoscale, as exemplified in several chapters in this book, comparable results on the fundamental mechanisms ruling abrasive wear are still lacking. Here, we refer the reader elsewhere [1] for an overview of nanowear on hard surfaces, and limit our discussion to the investigation of early stages of nanowear, as observed on polymer surfaces and layered materials within our ongoing collaboration. Apart from shedding light on fundamental mechanisms of surface wear, a better understanding of incipient wear processes on these model materials is important in applications ranging from the control of the environment pollution caused by nanoplastics to nanomachining. It is therefore necessary, as a first step, to briefly introduce these issues.

E. Gnecco (✉) · H. Khaksar
Marian Smoluchowski Institute of Physics, Jagiellonian University, 30-348 Krakow, Poland
e-mail: enrico.gnecco@uj.edu

J. J. Mazo
Departamento de Estructura de la Materia, Física Térmica y Electrónica and Grupo Interdisciplinar de Sistemas Complejos (GISC), Universidad Complutense de Madrid, 28040 Madrid, Spain

E. Gnecco and E. Meyer (eds.), *Fundamentals of Friction and Wear on the Nanoscale*, NanoScience and Technology, https://doi.org/10.1007/978-3-031-63065-1_12

Environmental plastic, notably pollution from microplastics (MPs) and nanoplastics (NPs), is a significant public concern. Open questions about the consequences of such pollution appear more and more frequently in newspapers and popular science magazines [2]. Due to their small size, NPs are easily dispersed through the atmosphere [3]. They have been found in various consumer goods, including seafood, milk, salts, fruits, vegetables, tap and bottled water, and even beverages [4–6]. As a result, NPs pose a potential hazard to both human health and the environment [7–9]. Numerous methodologies have been proposed to detect NPs. However, an adequate approach that can satisfactorily identify these particles is still lacking. Arguably, the visual method is the initial and rudimentary technique for detecting plastic particles, but this approach is obviously limited to macroscopic particles [10, 11]. As the particle size decreases, other factors, such as the total particle count, particle size distribution, morphology, color, chemical composition, and mass, become increasingly critical [4]. In this context, AFM is a very promising technique not only for detecting NPs at individual level, but also to observe how the particles are generated when a polymer surface is scratched by a nanotip, and to correlate the particle formation and detachment to the forces acting on the tip [12–15]. The experimental observations presented in Sect. 2 show that, if the polymer surfaces are initially flat, NPs are only developed in a second stage from the crest of wavy "ripple" structures formed as a result of the first scratches. For the formation of those primeval structures, a phenomenological model based on the Prandtl-Tomlinson (PT) mechanism for atomic-scale friction, has been proposed, which is also discussed below.

The investigation of nanowear on layered materials has also a strong link to environmental issues. Lubrication aims indeed to reduce energy consumption and greenhouse gases in the atmosphere. The Industrial Revolution in the eighteenth century would have not been possible without it. Its applications embrace automotive, aerospace, and shipbuilding industries, to name only the most straightforward examples. Traditional lubrication uses oily and fat fluids, which can ultimately damage our health and environment. It is also inapplicable in a vacuum or in specific sectors such as textile industry. It is therefore crucial to develop alternative lubrication techniques that avoid fluids and characterize their functioning under model conditions. Solid lubricants, specifically 2D materials like graphene, molybdenum disulphide, tungsten disulphide, and hexagonal boron nitride, are among the most suited replacements so far [16]. In the form of solid powders they already find commercial applications as additives to traditional fluid in engine oil and diesel fuel. They can also "lubricate" dry contacts and prevent wear by coating the surfaces they are aimed to protect. Also in this context, AFM can be successfully used to scratch the solid surfaces on the nanoscale, and image the resulting structures afterwards. It can also record the friction forces accompanying the scratch process and allow a rough estimation of the energy consumption in it. Starting from the pioneer results by Klemenz et al. on graphene on platinum [17], the topic has been more and more frequently addressed experimentally and numericall. Section 3 presents some illustrative examples from our research team.

2 Evolution of Compliant Polymer Surfaces Scraped by a Nanotip

The principle of AFM and its application to nanotribology is described elsewhere in the first chapter of this book. Here we will only focus on the use of AFM to scratch polymer surfaces, to image them before/during/after scratching, and to measure the forces accompanying the scratch processes in real time. Figure 1 illustrates the concept of AFM-based nanoscratching. The apex of the probing tip consists in a cone or a pyramid based on silicon, which is much harder than the surface to be scratched. In most commercially available setups, the tip is pulled by a spring k along a series of parallel scan lines, with a separation b among them (not shown). Alternatively, zigzag pathways can be also used, but the results obtained in this way [18] are more difficult to interpret. Other scan path geometries are possible using dedicated nanolithography scripts.

Let us assume that the surface is initially flat. If the tip is simply pushed into it without scanning, an indentation pit will be formed. The pit is surrounded by a rim. With loading forces F_N in the order of few hundreds of nN the rim has a rather circular shape, even if the probe has a pyramidal body: see Fig. 1b. Assuming that the deformation is plastic, the total volume of the material displaced from the pit is equal to the volume of the extruded rim. However, it is well-known that, in the deformation of polymeric materials, the solid viscosity also plays a key role. It means that the indentation depth will not reach its final value when F_N is applied, but rather increase asymptotically towards it. The same can be said for the height of the surrounding

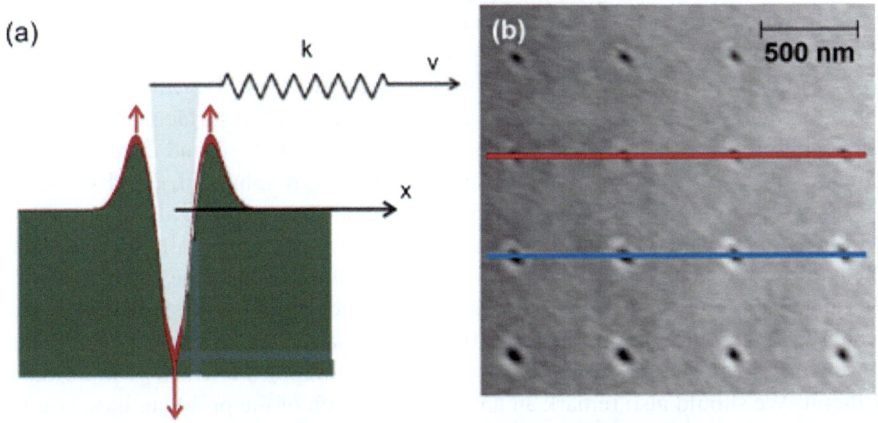

Fig. 1 **a** Sketch of the basic mechanism leading to nanowear on a compliant surface. A hard tip is indented into the surface. The resulting plastic deformation generates a rim all around it. At the same time, the tip is pulled laterally by a spring. **b** Nanoindentation pits on a polystyrene surface resulting from the application of a normal force of 0.4 μN (upper two rows) and 1.2 μN (lower two rows). The cross-sections along the red and blue lines (not shown) are consistent with the model discussed in the text. Adapted with permission from Ref. [19]

rim. Assuming, as a first approximation, that both pit and rim evolve according to a Gaussian profile, the time variation of the surface height at a distance r from the center of the indentation pit, $h(r)$, is approximately described by the equation

$$\frac{1}{N}\frac{dh}{dt} = -\exp\left(-\frac{r^2}{2\sigma^2}\right) + \alpha\exp\left(-\frac{(r-2\sigma)^2}{2\sigma^2}\right), \tag{1}$$

where N is the indentation rate, σ is the half-width of the two Gaussians and the factor 2 in the numerator has been chosen for consistency with the experimental observation. Moreover, the prefactor $\alpha \approx 0.20$ is obtained from the condition that the total volume is conserved [19].

Suppose now that the tip is driven laterally by the spring. The spring will elongate and, at the same time, the tip will be pulled to the right. However, the movement of the tip is hindered by the ongoing viscoplastic indentation, which tends to push the tip more and more deeply into the surface. The combination of the two processes can be described phenomenologically assuming that the lateral "energy landscape" seen by the tip precisely reproduces the evolving surface profile. The validity of this assumption is confirmed by the good agreement between its predictions and the results of nanoscratch tests performed by AFM: see below. In practice, it allows us to apply the well-known PT model for atomic-scale friction on a flat crystalline surface to the case of a time-varying tip-surface interaction (see Chap. 4 for more details).

In this way, the surface evolution is essentially ruled by four parameters: the indentation rate N, the tip width σ, the spring constant k and, also important, the velocity v of the free end of the spring. Depending on the ratios between these parameters two distinct regimes are expected. If N is sufficiently large and/or v and k are sufficiently low, the tip will remain pinned at its initial location for a certain time. However, when the spring elongation reaches a critical value (corresponding to a so-called *elastic instability*), the tip will suddenly cross the rim along the pull direction, and end up pinned again at a new location on the surface. The resulting stick–slip motion is accompanied by pile-ups of material all around the consecutive pinning sites. In the opposite case (low N and/or high values of v and k) the tip will move continuously driven by the stiff spring and, as a result, a shallow groove will be formed. The two outcomes have been confirmed numerically by estimating the equilibrium position of the tip at increasing values of the time t (Fig. 2). Note that there is no need of solving the equation of motion of the system in this way, which would require to introduce an effective tip mass m and a damping coefficient γ slowing down the tip in the slip phase, the estimation of which is notoriously difficult. We should also remark an analytical solution of the problem, based on the same model, which was recently given by Bradley and Shipman [21].

The PT mechanism so introduced can be also used to predict the shape of more complex nanowear patterns such as those formed by scanning a series of parallel lines one after the other (left to right). On each scan line the tip will encounter the structures resulting from the previous scan, which influences the "landing" position after slip. The smaller is the separation b, the larger is this influence. In this 3D case problem

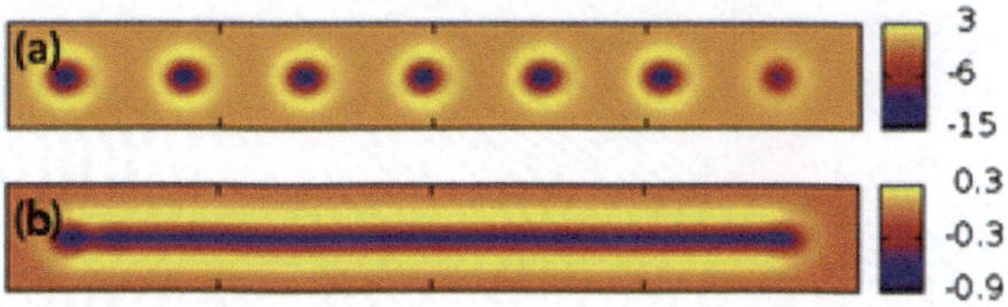

Fig. 2 **a** Numeric simulation of **a** a series of indentation pits, and **b** a continuous wear track obtained with the modified Prandtl-Tomlinson mechanism described in the text below and above a critical scan velocity. The color scale is given in arbitrary units. Adapted with permission from Ref. [20]

has been solved numerically by Mazo et al. [22]. As a result (Fig. 3a), the indentation rims were found to merge and evolve into a series of parallel ripples tilted by an angle θ with respect to the fast scan direction (x axis). This angle increases if b increases, up to 90° as $b \gtrsim \sigma$. At the same time, the corrugation of the ripples decreases. However, such a steady pattern is reached only after scanning few scan lines, as seen at the bottom of the figure. Moreover, on the left side of the scanned (square) area, the ripples tend to align at 90°, independently of b. The transition between the two orientation presents discontinuities, which are highlighted by the black rectangle in the figure. All these tendencies can be also recognized in corresponding nanowear tests performed by AFM, an example of which is given in Fig. 3b.

It is interesting to see how these conclusions remain valid if the orientation or the geometry of this area are changed. As an example, Fig. 4 shows the results of AFM tests with the parallel scan lines bounded by different geometric shapes. In all cases the ripples tend to follow the left edge profiles. This situation is kept if the edge is oriented close to the steady angle θ expected in the previous case. If this is not the case, the ripples tend to turn gradually towards θ. Please see Ref. [23] for similar results and numeric reproductions of those effects on circles and star-shaped patterns.

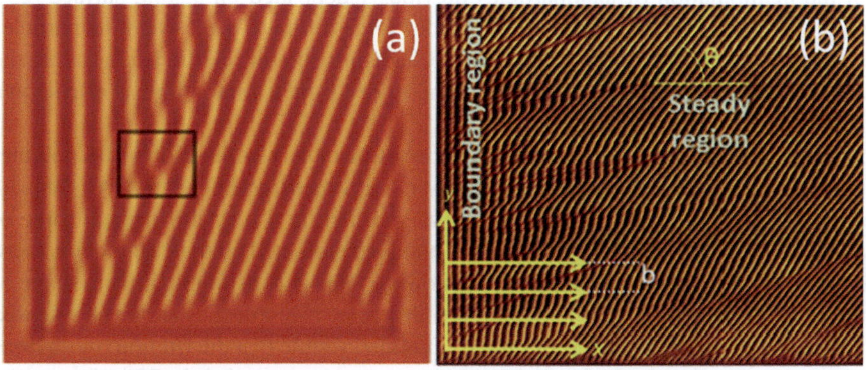

Fig. 3 **a** Ripple structures predicted numerically by the Prandtl-Tomlinson model applied to a surface profile evolving according to Eq. (1). **b** Corresponding results on a polystyrene surface as observed by AFM. Frame width: 10 μm. Adapted with permission from Ref. [22]

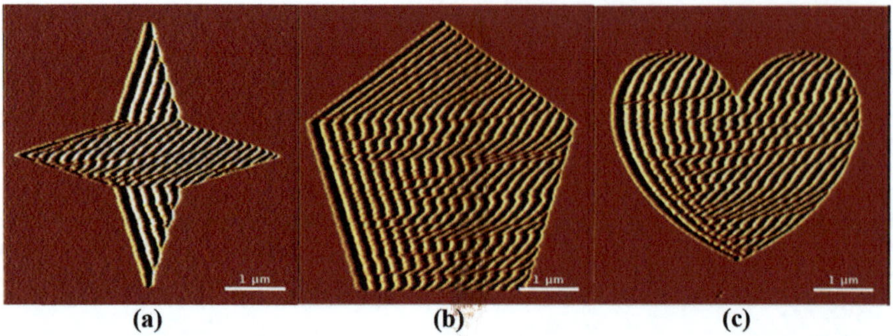

Fig. 4 Ripple patterns obtained experimentally by scanning **a** a diamond, **b** a pentagon, and **c** a hearth shaped area using AFM under loading conditions similar to those used in Ref. [23]

Suppose now that the scanning is repeated two, three, ... times. Based on the previous discussion, we would expect the ripple corrugation to grow without limit, or, assuming that N decreases with time, to increase toward an asymptotic value. However, this is not the case experimentally. Since the ripples originate from the merging of adjacent rims, their profile is not straight, but slightly wavy. The peaks on their tops grows faster than the rest of the surface structure. At the nanoscopic level, the molecular chains forming the polymer get more and more stretched. This is a well-known phenomenon in polymer physics, which is named *crazing* [24]. The process does not continue indefinitely. At a certain point, it is energetically convenient for the bulges to get released in the form of nanospheres. This is precisely what Hennig et al. have recently observed [3]. Once detached from the surface, the *nanoplastics* so formed can be found at the edges of the scanned area, where they align and partially pile up (Fig. 4). The pathways of the particles pushed by the tip from their nucleation sites up to their final locations on the edges can be recognized in the lateral force maps acquired while scratching. Similar pathways are well-known in the context of the manipulation of nanoparticles previously deposited on a solid surface [25]. Oppositely to that case, the lateral (friction) force does not peak when the particles start moving. In line with the crazing mechanism operating in the background, the release of the nanoplastics is indeed found to occur rather smoothly, as also shown in Ref. [3] (Fig. 5).

Finally, we have also investigated the formation of surface structures on a polymer blend, specifically polystyrene (PS) /poly n-butyl methacrylate (PnBMA) [26]. As seen in Fig. 6, not only ripples were observed, but, at a very low normal force of about 10 nN, also close-packed arrays of nanopits predicted by the PT model in these conditions. The ripples tend again to follow geometric constraints, i.e. the shape of the pre-existing ring and hole domains in this case. Nanoplastics are again released if the scratching is repeated (not shown). It is also visible that the friction oscillations accompanying the ripple formation are larger on the more brittle PS, whereas the less stiff PnBMA gets more damaged. Numeric simulation aimed to reproduce these results has not been attempted yet.

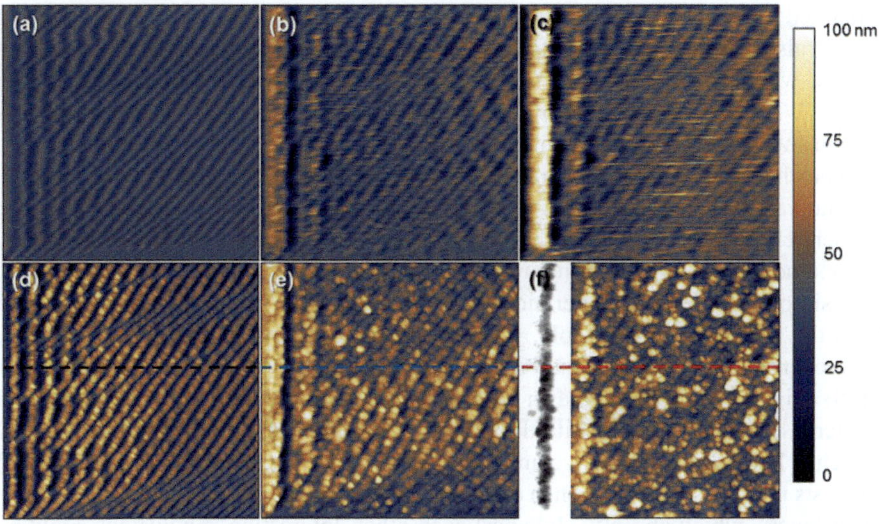

Fig. 5 **a–c** Surface structures observed when different regions of a PS surfaces are scratched for the first, third and fifth time respectively. **d–f** High resolution tapping-mode images of the same regions revealing the formation of nanoplastics out of the scratch process. Frame sizes: $5 \times 5\ \mu m^2$. Reproduced with permission from Ref. [3]

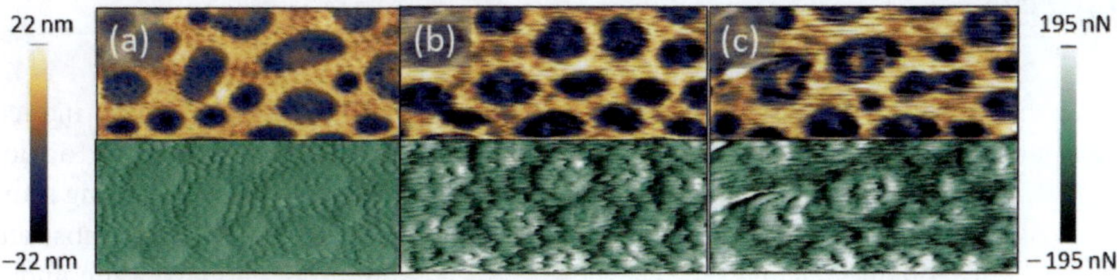

Fig. 6 **Upper row**: Surface topography of a PS/PMMA blend scratched by applying a normal force of 10, 50 and 90 nN respectively. **Lower row**: Corresponding lateral force images showing the resulting nanopits and ripples with enhanced contrast. Frame sizes: $1.8 \times 1\ \mu m^2$. Reproduced with permission from Ref. [26]

3 Exfoliation of Layered Surfaces Scraped by a Nanotip

Very different results are obtained when a layered material is scraped by a nanotip. In this case, silicon probes are unsuited since they are easily worn off well before damaging the surfaces. Not so surprisingly, diamond probes are much more efficient, provided that their apex is sharp enough. As a first example, Fig. 7a shows a series of wear grooves formed on a MoS_2 monolayer previously grown on amorphous SiO_2. The layer was scratched along the crystallographic armchair direction of MoS_2, and went peeled off along a zigzag direction at 30° and then folded again at −60°, as better seen in the corresponding AFM images in Fig. 7b. As a result the SiO_2 substrate remained exposed through a series of triangular patches. The detail highlighted in Fig. 7b also shows that flakes exfoliated on one side of the wear track can be multiply folded, whereas the other side of the track went structured by a periodic "tooth pattern". This feature possibly originates from the ripples formed on the amorphous silicon substrate, which was simultaneously scratched by the diamond tip.

Note that the wear structures in Fig. 7a were obtained with a normal force of 3 μN. Similar values were required for scratching multilayer MoS_2. In that case, however, the scratch process resulted in the formation of chips with linear size of about 100 nm and height of few tens of nm rolled out from the wear track. As detailed in Ref. [27] simple energetic considerations allowed to explain why the flakes originated from the monolayer tend to fold, while thicker layers tend to bend and form chips. Moreover, stick–slip was unambiguously recognized while scratching multilayer MoS_2 (Fig. 8a) but not monolayer MoS_2. The stick–slip repetition length was found to increase logarithmically with the scan velocity, although it remained well below the chip length (Fig. 8b). A comparison with TEM analysis by Tang et al. [28] suggests that the slip corresponds to the sudden rupture of the interface between two inner layers, which abruptly slide past each other. Interesting conclusions could be also drawn from the decreasing value of the lateral stiffness with the scan velocity (not shown) indicating a possible influence of the environmental humidity on the nanoscratch process.

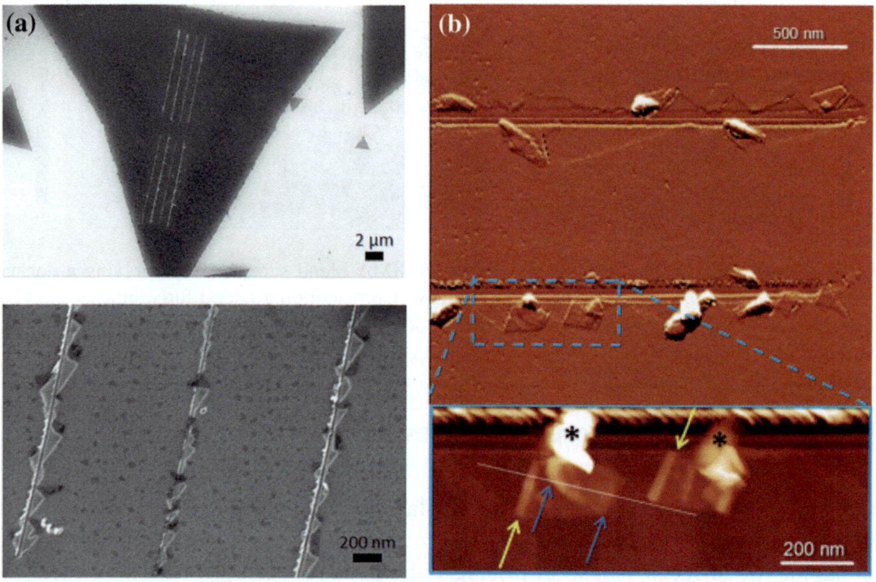

Fig. 7 **a** SEM images of a MoS_2 monolayer scratched eight times by a diamond tip. **b** Tapping mode AFM images highlighting the crack and fold directions of the layer, and a periodic tooth structure. While the upper part is a map of a phase signal, the magnified part below is a topography image. Adapted with permission from Ref. [27]

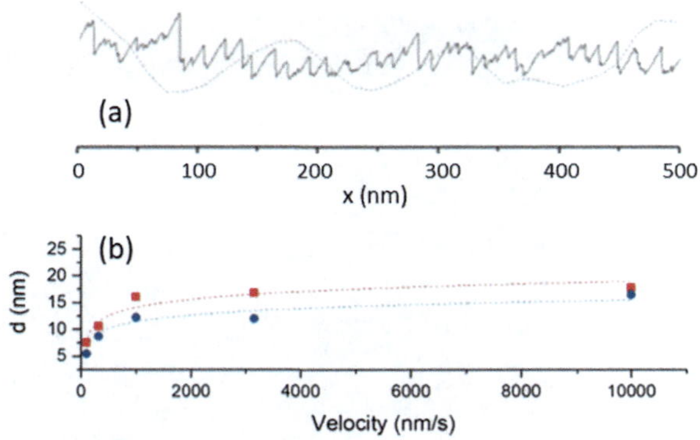

Fig. 8 **a** Stick–slip profile (in arbitrary units) observed while scratching a MoS_2 surface with a normal force of 2.5 μN using AFM. **b** Average stick–slip repetition distance as a function of the scan velocity. Adapted with permission from Ref. [27]

In a later study, the results on MoS_2 have been compared with those obtained on other layered materials [29]. Figure 9 shows the opposite response of a graphite and a mica surface. In the first case an extremely long chip was peeled off, flipped and fold onto the surface with an angle of rotation of about 120° with respect to the wear track. In the second case, the brittle surface was apparently pulverized by the diamond tip thus leaving a groove with alternating depths of 1 and 2 ML on the surface. Such an alternance was already observed in an early study by Kopta et al. and possibly related to thermally activated breakup and formation of Si–O bonds [30].

The previous examples are intended to illustrate the potential of AFM in the investigation of abrasive wear in its early stage. Other possibilities are offered by the combination of commercial nano-/microindenters for scratching and AFM for imaging the resulting structures. Please see Fig. 10 for an image of a microwear grove obtained with a normal force of 30 mN on amorphous SiO_2 as observed by AFM. The ripple pattern obtained on such a brittle material could be also related to the stick–slip motion of the indenter based on the assumption that the indenter has to overcome a well-defined static friction threshold to break up the glass, and it experiences a lower kinetic friction while cutting it in the slip phase [31]. This work is currently being extended on bulk layered materials.

4 Conclusions and Outlook

In this chapter, we have discussed two significant cases of nanowear of polymer surfaces and layered materials, as investigated by AFM. The surfaces could be scratched and imaged using the same probe. The ripple structures formed on polymers could be well reproduced by an extension of the well-known Prandtl-Tomlinson model for atomic-scale friction. Extrusion and release of nanoplastics is visible if the scratching is repeated. On layered materials the exfoliation of flakes and chips is the most common outcome. Some repetitive patterns are observed also in this case, but

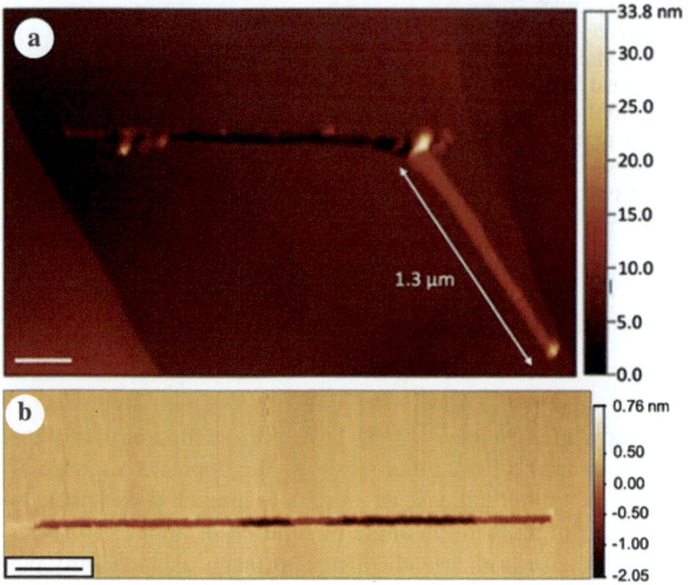

Fig. 9 Nanowear structures observed after scratching **a** a graphite and **b** on a mica surface left to right with a normal force of 3 μN. The scale bars correspond to 500 and 200 nm respectively. Adapted with permission from Ref. [29]

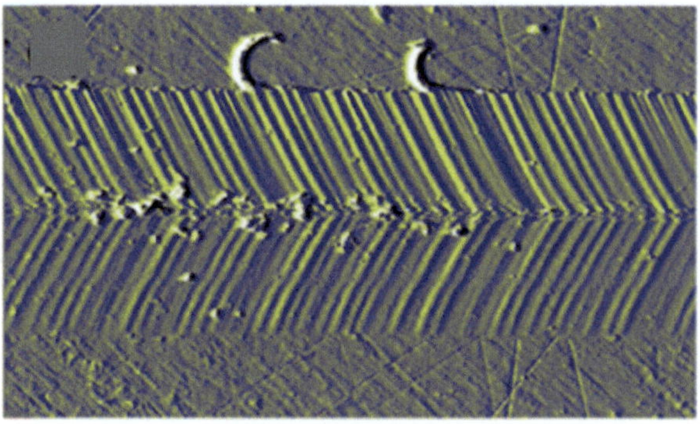

Fig. 10 Herringbone "ripple" pattern observed on amorphous SiO_2 previously scratched (left to right) with a normal force of 30 mN and a scan velocity of 10 μm/s. Frame size: 10 × 6 μm^2. Reproduced with permission from Ref. [31]

the problem is much more complex, periodicities are not easy to reveal, and, to the best of our knowledge, phenomenological models reproducing the observed results are still missing. The relevance of these topics, e.g. in environmental protection, is easy to recognize.

Acknowledgements We thank Pedro Martinez, Jana Hennig, an Alper Özogul for their collaboration on the modelling and experimental work presented in this chapter. The support of the Polish National Science Center (NCN) via the Opus Grant No. UMO/2021/43/B/ST5/00705 and of the

Strategic Program Excellence Initiative at the Jagiellonian University SciMat (Grant No. U1U/P05/NO/01.05). is gratefully acknowledged by EG and HK.

References

1. R. Colaco, From nano and microcontacts to wear of materials, in *Fundamentals of Friction and Wear on the Nanoscale*, 2 edn, eds. by E. Gnecco, E. Meyer (Springer, 2015)
2. C. Schwaferts, R. Niessner, M. Elsner, N.P. Ivleva, Methods for the analysis of submicrometer- and nanoplastic particles in the environment. TrAC, Trends Anal. Chem. **112**, 52–65 (2019)
3. J. Hennig, A. Litschko, J.J. Mazo, E. Gnecco, Nucleation and detachment of polystyrene nanoparticles from plowing-induced surface wrinkling. Appl. Surface Sci. Adv. **6**, 100148 (2021)
4. C. Vitali, R. Peters, H. G. Janssen, M.W. Nielen, Microplastics and nanoplastics in food, water, and beverages; part I. Occurrence. TrAC Trends Anal. Chem. 116670 (2022)
5. Y. Li, Z. Wang, B. Guan, Separation and identification of nanoplastics in tap water. Environ. Res. **204**, 112134 (2022)
6. R. Kumar, C. Manna, S. Padha, A. Verma, P. Sharma, A. Dhar, A. Ghosh, P. Bhattacharya, Micro (nano) plastics pollution and human health: How plastics can induce carcinogenesis to humans? Chemosphere **298**, 134267 (2022)
7. K. Blackburn, D. Green, The potential effects of microplastics on human health: what is known and what is unknown. Ambio **51**(3), 518–530 (2022)
8. W. Zhang, Q. Wang, H. Chen, Challenges in characterization of nanoplastics in the environment. Front. Environ. Sci. Eng. **16**, 1–3 (2022)
9. V. Kumar, E. Singh, S. Singh, A. Pandey, P.C. Bhargava, Micro-and nano-plastics (MNPs) as emerging pollutant in ground water: environmental impact, potential risks, limitations and way forward towards sustainable management. Chem. Eng. J. **459**, 141568 (2023)
10. V. Hidalgo-Ruz, M. Thiel, Distribution and abundance of small plastic debris on beaches in the SE Pacific (Chile): a study supported by a citizen science project. Mar. Environ. Res. **87**, 12–18 (2013)
11. R. Ahmed, A.K. Hamid, S.A. Krebsbach, J. He, D. Wang, Critical review of microplastics removal from the environment. Chemosphere **293**, 133557 (2022)
12. S. Zimmermann, J.L. Mead, F.T. von Kleist-Retzow, Probing friction and adhesion of individual nanoplastic particles. J. Phys. Chem. C **124**(44), 24145–24155 (2020)
13. J. Liu, H. Jiang, Q. Cheng, C. Wang, Investigation of nano-scale scratch and stick-slip behaviors of polycarbonate using atomic force microscopy. Tribol. Int. **125**, 59–65 (2018)
14. Y. Sun, Y. Yan, Z. Hu, X. Zhao, J. Yan, 3D polymer nanostructures fabrication by AFM tip-based single scanning with a harder cantilever. Tribol. Int. **47**, 44–49 (2012)
15. S. Friedrich, B. Cappella, Friction and mechanical properties of AFM-scan-induced ripples in polymer films. Front. Mech. Eng. **7**, 672898 (2021)
16. M. Marian, D. Berman, A. Rota, R.L. Jackson, A. Rosenkranz, Layered 2D nanomaterials to tailor friction and wear in machine elements—a review. Adv. Mater. Interfaces **9**, 2101622 (2022)
17. A. Klemenz, L. Pastewka, S.G. Balakrishna, A. Caron, R. Bennewitz, M. Moseler, Nano Lett. **14**, 7145 (2014)
18. R.H. Schmidt, G. Haugstad, W. Gladfelter, Langmuir **15**, 317 (1999)
19. E. Gnecco, P. Pedraz, P. Nita, F. Dinelli, S. Napolitano, P. Pingue, Surface rippling induced by periodic instabilities on a polymer surface. New J. Phys. **17**, 032001 (2015)
20. P.J. Martinez, E. Gnecco, J.J. Mazo, Numerical study of pattern formation in compliant surfaces scraped by a rigid tip. Phys. Rev. E **103**, 022802 (2021)

21. R.M. Bradley, P.D. Shipman, Theory of the oscillatory instability of a rigid tip scraped over a polymer surface. Phys. Rev. E **106**, 054803 (2022)
22. J.J. Mazo, P.J. Martínez, P. Pedraz, J. Hennig, E. Gnecco, Plowing-induced structuring of compliant surfaces. Phys. Rev. Lett. **122**, 256101 (2019)
23. J. Hennig, V. Feller, P.J. Martínez, J.J. Mazo , E. Gnecco, Locking effects in plowing-induced nanorippling of polystyrene surfaces. App. Surf. Sci. **594** 153467 (2022)
24. J. Rottler, M.O. Robbins, Growth, microstructure, and failure of crazes in glassy polymers. Phys. Rev. E **68**, 011801 (2003)
25. A. Rao, M.L. Wille, E. Gnecco, K. Mougin, E. Meyer, Trajectory fluctuations accompanying the manipulation of spherical nanoparticles. Phys. Rev. B **80**, 19340 (2009)
26. H. Khaksar, C. Ma, N. Janiszewska, K. Awsiuk, A. Budkowski, E. Gnecco, Nanoscale wear evolution on a polystyrene/poly (n-butyl methacrylate) blend. Wear **536–537**, 205160 (2024)
27. A. Özogul, F. Trillitzsch, C. Neumann, A. George, A. Turchanin, E. Gnecco, Plowing-induced nanoexfoliation of mono- and multilayer MoS_2 surfaces. Phys. Rev. Materials **4**, 033603 (2020)
28. D.M. Tang, D.G. Kvashnin, S. Najmei, Y. Bando, K. Kimoto, P. Koskinen, P.M. Ajayan, B.I. Yakobson, P.V. Sorokin, J. Lou, D. Golberg, Nanomechanical cleavage of molybdenum disulphide atomic layers. Nat. Commun. **5**, 3631 (2014)
29. A. Ozogul, E. Gnecco, M.Z. Baykara, Nanolithography-induced exfoliation of layered materials. Appl. Surf. Sci. Adv. **6**, 100146 (2021)
30. S. Kopta, M. Salmeron, The atomic scale origin of wear on mica and its contribution to friction. Langmuir **113**, 8249 (2000)
31. E. Gnecco, J. Hennig, E. Moayedi, L. Wondraczek, Surface rippling of silica glass surfaces scraped by a diamond indenter. Phys. Rev. Materials **2**, 115601 (2018)

Nanoscale Study of Antiwear Tribofilm Growth Using In Situ Atomic Force Microscopy Under Lubricated Contacts

Jitendra Soni, Himanshu Rai, and Nitya Nand Gosvami

Abstract We dedicate the current book chapter to highlighting the emerging nanoscale approach in tribology research, with a focus on the investigation of zinc dialkyldithiophosphate (ZDDP)-based antiwear tribofilms. Leveraging the in situ Atomic Force Microscope (AFM) technique, researchers have gained fundamental insights into the intricacy of tribochemical mechanisms within buried lubricated sliding contacts. The chapter covers a comprehensive overview of the complex tribological response of ZDDP at the sliding interface and its investigation through a detailed discussion on the AFM-based technique. Moreover, it reviews the profound influence of contact stresses, temperature, and substrate chemistry on tribofilm growth kinetics. It further incorporates the structural and chemical characterization-based inferences to deepen the understanding of tribofilm properties. The chapter underscores the unique ability of the technique in conjunction with the phenomena to create nanoscale structures and controlled imprints, emphasizing its significance in advancing tribology research and beyond.

1 Introduction

One of the primary factors contributing to failures in mechanical systems is the wear of the moving components such as brakes, bearings, gears, etc. [1–3] The surfaces of these moving components are often protected by thermal diffusion treatments [4], or wear-resistant coatings [5]; nevertheless, despite these protective measures, the thin protective layers ultimately wear out over time. Material wear typically begins when surfaces in mechanical contact experience friction under a load while being rubbed against each other. The introduction of a fluid lubricant film between sliding surfaces

J. Soni · H. Rai · N. N. Gosvami (✉)
Department of Materials Science and Engineering, Indian Institute of Technology Delhi, Hauz Khas, New Delhi 110016, India
e-mail: ngosvami@iitd.ac.in

H. Rai
Department of Control Engineering, Faculty of Electrical Engineering, Czech Technical University in Prague, Technicka 2, 16627, Prague 6, Czech Republic

E. Gnecco and E. Meyer (eds.), *Fundamentals of Friction and Wear on the Nanoscale*, NanoScience and Technology, https://doi.org/10.1007/978-3-031-63065-1_13

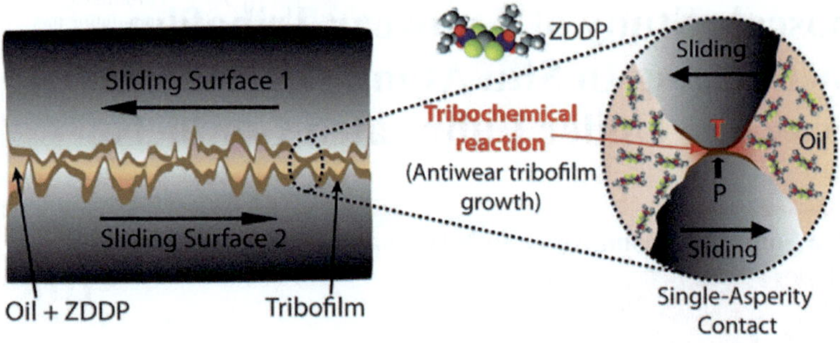

Fig. 1 Schematic depicting the formation of antiwear tribofilm at the single-asperity contact in the presence of ZDDP-containing lubricant. Reprinted with permission from Ref. [12]. Copyright (2019) MRS Bulletin

can substantially minimize friction and wear [6]. The significant contact pressure and shear forces occurring at the rough interface, where multi-asperities come in contact, frequently initiate tribochemical, mechanochemical, and catalytic reactions at sliding interfaces, particularly in the presence of reactive lubricant additives [7, 8]. Zinc dialkyldithiophosphate (ZDDP) is one such additive used in industrial lubricants which results in the nucleation and growth of protective and lubricious tribofilms between the contacting surfaces, thereby preventing wear. Figure 1 represents a schematic depicting the tribochemical reaction occurring at contacting asperities, facilitated by a lubricant containing ZDDP, between two surfaces sliding against each other. This phenomena is particularly evident in the boundary lubrication regime, where a protective tribofilm consisting of phosphate glass develops at the surfaces in contact, effectively preventing direct contact between them [9–12].

Understanding the complex tribochemical reactions of ZDDP during frictional sliding has posed a formidable challenge, primarily due to the obscured view of buried interfaces, which hinders direct observation. Given the limitations of existing techniques in direct observation, this chapter extensively covers in situ nanotribological investigations conducted using atomic force microscopy (AFM)-based techniques that can facilitate real-time monitoring of the development of tribochemical films. In the book chapter, we highlight the comprehensive insights based on our recently published review article [13] which serves as a foundational resource for the book chapter. The book chapter encompasses an in-depth discussion of the AFM-based in situ technique, providing a thorough exploration of the tribochemical reactions induced by antiwear additives such as ZDDP. Moreover, it aims to offer an overview of the fundamental mechanisms underlying the formation of ZDDP-based antiwear tribofilms. Additionally, the chapter emphasizes the significant role of substrate chemistry and various growth parameters, including contact pressure, shear forces, and temperature. These insights are further reinforced by inspecting structural and chemical characterization-based inferences scrutinized by the correlation of AFM results with chemical characterization techniques.

2 ZDDP Antiwear Tribofilm and Its Growth Mechanisms

For over 80 years, ZDDP, a phosphorus- and sulfur-containing compound, has been an important element in various engine oils [14]. Its initial use in the 1940s was focused on its antioxidative and corrosion-inhibiting properties, with its antiwear capabilities being overlooked for almost a decade. Subsequently, it gained widespread adoption in industrial lubricants and engine oils, leading to the development of more reactive secondary alkyl ZDDPs in the 1980s [14]. However, concerns were raised a decade later about accumulation within catalytic converters due to phosphorous- and sulfur-based by-products along with ash from combustion exhausts which gradually poison and eventually degrade their efficiency and lifespan over time [14–17]. Simultaneously, it was observed that although ZDDP demonstrated outstanding antiwear capabilities, it also had its drawbacks, such as contributing to micropitting damage on steel surfaces and reducing the lifespan of bearings [18–21]. As a result of these challenges, restrictions were placed on permissible concentration levels of ZDDP additives in lubricants. This compelled industries seeking alternative approaches including partial or complete replacement of ZDDPs or developing new additives without phosphorus or sulfur [14]. The pursuit for a substitute for ZDDP is crucial; nevertheless, its cost-efficiency and outstanding antiwear capabilities present a challenging obstacle.

Understanding ZDDP has been challenging due to limited insights into its interactions with sliding interfaces, compounded by the complexity arising from the multi-asperity nature of engineering surfaces. The obscured view of buried interfaces has further hindered investigations into these sliding interfaces. Researchers have undertaken numerous studies in recent decades and put forth various models and mechanisms to unravel the complexities of tribochemical reactions and the formation of ZDDP tribofilms. These mechanisms have significantly enhanced our understanding of tribofilm formation on diverse surfaces, including metals [23–25], ceramics [26–29], silicon [22], and diamond-like carbon (DLC) coatings [30]. While an initial hypothesis suggested a chemical interaction between ZDDP (or Zn^{2+}) and iron oxide wear debris (or Fe^{3+}) on steel surfaces based on the hard and soft acid–base (HSAB) reaction model [31], explanations for tribofilm formation on non-ferrous surfaces led to new challenges in understanding the mechanism [32–36]. Discovering ZDDP tribofilms on non-ferrous surfaces without Fe^{3+} ions challenged existing HSAB reaction model. Tribofilms on certain non-ferrous tribopairs, such as DLC/DLC, exhibit weak adhesion to the surface, making them prone to detachment [37, 38]. On the contrary, tribofilms on steel/steel tribopairs firmly adhere to the surface, even sustaining scratch tests [39]. An alternate model proposed the theory of localized flash temperature rise at surface asperities governed by moving heat source theory [40, 41]. Nevertheless, tribofilms have been observed even under conditions of low sliding speeds, where significant elevations in flash temperatures are not pronounced thus far [42]. The concept suggests that charged regions at contacting asperities could lead to emission of energetic particles like photons, electrons, ions, and X-rays explaining occurrence of tribochemical reactions. However, detecting

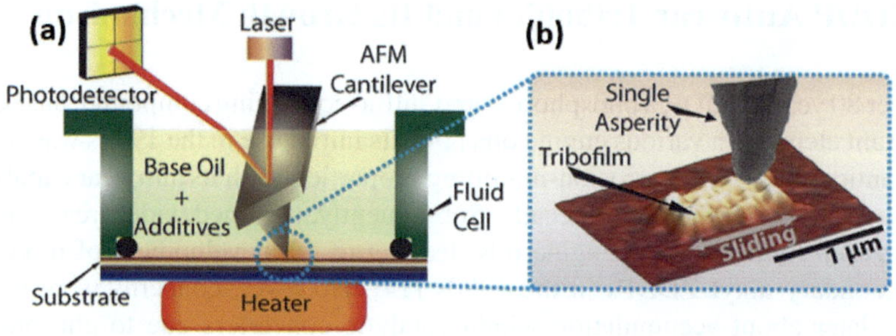

Fig. 2 Schematic illustrating the AFM-based in situ technique, featuring **a** the experimental setup comprising an AFM cantilever and substrate immersed in lubricant, confined by a liquid (fluid) cell and heater facilitating elevated temperatures of up to 200 °C; and **b** a single-asperity AFM probe scanning the surface to form tribofilm. Reprinted with permission from Ref. [22]. Copyright (2015) Science

triboemission in lubricated conditions has proven challenging so far [43–45]. The investigation led by Zhurkov [46] and Bell [47], examining the impact of applied force on chemical reactions, culminated in the development of the stress-assisted and thermal activation (SATA) model. The model is widely accepted among all the proposed mechanisms till date and significantly contributes to the understanding of the kinetics of tribofilm formation [40]. The SATA model conforms to the idea that tribochemical reaction rates follow the Arrhenius reaction kinetics and stress applied during sliding significantly reduces the activation energy required for ZDDP decomposition. Gosvami et al. [22] introduced an AFM-based technique, as illustrated in Fig. 2a and b, for real-time monitoring of in situ development of ZDDP tribofilms, providing essential visualization and identification of underlying physical mechanisms. In alignment with the SATA model, their research experimentally showed an exponential rise in tribofilm growth rate with contact pressure and temperature. Moreover, the study strongly endorsed the possibilities indicated by atomistic simulations [48], suggesting the bonding between zinc phosphate molecules under high contact pressure as a thermal or mechanochemical decomposition outcome of ZDDP. Subsequently, other studies have further shown that ZDDP tribofilm growth is indeed influenced by shear stress [40].

3 AFM-Based In Situ Nanoscale Technique to Study Tribofilm Growth

The AFM is an essential and advanced tool for characterizing in situ tribofilm formation, enabling concurrent analysis of surface morphology, friction, and wear on the substrate (see Fig. 2) [22]. Macroscopic experiments such as the mini traction machine (MTM) or pin-on-disc tests often induce substantial plastic deformation at the asperity level and mechanical blending of wear debris with the surface across a range of loads from milliNewtons (mN) or Newtons (N). Conversely, AFM-driven measurements operate within a relatively lower load spectrum, typically ranging from a few nanoNewtons (nN) to microNewtons (μN). This substantial reduction in load minimizes complications arising from extensive wear [22, 49]. Additionally, the AFM technique employs a small-sized probe (approximately 20 nm in diameter), facilitating topographical evaluations with exceptionally high resolution and determining localized coefficient of friction (COF) values. Such a level of precision is impractical to ascertain in macroscale tests.

AFM experiments are conducted on a smooth and pristine surface with nanoscopic roughness. Typically, the setup involves using sharp AFM probes made of materials like Si, DLC-coated, or diamond-coated Si that allow for imaging of surface features at nanometer scale resolutions. Concurrently, microsphere probes come into play to measure frictional forces across multiple asperities in the sliding region, as depicted in Fig. 3a. The scanning electron microscope (SEM) image in Fig. 3b displays a steel microsphere firmly attached to the end of the AFM cantilever. To establish a multi-asperity contact, several other hard material probes, such as alumina microspheres (30–50 μm in diameter), can also be attached to the end of the cantilever. When operated in lateral mode, the AFM ensures that the probe maintains contact with the surface and moves perpendicular to the axis of its cantilever while scanning across the surface. This technique allows for detection of both normal and lateral deflections which enables obtaining measurements related to topography and frictional forces, respectively.

The AFM-based technique emerges as a potent tool for directly observing the in situ growth of the tribofilm, particularly involving single-asperity nanoscale contacts within ZDDP-containing lubricants. It allows for simultaneous analysis of surface topography and physical properties while maintaining precise control over loading and sliding conditions. In contrast to ex situ methods, in situ AFM-based nanotribological investigations present several advantages while studying the antiwear properties of ZDDP. This approach enables concurrent measurements of various

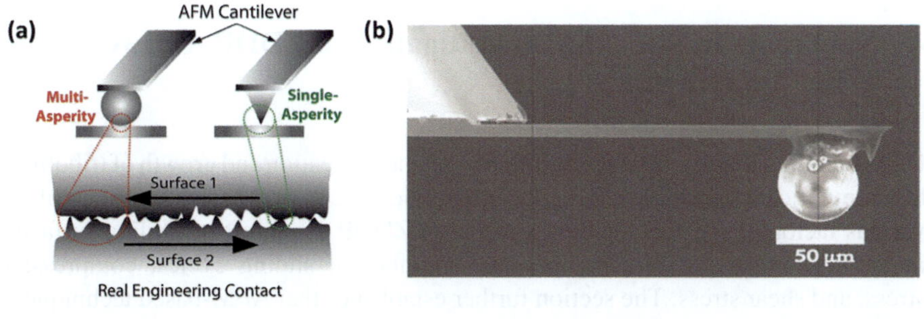

Fig. 3 **a** Schematic illustrating two surface contact scenarios: multi- and single -asperity contact portrayed by the microsphere probe attached to the free end of the cantilever (left) and the sharp AFM probe (right), respectively. **b** Electron micrograph displaying a standard cantilever with modifications featuring a steel colloid adhered to the free end of the cantilever. Reprinted with permission from Ref. [49]. Copyright (2018) Tribology Letters

local properties of the tribofilm, including changes in morphology, thickness, friction, wear, and adhesion. Conversely, ex situ volume and chemical analysis of tribofilm may introduce inaccuracies due to potential alterations during the removal of excess lubricant from the sample surface. Prior to the adoption of AFM-based technique, a significant gap existed in aligning with growth mechanisms and providing precise explanations for the patchiness and limited thickness of the tribofilm.

The fundamental insights derived from in situ nanoscale investigations using AFM can be extended to macro and atomic scales, providing a basis for designing components, devices, and systems with informed strategies to enhance tribological performance. One potential application of the AFM-based technique entails imprinting chemically bonded nanoscale patterns on diverse substrates using reactive ink molecules or discrete nanoparticles dispersed in a suitable fluid while the AFM probe is submerged [50, 51]. Khare et al. [50] introduced an innovative additive manufacturing technique called nanotribological printing (NTP) utilizing AFM, which facilitates the production of nanoscale structures and controlled patterns with customizable microstructure, thickness, and lateral dimensions, as illustrated in Fig. 4a. Figure 4b shows NTP using ZDDP dispersed in synthetic oil, resulting in a ZDDP-derived pattern imprinting the word "RESEARCH". The study further demonstrates the formation of amorphous, polycrystalline, and nanocomposite structures, along with sequential patterning with multi-material. Figure 4c and d illustrate layered patterns featuring a ZrO_2 film and a ZDDP-derived overlay developed by sequential patterning with ZrO_2- and ZDDP-containing base oil at temperatures of 25 and 100 °C, respectively. These patterns encompass layered heterostructures with distinct compositions for each layer and composite nanostructures with controllable loading fractions of two or more material chemistries. Notably, the resulting structures can be made electrically conductive by incorporating suitable nanomaterial within the tribofilm. The versatility of NTP in generating nanoscale patterns using standard AFM cantilever probes and operation modes establishes the AFM technique as a readily accessible and adaptable method for nanoscale additive manufacturing.

4 Nanoscale In Situ AFM Investigation of ZDDP Antiwear Tribofilm Growth

Understanding the key parameters influencing the nucleation and growth of tribofilms holds significant importance. This section provides a comprehensive summary of the various factors that affect the development of ZDDP tribofilms on sliding surfaces, including contact pressure, temperature, number of sliding cycles, compressive stress, and shear stress. The section further establishes that AFM-based technique is a promising method for studying antiwear tribofilms providing a robust validation for SATA model.

The application of the in situ AFM technique unveiled a profound impact of contact pressure and temperature on the morphology and growth of the tribofilm.

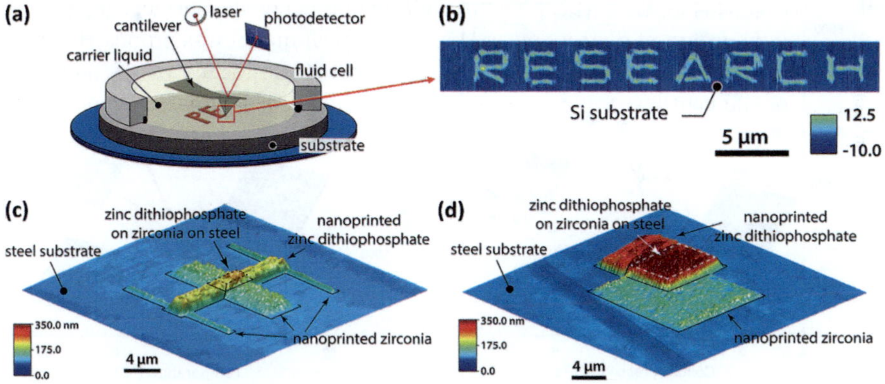

Fig. 4 Schematic illustrating the NTP process on a substrate. **a** Experimental setup consisting of an AFM cantilever immersed in an ink-containing fluid carrier within the fluid cell. **b** A ZDDP-derived pattern developed on a Si substrate. **c** and **d** Layered patterns comprising a ZrO_2 base pattern and a ZDDP-derived overlay developed by sequential patterning with ZrO_2- and ZDDP-containing PAO4 at 25 and 100 °C, respectively. Reprinted with permission from Ref. [50]. Copyright (2018) ACS Nano Letters

The technique revealed that the growth rate of the tribofilm exponentially increases with the increasing contact pressure and temperature [22]. Figure 5a and b illustrates the exponential increase in the growth rate of tribofilm with increasing contact pressure and temperature, respectively. As the pressure exceeds a critical threshold, it results in plastic deformation and wear of the tribofilm. This leads to the stabilization of the growth rate due to simultaneous removal and formation of the tribofilm. Additionally, these nanoscale investigations demonstrated that the mechanical hardness of the substrate and the number of sliding cycles exerted significant influence, while the sliding speed showed minimal impact [52, 53]. The growth rate apparently increases with the increase in sliding speed owing to the reduced time per cycle however, there was no significant impact of the shear rate on reaction kinetics. This suggests that each cycle contributes to the deposition of the tribofilm, with the extent depending on the contact pressure and temperature. These findings are consistent with the macroscopic experiments conducted by Clasen et al. [54], proposing that shear stress in the boundary lubrication regime is minimally influenced by the shear rate and primarily relies on COF and applied normal load on contacting asperities.

Typically, the thickness of ZDDP tribofilm directly correlates with the number of sliding cycles influenced by the contact pressure. Gosvami et al. [22] used the AFM-based technique to observe the growth kinetics of tribofilm, revealing two distinct phases in Fig. 6a. The initial phase showed a slow linear growth for approximately 1200 cycles with zero-order reaction kinetics, succeeded by a rapid growth characterized by first-order kinetics. The distinction in the stages revealed that the decomposition of ZDDP commences with a gradual and consistent reaction rate and follows intricate reaction pathways. Ultimately, the growth achieves a stable equilibrium over extended sliding periods, making simultaneous generation and elimination

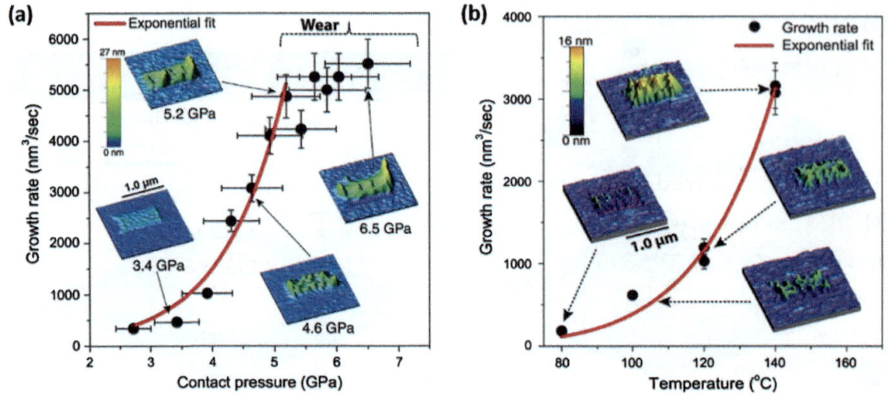

Fig. 5 Exponential growth of the tribofilm with the increasing **a** contact pressure, and **b** temperature. Reprinted with permission from Ref. [22]. Copyright (2015) Science

of the tribofilm. Dorgham et al. [53] also studied nanoscale techniques to investigate the effect of contact pressure on the growth kinetics. Their research noted a direct correlation between contact pressure and the reaction kinetics of ZDDP decomposition: increasing contact pressure exponentially decreased the number of sliding cycles required for transition from slow to rapid growth kinetics as depicted in Fig. 6b. Interestingly, their results could be fitted using first-order kinetics without an initial phase of zero-order growth—resulting in a stable equilibrium. Temperature variation exhibited exponential trends affecting the growth of tribofilm. Furthermore, the combined effects of contact pressure and temperature demonstrated significant 50-fold increase in the growth rate, reflecting their collective effectiveness in the ZDDP decomposition process. [55, 56]

In situ AFM technique fails to offer specific insights into the distinct impacts of stress components, such as compressive and shear stress. The interdependence between shear and compressive stress posed a challenge to manipulate one while maintaining the other constant. Consequently, the technique could not conclusively determine whether contact pressure or shear stress predominantly controlled the process. To overcome this constraint, an innovative method was introduced by Zhang et al. [40] and Fang et al. [57], focusing on the formation of ZDDP tribofilms under elastohydrodynamic (EHD) lubrication conditions. The approach aimed to investigate the influence of shear stress on the kinetics of ZDDP tribofilm growth on sliding surfaces. The study demonstrated that ZDDP tribofilm formation could occur without direct solid–solid contact, effectively preventing plastic deformation of the asperity and triboemissions, indicating independence from flash temperature. In a significant study, Zhang et al. [40] investigated ZDDP-containing base fluids with varying shear properties under EHD lubrication conditions and provided compelling evidence unequivocally demonstrating that tribofilm formation exclusively occurred in fluids generating high shear stress, but not in those exhibiting low shear stress at identical contact pressure levels. Furthermore, consistent with the SATA model, the

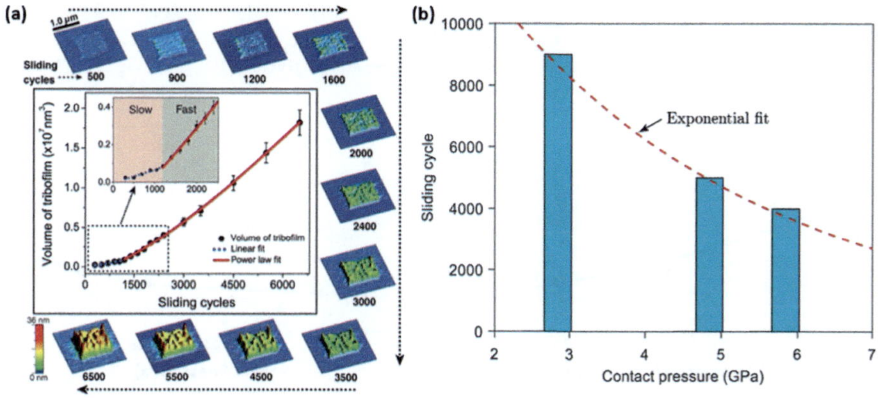

Fig. 6 **a** Evolution of the morphology and volume of tribofilm over successive sliding cycles observed through in situ AFM-based technique. Reprinted with permission from Ref. [22]. Copyright (2015) Science. **b** Number of sliding cycles required for the tribofilm to transition from slow to fast growth kinetics under varying contact pressures.Reprinted with permission from Ref. [53]. Copyright (2019) Tribology International

study affirmed that increasing atomic lateral force and temperature due to high shear stress reduced the activation energy barrier, enabling reactant molecules to surpass it towards transition state. Fang et al. [57] recently further elucidated the individual impacts of individual stress components on the kinetics of tribofilm formation. The study revealed that increasing compressive stress impeded tribofilm growth rate and inhibited ZDDP tribofilm formation with shear stress and temperature held constant, as depicted in Fig. 7a. Conversely, the growth rate increased with the increasing shear stress at a constant compressive stress (Fig. 7b), affirming that shear stress predominantly drives the formation of the tribofilm. Moreover, the increase in temperature expedited the growth rate, indicating that thermal factors significantly contribute to tribofilm formation.

The impact of compressive stress, shear stress, and temperature can be analyzed by the following modified form of the Eyring model [58]

$$k = A \cdot \exp\left(-\frac{(\Delta U_{act} - \tau\Delta V_{act} + P\Delta\Omega_{act})}{k_b T}\right) \quad (1)$$

where k is the rate constant, A is the pre-exponential factor, ΔU_{act} is the activation energy, k_b is Boltzmann constant, T is the temperature, ΔV_{act} and $\Delta\Omega_{act}$ are the activation volumes corresponding to the shear stress, τ and the compressive stress, P, respectively. Equation (1) elucidates the dynamic interplay of shear stress on ZDDP molecules, exerting a tensile force that promotes the stretching and breaking of Zn–S or S–R bonds. This mechanism accelerates the growth of the tribofilm significantly [57]. In contrast, compressive stress increases the activation barrier and obstructs the stretching and breaking of bonds. It potentially brings the atoms closer and hinders the crucial bond-breaking reactions needed to form phosphates

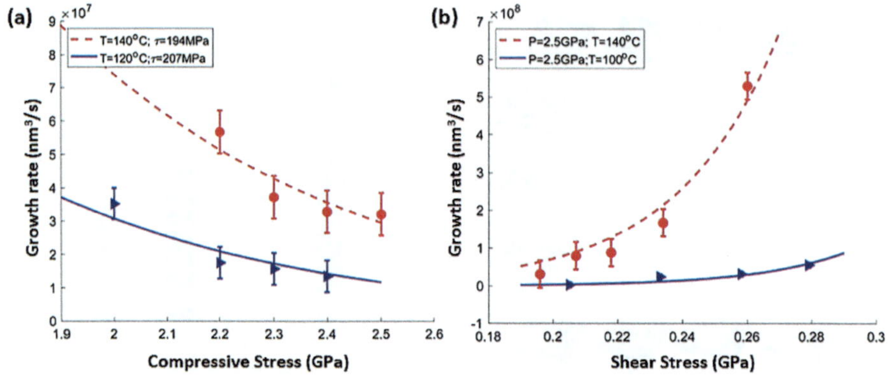

Fig. 7 Dependence of growth rate of tribofilm on the **a** compressive stress, and **b** shear stress. Reprinted with permission from Ref. [58]. Copyright (2023) Faraday Discuss

on sliding surfaces—an essential initial step in the tribofilm formation process. The positive activation volumes observed under both stress conditions indicate that the bond-breaking process is the limiting factor in tribofilm formation. Interestingly, a substantial activation volume for shear stress (0.18 nm^3) emphasizes its pivotal role in promoting tribofilm formation. In contrast, the smaller activation volume for compressive stress (0.01 nm^3) implies its relatively modest effect in impeding tribofilm formation.

In conclusion, the discussion emphasizes the exponential growth of ZDDP tribofilm with increasing contact pressure and temperature, which aligns seamlessly with the SATA model. These influential parameters lower the energy barrier and introduce a mechanochemical effect during ZDDP decomposition. The initial growth of the tribofilm is gradual but accelerates substantially after a certain number of sliding cycles, especially at higher contact pressure. Additionally, a comprehensive analysis of the individual impacts of stress components brings clarity that shear stress actively encourages bond breaking of the precursor molecule, facilitating the tribofilm formation. Conversely, compression inhibits the process and controls the overall reaction.

5 Structural, Mechanical, and Chemical Characterization of ZDDP Tribofilms

Interestingly, AFM is a versatile tool that extends beyond offering insights into surface topography by facilitating the precise measurement of crucial properties such as friction, wear, and adhesion. The AFM measurements of the ZDDP tribofilms, generated in a macroscale experiment [40, 59], revealed distinctive feature of ~ 200 nm thick, characterized by rough, patchy, and well-separated circular and elongated pads of ZDDP tribofilms. In an interesting study, Aktary et al. [60] employed

AFM to reveal that ZDDP tribofilms exhibited a pad-like structure. Initially, these films manifested as discrete patches, gradually transitioning into a cohesive structure characterized by distinct peaks and deep valleys. The origin of film patchiness posed a challenge for macroscale studies, making it difficult to determine whether it resulted from varying contact pressures across multi-asperity contacts. Nonetheless, the in situ AFM-based technique, focusing on single-asperity contact, proposed that inherent surface irregularities such as defects, roughness, and variations in atomic-level stress at these contacts contribute to the occurrence of random nucleation events [22]. The in situ tribofilm generation can also be demonstrated using multi-asperity AFM probes. Figure 8a illustrates a tribofilm generated and imaged in situ using a steel microsphere probe. However, the AFM image has a lower resolution due to the blunt nature of the probe, limiting the ability to resolve the fine details of the surface. Subsequently, the same tribofilm can be imaged using a sharp probe, as displayed in Fig. 8b. Sharp probes can capture fine details, revealing intricate surface structures and features with greater clarity. The sharp probe can reproduce the image obtained from the steel probe cantilever but with improved resolution, confirming the ease of locating the tribofilm for subsequent investigations utilizing alternate AFM modes [49].

The AFM technique proves invaluable in precisely quantifying nanomechanical properties such as elastic modulus and hardness of ZDDP tribofilms [60–62]. An intriguing finding surfaced during the examination of the mechanical characteristics of both pad and valley regions within the ZDDP tribofilm formed on a steel substrate

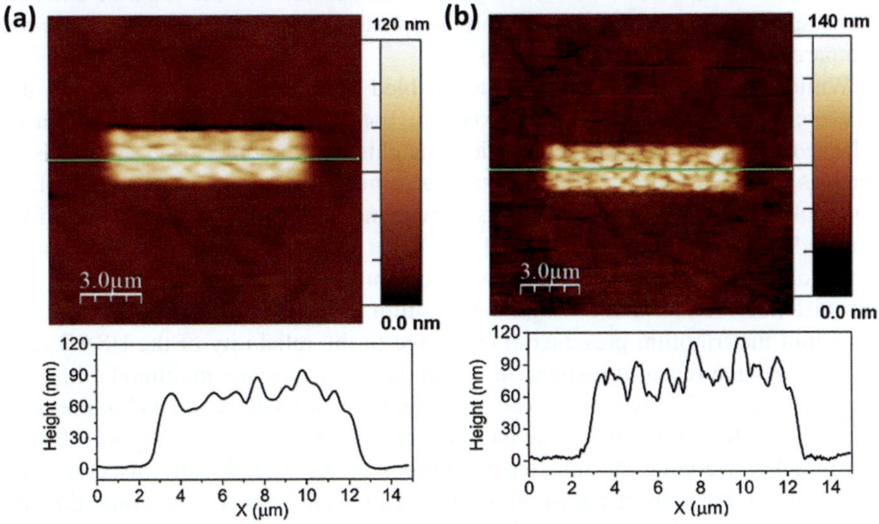

Fig. 8 **a** AFM topography image of the steel substrate captured by utilizing a steel probe cantilever, showcasing the in situ tribofilm formation. **b** Topographic image of the identical area utilizing a sharp Si probe cantilever. The height profile is displayed beneath each image, following the green color line. Reprinted with permission from Ref. [48]. Copyright (2018) Tribology Letters

using AFM [61, 62]. Graham et al. [62] observed that the thicker domains within the tribofilm, featuring larger pad areas, demonstrated an elastic modulus nearly matching that of the steel, approximately 220 GPa. Conversely, the smaller pads and valley areas in the tribofilm, distinguished by their reduced thickness, displayed a lower elastic modulus of approximately 30 GPa. The difference in modulus and hardness compared to steel provides sacrificial attributes to the tribofilm, effectively absorbing applied stresses at the interface and consequently reducing wear. Nicholls et al. [61], in accordance with Graham et al. [62], successfully demonstrated a compelling correlation between the length of polyphosphate chains and the elastic response of the film. The research predominantly associated the higher stiffness observed in larger pads with the presence of longer phosphate chain lengths, in contrast to the smaller pads and valley regions. The hypothesis proposes that shorter-chain polyphosphates undergo polymerization under hydrostatic pressure, resulting in the generation of longer, interconnected chains [63]. The stiffer long-chain polyphosphates are primarily located in the upper regions of tribofilm asperities, where pressures are at their maximum. While nanoindentation techniques could also be applied to evaluate nanomechanical properties, Nichollas et al. [61, 64] also highlighted a drawback of the nanoindentation technique, noting that the relatively large radius of the Berkovich tip leads to the averaging of mechanical responses over a broader area. Furthermore, the thinness of the film poses challenges for precise mechanical property measurements using the nanoindenter tip. However, AFM, with its ability to perform indentations at substantially lower loads and penetration depths, facilitates more precise measurements and discerns quasi-static behavior on thinner antiwear films. In essence, AFM investigations substantiate the superior antiwear characteristics attributed to longer polyphosphate chains in ZDDP tribofilms when compared to their shorter counterparts.

While the in situ AFM technique enabled the simultaneous generation and measurement of ZDDP tribofilm thickness on both ferrous and non-ferrous surfaces, it lacked the ability to provide chemical insights. On the other hand, ex situ EDX analysis provided valuable insights into the elemental composition of the tribofilm. The analysis detected prominent peaks corresponding to Zn, P, and S within the film, alongside smaller peaks adjacent to it [22]. Auger electron microscopy (AES), recognized for its enhanced surface sensitivity in contrast to EDX, subsequently verified the existence of Zn, P, and S within the tribofilm. Interestingly, the lack of displaced Fe within the tribofilm presented a challenge to the reliability of the HSAB reaction model. In a recent investigation, a comparable nanoscale multimodal imaging approach using AFM was employed, combining in situ and ex situ techniques with nano-IR and ToF–SIMS [65]. The methodology employed in this study was distinctive in acknowledging the inherent limitations of AFM in directly discerning the chemical composition of the tribofilm during its formation. To overcome this, the research introduced a novel workflow utilizing nano-IR, effectively monitoring the dynamic chemical evolution of the tribofilm. This innovative approach provided a comprehensive depiction of the nanoscale changes in the topography of the tribofilm induced by a phosphorus additive in the base oil. Furthermore, it delved into the examination of local chemical transformations occurring within the tribofilm.

6 Role of Substrate Chemistry on Tribofilm Formation and Properties

Although antiwear tribofilms derived from ZDDP offer considerable protection against severe wear by potentially averting abrasion, adhesion or plastic deformation at the sliding contact, their efficient growth and performance hinge on the interaction between ZDDP and solid surfaces during sliding. These tribofilms have been noted on a range of surfaces, primarily on ferrous-based surfaces such as steel, and also on non-ferrous surfaces including Si, Al, Mg alloys, ceramics, and DLC coatings. Nevertheless, earlier investigations have suggested that the formation of ZDDP tribofilms is significantly influenced by the intrinsic characteristics of the solid surfaces, including factors such as roughness [66], hardness [67], and chemical composition [68, 69]. Iron and steel surfaces are renowned for producing exceptionally durable tribofilms that exhibit strong surface adhesion and remain unaffected even after undergoing a scratch test. Conversely, non-ferrous tribopairs, such as DLC/DLC or Al-based materials, generate less resilient tribofilms that are prone to detachment from the surface when subjected to rubbing or ultrasonic treatment [39, 52]. Moreover, the HSAB model effectively elucidated the process of ZDDP tribofilm formation on ferrous surfaces by proposing an exchange of Zn^{2+} ions from the ZDDP molecule and Fe^{3+} ions from the substrate, followed by their interaction with phosphates. However, the applicability of this model in explaining the growth of ZDDP tribofilms on non-ferrous tribopairs remained uncertain and was found to be inconsistent with observations on non-ferrous surfaces. Studies conducted on ceramic surfaces such as tungsten carbide or sapphire substrates demonstrated that, despite experiencing high contact stress, these surfaces did not display visible wear or plastic deformation. However, stable tribofilms were observed on these surfaces [57], indicating that unlike what the HSAB model suggests, the formation of the tribofilm did not necessitate any cation exchange involving sliding induced wear debris.

Several research studies have focused on Al-Si alloys and reported the presence of strong tribofilms with pad-like structures on the harder Si inclusions, akin to those observed on steel surfaces [34]. Remarkably, nanoscale investigations utilizing the AFM-based in situ technique unveiled intriguing revelations: within the Al-Si (ADC12) alloy, minimal wear was observed, allowing for the development of ZDDP tribofilms in both the softer Al matrix and the harder Si phase at elevated temperatures [70]. Moreover, the Al matrix generated thinner tribofilms and displayed increased roughness compared to the Si phase, as depicted in Fig. 9. However, specifically within the Al matrix phase, nanoscale examination revealed considerable wear and material accumulation along the wear track edges. This behavior was attributed to the transfer of reactive species readily generated on the Si phase to the Al matrix. The author suggested that as the AFM probe raster-scan along the phase junction, the presence of the Si phase might induce the growth of the tribofilm on the adjacent Al matrix phase. This could also be due to the increased hardness of the soft Al matrix adjoining the harder Si phase [71]. Nevertheless, there were no discernible changes

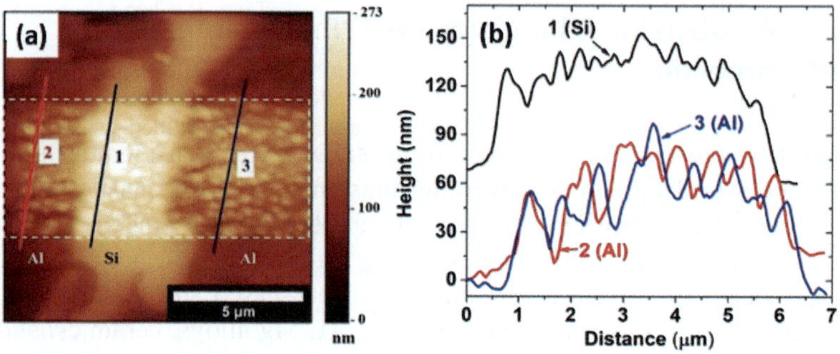

Fig. 9 **a** AFM image shows the progressive formation of a tribofilm within the experimental region, scanned for 500 cycles over both the Al matrix and Si phase concurrently. **b** Topographical height profiles depicting the thickness of the tribofilm on both the Al matrix and Si phase. Reprinted with permission from Ref. [70]. Copyright (2020) Tribology International

in the mechanical properties observed in the vicinity of the Si phase [70]. Indeed, Gosvami et al. [52] demonstrated the impact of strain hardening on tribofilm growth during extended rubbing on a pure Al surface. Initially, the soft Al surface underwent substantial wear; nevertheless, due to the eventual strain hardening process, the surface attained sufficient hardness to facilitate the development of a tribofilm. A similar examination was conducted on AZ91 Mg alloy which demonstrated that the harder $Mg_{17}Al_{12}$ phase facilitated tribofilm growth, whereas the softer α-Mg matrix experienced substantial wear [72].

7 Impact of Other Additives on the Formation of ZDDP Tribofilms

In real-world applications, ZDDP is commonly blended with supplementary additives, notably friction modifiers, dispersants, detergents, and viscosity modifiers. The interaction between these additives and ZDDP can result in synergistic effects that amplify ZDDP's performance or lead to unfavorable consequences. For instance, the incorporation of molybdenum dialkyl dithiocarbamate (MoDTC) with ZDDP has been noted to impede the growth of ZDDP tribofilms due to the competitive adsorption phenomena between ZDDP and MoDTC [73]. On the other hand, in a macroscale investigation, Renguo et al. [74] reported a synergistic impact by combining an optimal concentration of MoDTC with ZDDP [75], resulting in the conversion of ZDDP into a protective zinc phosphate glass tribofilm with antiwear properties. Simultaneously, MoDTC underwent decomposition, forming a MoS_2 tribolayer, contributing to a reduced COF on the surface [76–80]. Similarly, the blending of nanodiamond (ND) additives with ZDDP exhibited a synergistic effect [81]. The embedment of ND into the tribofilm enhanced its adhesion, thickness and stiffness, consequently improving the load-bearing capacity of the substrate.

This section thoroughly explores the nanoscale investigation of additional lubricant additives and their influence on the properties and efficacy of ZDDP tribofilm. A sequence of nanotribological investigations has been carried out utilizing the AFM-based technique to explore the combined effect of ZDDP and other additives [73, 82, 83]. A recent localized examination focused on individual phases within ADC12 alloy using a base oil containing ZDDP as well as MoDTC additives [73]. The investigation unveiled a comparatively thinner tribofilm in the Si phase; however, there was a significant decrease in the COF compared to ZDDP alone. Intriguingly, the Al matrix, away from the Si phase, displayed reduced friction, accompanied by the formation of a thin tribofilm within the Al matrix. This underscores the enhanced antiwear properties achieved when combining ZDDP with MoDTC, contrasting with the elevated wear witnessed in scenarios using ZDDP alone [74]. Therefore, incorporating MoDTC with ZDDP enhances antiwear properties specifically for softer surfaces. In a comparable investigation on ADC12 alloy, incorporating tert-butoxy functionalized h-BN nanosheets (h-BN-TBO) into ZDDP-containing lubricant resulted in a significant enhancement of the tribological properties due to the synergistic interplay between the combined additives [82]. This combination facilitated the formation of durable antiwear tribofilms, enhancing lubricating properties across the Si and Al phases. An additional significant nanoscale tribological investigation conducted on the AZ91 Mg alloy, utilizing a lubricant infused with an acrylamide additive instead of ZDDP, resulted in the formation of nanometer-thick tribofilms on both precipitate and matrix surfaces. The findings underscore its potential as a feasible alternative to ZDDP, providing efficient lubrication for Mg-based alloys [83]. Dorgham et al. [53] conducted an investigation utilizing an ashless dialkyl dithiophosphate (DDP) additive. Their findings indicated that the breakdown of the DDP molecule adhered to complex and fractional-order reaction kinetics, implying the creation of phosphate glass and the generation of volatile products. Notably, the growth rate of tribofilm exhibited exponential growth in relation to contact pressure, whereas its response to temperature followed a linear trend. Conversely, the growth rate of the tribofilm based on ZDDP showed an exponential rise with both contact pressure and temperature. Additionally, the lower energy barrier for DDP decomposition, as compared to ZDDP, suggested that DDP is more sensitive to contact pressure than ZDDP. Unlike sulfur- and phosphorous-based additives, Khare et al. [84] investigated the potential of ZrO_2 nanoparticles as a substitute for ZDDP. Using the in situ AFM technique, the research uncovered that unlike ZDDP-based tribofilms, tribofilms formed by functionalized ZrO_2 nanoparticles, follow the growth process dictated by sintering rate law kinetics. This growth mechanism involves stress-activated tribosintering, strongly supporting the idea that tribofilms resulting from nanoparticle antiwear additives develop through stress-induced sintering of nanoparticles, possibly entailing the elimination of initially attached dispersing ligands. Furthermore, the research demonstrated the extension of the ZrO_2 tribofilm over a temperature spectrum from −25 to 100 °C, while ZDDP additives faced challenges in forming robust antiwear tribofilms at room temperature.

In summary, research on ZDDP tribofilms has expanded to explore interactions with various additives, uncovering both synergistic and adverse effects. The incorporation of ZDDP with compounds like MoDTC, ND, and h-BN demonstrates encouraging synergies, improving both antiwear and friction-reduction characteristics. Recent nanotribological investigations underscore the benefits of ZDDP blends, showcasing enhanced performance across diverse surfaces. Significantly, the interaction between ZDDP and other additives offers an avenue for finely tuning lubricant compositions, optimizing them for superior wear protection and operational efficiency.

8 Conclusions and Future Perspectives

The present chapter aims to comprehensively review AFM-based nanotribological investigations involving the tribochemical decomposition of essential lubricant additives such as ZDDP, which enhance surface tribological properties by generating a protective antiwear film during sliding contact. Understanding the decomposition mechanisms of ZDDP has advanced significantly with the emergence of the AFM-based in situ technique, allowing real-time monitoring of tribofilm growth and physical characteristics. This technique involves explicitly examining nanoscale contacts at single-asperities within ZDDP-containing lubricants. Before the development of the technique, there was a considerable gap in validating the growth mechanisms and providing accurate explanations for the physical attributes of the tribofilm. In line with the SATA model, the technique experimentally demonstrated tribofilm growth exhibiting exponential rates concerning contact pressure and temperature. A detailed analysis of the distinct impacts of shear and compressive stress confirms that shear stress facilitates bond-breaking, thereby accelerating growth rates. In contrast, compressive stress increases the activation barrier, decelerating the growth rate.

AFM exhibits limitations in providing direct insights into the chemical composition of the tribofilm during its formation; however, integration of nano-IR comprehensively captures the local chemical transformation within the contact during the growth of the tribofilm. While the technique holds significant promise in comprehending the core aspects of tribochemical reactions, a notable research gap persists in addressing the detrimental impacts of ZDDP on vehicles exhaust systems and the environment. Considering this, the comprehensive discussion presented in this chapter regarding the in situ growth of antiwear tribofilms through AFM could pave the way for future research avenues. This discussion may inspire researchers to delve deeper into the fundamental comprehension of tribochemical reactions, potentially aiding in further enhancing the tribological properties of mechanical components without posing environmental challenges.

References

1. D. Berman, A. Erdemir, Achieving ultralow friction and wear by tribocatalysis: enabled by *In-Operando* formation of nanocarbon films. ACS Nano. **15**(12), 18865–18879 (2021). https://doi.org/10.1021/acsnano.1c08170
2. K. Holmberg, P. Andersson, A. Erdemir, Global energy consumption due to friction in passenger cars. Tribol. Int. **47**, 221–234 (2012). https://doi.org/10.1016/j.triboint.2011.11.022
3. E.R. Booser, in *Tribology Data Handbook* (CRC Press, 1997). https://doi.org/10.1201/9780367802509
4. T. Morita, Y. Hirano, K. Asakura, T. Kumakiri, M. Ikenaga, C. Kagaya, Effects of plasma carburizing and DLC coating on friction-wear characteristics, mechanical properties and fatigue strength of stainless steel. Mater. Sci. Eng. A **558**, 349–355 (2012). https://doi.org/10.1016/j.msea.2012.08.011
5. K. Jacques, N. Murthy, S. Dixit, D. Berman, S. Berkebile, Method for tribological experiment to study scuffing initiation on AISI 52100 steel and hard ceramic coatings. Tribol. Int. **160**, 107001 (2021). https://doi.org/10.1016/j.triboint.2021.107001
6. A. Abdelbary, L. Chang, Introduction to engineering tribology, in *Principles of Engineering Tribology* (Elsevier, 2023), pp. 1–32. https://doi.org/10.1016/B978-0-323-99115-5.00013-X
7. M.F. Ashby, J. Abulawi, H.S. Kong, Temperature maps for frictional heating in dry sliding. Tribol. Trans. **34**(4), 577–587 (1991). https://doi.org/10.1080/10402009108982074
8. T.D.B. Jacobs, R.W. Carpick, Nanoscale wear as a stress-assisted chemical reaction. Nat. Nanotechnol. **8**(2), 108–112 (2013). https://doi.org/10.1038/nnano.2012.255
9. Y. Junbin, D. Junxiu, Tribocatalysis reaction during antiwear synergism between borates and Sn(IV) compounds in boundary lubrication. Tribol. Int. **29**(5), 429–432 (1996). https://doi.org/10.1016/0301-679X(95)00101-9
10. T. Onodera, K. Kawasaki, T. Nakakawaji, Y. Higuchi, N. Ozawa, K. Kurihara, M. Kubo, Tribocatalytic reaction of polytetrafluoroethylene sliding on an aluminum surface. J. Phys. Chem. C **119**(28), 15954–15962 (2015). https://doi.org/10.1021/acs.jpcc.5b01370
11. R. Larson, The performance of zinc dithiophosphates as lubricating oil additives. Ind. Lubr. Tribol. **10**(8), 12–19 (1958). https://doi.org/10.1108/eb052531
12. T.D.B. Jacobs, C. Greiner, K.J. Wahl, R.W. Carpick, Insights into tribology from in situ nanoscale experiments. MRS Bull. **44**(06), 478–486 (2019). https://doi.org/10.1557/mrs.2019.122
13. J. Soni, N.N. Gosvami, Recent advancements in understanding of growth and properties of antiwear tribofilms derived from zinc dialkyl dithiophosphate additives under nanoscale sliding contacts. Langmuir **40**(7), 3301–3319 (2024). https://doi.org/10.1021/acs.langmuir.3c02512
14. H. Spikes, The history and mechanisms of ZDDP. Tribol. Lett. **17**(3), 469–489 (2004). https://doi.org/10.1023/B:TRIL.0000044495.26882.b5
15. R.I. Taylor, Tribology and energy efficiency: from molecules to lubricated contacts to complete machines. Faraday Discuss. **156**, 361 (2012). https://doi.org/10.1039/c2fd00122e
16. G.C. Smith, Surface analytical science and automotive lubrication. J. Phys. D Appl. Phys. **33**(20), R187–R197 (2000). https://doi.org/10.1088/0022-3727/33/20/201
17. H. Spikes, Low- and zero-sulphated ash, phosphorus and sulphur anti-wear additives for engine oils. Lubr. Sci. **20**(2), 103–136 (2008). https://doi.org/10.1002/ls.57
18. C. Benyajati, A.V. Olver, C.J. Hamer, An experimental study of micropitting, using a new miniature test-rig, in *Tribology Series,* vol. 43 (Elsevier, 2003), pp. 601–610. https://doi.org/10.1016/S0167-8922(03)80088-3
19. E. Lainé, A.V. Olver, T.A. Beveridge, Effect of lubricants on micropitting and wear. Tribol. Int. **41**(11), 1049–1055 (2008). https://doi.org/10.1016/j.triboint.2008.03.016
20. M.N. Webster, C.J.J. Norbart, An experimental investigation of micropitting using a roller disk machine. Tribol. Trans. **38**(4), 883–893 (1995). https://doi.org/10.1080/10402009508983485
21. V. Brizmer, H.R. Pasaribu, G.E. Morales-Espejel, Micropitting performance of oil additives in lubricated rolling contacts. Tribol. Trans. **56**(5), 739–748 (2013). https://doi.org/10.1080/10402004.2013.790097

22. N.N. Gosvami, J.A. Bares, F. Mangolini, A.R. Konicek, D.G. Yablon, R.W. Carpick, Mechanisms of antiwear tribofilm growth revealed in situ by single-asperity sliding contacts. Science **348**(6230), 102–106 (2015). https://doi.org/10.1126/science.1258788
23. M.L. Suominen Fuller, K.L. De Jong, M. Kasrai, G.M. Bancroft, Electroless generation of phosphate films on metals from zinc dialkyldithiophosphates. Chem. Mater. **12**(5), 1300–1304 (2000). https://doi.org/10.1021/cm990636t
24. M. Mee, A.A. Torrance, A study of the lubricated wear of plain bearing materials in a ZDDP-containing lubricant. Wear **128**(2), 201–217 (1988). https://doi.org/10.1016/0043-1648(88)90185-8
25. D. Yang, J. Zhou, Q. Xue, Study of the tribochemical behavior of Al Ion-implanted pure iron lubricated with ZDDP. Surf. Coat. Technol. **102**(3), 223–232 (1998). https://doi.org/10.1016/S0257-8972(97)00693-2
26. J.S. Sheasby, T.A. Caughlin, W.A. Mackwood, A comparison of the boundary lubrication of 52100 steel, zirconia and silicon nitride by S, P, S/P and zinc dialkyl dithiophosphate additives. Wear **196**(1–2), 100–109 (1996). https://doi.org/10.1016/0043-1648(95)06862-7
27. B. Mingwu, Z. Xushou, Q. Shangkui, Tribological properties of silicon nitride ceramics coated with molybdenum films under boundary lubrication. Wear **169**(2), 181–187 (1993). https://doi.org/10.1016/0043-1648(93)90296-X
28. T. Haque, A. Morina, A. Neville, R. Kapadia, S. Arrowsmith, Non-ferrous coating/lubricant interactions in tribological contacts: assessment of tribofilms. Tribol. Int. **40**(10–12), 1603–1612 (2007). https://doi.org/10.1016/j.triboint.2007.01.023
29. Y. Tanita, T. Mine, K. Nakajima, Tribological reaction generated on ceramic—steel couples under boundary lubrication. J. Tribol. **112**(4), 637–642 (1990). https://doi.org/10.1115/1.2920309
30. T. Haque, A. Morina, A. Neville, R. Kapadia, S. Arrowsmith, S. Tung, B. Kinker, M. Woydt, S.W. Dean, Study of the ZDDP antiwear tribofilm formed on the DLC coating using AFM and XPS techniques. J. ASTM Int. **4**(7), 100937 (2007). https://doi.org/10.1520/JAI100937
31. J.M. Martin, Antiwear mechanisms of zinc dithiophosphate: a chemical hardness approach. Tribol. Lett. **6**(1), 1–8 (1999). https://doi.org/10.1023/A:1019191019134
32. B. Vengudusamy, J.H. Green, G.D. Lamb, H.A. Spikes, Tribological properties of tribofilms formed from ZDDP in DLC/DLC and DLC/steel contacts. Tribol. Int. **44**(2), 165–174 (2011). https://doi.org/10.1016/j.triboint.2010.10.023
33. F.-X. Wang, Y.-Q. Cheng, D.-H. Guan, On the tribological behavior and surface analysis of a sliding PSZ ceramic-steel pair. J. Tribol. **117**(3), 548–552 (1995). https://doi.org/10.1115/1.2831289
34. Y. Shimizu, H.A. Spikes, The influence of aluminium-silicon alloy on ZDDP tribofilm formation on the counter-surface. Tribol. Lett. **65**(4), 137 (2017). https://doi.org/10.1007/s11249-017-0915-8
35. M. Burkinshaw, A. Neville, A. Morina, M. Sutton, ZDDP and its interactions with an organic antiwear additive on both aluminium-silicon and model silicon surfaces. Tribol. Int. **69**, 102–109 (2014). https://doi.org/10.1016/j.triboint.2013.09.001
36. X. Zhao, J. Liu, B. Zhu, H. Miao, Z. Luo, Effects of antiwear additives on the friction and wear of Si3N4/steel sliding contacts. Wear **201**(1–2), 99–105 (1996). https://doi.org/10.1016/S0043-1648(96)07215-8
37. M. Ueda, A. Kadiric, H. Spikes, ZDDP Tribofilm formation on non-ferrous surfaces. Tribol. Online **15**(5), 318–331 (2020). https://doi.org/10.2474/trol.15.318
38. B. Vengudusamy, J.H. Green, G.D. Lamb, H.A. Spikes, Durability of ZDDP tribofilms formed in DLC/DLC contacts. Tribol. Lett. **51**(3), 469–478 (2013). https://doi.org/10.1007/s11249-013-0185-z
39. S. Equey, S. Roos, U. Mueller, R. Hauert, N.D. Spencer, R. Crockett, Reactions of zinc-free anti-wear additives in DLC/DLC and steel/steel contacts. Tribol. Int. **41**(11), 1090–1096 (2008). https://doi.org/10.1016/j.triboint.2008.03.004
40. J. Zhang, H. Spikes, On the mechanism of ZDDP antiwear film formation. Tribol. Lett. **63**(2), 24 (2016). https://doi.org/10.1007/s11249-016-0706-7

41. I.L. Ryhming, On temperature and heat source distributions in sliding contact problems. Acta Mech. **32**(4), 261–274 (1979). https://doi.org/10.1007/BF01379011
42. H. Fujita, H.A. Spikes, The formation of zinc dithiophosphate antiwear films. Proc. Inst. Mech. Eng. J: J. Eng. Tribol. **218**(4), 265–278 (2004). https://doi.org/10.1243/1350650041762677
43. C.G. Camara, J.V. Escobar, J.R. Hird, S.J. Putterman, Correlation between nanosecond x-ray flashes and stick-slip friction in peeling tape. Nature **455**(7216), 1089–1092 (2008). https://doi.org/10.1038/nature07378
44. K. Nakayama, Triboemission of charged particles and resistivity of solids. Tribol. Lett. **6**(1), 37–40 (1999). https://doi.org/10.1023/A:1019139002295
45. C.K. Kajdas, Importance of the triboemission process for tribochemical reaction. Tribol. Int. **38**(3), 337–353 (2005). https://doi.org/10.1016/j.triboint.2004.08.017
46. S.N. Zhurkov, Kinetic concept of the strength of solids. Int. J. Fract. Mech. **1**(4), 311–323 (1965). https://doi.org/10.1007/BF03545562
47. G.I. Bell, Models for the specific adhesion of cells to cells: a theoretical framework for adhesion mediated by reversible bonds between cell surface molecules. Science **200**(4342), 618–627 (1978). https://doi.org/10.1126/science.347575
48. N.J. Mosey, M.H. Müser, T.K. Woo, Molecular mechanisms for the functionality of lubricant additives. Science **307**(5715), 1612–1615 (2005). https://doi.org/10.1126/science.1107895
49. N.N. Gosvami, J. Ma, R.W. Carpick, An in situ method for simultaneous friction measurements and imaging of interfacial tribochemical film growth in lubricated contacts. Tribol. Lett. **66**(4), 154 (2018). https://doi.org/10.1007/s11249-018-1112-0
50. H.S. Khare, N.N. Gosvami, I. Lahouij, Z.B. Milne, J.B. McClimon, R.W. Carpick, Nanotribological printing: a nanoscale additive manufacturing method. Nano Lett. **18**(11), 6756–6763 (2018). https://doi.org/10.1021/acs.nanolett.8b02505
51. A. Dorgham, C. Wang, A. Morina, A. Neville, 3D tribo-nanoprinting using triboreactive materials. Nanotechnology **30**(9), 095302 (2019). https://doi.org/10.1088/1361-6528/aaf70c
52. N.N. Gosvami, I. Lahouij, J. Ma, R.W. Carpick, Nanoscale in situ study of ZDDP tribofilm growth at aluminum-based interfaces using atomic force microscopy. Tribol. Int. **143**, 106075 (2020). https://doi.org/10.1016/j.triboint.2019.106075
53. A. Dorgham, P. Parsaeian, A. Azam, C. Wang, A. Morina, A. Neville, Single-asperity study of the reaction kinetics of *P*-based triboreactive films. Tribol. Int. **133**, 288–296 (2019). https://doi.org/10.1016/j.triboint.2018.11.029
54. C. Clasen, H. Pirouz Kavehpour, G.H. McKinley, Bridging tribology and microrheology of thin films. Appl. Rheol. **20**(4), 45049 (2010). https://doi.org/10.3933/APPLRHEOL-20-45049
55. N.N. Gosvami, P.C. Nalam, A.L. Exarhos, Q. Tam, J.M. Kikkawa, R.W. Carpick, Direct torsional actuation of microcantilevers using magnetic excitation. Appl. Phys. Lett. **105**(9), 093101 (2014). https://doi.org/10.1063/1.4894737
56. Z. Yin, M. Kasrai, M. Fuller, G.M. Bancroft, K. Fyfe, K.H. Tan, Application of soft x-ray absorption spectroscopy in chemical characterization of antiwear films generated by ZDDP Part I: the effects of physical parameters. Wear **202**(2), 172–191 (1997). https://doi.org/10.1016/S0043-1648(96)07272-9
57. L. Fang, S. Korres, W.A. Lamberti, M.N. Webster, R.W. Carpick, What stress components drive mechanochemistry? A study of ZDDP tribofilm formation. Faraday Discuss. **241**, 394–412 (2023). https://doi.org/10.1039/D2FD00123C
58. I.M. Ward, Review: the yield behaviour of polymers. J. Mater. Sci. **6**(11), 1397–1417 (1971). https://doi.org/10.1007/BF00549685
59. K. Topolovec-Miklozic, T.R. Forbus, H.A. Spikes, Film thickness and roughness of ZDDP antiwear films. Tribol. Lett. **26**(2), 161–171 (2007). https://doi.org/10.1007/s11249-006-9189-2
60. M. Aktary, M.T. McDermott, G.A. McAlpine, Morphology and nanomechanical properties of ZDDP antiwear films as a function of tribological contact time. Tribol. Lett. **12**(3), 155–162 (2002). https://doi.org/10.1023/A:1014755123184
61. M.A. Nicholls, P.R. Norton, G.M. Bancroft, M. Kasrai, T. Do, B.H. Frazer, G. De Stasio, Nanometer scale chemomechanical characterization of antiwear films. Tribol. Lett. **17**(2), 205–216 (2004). https://doi.org/10.1023/B:TRIL.0000032447.32442.6a

62. J.F. Graham, C. McCague, P.R. Norton, Topography and nanomechanical properties of tribochemical films derived from zinc dialkyl and diaryl dithiophosphates. Tribol. Lett. **6**(3/4), 149–157 (1999). https://doi.org/10.1023/A:1019124026402
63. M.A. Nicholls, G.M. Bancroft, P.R. Norton, M. Kasrai, G. De Stasio, B.H. Frazer, L.M. Wiese, Chemomechanical properties of antiwear films using x-ray absorption microscopy and nanoindentation techniques. Tribol. Lett. **17**(2), 245–259 (2004). https://doi.org/10.1023/B:TRIL.0000032451.97552.3d
64. M.A. Nicholls, T. Do, P.R. Norton, M. Kasrai, G.M. Bancroft, Review of the lubrication of metallic surfaces by zinc dialkyl-dithiophosphates. Tribol. Int. **38**(1), 15–39 (2005). https://doi.org/10.1016/j.triboint.2004.05.009
65. M. Lorenz, A.A. Pawlicki, H.E. Hysmith, K. Cogen, H. Thaker, O.S. Ovchinnikova, Direct multimodal nanoscale visualization of early phosphorus-based antiwear tribofilm formation. ACS Appl. Mater. Interfaces **14**(30), 35157–35166 (2022). https://doi.org/10.1021/acsami.1c16761
66. K.J. Kubiak, T.G. Mathia, M. Bigerelle, Influence of roughness on ZDDP tribofilm formation in boundary lubricated fretting. Tribol. Mater. Surfaces Interfaces **6**(4), 182–188 (2012). https://doi.org/10.1179/1751584X12Y.0000000020
67. Y.-R. Li, G. Pereira, M. Kasrai, P.R. Norton, The effect of steel hardness on the performance of ZDDP antiwear films: a multi-technique approach. Tribol. Lett. **29**(3), 201–211 (2008). https://doi.org/10.1007/s11249-008-9297-2
68. J. Jelita Rydel, K. Pagkalis, A. Kadiric, P.E.J. Rivera-Díaz-del-Castillo, The correlation between ZDDP tribofilm morphology and the microstructure of steel. Tribol. Int. **113**, 13–25 (2017). https://doi.org/10.1016/j.triboint.2016.10.039
69. K. Pagkalis, H. Spikes, J. Jelita Rydel, M. Ingram, A. Kadiric, The influence of steel composition on the formation and effectiveness of anti-wear films in tribological contacts. Tribol. Lett. **69**(2), 75 (2021). https://doi.org/10.1007/s11249-021-01438-6
70. P. Mittal, Y. Maithani, J.P. Singh, N.N. Gosvami, In Situ microscopic study of tribology and growth of ZDDP antiwear tribofilms on an Al–Si alloy. Tribol. Int. **151**, 106419 (2020). https://doi.org/10.1016/j.triboint.2020.106419
71. A. Lombardi, F. D'Elia, C. Ravindran, B.S. Murty, R. MacKay, Analysis of the secondary phases in the microstructure of 319 type al alloy engine blocks using electron microscopy and nanoindentation. Trans. Indian Inst. Met. **64**(1–2), 7–11 (2011). https://doi.org/10.1007/s12666-011-0002-z
72. D. Kumar, J. Jain, N.N. Gosvami, In Situ study of role of microstructure on antiwear tribofilm formation on AZ91 magnesium alloy under zinc dialkyldithiophosphate containing lubricant. Adv. Eng. Mater. **22**(8), 2000335 (2020). https://doi.org/10.1002/adem.202000335
73. P. Mittal, H. Rai, N.N. Gosvami, Microscopic tribology of ADC12 alloy under lubricant containing ZDDP and MoDTC using in situ AFM. Tribol. Lett. **69**(2), 35 (2021). https://doi.org/10.1007/s11249-021-01408-y
74. R. Lu, S. Shiode, H. Tani, N. Tagawa, S. Koganezawa, A study on the tribofilm growth and tribological properties of tribofilms formed from zinc dialkyl dithiophosphate (ZDDP) and molybdenum dialkyl dithiocarbamate (MoDTC). Tribol. Online **13**(3), 157–165 (2018). https://doi.org/10.2474/trol.13.157
75. H. Okubo, M. Yonehara, S. Sasaki, In Situ raman observations of the formation of MoDTC-derived tribofilms at steel/steel contact under boundary lubrication. Tribol. Trans. **61**(6), 1040–1047 (2018). https://doi.org/10.1080/10402004.2018.1462421
76. C. Grossiord, K. Varlot, J.-M. Martin, T. Le Mogne, C. Esnouf, K. Inoue, MoS2 single sheet lubrication by molybdenum dithiocarbamate. Tribol. Int. **31**(12), 737–743 (1998). https://doi.org/10.1016/S0301-679X(98)00094-2
77. Y. Yamamoto, S. Gondo, Friction and wear characteristics of molybdenum dithiocarbamate and molybdenum dithiophosphate. Tribol. Trans. **32**(2), 251–257 (1989). https://doi.org/10.1080/10402008908981886
78. J.M. Martin, T. Le Mogne, C. Grossiord, T. Palermo, Tribochemistry of ZDDP and MoDDP chemisorbed films. Tribol. Lett. **2**(3), 313–326 (1996). https://doi.org/10.1007/BF00173135

79. J. Graham, H. Spikes, S. Korcek, The friction reducing properties of molybdenum dialkyldithiocarbamate additives: Part I—factors influencing friction reduction. Tribol. Trans. **44**(4), 626–636 (2001). https://doi.org/10.1080/10402000108982504
80. J. Graham, H. Spikes, R. Jensen, The friction reducing properties of molybdenum dialkyldithiocarbamate additives: Part II—durability of friction reducing capability. Tribol. Trans. **44**(4), 637–647 (2001). https://doi.org/10.1080/10402000108982505
81. B. Acharya, C.M. Seed, J. Krim, Shear activation of ZDDP reaction films in the presence and absence of nanodiamonds. Appl. Surf. Sci. Adv. **7**, 100214 (2022). https://doi.org/10.1016/j.apsadv.2022.100214
82. P. Mittal, H. Rai, S. Kumari, O.P. Khatri, N.N. Gosvami, Efficient friction and wear reduction of Al-Si alloy via tribofilms generated from synergistic interaction of ZDDP and chemically functionalized h-BN additives. Appl. Surf. Sci. **595**, 153520 (2022). https://doi.org/10.1016/j.apsusc.2022.153520
83. D. Kumar, J. Jain, N.N. Gosvami, Nanometer-thick base oil tribofilms with acrylamide additive as lubricants for AZ91 Mg alloy. ACS Appl. Nano Mater. **3**(10), 10551–10559 (2020). https://doi.org/10.1021/acsanm.0c02644
84. H.S. Khare, I. Lahouij, A. Jackson, G. Feng, Z. Chen, G.D. Cooper, R.W. Carpick, Nanoscale generation of robust solid films from liquid-dispersed nanoparticles via in situ atomic force microscopy: growth kinetics and nanomechanical properties. ACS Appl. Mater. Interfaces **10**(46), 40335–40347 (2018). https://doi.org/10.1021/acsami.8b16680

In Operando Formation of Layered Materials for Friction Reduction

Mauro Ferrario and Maria Clelia Righi

Abstract Friction and wear result in massive energy and environmental costs. The technologies nowadays available to reduce these costs are based on materials and intense research efforts are being devoted to improving the efficiency of lubricants. Among them 2D materials have emerged as promising alternative to liquid lubricants as they can provide extremely low friction at a potentially much lower environmental impact as they do not require the use of petroleum oils. Moreover, they are particularly suited for lubricating tribological systems where the use of liquid lubricants is not possible, such as those operating in vacuum, high-temperature or at the nanoscale. While the friction coefficients provided by the 2D materials can reach super-low values in mild conditions, higher pressures applications often suffer from the need of replenishment into the wear tracks. A smart solution to overcome this problem is represented by the possibility to synthesize the slippery layers in operando conditions through tribochemical reactions involving molecules made available in the tribological environment as gases, powders or additives in liquid media. The present chapter offers an overview on the state-of-the art knowledge on mechanochemical/tribochemical synthesis and the *in-silico* experiments based on ab initio molecular dynamics that can be performed to monitor in real time the formation of 2D tribofilms. Two case studies are also described that concerns the tribological synthesis of graphene and transition metal dichalcogenides layers.

1 Layered Materials as Efficient Solid Lubricants

Lubricants save vast amounts of energy and material resources as they reduce friction and wear, also increasing the lifetime of devices. However, the climate crisis calls for solutions not dependent on non-renewable fossil resources, from which most of the current liquid lubricants are extracted. Solid lubricants are increasingly coming back into the focus of research community mainly because they can provide

M. Ferrario · M. C. Righi (✉)
Department of Physics and Astronomy, University of Bologna, 40127 Bologna, Italy
e-mail: clelia.righi@unibo.it

E. Gnecco and E. Meyer (eds.), *Fundamentals of Friction and Wear on the Nanoscale*, NanoScience and Technology, https://doi.org/10.1007/978-3-031-63065-1_14

extremely low friction at potentially much less environmental impact. They are typically essential when the operating conditions of machine elements do not allow the use of liquid lubricants, such as in a high temperature or vacuum environment, or when contamination is critical.

Most solid lubricants have a layered or lamellar structure, like graphite, MoS_2, and hexagonal boron nitride (h-BN). Typically, their mechanism of function is explained as a result of the ability of crystalline layers to undergo easy interlayer shear [1]. However, it has been shown by first principles calculations that it is their chemical inertness the key property of layered materials for the friction-reduction function. The flakes exfoliated during rubbing reach the most reactive parts of the surfaces and, as atomic-thin coatings, quench down the surface energy. The fully-saturated shells of the layer basal plane, make the interaction of covered microasperities (or the interaction of a covered-microasperity with other interfacial layers) highly repulsive at short range due to purely quantum–mechanical effects, i.e., those arising from the Pauli exclusion principle. The interfacial spacing is thus kept larger than the one required for establishing chemical interactions across the interface, with a consequent reduction of adhesion and friction [2, 3].

Offering the advantages of being solid, the layered materials can be applied to open systems and can also be used in poles apart applications, such as nanoscale devices and aircrafts, everywhere liquid lubrication is not appropriate. On the other hand, the 2D materials present the drawback of a difficult replenishment once removed from the wear track, being their peeling off from the substrate often observed in harsh tribological conditions. This disadvantage can be overcome by the in-operando synthesis of the layered materials from molecules in the form of gases, powders or liquid media made available in the tribological environment. The spontaneous formation of graphitic structure catalyzed by fresh metal surfaces exposed by wear, is, e.g., observed in the metalon-metal hip replacement, where carbon material with minimal nitrogen and low hydrogen content is retrieved directly from patients [4], or when a hydrocarbon contaminant forms on the surfaces of the electrodes and on the conductive bar of MEMS and NEMS devices after many switching cycles [5].

2 Mechanochemical and Tribochemical Synthesis

The synthesis of layered material within a sliding interface is realized through tribochemical reactions, which similarly to mechanochemical reactions are activated by mechanical stresses. The modern study of mechanochemical reactions goes back to the late nineteenth century with the work and foundational discoveries of M. Carey Lea, the father of mechanochemistry [6], who was the first to systematically demonstrate the effects of mechanical forces on activation energies as early as 1892.

Several critical reviews, surveys, and lectures view mechanochemical reactions as the viable and sustainable synthesis methodology for future manufacturing endeavors, at the root of the development of a new generation of chemistry [7–10]. Mechanochemical reactions are becoming very attractive thanks to their inherent green and sustainable nature and, first of all, to the opportunity to reduce environmentally unfriendly solvents. The last ten years have witnessed a surge of interest in mechanochemistry, mainly related to the development of ingenious methologies for solvent-free material synthesis, with a large number of inorganic and organic mechanochemical reactions already classified on the base of the resulting products [10–12]. In the field of nanoelectronics the production of nanocrystalline compunds by (high-energy) ball-milling, avoiding toxic precursors present in traditional synthesis, have also generated a lot of interest [13]. The strain-based energy input enables novel reaction pathways resulting in materials having unique electronic properties. Transition electron microscopy (TEM), electron diffraction and spectroscopic methods in conjunction with in situ nanoindentation have been used to investigate how the mechanical energy input drive the nucleation of the new phase at the interface [14].

With respect to mechanochemical reactions, tribochemical reactions can be seen as those induced at a sliding interface that typically result in the chemical formation of "tribofilms", i.e. highly desirable intercalated layers that avoid direct microasperities contact and thus reduce friction and wear [15, 16]. Of the several factors that can be used to distinguish tribochemistry from mechanochemistry the most notable ones are the persistent, load induced, molecular confinement at the buried interface and the shear induced intermixing.

While mechanically induced synthesis, primarily through the advancement of traditional milling methodologies, is currently adopted in numerous fields of application, tribochemical synthesis is almost always restricted to the tribofilm formation from lubricant additives present in engine oils, such as zinc dialkyldithiophosphate (ZDDP) [17], and the development of solid lubricant coatings, e.g. polytetrafluoroethylene (PTFE) [18].

The kinetic of mechanochemical/tribochemical reactions highly differs from the kinetic of thermally activated reactions at the open surface in static conditions. Typically the rates are accelerated by the combined action of both pressure and shear [19]. It is still rather difficult to separate the action of each mechanical force by experiment, identifying the specific role of load and shear in reducing the activation energy. On the bases of empirical observations, it has been suggested that the shear stress is a more important driving factor for a tribochemical reaction than the normal stress [20]. However, no systematic study has been performed so far, due to the difficulty in decoupling the two stress components experimentally. Simulation could play an important role here. Phenomelogical theories are limited to qualitative representations of the mechanochemical/tribochemical processes and of the reaction kinetics. Computational approaches, on the contrary, allows one to monitor in real time the atomistic processes occurring at the buried interface and gain insights into the reaction paths.

There are a variety of factors that can play important roles in driving tribochemical reactions. Some of those reported in the literature are frictional heating [15], formation of defects and triboemission. Frictional heating simply refers to the local rise in temperature as a result of the sliding contact between two surfaces that mostly occurs through contacting asperities and produces high local temperatures.

Heating evolves on very short timescales, and temperature reaches a maximum, usually referred to as the flash temperature, when all the frictional power is dissipated through a single asperity [21]. Other important factors in enhancing the reaction rates of mechanochemical processes are defects that can be created during rubbing, particularly surface wear. Bulk material is separated from the oxygen-rich atmosphere by heavily oxided surface layers which usually remain largely chemically inert. Wearing away these exterior layers exposes the fresh material underneath and results in accelerated tribochemical reactions [15], as it was found for the case of the enhanced decomposition of widely used lubricant additive ZDDP on native iron surfaces exposed through surface wear [22]. More recent examples in literature also corroborate these observations, such as how freshly scratched gold surfaces resulted in enhanced chemisorption [23], as well as how nascent steel surfaces could induce the formation of iron phosphide tribofilms [24, 25].

The intense mechanical forces endured by reactants often result in triboemissions, i.e. the emission of particles, electrons or photons and other particles, from the sliding interface [26]. Exoelectron triboemisssions from within the wear track have been proposed as a driving force for lubricant breakdown and tribofilm formation [27, 28]. Finally, it is important to notice that ab initio molecular dynamics simulations revealed that tribochemical reaction can be activated at the sliding buried interface in extremely short times, e.g. few picoseconds, even in the absence of flash temperatures and triboemission. These results suggest that other effects, related to the perturbation of the electronic structure of the confined molecules, may play a relevant role in promoting tribochemical reactions [29, 30].

The most widely studied example of tribochemical synthesis is from ZDDP complexes, used as an wear resistent additive to engine oil [31]. Several hypothesis have been made to describe the process of ZDDP decomposition and fragment recombination in zincphosphate films. Some insights have been obtained in a simplified, single asperity geometry, by using an atomic force microscopy to monitor the growth rate of ZDDP tribofilms as a function of load and shear [32]. Eventually, tribologists attempted to replicate the tribofilm by introducing polyphosphate additives directly; these more environmentally friendly lubricants, e.g. without heavy metal ions, adopted the wear reducing properties of normal ZDDP [33]. When tested experimentally, these new glassy polyphosphate lubricants at a steel/steel contact demonstrated high thermal stability, a 60% decrease in friction, and a 59% reduction in wear volume [34].

Recent attention has been given to tribofilms formed from carbon based molecules, which have been shown to possess admirable tribological properties. The conventional way to synthesize diamond-like carbon films, which have excellent friction and wear reduction properties, is chemical vapour deposition [35–36]. Erdemir and coworkers [37] showed that a tribocatalytic synthesis of DLC can also be realized directly by rubbing the hydrocarbon molecules of the base oil confined within catalytic surfaces.

3 Monitoring in Real Time the Tribochemical Synthesis of Layered Materials Through Ab Initio Molecular Dynamics

To design effective lubricant additives or solid lubricants, is highly desirable to understand how tribofilm are formed and function. However, monitoring in real time the chemical reactions occurring at the buried sliding interface is nowadays practically impossible by experiments. Typically, by postmortem analysis of the wear tracks one can gain information on the reaction products, but very little can be inferred on the reaction paths and mechanisms. Atomistic simulations play a crucial role here as they allow to open a window on the complex scenario of tribochemical processes and "observe" what happens during sliding. Moreover, simulations based on a quantum mechanical approach can provide fundamental understanding on the effects of molecular confinement and mechanical stresses on reaction kinetics.

Indeed, at present, sistematic Density Functional Theory (DFT) simulations are, although more computationally expensive than force-fields based ones, the most accurate tools we can use to rationalize the experimental observations and address their most important aspects, such as: how mechanical energy is transferred into chemical changes; which are the relevant microscopic states of the samples after being locally hit; which are the kinetics pathways that influence most the mechanochemical transformation. Ab initio molecular dynamics (AIMD) are in particular better suited to follow in real time the changes in the electronic structure due to extreme tribological conditions. Their use has, for example, permitted to reveal how dissociation under tribological conditions of lubricant additives can passivate exposed metallic surface and improve significantly friction and wear properties [29, 38]. The DFT based atomistic simulations here reported share many common aspects, like being performed within the generalized gradient approximation in the Perdew – Burke – Ernzerhof parametrization [39], Grimme's DFT-D2 dispersion correction [40], the use of ultrasoft pseudopotentials [41], the plane-wave expansion of spin-polarized electronic wave functions. More specifically, the tribo-molecular dynamics simulations, either in the Car Parrinello(CP) scheme [42] or in the Born – Oppenheimer(BO) scheme, fully quantum or, to keep simulations computationally more affordable, with the Quantum Mechanics/Molecular Mechanics (QM/MM) approach [43], were performed using modified versions of the CP and PWSCF modules included in the Quantum ESPRESSO computational suite [44]. These were realized applying a constant normal stress together with a constant, relative, sliding velocity of the two surfaces, while temperature control was being enforced by thermostats acting only on the thermal velocities of the substrates [29, 45]. The LAMMPS package [46] was in some cases employed for the initial preparation of the atomic configurations and together with Quantum Espresso for the QM/MM scheme.

4 Case Studies: Graphene and Transition Metal Dichalcogenides Formed In Situ by Tribochemical Reactions

4.1 Graphene from Methane

It has been reported experimentally that methane (CH_4) can be tribologically converted to graphene, nano-onion and disordered carbons on the sliding surfaces of coatings containing catalytically active particles of Ni or VN under atmospheric gas pressure and at room temperature [47]. A continuous extraction of graphene and other carbon nanostructure from CH_4 permits the in-operando formation of a tribofilm substantially lowering friction and providing a remarkable reduction in wear. With Ni and Cu playing the role of catalysts for the methane dehydrogenation, it is the confinement generated by the applied normal stress and the mechanical energy transferred through the shear stress that drive the tribochemical formation of the mentioned carbon nanostructures. AIMD simulations of the tribological events taking place at the sliding contact interface have been key to understandand the molecular-level mechanisms behind the process leading to the tribofilm formation.

The reaction energy, E_R, for the complete dissociation of a CH_4 molecule on a metal surface can be calculated as the difference of the chemisorbed fragments and the physisorbed molecule on the same surface. For methane the value of the dissociation energies are positive, indicating an endothermic reaction on both the Ni (111) surface, $E_R = 0.02$ eV, and the VN (100) surface, $E_R = 0.89$ eV, with Ni being the more favorable catalyst. Despite being not energetically favorable at ordinary conditions the dehydrogenation reaction can easily occur, even at high rates, when mechanical stresses are applied to produce molecular confinement, sliding and localised temperature increases.

Harsh conditions, chosen to model temperatures and stresses occurring in the real contact area [48], were considered in the AIMD simulations by applying a 5GPa constant normal stress together with a constant, relative, 200 m/s sliding velocity of two opposite metal slabs along [11-2] whose temperature was kept constant at 1000 K, while leaving the atoms of the CH_4 molecules evolve only under the actions of the interaction forces.

Confinement under applied load and shear was already observed to favour dissociation for water molecules at diamond interfaces [29, 49] and for organophosphorus additives at iron interfaces [30, 50]. AIMD simulations performed for CH_4 at room temperatures shows the molecular dissociation process starting immediately in the dynamical simulations as soon as molecoles begin colliding with the Ni surface, events which become frequent in the presence of applied normal and shear stresses.

The formation of carbon nanostructures is also favoured by the rapid diffusion of the dissociated hydrogen atoms into the Ni bulk. The concentration of carbon atoms at the interface increases with the formation of hydrocarbon clusters with low hydrogen content adsorbed on the metal surface. As soon as the interfacial concentration of H decreases short chains with two or more carbon atoms start forming, bonded to Ni atom on the surface. This is in agreement with in-operando near-ambient-pressure X-ray photo-electron spectroscopy identification of reactive carbon intermediates, C_n, during the thermal decomposition of CH_4 on Ni(111) [51]. Ni_2C reconstruction and other carbide structures often found during graphene growth on Ni (111) were not observed in the simulations, a finding that could be correlated with the presence of the load conditions and of a sliding interface that can act as a limiting factor to free carbon diffusion on and below the Ni surface.

Some insights could be gained by comparing the snapshots from AIMD for two different concentration of carbon atoms, corresponding to coverages of 50% (A) and 100% (B) of the Ni interface. At the lower coverage, after the initial formation of independent hydrocarbon chains, with decreasing hydrogen content every interfacial C atom become included in chains that start cross-linking to each other, until a single disordered sparse tribofilm is formed and gradually becomes thinner and thinner under the concerted action of the normal and shear stresses. The rubbing process promotes the change in the carbon atom hybridization to form sp to distorted sp^2 planar structures precursor of carbon rings. Indeed at the higher coverage the formation of carbon rings with the growth of graphene flakes could be observed. The change in the structure is best illustrated by comparing the radial distribution functions for the A and B configurations with the one for graphene, see Fig. 1. The shift of the peak of the C−C radial distribution functions from ~ 1.43 to ~ 1.47 Å for the 100% coverage case clearly indicates the rehybridization from sp to sp^2 that accompanies the structural change from interconnected carbon chains to 2D carbon structures and provides further evidence that graphene is being formed under the simulated tribological conditions. The coordination mechanism leading to graphene flake growth can be fully appreciated in the movies avaliable as Supporting Information in the article by Ramirez et al. [47]

4.2 *Transition Metal Dichalcogenides (TMD)*

4.2.1 *MoS_2 from MoDtC*

Crystallization of MoS_2 can take place in situ starting from precursors such as molybdenum dithiocarbamates (MoDTCs), a group of organo-molybdenum lubricant additives which can be transformed into MoS_2 in the presence of high temperatures and mechanical stresses. The mechanism underlying the transformation of MoDTC in MoS_2 is a complex series of tribochemical steps that have been the subject of many debates in the recent past [52, 53]. Sulfur and oxygen present at the interface play an important role in this reaction: MoDTC compounds undergo quick oxidation in air, causing the chemical structure and properties of these complexes to change [54]. The complex chemistry of these compounds has been elucidated by means of DFT calculations [55] focusing first on the properties of the isolated MoDTC molecule, identifying the ligand position between Mo and C as the most favorable site for

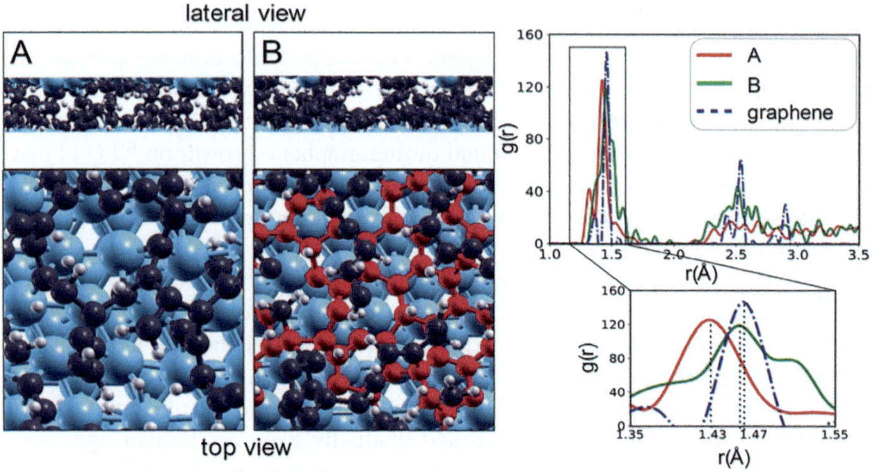

Fig. 1 Snapshot of simulated system containing 12 (A, 50% coverage) and 24 (B, 100% coverage) CH_4 molecules intercalated between metal slabs of area corresponding to $4 \times 3\sqrt{3}$ cells. On the right the C–C radial distribution functions compared with graphene (from Ramirez et al. [47])

oxygen in the molecule. Indeed, the dissociation of MoDTC with oxygen in ligand position results in the release of the central molecular unit containing Mo and S atoms with the correct stoichiometry to form MoS_2, contrary to what happens with oxygen in the terminal position, on top of Mo, hindering the transformation of the MODTC complexes into MoS_2. Moreover, MoDTC is not able to chemisorb and decompose on passivated surfaces. An activation time for the additive functioning is, thus, required, which is necessary to remove the superficial oxide layer by rubbing, consistently with the run-in period typically observed experimentally [56, 57].

The complex tribochemical reaction from MoDTC to the formation of the tribofilm can be separated into two main steps, the first one being the formation of MoS_x fragments, containing Mo and S atoms from the central core of MoDTC, and the second one the conversion of different units of MoS_x into a crystalline MoS_2 layer [58, 59]. Simulations are particularly useful to clarify the debated aspects of this complicated reactivity. Several authors [60, 61] simulated the crystallization of MoS_2 by using different reactive force fields such as AIREBO [62] and ReaxFF [63], but these are designed to model the formation and breaking of chemical bonds only in standard conditions and, as it turns out, they cannot reproduce well the chemical properties of the species under the effect of confinement and stresses, whereas DFT based approaches do provide the necessary accuracy. In the study of Peeters et al. [64] it was shown how, by means of DFT calculations of structural optimization under load, one can monitor the way compression destabilizes the geometry of these MoS_x units and how, introducing the effects of temperature and shear stress on the destabilized MoS_x units dynamic ab-initio calculations, as sketched in Fig. 2, can to provide an *in-situ* microscopic picture of the evolution of the process.

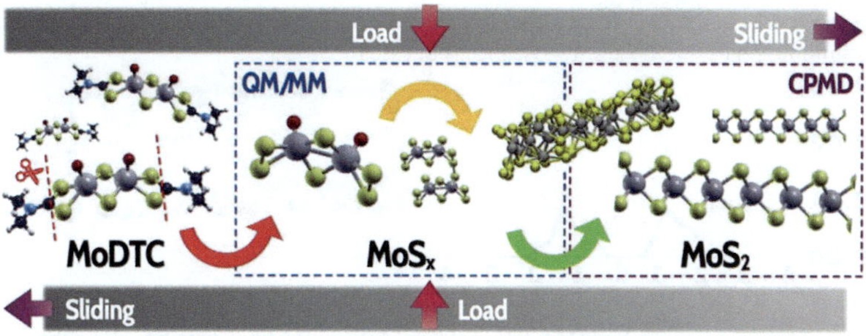

Fig. 2 A cartoon description of the adopted simulation route from MoDTC to crystalline MoS_2 (from Peeters et al. [64])

Starting with a system of twelve units originated from the central core of MoDTC and enclosed by two crystalline MoS_2 layers, they first performed ab-initio geometry optimizations under load comparing two different kinds of MoS_x fragments: units containing only molybdenum and sulfur and units with two oxygen atoms in terminal positions. The effect of increasing the load progressively destabilized the molecular units until the original MoS_x geometry is completely lost, apart from the persistence of Mo atoms preferrably clustering in the most central zone.

Further reorganization was then obtained introducing temperature and shear stress effects in dynamical simulations performed with a QM/MM approach, where the confining and sliding crystalline MoS_2 layers where described classically. After an initial thermalization of the whole system at 300 K shear stresses imposed on the inner MoS_x units by applying relative velocity kicks to the Mo atoms of the external layers, lead to the formation of an inner amorphous Mo, S rich layer catalyzed by the mechanical energy transferred during the sliding, with peculiar atomic conformations, as the one shown Fig. 3, pictorially by the final atomic configuration and more quantitatively by the Mo and S densities along the z direction, perpendicular to the crystalline MoS_2 surfaces. Indeed, Mo and S atoms showed in all the dynamics a strong tendency to increase their coordination. Comparison of systems with different stoichiometry indicated that the concentration of oxygen atoms can only influence the kinetic of the process, but the presence of oxygens do not hinder the formation of the inner, amorphous, Mo and S rich layer.

The final step of MoS_2 crystallization from the amorphous Mo, S layer was simulated by means of CP molecular dynamics. A set of dynamical simulations, all starting from the same initial configuration equilibrated at 300 K, were performed applying different tribological conditions, realized varying the values of the thermostat temperatures and of the imposed normal and shear stresses. As it is clearly illustrated by the set of successive atomic configurations in Fig. 4, a complete crystallization of the inner amorphous layer was successfully obtained, in the one case in which both the normal and shear stress were increased together to 5.6 GPa and 5.32 GPa respectively. Crystallization was also almost reached in another case in presence of higher

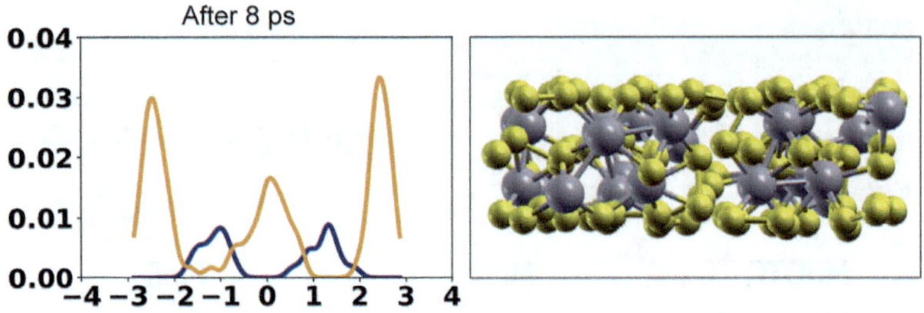

Fig. 3 Formation of an amorphous Mo, S layer from MoS_x units: final densities along the zdirection, Mo (blue), S(yellow) and atomic configurations after 8 ps of dynamics under imposed sliding (from Peeters et al. [64])

temperatures that however increases defect formations rather than favouring the crystallization process. More generally, the increase of temperature apparently did not lead to any improvement in the kinetic of the transformation. In all runs the amorphous MoS_2 layer started rearranging immediately toward configurations characterized by an increased confinement of the Mo atoms in the central zone, somewhat closer to the crystalline arrangment but yet with a varied degree of disorder.

Higher normal stresses induce a more pronounced rearrangment, but also imply a higher friction deceleration, so that they must be accompayned by a corresponding increase of the shear stresses to overcome the complete slowing down of the sliding motion. Remarkably, the imposition of alternate compression/decompression stages turned out to be less efficient than the persistent application of higher loads, indicating that the crystallization path is an activated one and harsh tribological conditions must be maintaned for period long enough to promote the creation of a subcritical crystalline cluster.

Some further insights could be gained by monitoring, along the time evolution, the energy difference per unit area, ΔE, between inner layer and the average of the energies of the two crystalline layers confining the amorphous one. In all cases, from an initial value of just above 3 J/m^2, the time averages of ΔE only showed

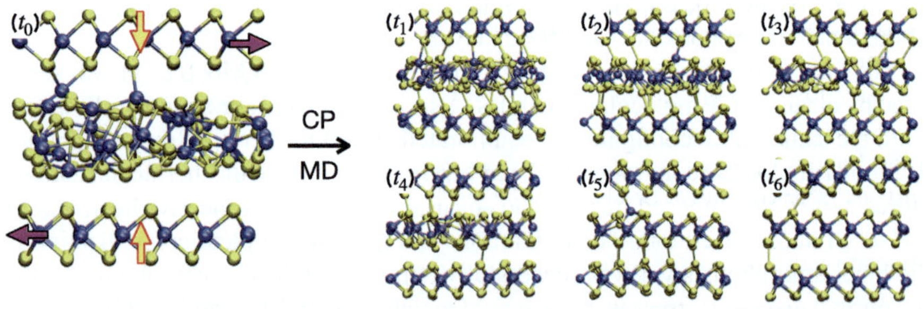

Fig. 4 Snapshots from the CP dynamical tribocrystallization, (t_0) initial configuration, (t_6) final configurational outcome (from Peeters et al. [64])

minor decreases, between 10 and 20%, but for the successful crystallization case where ΔE, after some small oscillations around the initial value, started to decrease uniformly with time until finally stabilized down to a very small value after about 20 ps. The initial waiting period can be related to the time needed for the formation of a critical crystalline cluster triggered by the concurrent action of the high normal and shear stresses, then after reaching the critical size, the degree of crystallinity grows uniformly until the phase transformation of the whole layer is completed. Overall the QM/MM and Car-Parrinello simulations of the crystallization of the amorphous inner layer showed that the formation of a beneficial, crystalline tribolayer from MoDTC molecules can occur spontaneously provided the system is subjected to the continued action of a sufficiently high shear stress concurrent to a sufficiently high degree of confinement realized by imposing at the same time a high enough normal stress.

4.2.2 *$MoSe_2$ from Se Nanopowder*

The formation of crystalline TMDs has also been observed for the chalcogen element Se, uniquely used in form of a nano-powder that is sprinkled onto a sliding metal surface. The tribochemical reactions occurring in situ result in the conversion of the powder and metal debris into a layered material that is responsible for the establishment of conditions of very low friction and wear. The use of a chalcogen powder has a number of technological advantages: it prevents the degradation of the lubricant properties by the interaction with air humidity; the tribological contact can be easily replenished by feeding nano-powders into the contact; the approach may also be applied at high temperature conditions or at the nanoscale and it does not need environmental-unfriendly liquid solvents to vehiculate the lubricant additive. This solid–solid state *in-operando* formation of the self-lubricating TMD tribofilms follows a rather specific route. The underlying mechano-synthesis has been unravelled by AIMD simulations that, accurately describing the formation/dissociation and hybridization-changes in the chemical bonds, permitted to explore the reaction pathways involved in the *in-operando* formation of $MoSe_2$ driven by the enhanced reactivity triggered by the applied mechanical stresses [65].

The formation of TMDs from their metal and chalcogenides constituents is energetically favorable: the *ab-initio* calculated energy of a $MoSe_2$ unit in the TMD layer is 2.0 eV lower than the sum of the energies of a Mo and two Se atoms in their bulk structures, meaning that the energetics provides the required thermodynamic driving force for the spontaneous transformation of the metal and chalcogenide elements into the TMD layers. The occurrence of such a transformation starting from Se nanopowder however requires the presence of an activating agent, the mechanical imposition of normal and shear stresses, to overcome the energy barriers of the reactions. It must be noted that, in order to observe the full formation of TMD from Mo and Se particles, AIMD simulations needed to be run for a considerable (for DFT calculations) length of time, i.e. up to hundreds of ps. The simulations also clearly emphasized that the presence of metallic debris is a prerequisite for the tribochemical reaction with the Se nanopowder, ultimately resulting in the formation of crystalline TMD layers. These metallic debris particles are most likely extracted from the substrate, thanks to wear-induced defects and debris, during the early stages of surface sliding. The different degree of Mo-Se intermixing observed in the dynamical simulations shows that is much more difficult to extract Mo atoms from an oxidized nanoparticle or from an undefected metallic surface. These findings are confirmed by the values of the calculated binding energies of the metal atom within the three different precursors considered by the authors. The highest energy value found was the one needed to extract a metal atom from an oxidized nanoparticle, 10.1 eV, followed by the case of an undefected metallic surface, 8.7 eV, while 6.5 eV are sufficient to extract an atom from the metallic nanoparticle. Although the oxidic wear debris formed during rubbing do not directly facilitate the formation of TMD layers, they can in turn accelerate abrasive wear, since oxidized surfaces might be more easily damaged by rubbing, leading to the formation of surface defects and to the extraction of metallic wear particles, which are then converted rapidly into TMD layers under shear and load once brought into contact with the Se nanoparticles.

Figure 5 summarizes the conditions and outcomes of three simulations with different system compositions, with starting configurations optimized under a 10 GPa load. The high applied load induces the weak Se-Se bonds to rearrange and adhere to the metallic nanoparticles and the boundary TMD surfaces or to the metallic surfaces.

The dynamical evolutions clearly show that normal and shear stresses promote the *in-operando* formation of a crystalline MoSe interlayer, provided that the mixing of Se nanoparticles and metal debris is not hindered by the presence of oxygen, as in the case of a molybdenum-oxide nanoparticle. In the reference system with the Se nanoparticle confined between two ideal (110) Mo surfaces the cohesive energy of the atoms in the metallic surfaces is too high for them to be detached and no further structural change is observed, after the initial chemisorption under load of Se atoms onto the metal surface, with the disaggregation of the Se nanoparticles occurring immediately during the initial static relaxation. The evolution of the pure metal nanoparticle embedded in the chalcogenide powder, shows that the crystallization process, as for the previous case of MoS_2, can be subdivided into two major steps, the first one being the quick creation of a positionally disordered intercalated layer, where sliding induces diffusive motions that quickly lead to atom intermixing with the establishment of Mo-Se bonds in the trigonal prismatic coordination typical of TMDs, with the Se atoms diffusing preferentially in the most central zone. Then the re-organization of the amorphous layer progresses on a slower timescale until an almost defect-free crystalline TMD layer is formed.

The kinetic of the formation of the TMD layer could be monitored by means of the structure factor $S(G)$ of the Mo atoms $S(G) = \sqrt{\left|\sum_{j=1}^{N} e^{i\boldsymbol{G}\cdot\boldsymbol{r}_j}\right|^2/N}$ where $\boldsymbol{G}$ is the summation of the two (lateral) reciprocal lattice basis vectors of the crystalline $MoSe_2$ layer and $\boldsymbol{r}_j$ is the real space position of the j-th Mo atom belonging either to the debris particle or to the Mo (110) surface. $S(G)$ has a limit value of ~ 1/N in

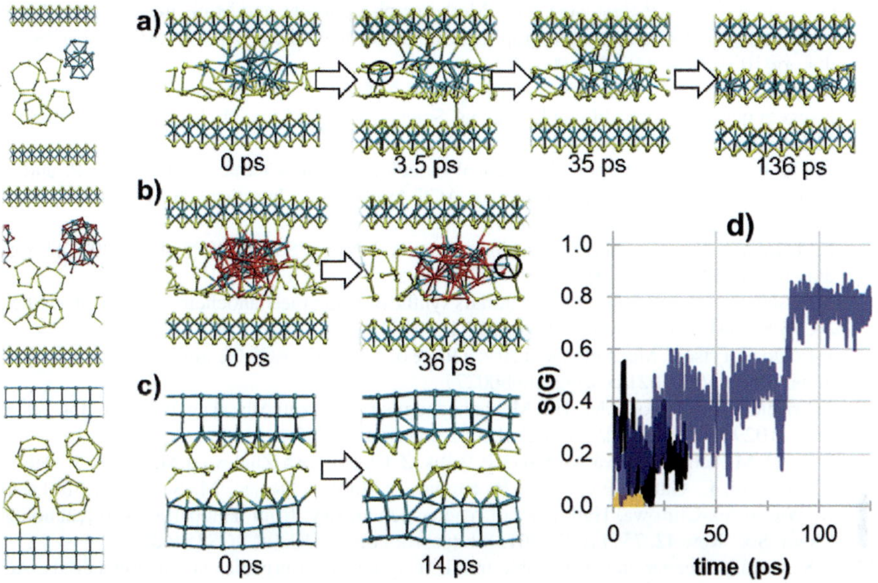

Fig. 5 Dynamical evolution of the explored tribological systems: Mo_{20} debris particle (top) and $Mo_{20}O_{40}$ debris particle (middle) together with a Se_{40} nanoparticle between two TMD boundary layers; Se_{64} nanoparticle within two Mo(110) surfaces (bottom). The black circles in panels highlight the detachment of Mo atoms from the debris particles. Color code: Mo in cyan, Se in yellow, O in red. Side panel, time evolution of the structure factor S(G): (blue) for the Mo_{20} debris particle; (black) for the $Mo_{20}O_{40}$ debris particle and (yellow) for the Mo(110) surface (from Grützmacher [65])

case of positional disorder while should be very close to 1 when the atomic positions are fully commensurable with the $MoSe_2$ lattice [66]. The $S(G)$ values calculated summing over the atoms of the Mo_{20} nanoparticle increase progressively to end up oscillating around 0.8 signaling that a high degree of crystallinity has been reached, contrary to what happens with the sum over the Mo atoms of the metal-oxide particle that keeps oscillating around the initial value. The reference system $S(G)$ calculated summing over the atoms of the Mo (110) surface remains instead always very close to zero as the periodicity of the surface is not commensurable with the one of the $MoSe_2$ lattice, the very small values implying that the surface structure is only perturbed in a minor way by the Se chemisorption.

References

1. J.C. Spear, B.W. Ewers, J.D. Batteas, 2D-nanomaterials for controlling friction and wear at interfaces. Nano Today **10**, 301–314 (2015). https://doi.org/10.1016/j.nantod.2015.04.003
2. M.C. Paolo Restuccia, Righi, Tribochemistry of graphene on iron and its possible role in lubrication of steel, Carbon **106**, 118–124 (2016). https://doi.org/10.1016/j.carbon.2016.05.025

3. G. Losi, M. Cutini, P. Restuccia et al., Modeling phosphorene and MoS_2 interacting with iron: lubricating effects compared to graphene. J. Nanostruct. Chem. **13**, 497–505 (2023). https://doi.org/10.1007/s40097-022-00478-1
4. Y. Liao, R. Pourzal, M.A. Wimmer, J.J. Jacobs, A. Fischer, L.D. Marks, Graphitic tribological layers in metal-on-metal hip replacements. Science **334**, 1687–1690 (2011). https://doi.org/10.1126/science.1213902
5. Y. Qi, J. Yang, A.M. Rappe, Theoretical modeling of tribochemical reaction on Pt and Au contacts: mechanical load and catalysis. ACS Appl. Mater. Interfaces. **8**, 7529–7535 (2016). https://doi.org/10.1021/acsami.5b12350
6. L.M. Takacs, Carey lea, the first mechanochemist. J. Mater. Sci. **39**, 4987–4993 (2004). https://doi.org/10.1023/B:JMSC.0000039175.73904.93
7. T. Friščić, New opportunities for materials synthesis using mechanochemistry. J. Mater. Chem. **36**, 7599–7605 (2010). https://doi.org/10.1039/C0JM00872A
8. J.-L. Do, T. Friščić, Mechanochemistry: a force of synthesis. ACS Central Sci. **3**, 13–19 (2017). https://doi.org/10.1021/acscentsci.6b00277
9. T. Friščić, C. Mottillo, H.M. Titi, Mechanochemistry for synthesis. Angew. Chem. Int. Ed. **59**, 1018–1029 (2020). https://doi.org/10.1002/anie.201906755
10. P. Baláž, M. Achimovičová, M. Baláž, P. Billik, Z. Cherkezova-Zheleva, J.M. Criado, F. Delogu, E. Dutková, E. Gaffet, F.J. Gotor, R. Kumar, I. Mitov, T. Rojac, M. Senna, A. Streletskii, K. Wieczorek-Ciurowa, Hallmarks of mechanochemistry: from nanoparticles to technology. Chem. Soc. Rev. **42**, 75717637 (2013). https://doi.org/10.1039/C3CS35468G
11. R.S. Varma, Greener and sustainable trends in synthesis of organics and nanomaterials. ACS Sustain. Chem. Eng. **4**, 5866–5878 (2016). https://doi.org/10.1021/acssuschemeng.6b01623
12. T.K. Achar, A. Bose, P. Mal, Mechanochemical synthesis of small organic molecules. Beilstein J. Organic Chem. **13**, 1907–1931 (2017). https://doi.org/10.3762/bjoc.13.186
13. P. Baláž, M. Hegedüs, M. Achimovičová, M. Baláž, M. Tešinský, E. Dutková, M. Kaňuchová, J. Briančin, Semi-industrial green mechanochemical syntheses of solar cell absorbers based on quaternary sulfides. ACS Sustain. Chem. Eng. **6**, 2132–2141 (2018). https://doi.org/10.1021/acssuschemeng.7b03563
14. D. Huitink, L. Peng, R. Ribeiro, H. Liang, *In situ* observation of stress-induced Au–Si phase transformation. Appl. Phys. Lett. **94**, 183111 (2009). https://doi.org/10.1063/1.3132060
15. T.E. Fischer, Tribochemistry. Annu. Rev. Mater. Res. **18**, 303–323 (1988). https://doi.org/10.1146/annurev.ms.18.080188.001511
16. S.M. Hsu, J. Zhang, Z. Yin, The nature and origin of tribochemistry. Tribol. Lett. **13**, 131–139 (2002). https://doi.org/10.1023/A:1020112901674
17. H. Spikes, The history and mechanisms of ZDDP. Tribol. Lett. **17**, 469–489 (2004)
18. K.L. Campbell, M.A. Sidebottom, C.C. Atkinson, T.F. Babuska, C.A. Kolanovic, B.J. Boulden, C.P. Junk, B.A. Krick, Ultralow wear PTFE-based polymer composites the role of water and tribochemistry. Macromolecules **52**, 5268–5277 (2019)
19. V.V. Boldyrev, Mechanochemistry and mechanical activation of solids. Usp. Khim. **75**, 203–216 (2006)
20. H. Spikes, Stress-augmented thermal activation: tribology feels the force. Friction **6**, 1–31 (2018)
21. J.F. Archard, The temperature of rubbing surfaces. Wear **2**, 438–455 (1955)
22. R.C. Watkins, The antiwear mechanism of zddps. 2. Tribol. Int. **15**, 13–15 (1982)
23. S. Mori, Y. Shitara, Tribochemical activation of gold surface by scratching. Appl. Surf. Sci. **78**, 269–273 (1994)
24. M.I. De Barros-Bouchet, M.C. Righi, D. Philippon, S. Mambingo-Doumbe, T. Le-Mogne, J.M. Martin, A. Bouffet, Tribochemistry of phosphorus additives: experiments and first-principles calculations. RSC Adv. **5**, 49270–49279 (2015). https://doi.org/10.1039/C5RA00721F
25. D. Philippon, M.I. De Barros-Bouchet, T. Le Mogne, O. Lerasle, A. Bouffet, J.M. Martin, Role of nascent metallic surfaces on the tribochemistry of phosphite lubricant additives. Tribol. Int. **44**, 684–691 (2011)

26. C. Kajdas, K. Hiratsuka, Tribochemistry, tribocatalysis, and the negative-ion-radical action mechanism. Proc. Inst. Mech. Eng. Part J J. Eng. Tribol. **223**, 827–848 (2009)
27. C.K. Kajdas, Importance of the triboemission process for tribochemical reaction. Tribol. Int. **38**, 337–353 (2005)
28. Y. Wang, N. Yamada, J.X. Xu, J. Zhang, Q. Chen, Y. Ootani, Y. Higuchi, N. Ozawa, M.I.D. Bouchet, J.M. Martin, et al., Triboemission of hydrocarbon molecules from diamond-like carbon friction interface induces atomic-scale wear. Sci. Adv. **5**, eaax9301 (2019)
29. G. Zilibotti, S. Corni, M.C. Righi, Load-induced confine- ment activates diamond lubrication by water. Phys. Rev. Lett. **111**, 146101 (2013)
30. S. Loehlé, M.C. Righi, Ab initio molecular dynamics simulation of tribochemical reactions involving phosphorus additives at sliding iron interfaces. Lubricants **6**, 31 (2018). https://doi.org/10.3390/lubricants6020031
31. G.M. Bancroft, M. Kasrai, M. Fuller, Z. Yin, K. Fyfe, K.H. Tan, Mechanisms of tribochemical film formation: stability of tribo- and thermally-generated ZDDP films. Tribol. Lett. **3**, 47–51 (1997)
32. N.N. Gosvami, I. Lahouij, J. Ma, R.W. Carpick, Nanoscale in situ study of ZDDP tribofilm growth at aluminum-based interfaces using atomic force microscopy. Tribol. Int. **143**, 106075 (2020)
33. S.H. Wan, A.K. Tieu, Q. Zhu, H.T. Zhu, S.G. Cui, D.R.G. Mitchell, C. Kong, B. Cowie, J.A. Denman, R. Liu, Chemical nature of alkaline polyphosphate boundary film at heated rubbing surfaces. Sci. Rep. **6**, 26008 (2016)
34. A.K. Tieu, N. Kong, S.H. Wan, H.T. Zhu, Q. Zhu, D.R.G. Mitchell, C. Kong, The Influence of alkali metal polyphosphate on the tribological properties of heavily loaded steel on steel contacts at elevated temperatures. Adv. Mater. Interfaces **2**, 1500032 (2015)
35. A. Erdemir, C. Donnet, Tribology of diamond-like carbon films: recent progress and future prospects. J. Phys. D Appl. Phys. **39**, R311–R327 (2006)
36. H.X. Wu, A.M. Khan, B. Johnson, K. Sasikumar, Y.W. Chung, Q.J. Wang, Formation and nature of carbon-containing tribofilms. ACS Appl. Mater. Interfaces **11**, 16139–16146 (2019)
37. A. Erdemir, G. Ramirez, O.L. Eryilmaz, B. Narayanan, Y.F. Liao, G. Kamath, S. Sankaranarayanan, Carbon-based tribofilms from lubricating oils. Nature **536**, 67–71 (2016)
38. M. Clelia Righi, Understanding the tribochemistry of lubricant additives by *Ab initio* calculations: the case of phosphites, in: *Advanced Analytical Methods in Tribology, Microtechnology and MEMS*, ed. by M. Dienwiebel, M.I. De Barros Bouchet (Springer, Cham, 2018). https://doi.org/10.1007/978-3-319-99897-8_12
39. J.P. Perdew, K. Burke, M. Ernzerhof, Generalized gradient approximation made simple. Phys. Rev. Lett. **77**, 3865–3868 (1996). https://doi.org/10.1103/PhysRevLett.77.3865
40. S. Grimme, Semiempirical GGA-type density functional constructed with a long-range dispersion correction. J. Comput. Chem. **27**, 1787–1799 (2006)
41. D. Vanderbilt, Soft self-consistent pseudopotentials in a generalized eigenvalue formalism. Phys. Rev. B **41**, 7892–7895 (1990)
42. R. Car, M. Parrinello, Unified approach for molecular dynamics and density-functional theory. Phys. Rev. Lett. **55**, 2471–2474 (1985)
43. A. Warshel, M. Levitt, Theoretical studies of enzymic reactions: dielectric, electrostatic and steric stabilization of the carbonium ion in the reaction of lysozyme. J. Mol. Biol. **103**, 227–249 (1976)
44. P. Giannozzi, S. Baroni, N. Bonini, M. Calandra, R. Car, C. Cavazzoni, D. Ceresoli, G.L. Chiarotti, M. Cococcioni, I. Dabo et al., QUANTUM ESPRESSO: a modular and open-source software project for quantum simulations of materials. J. Phys. Condens. Matter **21**, 395502 (2009)
45. P. Restuccia, M. Ferrario, M.C. Righi, Quantum mechanics/molecular mechanics(QM/MM) applied to tribology: real-time monitoring of tribochemical reactions of water at graphene edges. Comput. Mater. Sci. **173**, 109400 (2019)
46. S. Plimpton, Fast parallel algorithms for short-range molecular dynamics. J. Comput. Phys. **117**, 1–19 (1995)

47. G. Ramirez, O.L. Eryilmaz, G. Fatti, M.C. Righi, J. Wen, A. Erdemir, Tribochemical conversion of methane to graphene and other carbon nanostructures: implications for friction and wear. ACS Appl. Nano Mater. **3**, 8060–8067 (2020) https://doi.org/10.1021/acsanm.0c01527
48. M. Kalin, Influence of flash temperatures on the tribological behaviour in low-speed sliding: a review. Mater. Sci. Eng. A **374**, 390–397 (2004)
49. S. Kajita, M.C. Righi, A Fundamental mechanism for carbon- film lubricity identified by means of Ab Initio molecular dynamics. Carbon **103**, 193–199 (2016)
50. S. Loehlé, M.C. Righi, First principles study of organophosphorus additives in boundary lubrication conditions: effects of hydrocarbon chain length. Lubr. Sci. **29**, 485–491 (2017). https://doi.org/10.1002/ls.1382
51. D.I. Miakota, R.R. Unocic, F. Bertoldo, G. Ghimire, S. Engberg, D. Geohegan, K.S. Thygesen, S. Canulescu, Nanoscale **14**, 9485 (2022)
52. C. Grossiord, K. Varlot, J.-M. Martin, T.L. Mogne, C. Esnouf, K. Inoue, MoS_2 single sheet lubrication by molybdenum dithiocarbamate. Tribol. Int. **31**, 737–743 (1998)
53. D.N. Khaemba, A. Neville, A. Morina, New insights on the decomposition mechanism of molybdenum DialkyldiThioCarbamate (MoDTC): a Raman spectroscopic study. RSC Adv. **6**, 38637–38646 (2016)
54. M. De Feo, C. Minfray, M.I. De Barros Bouchet, B. Thiebaut, J.M. Martin, MoDTC friction modifier additive degradation: correlation between tribological performance and chemical changes. RSC Adv. **5**, 93786–93796 (2015)
55. S. Peeters, P. Restuccia, S. Loehle, B. Thiebaut, M.C. Righi, Characterization of molybdenum dithiocarbamates by first-principles calculations. J. Phys. Chem. A **123**, 7007–7015 (2019)
56. S. Peeters, P. Restuccia, S. Loehle, B. Thiebaut, M.C. Righi, Tribochemical reactions of MoDTC lubricant additives with iron by quantum mechanics/molecular mechanics simulations. J. Phys. Chem. C **124**, 13688–13694 (2020)
57. Peeters, S.; Charrin, C.; Duron, I.; Loehl´e, S.; Thiebaut, B.; Righi, M. C. Importance of the catalytic effect of the substrate in the functionality of lubricant additives: the case of molybdenum dithiocarbamates. *Materials Today Chemistry* **21**, 100487 (2021).
58. D.N. Khaemba, A. Neville, A. Morina, A methodology for Raman characterisation of MoDTC tribofilms and its application in investigating the influence of surface chemistry on friction performance of MoDTC lubricants. Tribol. Lett. **59**, 38 (2015)
59. Khaemba, D. N. Raman Spectroscopic Studies of Friction Modifier Molybdenum DialkyldiThioCarbamate (MoDTC). Ph.D. thesis, University of Leeds, 2016; http://etheses.whiterose.ac.uk/12923/
60. P. Nicolini, R. Capozza, P. Restuccia, T. Polcar, Structural Ordering of Molybdenum Disulfide Studied via Reactive Molecular Dynamics Simulations. ACS Appl. Mater. Interfaces. **10**, 8937–8946 (2018)
61. R. Chen, A. Jusufi, A. Schilowitz, A. Martini, Formation of MoS2 from elemental Mo and S using reactive molecular dynamics simulations. J. Vac. Sci. Technol., A **38**, 022201 (2020)
62. S.J. Stuart, A.B. Tutein, J.A. Harrison, A Reactive Potential for Hydrocarbons with Intermolecular Interactions. J. Chem. Phys. **112**, 6472–6486 (2000)
63. A.C.T. van Duin, S. Dasgupta, F. Lorant, W.A. Goddard, ReaxFF: A Reactive Force Field for Hydrocarbons. J. Phys. Chem. A **105**, 9396–9409 (2001)
64. Stefan Peeters, Gabriele Losi, Paolo Restuccia, M.C. Righi, Unraveling the mechanism to form MoS2 lubricant layers from MoDTC by ab initio simulations, *Applied Surface Science* **606**, 154880 (2022). https://doi.org/10.1016/j.apsusc.2022.154880
65. P. G. Grützmacher, M. Cutini, E. Marquis, M. Rodríguez Ripoll, H. Riedl, P. Kutrowatz, S. Bug, C.-J. Hsu, J. Bernardi, C. Gachot, A. Erdemir, M. C. Righi, Se Nanopowder Conversion into Lubricious 2D Selenide Layers by Tribochemical Reactions. *Adv. Mater.* **35**, 2302076 (2023). https://doi.org/10.1002/adma.202302076
66. M. Reguzzoni, M.C. Righi, Size dependence of static friction between solid clusters and substrates. Phys. Rev. B **85**, 201412 (2012). https://doi.org/10.1103/PhysRevB.85.201412

Dissipation at Large Separations

Marcin Kisiel, Markus Langer, Urs Gysin, Simon Rast, Dilek Yildiz, Ernst Meyer, and Dong-Weon Lee

Abstract When two macroscopic bodies slide in contact, energy is dissipated due to friction. Sometimes it is desired, like in case of brakes in the bicycle, sometimes unwelcome—when you ask yourself why your automated coffee machine broke for the third time. In nanoscale, a tiny friction force is present when bodies in relative motion are separated by a few nanometer gap. This non-contact form of friction might be successfully measured by highly sensitive cantilever oscillating like a tiny pendulum over the surface. The elusive non-contact friction might arise due to vdW interaction, which is mediated by the long-range electromagnetic field or in many cases by fluctuations of static surface charges arising from material inhomogeneities. The huge dissipation might also originate from hysteretic switching of the studied material under the external action of the oscillating probe. In this chapter several experiments reporting on non-contact friction are discussed. First the Joule dissipation channel is discussed. Next we report on non-contact friction measurement over metal—superconductor transition, which allows to distinguish between phononic and electronic contribution to friction. Energy dissipation over a phase transition is further demonstrated on $SrTiO_3$ crystal undergoing structural change. Next the non-contact friction due to switching of the charge density wave is discussed. Finally a energy dissipation due to single electron charging is reported on oxygen deficient $SrTiO_3$ and topologically protected Bi_2Te_3 crystals. Interestingly the energy losses due to the single electron charging on Bi_2Te_3 surface are observed due to the protected character of the surface.

M. Kisiel (✉) · M. Langer · U. Gysin · S. Rast · D. Yildiz
Institute of Physics, University of Basel, Klingelbergstrasse 82, CH-4056 Basel, Switzerland
e-mail: Marcin.Kisiel@unibas.ch

D.-W. Lee
MEMS and Nanotechnology Laboratory, Chonnam National University, Gwangju, South Korea

E. Meyer
Department of Physics, University of Basel, Basel, Switzerland

E. Gnecco and E. Meyer (eds.), *Fundamentals of Friction and Wear on the Nanoscale*, NanoScience and Technology, https://doi.org/10.1007/978-3-031-63065-1_15

1 Introduction

In a closed system, friction forces transfer work to heat. Energy dissipation is a non-reversible process, which is well known from everyday life [1, 2]. The mechanisms, which lead to these irreversible processes, are complex and still poorly understood. The vibration amplitude of a damped oscillator [3–5] decays in time that is equivalent to the fact that the kinetic energy is converted into heat. The energy transfer lasts until the cantilever system reaches its thermodynamic equilibrium. In this steady state stationary fluctuations from the mean value $<x>$ are observed. Both decay time τ and equilibrium fluctuations $x(t)$ contain information about the dissipative process. The equation motion of a linear damped harmonic oscillator can be described by the knowledge of the spring constant k, the eigenfrequency ω_0 and the damping coefficient Γ, the effective mass m_{eff} and the external force f_{ext}:

$$m_{eff}\frac{d^2x}{dt^2} + \Gamma\frac{dx}{dt} + kx = f_{ext}(t) \tag{1}$$

The external force f_{ext} can be regarded as sum of a non-stochastic force and a stochastic force. Since (1) is a linear differential equation, both contributions can be treated separately. In both cases the same parameters set (ω_0, Γ, m_{eff}) is required to describe the system. All quantities (fluctuating and non-fluctuating) derived from the cantilever movement depend on the same parameter set. An experimentally accessible quantity to describe the dissipation process of a vibrating cantilever at its resonance frequency is its decay time τ. The quality factor Q is given by:

$$Q = \frac{\tau\omega_0}{2} \tag{2}$$

The knowledge of the spring constant k and decay time τ permits us to calculate the damping coefficient or so to say noncontact friction—Γ.[1] Γ is the proportionality constant between the non-conservative friction force F and the velocity v of the oscillator.

$$F = -\Gamma \cdot v_{rms} \tag{3}$$

where the friction coefficient is given by:

$$\Gamma = \frac{k}{\omega_0 Q} \tag{4}$$

For an extremely soft free cantilever ($k = 130$ μN/m) with a resonance frequency of 5.8 kHz and a vibration amplitude of $x_0 = 20$ nm, a Q of 240,000 and a $\Gamma = 10^{-14}$ kg/s we obtain friction force $F = \Gamma\frac{\omega_0 x_0}{\sqrt{2}} = 7.6aN$. The dissipated power

[1] Although, the term friction is commonly related to the situation of two surfaces in contact, we will stick to it in the rest of this chapter always referring to the noncontact form of friction.

$P = \Gamma \cdot {v_{rms}}^2 = 4.0 \times 10^{-21}$ Watt (2.5×10^{-2} eV/s or 4.2 μeV/cycle). For non-contact force sensors with a resonance frequency of 300 kHz, a $\Gamma = 10^{-11}$ kg/s and typical oscillation amplitude of 1 nm, the friction force is $F = 1.3 \times 10^{-14}$ N, which corresponds to a dissipated power of 1.7×10^{-17} Watt (110 eV/s or 367 μeV/cycle). The experimentally determined friction coefficient Γ might be a superposition of different friction coefficients Γ_i

$$\Gamma = \Gamma_0 + \Gamma_1 + \Gamma_2 ... + \Gamma_N \tag{5}$$

$\Gamma(T, p, x,)$ represents a quantity which depends on the temperature T, the pressure p, the material properties and external fields (e.g. magnetic or electrical fields). As the cantilever approaches the sample surface the electromagnetic field between tip and surface has an impact on the friction coefficient. This leads to a distance dependent friction coefficient at small tip sample separations [1, 6–8].

The internal friction of the cantilever Γ_0 describes the friction losses which occur by simply bending the cantilever. Internal friction has to be measured under vacuum conditions to minimize the viscous damping due to inelastic scattering between the vibrating cantilever and gas molecules. At pressures below 10^{-6} mbar viscous damping can be neglected [9]. The internal friction depends mainly on the material properties and the sensor geometry. There are two possibilities to lower the kinetic energy of a vibrating cantilever dissipation and sound wave scattering. The cantilever can be regarded as a one side clamped bar. In reality the cantilever is a micro-mechanically etched silicon bar, which is connected to a support. If the junction between bar and support is not ideally constructed, there exists the possibility that sound waves are scattered from the cantilever to the support, which is reducing the kinetic energy of the cantilever. This is not a dissipative process! On the other hand kinetic energy is transferred into heat. It is experimentally a proven fact that a rectangular bar cantilever scatters a minimum energy into the support. The dissipative energy losses can be induced due to stress and strain acting on the cantilever. Experimentally it is hard to distinguish between the two mechanisms.

The internal friction Γ_0 is omnipresent and determines if an additional friction coefficient $\Gamma_R = \Gamma_1 + \Gamma_2 + \Gamma_3 ... + \Gamma_N$ is detectable at all. Assuming the errors $\delta\Gamma$ and $\delta\Gamma_0$ are Gaussian distributed we find for Γ_R:

$$\Gamma_R = \Gamma - \Gamma_0 \pm (\delta\Gamma + \delta\Gamma_0) \tag{6}$$

Γ_R is limited by the sum of the statistical errors $\delta\Gamma + \delta\Gamma_0$. For an ultrasensitive cantilever (cf. Fig. 1) an internal friction coefficient of $\Gamma_0 = 1.47 \times 10^{-14}$ kg/s can be obtained under UHV conditions.

Approaching the cantilever close to a surface opens new dissipation channels. Noncontact friction is accurately measured by means of a very soft and sensitive cantilever, which however might jump into contact, when the attractive force gradient is larger than the spring constant. That happens on almost all surfaces because the attractive van der Waals forces and the only exception is the operation under liquid

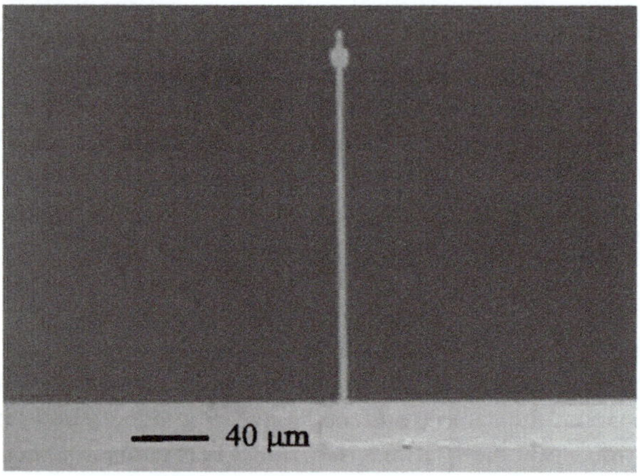

Fig. 1 Ultrasensitive single crystalline silicon force sensor with a length of 200 μm, a width of 5 μm and a thickness of 170 nm [10]. The spring constant is $k = 128$ μN/m and the eigenfrequency $f_0 = 5.8$ kHz. A quality factor Q = 240,000 can be obtained after annealing under UHV conditions for several hours. An internal friction coefficient Γ_0 in the order of $(1.47 \pm 0.27) \times 10^{-14}$ kg/s is achieved. With a vibration amplitude of 20 nm the dissipated power is: 4.0×10^{-21} Watt (2.5×10^{-2} eV/s or 4.2 μeV/cycle)

environment, where attractive forces are greatly reduced [11]. An easy way to overcome this obstacle is to oscillate the cantilever in the so-called pendulum geometry [12, 13]. This way the cantilever is oriented perpendicularly to the surface and oscillates like a tiny pendulum without snapping into the contact with the measured sample (Fig. 2).

There are several methods to determine the friction coefficient Γ: ring-down measurements, the measurement of the power spectral density S(ω) of the cantilever

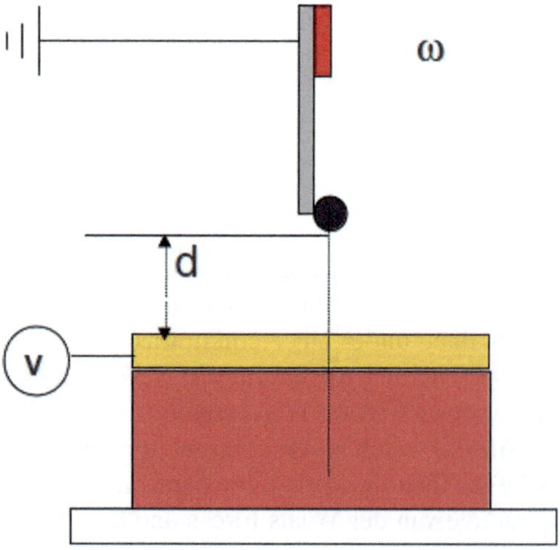

Fig. 2 Experimental setup: The cantilever is mounted perpendicularly to the surface to avoid a jump into contact. The distance d between force sensor and surface can be varied and a bias voltage V between cantilever and surface can be applied

fluctuations [14, 15] and the measurement of the excitation voltage A_{exc} needed to maintain constant amplitude cantilever's oscillation.

The smallest possible friction coefficients Γ can be sensed by using soft cantilever with a smallest possible constant k, a highest possible eigenfrequency and quality factor Q (Eq. 4). Cantilevers with these features are very force sensitive. The minimal detectable force F_{min} is given by:

$$F_{min} = \sqrt{\frac{2k_B T \Gamma \Delta\omega}{\pi}} \tag{7}$$

where k_B is the Boltzmann constant, T and $\Delta\omega$ are the temperature and the bandwidth of the measurement. A change of the friction coefficient and the temperature is always accompanied by a change of the minimal detectable force (Eq. 7).

Quantities like the frequency noise $\delta\omega$ of a cantilever which oscillates with an amplitude x_0 at its resonance frequency ω_0 depend on the friction coefficient Γ [16].

$$\delta\omega = \frac{\omega}{x_0 k}\sqrt{\frac{2k_B T \Gamma \Delta\omega}{\pi^3}} \tag{8}$$

To obtain the best possible frequency resolution of a self-driven cantilever the frequency noise $\delta\omega$ has to be minimized. Therefore, the measurement has to be performed at low temperatures, where thermal fluctuations are small and low Γ are found. The oscillator frequency noise $\delta\omega$ is influenced by Γ and the temperature T. With Eq. (7) the power spectral noise density $S(\omega)^2\Delta\omega = \delta\omega^2$ can be estimated in a narrow bandwidth

$$S(\omega) = \frac{2k_B T \omega_0^2 \Gamma}{\pi^3 k^2 x_0^2} \tag{9}$$

This contribution is structured as follows: First we investigate the internal friction of the cantilever Γ_0. In a second part we discuss different dissipation processes between moving bodies separated by a nanometer sized distance d. Experiment on superconducting Nb surface allowed to distinguish between different dissipation channels—electronic and phononic. The last part reports on experiment on $NbSe_2$ surface, where the non-contact friction is caused by the hysteretic processes induced in the sample.

2 Internal Friction of the Cantilever

The temperature dependence of resonance frequency of cantilevers is rather well understood [16]. Geometry changes due to thermal expansion can be neglected. However, the temperature dependence of the Young's modulus of Silicon, $E(T)$, is given by the Wachtman-formula:

$$E(T) = E_0 - B \cdot Texp\left(-\frac{T_0}{T}\right), \tag{10}$$

where T_0 is related to the Debye-temperature of the sensor material. With (Eq. 10) the temperature dependent resonance frequency can be calculated:

$$\omega_n = \alpha_n^2 \frac{t}{L^2} \sqrt{\frac{E}{12\rho}} \tag{11}$$

where $\alpha_1 = 1.875$ for the first eigenmode. t is the thickness, L the length and ρ the mass density. The experimental frequency vs. temperature data are well fitted with $T_0 = 317$ K [17]. According to $T_0 =$ D/2 a Debye temperature of $D = 634$ K is determined, which is in good agreement with literature values of $D = 645$ K for silicon.

In contrast, the damping of cantilevers is still rather poorly understood. Several contributions have to be distinguished:

1. Damping due to thermoelastic effects.
2. Damping due bulk losses.
3. Damping due to surface losses.
4. Damping due to acoustic emission into the bulk.
5. Losses due to the clamping.
6. Viscous damping due to the presence of gases or liquids.

As far as ultra-sensitive measurements under ultrahigh vacuum conditions are concerned, the influence of viscous damping at pressures below 10^{-6} mbar can be neglected. The influence of clamping can be optimized by rigid holders and the exclusion of glues with high damping rates. Damping due to acoustic emission is also found to be negligible in most practical cases [18]. Therefore, the first three mechanisms are the most important ones.

2.1 Thermo-Elastic Damping

The conduction of heat is an important energy loss mechanism. Periodical compression and expansion of oscillating micromechanical elements is associated with heat flow between compressed and expanded areas. The Zener-model is a continuum model of this thermo-elastic damping mechanism [19]. The internal friction is given by:

$$Q^{-1} = \frac{\alpha^2 TE}{\rho c_p} \frac{\omega\tau}{1 + (\omega\tau)^2} \tag{12}$$

where α is the thermal expansion coefficient, c_p the specific heat capacity and ρ is the mass density. The relaxation time τ is given by:

$$\tau = \frac{t^2}{\pi^2}\frac{\rho c_p}{\kappa}, \tag{13}$$

where κ is the thermal conductivity. Typical parameters for silicon at room temperature are $E = 1.68$ GPa, $\alpha = 2.54 \times 10^{-6}$ K^{-1}, $\rho = 2.33 \times 10^3$ kgm^{-3}, $c_p = 711$ J kg^{-1} K^{-1} and $\kappa = 150$ Wm^{-1} K^{-1}. The temperature dependence of the Young's modulus is small compared to the strong variations of thermal expansion (zero crossings at 20 K and 125 K). The thermal conductivity in the bulk varies between 100 and 5000 Wm^{-1} K^{-1} [20]. One should also take into account that the thermal conductivity is reduced below 100 Wm^{-1} K^{-1} due to phonon-boundary scattering for thickness of the order of microns at temperatures below 30 K [21].

At present, many experimental data indicate that thermo-elastic damping is the dominant loss mechanism at room temperature. At temperature below 200 K other channels start to dominate, which may be related to bulk or surface losses.

2.2 *Bulk and Surface Losses*

The scattering of elastic waves with defects on the surface or in the bulk is an important loss mechanism. The oscillation of the cantilever leads to a time dependent local stress field changing the energy landscape of the defects. Instabilities of these defects may occur, where atoms jump from one equilibrium position to another position. The energy difference between equilibrium positions is the activation energy. Therefore, damping vs. temperature curves show activation peaks, also called Debye peaks. So far, most of the experimental work is limited to crystalline silicon cantilevers which exhibit the highest Q-factor of available cantilevers. Typical Q-factors are between 10,000 up to 900,000. Comparable cantilevers made of Si_3N_4 or SiO_2 show much smaller Q-factors of 100–1000. Therefore, we conclude that bulk losses are dominant for these amorphous structures. In the case of silicon, bulk or surface losses may become dominant at temperatures below 200 K. At 160 K a peak is observed, which may be related to such an activation peak with an activation energy of 0.25 eV. Unfortunately, the nature of these defects in silicon is still poorly understood. Simple defects, such as vacancies or interstitials are ruled out because of their high activation barriers [17]. Recently, it has been observed that the 160 K peak can be reduced strongly by annealing under vacuum conditions [22]. It is also observed that the peak does not shift with the resonance frequency, which is not in agreement with the simple activation energy model. The authors suggest that the 160 K peak is related to an adsorbate layer. Another peak at 30 K shifts with the resonance frequency and seems to be in better agreement with a Debye peak [22].

Coating of cantilevers leads to a strong increase of dissipation. The polycrystalline nature of these metallic films implies grain boundaries, where increased phonon scattering leads to an increase of damping losses. Also the surface quality, that is the presence of silicon oxide layer or adsorbates, such as H_2O or hydrocarbons, lead to rather large damping losses. Yang et al. annealed extremely small cantilevers

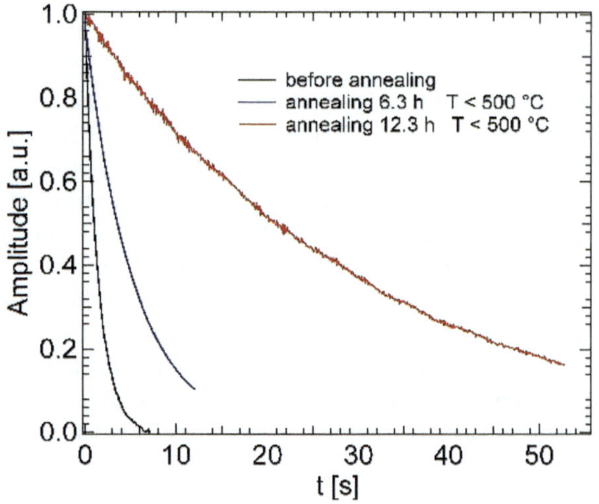

Fig. 3 Ring down measurement of a cantilever before and after annealing under UHV-conditions. The initial quality factor of 62,000 could be improved by an order of magnitude after 6 h annealing. Further annealing improved the quality factor to 1,240,000. The annealing temperature was too low to remove the oxide layer. Thus, the removal of weakly bound molecules, such as H_2O, OH or hydrocarbons, improves the quality factor

(length < 80 μm) [23]. They used rather high annealing temperatures (1000 °C), which was sufficient to remove the oxide layers. It has been shown that annealing at temperatures below 600 °C of rather large silicon cantilevers (length of 400 or 500 μm and thickness of 0.5–1.5 μm under ultrahigh vacuum (UHV) conditions also leads to a reduction of dissipation [24]. In the case reported in Fig. 3 the quality factor of 62,000 could be improved by an order of magnitude after 6 h annealing at temperatures below 600 °C. Further annealing improved the quality factor to 1,200,000. The annealing temperature was too low to remove the oxide layer and thus results in removal of water or other weakly bound molecules, such as H_2O, OH or hydrocarbons. Alternatively, defects on the surface or in the bulk of the cantilever may be reduced by the annealing procedure. It is also known that long term annealing leads to negligible amount of localized charges at the end of the probing tip.

3 Dissipation at Large Separations

In most of the cases the dissipation between two moving bodies separated by a distance d is due to an electromagnetic interaction. The range of the interacting force determines the distance dependency of the friction coefficient Γ. Therefore, it is possible to distinguish between long-range and short-range friction forces. The mechanism of the long-range, electrostatic dissipation known also as joule dissipation is well understood [25]. It is due to creation of local electric currents in the conducting sample by an oscillating tip. Static electric fields between two different surfaces can exist without any externally applied voltage due to different work functions of different orientations of the crystallites of a polycrystalline surface. The work function (contact potential) measured by applying a bias voltage between tip and sample is distance dependent [24]. This is due to the fact that the cantilever senses at large separations an average patch force resulting of several different oriented crystallites. At smaller tip-sample distances of the order of few nm and for compensated contact potential the phononic channel of energy dissipation might become dominant. In this case oscillating tip induced time dependent elastic deformation and the energy is lost to the creation of acoustic, longitudinal phonons. In this section electronic and phononic contribution to non-contact friction is discussed while oscillating the tip over the underlying crystal in the presence of the phase transition.

3.1 Dissipation Due to Electromagnetic Interaction

The dissipative force can be minimized by compensating the local electric field. Electrostatic dissipation was previously observed by Denk et al. [7]. Later Stipe et al. [8] observed electrostatic dissipation at separations of 1–200 nm by using ultrasensitive force sensors. A gold tip was attached to an ultra-sensitive cantilever in the pendulum geometry. The measurements of Stipe et al. were performed under high vacuum conditions. Friction coefficients of the order of 10^{-13} kg/s between tip and metal substrate were observed. The distance dependence of the friction coefficient was fitted by a power law $\Gamma \propto d^{-n}$ with an exponent $n = 1.3 \pm 0.2$. Volokitin et al. [26, 27] have calculated that dissipation at these large separations is consistent with the exponent $n = 1.5$ expected for the friction caused (on a spherical tip) by clean metal surface.

Rast et al. performed dissipation measurements under ultra high vacuum conditions at 7 K. They employed an ultrasensitive cantilever with a Co-Sm magnetic tip [24, 28]. The front edge of the tip is flat and has an area of 1.56 μm^2. As a substrate irradiated quartz was used (Suprasil 300) with 20 nm gold film. The friction coefficient is calculated according to a formula:

$$\Gamma = \Gamma_0 \left(\frac{A_{exc}(d)}{A_{exc,0}} - \frac{f(d)}{f_0} \right), \tag{14}$$

where Γ_0 is the friction due to intrinsic losses of the cantilever, measured at large tip-sample separation, $A_{\text{exc}}(d)$ and $f(d)$ are the distance dependent excitation amplitude (as measured by the excitation voltage needed to excite the cantilever at constant oscillation amplitude A) and frequency of the cantilever and the suffix zero refers to the free cantilever. As shown in Fig. 4 both conservative forces and dissipative forces increase with applied bias voltage at separations larger than 250 nm. Both obey to a quadratic power law $\Gamma, F \propto (V_{bias}{}^2)$. The quadratic behaviour of Γ versus bias voltage is direct evidence of the electronic origin of friction [26, 27, 29]. The friction coefficient has its minimum, where the conservative force is minimal and changes approximately by 2×10^{-13} kg/s when 1 V bias voltage is applied. Chumak et al. and Volokitin et al. [26, 27, 30] calculated the friction coefficient between a metallic tip and a flat metallic surface. They concluded that time dependent electric

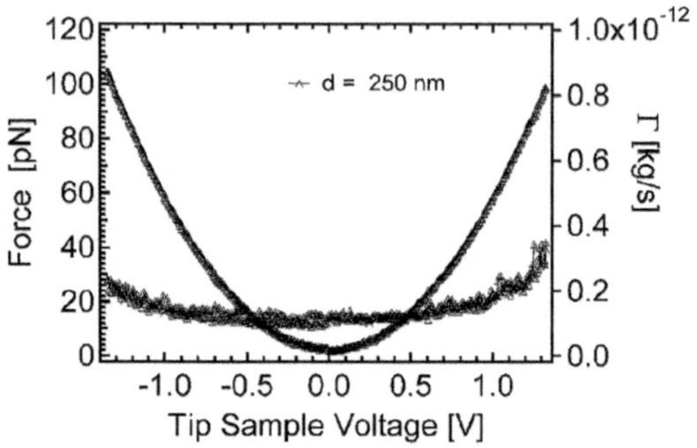

Fig. 4 Voltage-dependence of the normal force and friction coefficient at a separation of d = 250 nm of a 20 nm thick gold film. The parabolic force dependence is related to capacitive forces. The friction coefficient increases with applied voltage. The quadratic behavior of the normal force and the quadratic behavior of friction coefficient in respect to bias voltage indicate that an electrical field is involved in the process

field induces local electric currents in the sample and cantilever. Thus the induction of the current is responsible for an energy dissipation.

Compensating the contact potential allows us to investigate the friction force which is not dominated by Joule losses. Figure 5 shows the distance dependence of the friction coefficient as a function of distance with compensated contact potential. At tip sample separations which are larger than 30 nm the friction force is dominated by the internal friction of the force sensor. At this separation the friction coefficient is smaller than $< 10^{-14}$ kg/s—the internal friction of the force sensor and the strength of the electromagnetic field between tip and sample is too small to create a measurable friction force. At separations below 30 nm the friction coefficient is larger than the internal friction of the cantilever and an non-reversible energy flow between tip and sample is observed. Volokitin (see previous chapter in this book) calculated the friction coefficient of a tip separated by a distance d generated by a van der Waals friction force in dielectrics. For a cylindrical tip with radius R and width w the calculated friction coefficient scales with $d^{-1/2}$, which fits to our data.

Usually the cantilever oscillations are controlled by means of phase-locked loop (PLL) feedback system, where the frequency ω caused by the tip sample interaction is measured. The frequency noise determines frequency resolution of the measurement (Eq. 8). Figure 6 illustrates the power spectral density $S(\omega)$ of the frequency noise of a self-oscillating cantilever with an amplitude of 10 nm close to a CaF_2 substrate at room temperature under high vacuum conditions. The power spectral density is measured by sampling the frequency noise of the FM-demodulated signal of the cantilever vibration amplitude. At smaller tip sample separations the noise floor is increasing several orders of magnitude due to the distance dependent change of the friction coefficient Γ (Eq. 9). The noise floor $S(\omega)$ is changing more at low frequencies (1–10 Hz) than at frequencies above 10 Hz.

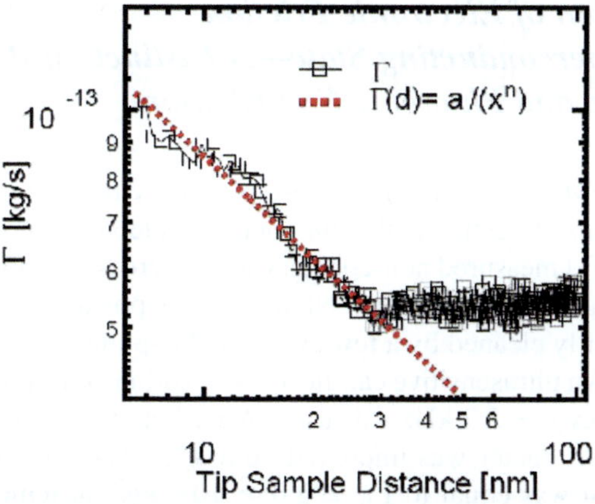

Fig. 5 Illustration of the friction coefficient with compensated contact potential at separations between 1 and 100 nm. At a separation larger than 30 nm the friction is governed by the internal friction of the cantilever. Therefore, the friction coefficient of the long-range interaction has to be smaller than $\Gamma << 10^{-14}$ kg/s

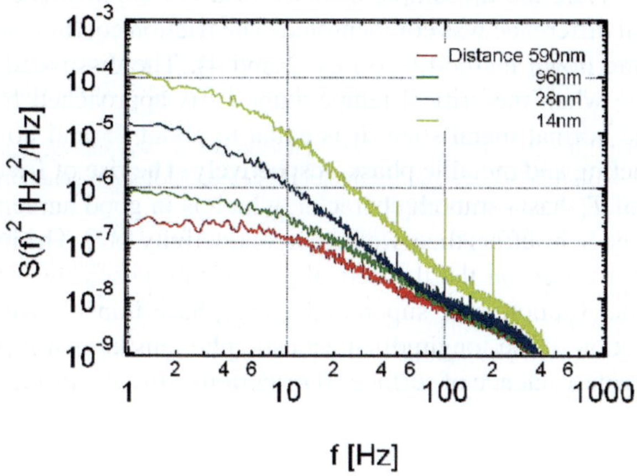

Fig. 6 Illustration of the power spectral density $S(f)$ of the frequency noise of a self-oscillating cantilever with an amplitude of 10 nm close to a CaF_2 substrate under ambient conditions. At large tip sample separations the frequency noise corresponds to the frequency noise of the free cantilever. The values of $S(f)$ decrease at higher frequencies since a lowpass filter limits the frequency band. As smaller the distance between tip and surface is, as higher is the power spectral density. With decreasing distance between tip and surface, Γ increases which is reflected in an increase of $S(f)$ (Eq. 9.)

3.2 Suppression of Electronic Friction in the Superconducting State—A Distinction Between Electronic and Phononic Contribution

The pendulum AFM tip oscillating in close proximity to a substrate might induce phononic excitations. To estimate the phononic and electronic contributions to friction, Kisiel et al. [31] measured non-contact friction across the superconducting transition of a Nb sample. A 150 nm thick Nb film was deposited on a Si(100) substrate and was subsequently cleaned by a few cycles of Ar sputtering and annealing under UHV conditions. An ultrasensitive cantilever with spring constant $k = 30$ mN/m and resonance frequency $f = 5.3$ kHz was used. After long term annealing under UHV condition the quality factor was improved up to $Q = 500{,}000$. The corresponding internal dissipation was equal to $\Gamma_0 = 2.0 \times 10^{12}$ kg/s at temperature T = 6 K. The cantilever end was exposed to focus ion beam (FIB) to form a sharp tip with spherical apex, approximately 50 nm in diameter (inset in Fig. 7). The oscillation amplitude $A = 5$ nm was constant during the measurement. The temperature dependence of the friction coefficient over critical temperature of Nb ($T_c = 9.2$ K) is shown in Fig. 7. Here the tip sample distance of d = 0.5 nm was constant and the contact potential difference was compensated. The friction coefficient was measured by means of ring down method (see Eqs. 2 and 4). The dissipated power rise by a factor of three when the critical temperature T_c is approached from below and levels off in the normal metal state. It is equal to about 25 and 80 µeV/cycle for the superconducting and metallic phase, respectively. The rise of friction coefficient in the vicinity of T_c has a smooth character, which is in good agreement with BCS (Bardeen-Cooper-Schrieffer) theory of superconductivity [32]. The interpretation of the experiment is based on the theoretical analysis proposed by Persson [33, 34], who realized that friction over superconducting phase transition is analog to the acoustic attenuation of the longitudinal acoustic phonons known from BCS theory. The temperature dependence of surface contribution to the electronic friction has the form:

$$\frac{\Gamma_{surf}(T)}{\Gamma_{surf}(T_c)} \simeq \frac{2}{exp(\Delta(T)/k_B T) + 1} \tag{15}$$

where $\Delta(T)$ is the temperature dependent energy band gap:

$$\Delta(T) = C \cdot k_B T_c (1 - (T/T_c))^{0.5} \tag{16}$$

The factor $C = 3.52$ is the same for all BCS superconductors and determines the ratio of the energy gap at $T = 0\,K$ to the thermal energy at T_c. The data shown in Fig. 7 agree with Eq. (15), with the fit parameter $C = 3.8 \pm 0.7$. The cantilever tip oscillating very close to the Nb-film produce the surface acoustic longitudinal waves due to time dependent mechanical stresses acting on the sample. The energy is lost to the emission of phonons. The acoustic wave can interact only with the normal

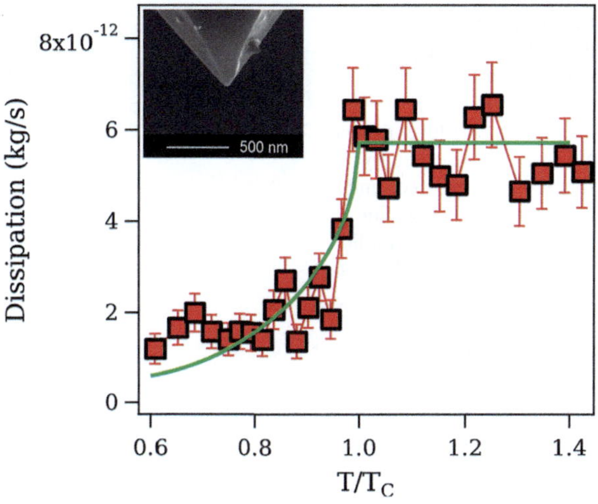

Fig. 7 Temperature dependence of the non-contact friction coefficient of the Nb surface. The friction coefficient increases as the critical temperature $T_c = 9.2$ K is approached and levels off in the Nb normal metal state. The green line is a fit of equation Eq. (15) to the measured data. Inset: the cantilever tip with spherical apex prepared by means of focused ion beam

electrons near the Fermi surface. Close to T_C the electron population is gradually growing and the acoustic wave attenuation rises rapidly. Thus the electronic friction caused by electron phonon interaction vary with the temperature according to the formula (15).

Further support to this hypothesis comes from voltage V and distance d dependent friction. In Fig. 8 the $\Gamma(V)$ is shown in normal, metallic state (T = 13 K) and superconducting state (T = 5.8 K), both acquired at the same tip sample distance d = 0.5 nm. In the normal state data follow a quadratic dependence $\Gamma(V) \propto V^2$, as expected for the electronic friction [26, 27], whereas in the superconducting state $\Gamma(V) \propto V^4$. Volokitin et al. [26, 27] calculated that, when an spherical tip is oscillating laterally above the elastic surface the friction $\Gamma \propto F^2$. Since $F \propto V^2$, the friction coefficient has to vary as a fourth power of the voltage, which is indeed the case in the superconducting state. The dependence of the friction coefficient Γ on distance d is shown in Fig. 9, again for normal and superconducting state of Nb film. In both cases the contact potential was compensated and equal about $V_{CPD} = 150 \pm 30$ mV. The distance was swept in between $0 < d < 3$ nm. As expected, the friction coefficient has much steeper distance dependence in the superconducting state as compared to the metallic state. The fit of the negative power function $\Gamma \propto d^{-n}$ to the experimental data gives $n = 1.0 \pm 0.1$ and $n = 3.8 \pm 0.3$ for normal and superconducting state, respectively. In normal state the distance dependence is in good agreement with the results obtained by Stipe et al. ($n = 1.3$) [29]. In the case of phononic friction, Lifshitz theory of van der Waals interaction [35] predicts that elastic stress leads to a vdW force $F(d) \propto d^{-2}$. Thus the phononic friction should vary as $\Gamma_{ph} \propto d^{-4}$ and the experimental value is in excellent agreement with this prediction.

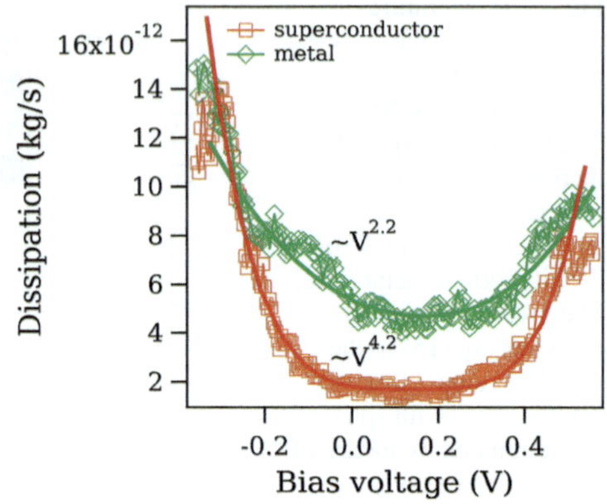

Fig. 8 The voltage dependence of the non-contact friction coefficient between oscillating cantilever tip and Nb surface, measured in superconducting (red) and normal (green) state. The dessipation has a parabolic dependence in metallic state, while $\Gamma \propto V^4$ for superconducting state of Nb. The tip sample distance is constant d = 0.5 nm

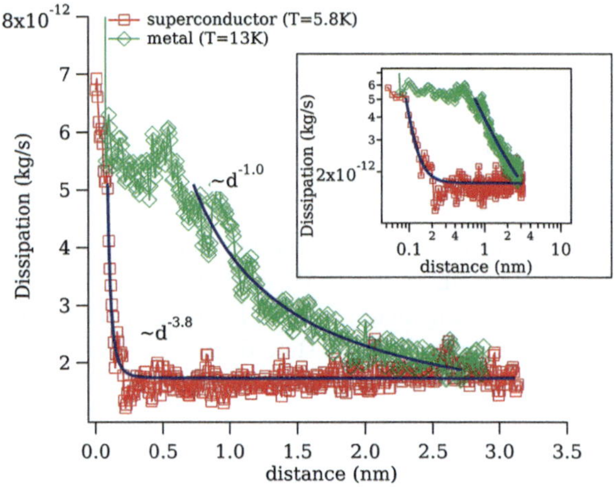

Fig. 9 Distance dependence of the non-contact friction coefficient for Nb superconducting (red) and Nb normal metal (green) state. In normal state clear rise of dissipation is visible few nanometers away from the surface, otherwise as in the case of superconducting Nb state. The data are well fitted by inverse power law dependence $\Gamma \propto d^{-1}$ and $\Gamma \propto d^{-4}$ respectively for normal and superconducting Nb state

3.3 The Energy Dissipation Across the Structural Phase Transition in $SrTiO_3$

An impact of the second order phase transition on cantilever energy dissipation was further demonstrated for $SrTiO_3$ (STO) surface [36]. STO single crystal undergoes antiferrodistortive transition just above 100 K. Close to this temperature, the high temperature ideal cubic perovskite crystal structure becomes unstable displacement of the ions, leading structural transition to orthorhombic unit cell. A phenomena is originally known from neutron scattering experiments [37], where inelastic neutron spectra showed, a strikingly narrow peak ("central peak") centered close to zero frequency. Here, we show that noncontact force microscope dissipation bears clear evidence of the antiferrodistortive phase transition of STO as proposed by molecular dynamics classical simulations [38]. Benassi et al. [38] showed that critical fluctuations should leave a footprint in the frictional dissipation of external mechanical probes such as AFM, when the temperature crosses the phase transition in the underlying bulk. The probe consisted of a soft cantilever (ATEC-CONT from Nanosensors) with spring constant k = 120 mN/m, the resonance frequency f = 11 kHz, quality factor $Q = 9.0 \times 10^5$ and a friction coefficient $\Gamma_0 = 2 \times 10^{-12}$ kg/s (at 6 K) and was operated in the pendulum geometry where the tip vibrations are parallel to the sample surface. Figure 10a shows the dissipated power as a function of temperature taken at different spots of the STO surface. Data at distance d = 12 nm corresponds to $\Delta f = -10$ Hz frequency shift and show the dissipation peak that is barely visible. Spectra taken at closer distances show a narrow dissipation peak within the temperature range 114 K < T < 118 K. The position of the dissipation peak depends on chosen surface spot and thus from local doping. Since STO is an insulator the coupling of the neutral peak at very close distances must be phononic. Similarly as in the case spherical tip oscillating above superconducting niobium surface discussed in previous section, dissipation due to creation of phonons should vary as d^{-4}. The experimental distance dependence shown is Fig. 10b is best fit by d^{-p} with p = 4.2, which is in excellent agreement with the expected value. Rise of dissipation peak at the phase transition is due to creation of acoustic phonons, corresponding to the oscillating strain wave under the tip. Thus, the structural phase transition is indeed revealed by non-contact AFM energy dissipation as predicted [38]. Pendulum AFM dissipation picks up precisely the central peak fluctuations, here responsible for the dissipation at the extremely low AFM pendulum frequency of 11 kHz. Due to the extremely low frequency, far away tip cannot directly couple to the primary atiferrodistortive order parameter, which upper bound was set to 6 MHz by neutron scattering experiments. Instead the rise of phononic dissipation at the phase transition is due to slowly varying tip-induced strain. Thus the AFM experiment showed that slow critical central peak fluctuations must involve a large component of strain, which is the secondary order parameter of the structural transition.

The central peak AFM dissipation is further analysed assuming that the external tip is a very weak perturbation of the underlying STO. Thus, rarely appropriate in friction, linear response approximation of dissipated power is assumed [36]:

$$\Gamma(\omega, T) = \Gamma_0 + \alpha k_B T \cdot \Im\chi(\omega, T) \tag{17}$$

where Γ_0 is the free cantilever dissipation, that is constant within the considered temperature T range. χ is the a lattice susceptibility, α is the positive distance dependent constant, ω is the frequency and k_B is the Boltzmann constant. Since AFM tip perturbation affects a large portion of the underlying sample, we can approximate its response by means of the bulk response of the material, as shown in Fig. 11a. Thus overall shape of measured dissipation peak might be theoretically modeled. Predicted critical form of AFM dissipation is given by the following formula:

$$\Gamma = \Gamma_0 + \frac{U}{1 + V \cdot t^{2\gamma}} \tag{18}$$

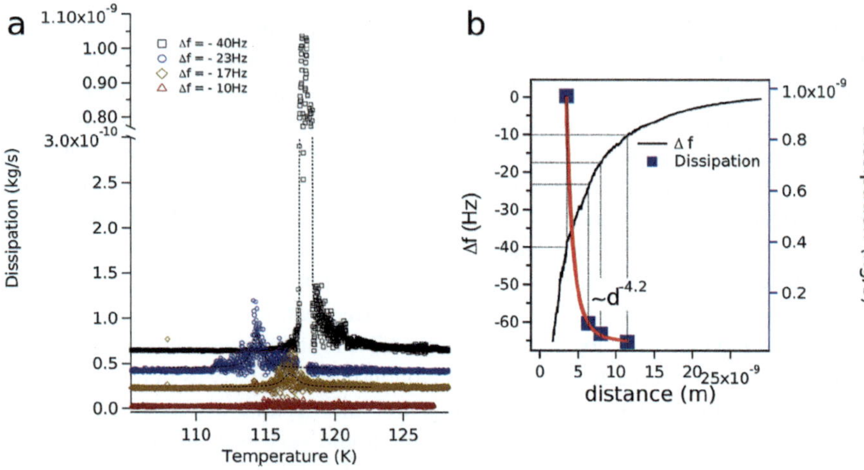

Fig. 10 **a** The experimental energy dissipation on $SrTiO_3$ surface as a function of temperature. Data were taken at four different surface spots and different tip sample distances d and the sharp peak corresponds to the critical temperature of the sample in the region under the tip. **b** is distance dependence of the dissipation (blue points) and frequency shift (black), taken as the maximum of the peak shown in **a**. A fit $\Gamma \sim d^{-p}$ to the experimental data is shown in red. An exponent $p = 4$ is expected for phononic dissipation is appropriate for coupling to acoustical surface fluctuations of an insulating bulk material

where U and V are positive constants and $t = |T - T_c|/T_c$ is the reduced temperature. Here T_c and γ are critical temperature and critical exponent of the transition. Figure 11b shows the experimental data $\Gamma - \Gamma_0$ for frequency shift $\Delta f = -40$ Hz and tip-sample distance $d = 3.5$ nm (black dots), however the y-axis is logarithmic.

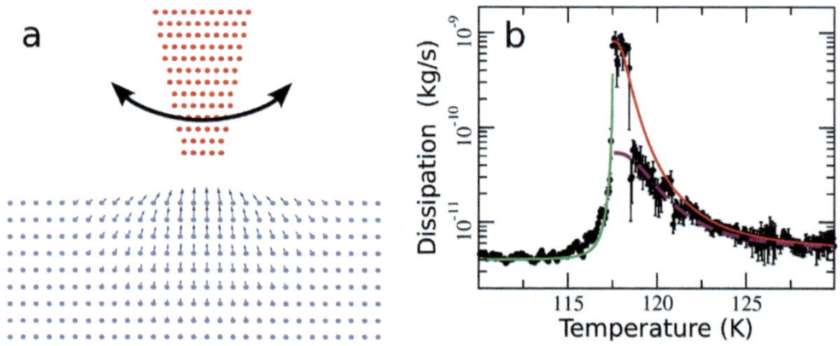

Fig. 11 **a** Simulated AFM tip perturbing a semi-infinite harmonic solid crystal through a vdW interaction. The red dots represent the tip atoms, whereas the blue dots are crystal atoms and arrows show the atomic displacement from relaxed positions. **b** experimental dissipation versus temperature measured across the critical temperature of STO—$T_c = 117.6$ K (black dots). Green and red solid lines are the theoretical fit according to Eq. (18) central peak frequency $\omega = 156$ kHz and pink dashed line similar fit for slightly different frequency $\omega = 44$ kHz. Both fits to the data yield the same value of critical exponent $\gamma = 1.24$

Red and green lines, fit above and below critical temperature $T_c = 117.6$ K according to Eq. (18) and pink dashed line, fit above T_c excluding high plateau data. The values of central peak fluctuation is $\omega = 156$ kHz and $\omega = 44$ kHz for solid and dashed lines, respectively. In both fit the critical exponent is fixed and equal to $\gamma = 1.24$, which confirms that critical behavior of STO is in the 3D-Ising universality class. Thus, the mechanical AFM dissipation not only rises at critical temperature, but the shape of the dissipation peak reflects the character of the criticality of the structural phase transition.

3.4 *The Noncontact Friction Due to Phase Slips of the Charge Density Wave (CDW) in $NbSe_2$ Sample*

So far we reported on dissipation mechanisms having the form of viscous drag as given by the equation Eq. (3). It this section we report on data when dissipation is produced by cycle of hysteretic processes induced in the sample by oscillating cantilever tip [39]. Apart from provoking such conventional forms of non-contact energy transfer as phonon and Joule ohmic dissipation, the external perturbation caused by an oscillating tip might push a finite quantum system, towards a transition or a level crossing with subsequent relaxation of the system. A level crossing implies a dissipation channel for the external tip provoking the change. The sample of interest is $NbSe_2$—an intercalated dichalcogenide compound with bulk charge density wave (CDW) state accompanied by periodic lattice distortion (PLD). The CDW materials have been an object of scientific interest since mid-1970s because their unusual properties; namely nonlinear conductivity, huge dielectric constants, elastic and thermal properties, etc. [40]. Most of them are related to the CDW motion under the external force. ATEC-CONT cantilever from Nanosensors was used ($k = 120$ mN/m, $f = 12$ kHz, $Q = 9.0 \times 10^5$). The probe was operating in pendulum geometry, however otherwise as in the case of conventional pendulum AFM experiments, the cantilever tip was chosen to be asymmetric (see inset Fig. 12). Due to the tip design the cantilever lateral oscillatory motion also implies a normal action.

The lateral oscillation amplitude $A = 5$ nm was kept constant by means of PLL. Due to tip asymmetry the normal oscillation amplitude was equal to $A_{norm} = 180$ pm. The measurement of friction versus distance $\Gamma(d)$ shows striking multiple of dissipation peaks arising at few nanometer distance from $NbSe_2$ surface, as shown in Fig. 12. Here the CPD was compensated. The friction coefficient was calculated according to Eq. (14). The presented result is contrary to the most of the non contact friction experiments, discussed so far in this chapter. Typically Γ increases smoothly as the tip approaches the sample and eventually saturates when the tip is close to contact with the surface. The amount of dissipated power (up to $P = 2$ meV/cycle) is much larger as compared to the experiment on Nb surface, for instance. The dissipated power $P(z, V)$ versus tip sample distance z and tip sample bias voltage V is shown in Fig. 13. The bright features correspond to high dissipation maxima.

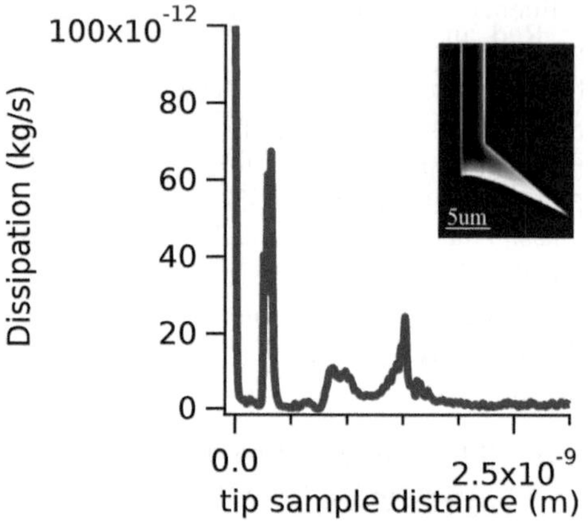

Fig. 12 The energy dissipation as a function of tip $NbSe_2$ sample distance for compensated contact potential difference. Three dissipation spikes positioned few nanometers above the surface are clearly visible. Inset: the pendulum AFM cantilever tip used in the experiment

Three dominant of them are superimposed with dashed lines which correspond to contours of constant cantilever oscillation frequencies—f_0—22 Hz, f_0—30 Hz and f_0—120 Hz—as tip approaches the sample's surface. Within each dissipation branch the amount of energy loss stays constant, meaning it doesn't depend on bias voltage V. The huge non contact friction increase is observed even after careful compensation of CPD between tip and sample. In other words no matter the character of the interaction force is van der Waals ($V = V_{CPD}$) or electrostatic ($V \neq V_{CPD}$). Moreover, $z(V)$ dependence of the particular dissipation branch has a parabolic behaviour. In fact, for capacitively coupled conical tip and sample the force varies as $F \propto V^2/z$, meaning that the each dissipation peak always follow the same tip-sample interaction force. The above observations mean that the effect is force controlled rather than voltage controlled. At this stage, it is necessary to mention that the similar dissipation peaks might occur if trapped charges are present at the end of the oxidized tip and they do not remain constant on the timescale of the experiment. That might be for instance due to charge leakage across the oxide layer [41]. In order to exclude this process the whole cantilever bar was annealed up to 700 °C, which results in removal of static charges from the probing tip. Moreover, to confirm further the universal character of the effect the experiment was repeated with an alternative instrument—the tuning fork AFM furnished with the metallic tip made of tungsten. In this setup the oscillation direction was perpendicular to the surface. The frequency, spring constant, quality factor and oscillation amplitude were equal to $f_0 = 25$ kHz, k = 2000 N/m, Q = 25,000 and A = 200 pm, respectively. Figure 14 shows dependence of non contact friction versus tuning fork tip-sample interaction force for several values of bias voltage ($-0.4\ V \leq V \leq 1.5\ V$). Again the friction measurements systematically show the existence of three dissipation maxima positioned at different interaction forces $F_{int} = -6.4, -8.2, -11.8$ nN, as the tip approaches the surface. In order to understand the origin of dissipation spikes a theoretical model was proposed, where the CDW is considered to be an elastic medium, perturbed locally by the attractive potential of the cantilever tip [42, 43]. The total energy is estimated according to equation:

$$E[\phi(x)] = \int\left[(\nabla\phi(x))^2 + V(x)\rho(x)\right]dx \tag{19}$$

where first term is an elastic energy and the second stands for perturbation. The $\rho(x) = \rho_0 \cos(Qx + \varphi(x))$ is an electron charge, where $\rho_0 = const$ and $\varphi(x)$ are CDW amplitude and CDW phase. The energy given by Eq. (17) is next minimized in order to find the preferential shape of the phase. This is shown in Fig. 15. For a given perturbation the charge peak below the tip displaces (red solid line) and the resulting CDW phase is getting distorted (red dashed line). As the force reaches critical value the central peak disappears and an extra 2π phase is pumped locally below the cantilever tip (green dashed line). This mechanism is provided by the "phase slip". The energies for the two discussed phase deformations are shown in Fig. 15 and a

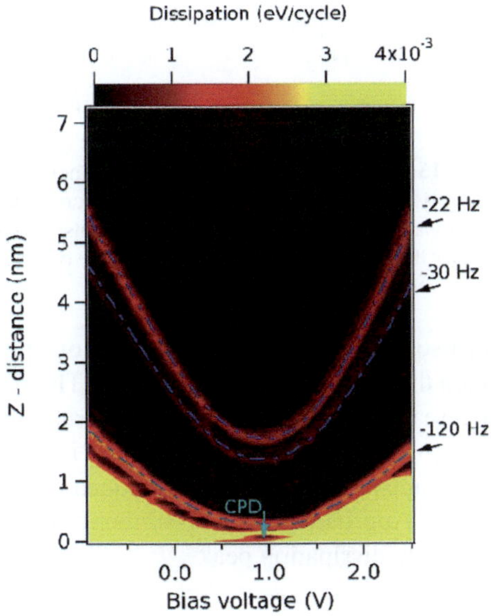

Fig. 13 Energy dissipation between $NbSe_2$ surface and oscillating pendulum AFM cantilever tip versus tip sample voltage and tip sample distance. Bright features correspond to multiplet of dissipation spikes observed few nanometers above the surface. The peaks always follow the same cantilever frequency $f = f_0 - 22$ Hz, $f = f_0 - 30$ Hz, $f = f_0 - 120$ Hz and constant frequency contours are shown with dashed lines

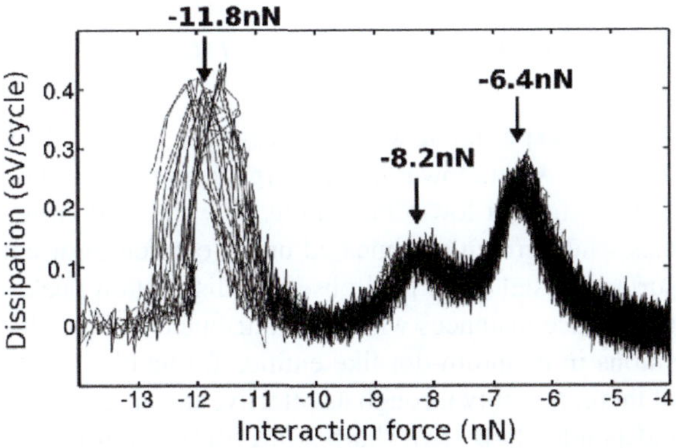

Fig. 14 The dissipation versus tip sample interaction force measured by means of tuning fork sensor. Many curves for different bias voltages are shown and the dissipation maxima, positioned always at the same interaction force are clearly visible

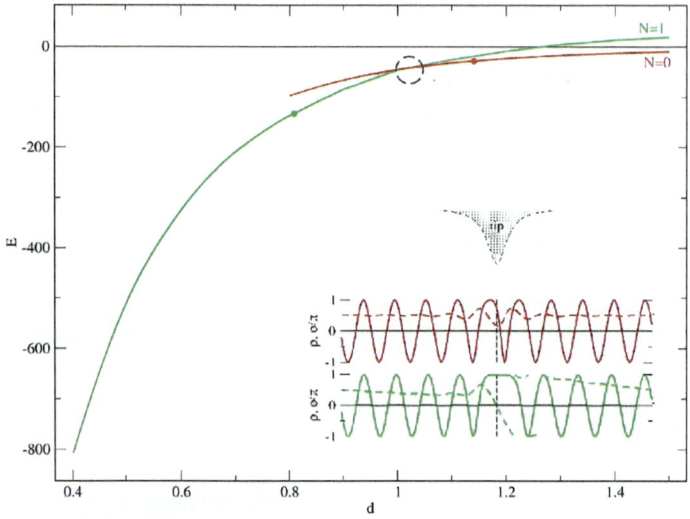

Fig. 15 The calculated energy E as a function of distance d for locally perturbed elastic charge density wave. The tip oscillation around crossover point (marked by a circle) causes hysteresis in tip dynamics. Inset shows the charge density (solid lines) and phase (dashed lines) of the CDW under the tip perturbation respectively for two different energy-distance curves

crossover energy after which the second phase deformation becomes more favourable then the first is marked by a circle. The tip oscillation near the threshold causes that phase motion is quasi-periodic, alternating between rapid advances by 2π and a slow creep of the phase. Since the process is hysteretic, it implies dissipation. The discussed mechanism works in broad frequency range, even if the perturbation frequency is small. Approaching the tip further down, there can be another phase slip explaining multiple dissipation peaks.

3.5 The Mechanical Dissipation from Charge Transitions in Oxygen-Deficient $SrTiO_3$

Another experiment where energy loss is caused by a oscillating tip provoking a quantum system towards the energy level transition is performed on oxygen deficient STO surface at low temperature (T = 5 K). Pristine STO develops oxygen vacancies when grown or annealed under oxygen-poor conditions. On the reduced STO surface Kisiel et al. [44] observed dissipation peaks at selected voltage-dependent tip-surface distances which are attributed to tip-induced charge and spin state transitions in quantum-dot-like entities formed by single oxygen vacancies or a group of them, possibly through a collective mechanism. Energy dissipation was measured independently by two sharp AFM tips oscillating in pendulum geometry at a range of lateral amplitudes 100 pm to 5 nm, for stiff (k = 50 N/m) and soft (k = 0.1 N/m) cantilever sensor, respectively. Depending on the choice of lateral amplitude performed AFM experiments have a local or non-local character. First the sample was characterized by means of Scanning Tunneling Microscopy (STM). Surface morphology and chemical composition of STO are known to change significantly under high-temperature annealing. The atomic resolution STM image (Fig. 16a) of the long-term-annealed (T = 1000 °C) surface among other features shows dark features, which considering that annealing was carried out at oxygen-poor conditions, more likely represent oxygen vacancies (V_O) at the surface. The dangling bonds of the Ti atoms bordering V_O vacancies are known to host the two electrons, or fractions thereof, left behind by the departed O atom. Depending on the local value of the electron chemical potential the charge transition is expected from the 1/2 doublet state of a q = −1e charged single-hole vacancy, and the two-hole q = −2e vacancy state [45]. Thus an isolated V_O at the surface of STO might be regarded as a quantum dot with a discrete set of charge states. A dissipation peak would arise in the tip oscillation from a periodic transition in the vacancies, when changing the local chemical potential by the tip. Such dissipation was observed at the crossing of electron energy levels in single-electron InAs quantum dots [46]. Such dissipation peak are also observed for reduced STO surface as shown in Fig. 16b. The dissipation spectra were taken at 15 nm tip-sample nanometer distance with a stiff cantilever sensor oscillating at amplitude of A = 1 nm, before (blue) and after (black) long-term annealing the sample up to about T = 1000 °C. Two distinct dissipation peaks located at $V = V_{CP} \pm 3.25$ V located symmetrically around the contact potential voltage $V_{CP} = -0.7$ V are a signature of single electron charging. A slight change of $V_{CP} \approx 0.5$ V between non-reduced and reduced surface was observed.

To clarify the tip-induced V_O transition mechanism tip-sample distance dependence of the dissipation was further studied with stiff cantilever tip (k = 50 N/m) and soft sensor (k = 0.1 N/m). Depending on the chosen sensor Kisiel et al. [44] performed local and non-local dissipation maps, shown in Fig. 17a, b for stiff and soft sensors, respectively. For distances d spanning a range of 15 nm the bias V dependence of the AFM non-contact friction coefficient was measured, thus obtaining a full Γ(V,d) map. The local dissipation data shown in Fig. 17a show the sharp dissipation peaks marked with arrows both reach $\Gamma = 5 \times 10^{-10}$ kg/s, maintaining roughly the same intensity independently of the tip-sample distance. The peaks are observed at non-zero biases with respect to V_{CP} and shift towards higher bias voltages with increasing tip-sample distance, which indicates that the effect is voltage rather than force controlled, similar to the case of quantum dots [46] and in analogy to quantum dots the amount of dissipated energy is in the order of 10–20 meV/cycle. Stiff sensors showed two dissipation peaks, whereas soft non-local sensors exhibit a whole family of dissipation maxima (Fig. 17b), due to the larger oscillation amplitudes and thus larger size of the tip shadow potential projected on the sample. A substantial change of V_{CP} value (ΔCPD = 0.1 V) indicated by the elbow shape of the arrow in Fig. 17b was observed in the transition while switching from one to the next dissipation maxima trace, which indicates that oscillating tip causing a persistent accumulation of charge below it.

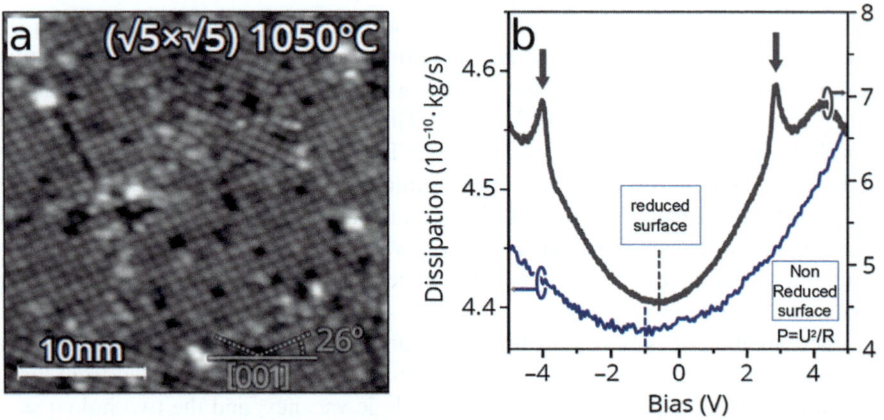

Fig. 16 **a** STM topography of the reduced STO surface obtained after annealing to T = 1050 °C. Dark features visible after high-temperature annealing are related to oxygen vacancies. **b** dissipation peak formation upon high-temperature annealing of reduced STO. Bias voltage dependence of friction coefficient Γ between the metallic tip oscillating with amplitude A = 1 nm and $SrTiO_3$ sample annealed at two different temperatures of 850 °C (black) and 950 °C (blue). Higher annealing temperatures lead to larger dissipation and pronounced dissipation peaks marked by arrows

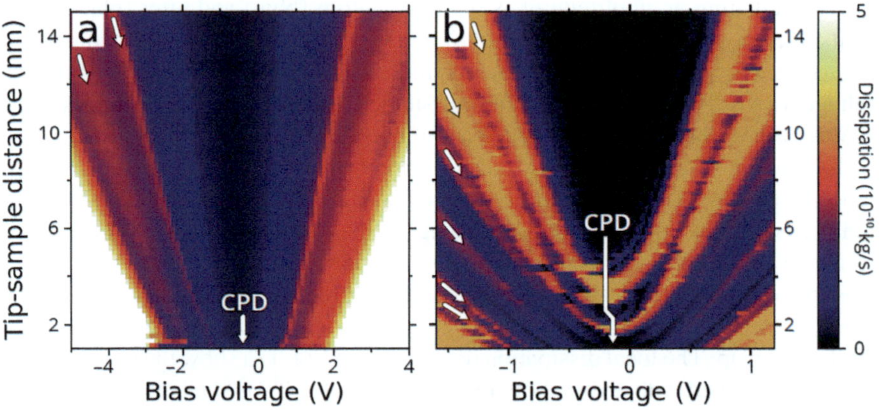

Fig. 17 Dissipation maps versus tip-sample distance and tip-sample voltage for local (**a**) and non-local (**b**) cantilever probes oscillating with amplitudes A = 1 nm and A = 5 nm. The arrows indicate the position of dissipation peaks

Based on the results the authors develop a phenomenological mechanism of charge transition of V_O on STO surface shown schematically in Fig. 18. The distant tip projects a potential shadow on the underlying STO surface which locally changes the chemical potential. If one or more oxygen vacancies happen to lie in the potential shadow of the tip, the vacancies closer to the tip experience different chemical

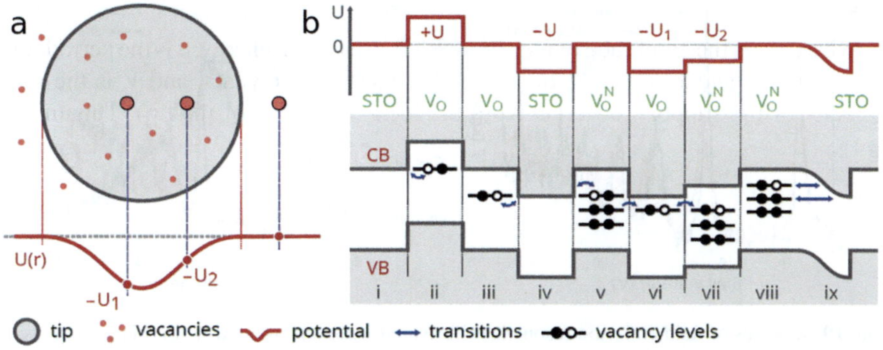

Fig. 18 Electronic transitions in oxygen vacancies caused by the potential of an oscillating AFM tip. **a** the AFM tip modifies locally the chemical potential at the surface within the shadow potential enclosing one or several vacancies. **b** several possible paths of vacancy charging, where scenario (ii) and (iii) correspond to electron exchange between single vacancy and conduction band of $SrTiO_3$, (v), (vi), (vii)—electron exchange between group of vacancies, and (viii), (ix)—electron exchange between vacancies and surface 2DEG

potential as compared to those lying at periphery. At the certain value of the potential the vacancy energy level can become aligned with an empty electron reservoir, thus provoking the electron to leave the vacancy and causing the transition. Some of the possible transitions are shown in Fig. 18. The electrons might be exchanged between vacancies themselves (v, vi, vii), vacancy and the conduction band of STO (ii, iii) or vacancy and 2D electron gas excising on the reduced STO surface (viii, ix). The work [44] demonstrated the presence of oxygen vacancies which behave like quantum-dot-like entities and the ease with which they can take or give electrons is important for the interfaces subsequently formed on STO substrate.

3.6 *The Mechanical Dissipation via Image Potential States on a Topological Insulator*

Whereas, Joule dissipation omnipresent in nowadays electronics is in general well understood the energy loss on topologically protected surface is less studied. The layered compound Bi_2Te_3 is a model topological insulator with prevented electron back-scattering and reduced electrical resistance at low temperatures due to the effect of weak anti-localization [47]. Due to protected character of the surface a suppression of Joule dissipation was observed by pendulum AFM [48] nano-mechanical dissipation. Instead a different type of mechanical energy loss was found, which was related to single electron tunneling into the image potential states that are localized few nm above the Bi_2Te_3 crystal surface. Image potential states (IPSs) on metallic surfaces resembling Rydberg series with the energy states lying slightly below the vacuum level. There are weakly coupled to the bulk in comparison to Shockley

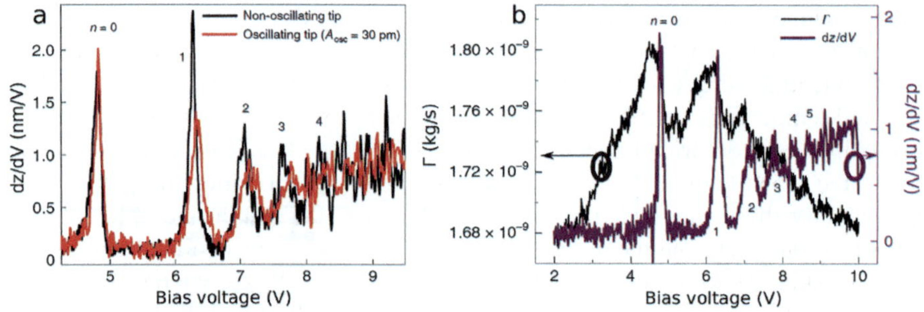

Fig. 19 **a** Series of field emission resonances measured with non-oscillating tip (black) and oscillating tip with an amplitude A = 30 pm (red). Number n indicates the quantum number n. The tip oscillation broadens the IPS peak. **b** simultaneously measured IPS by means of STS (violet) and AFM dissipation signal (black) performed by oscillating tip. Every field emission resonance peak is accompanied with changes in dissipation spectrum

surface states and in Scanning Tunneling Spectroscopy (STS) they are detected as Gundlach oscillations, which is a phenomenon of field emission resonance through IPSs in the tip–sample gap [49]. They are located a few nanometres away from the surface with an increasing tendency for higher quantum numbers. Yildiz et al., used conducting AFM tip to detected Reydberg series of IPS on Bi_2Te_3 surface by means of STS. The STS spectroscopy was performed with an non-oscillating tip and with oscillating tip (see Fig. 19a). The oscillation amplitude was chosen to be equal A = 30 pm.

The tip oscillations broadens the the IPS for larger bias voltages, due to reduced sensitivity at larger tip—sample distances and intermixing of the states of different quantum number n, by the tip-induced oscillating tunneling barrier. The tip oscillations allows to detect mechanical dissipation signal shown in Fig. 19b, together with simultaneously acquired IPS spectrum. The rise of the dissipation signal and the correspondence to the series of IPSs shows that field emission resonances affect the mechanical nanodissipation. Then, the tip is retracted 5 nm away from the surface and the measurement is switched to the AFM mode of operation where no tunneling current is detected. The tip oscillated with 300–400 pm lateral amplitude. After retraction, the sample bias was swept between 10 and −10 V while the tip was grounded, and dissipation Γ and frequency shift Δf spectra are measured. The results are shown in Fig. 20a. Yildiz et al. observed the first peak in the dissipation data located at V = ± 7.4 V and a second peak at V = ± 8.5 V. Both are symmetric with respect to the V_{CP}. Distance-dependent dissipation spectra measured at sample bias V ≤ 5 V are shown in Fig. 20b. The spectra show a $d^{-3/2}$ power law, in agreement with the theory of non-contact friction on a thin metallic film on an insulating/semiconducting bulk substrate [26, 27, 50]. Volokitin et al. [50] predicted non-contact friction is given by the formula (19.37) therein:

$$\Gamma = \frac{(4\pi\varepsilon_0)^2 w V^2 R^{0.5}}{2^{4.5}\pi\sigma d_f d^{3/2}} \tag{20}$$

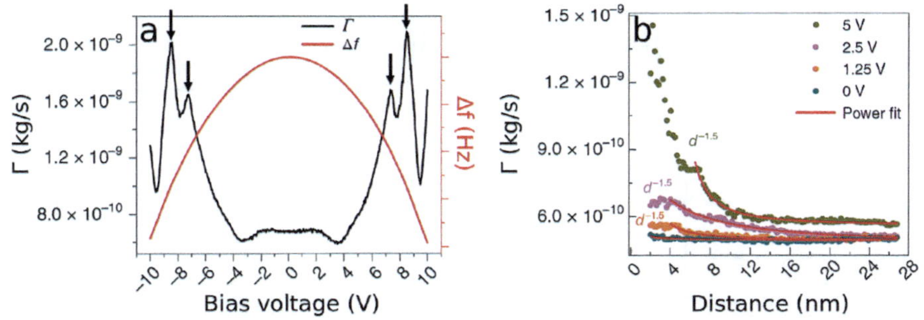

Fig. 20 Voltage (**a**) and distance (**b**) dependent AFM dissipation measured at T = 5 K. The voltage-dependent dissipation taken at constant distance d = 5 nm shows a series of dissipation peaks marked by arrows positioned symmetrically with respect to V_{CP}. Dissipation spectra measured at bias voltages V < 5 V follow a $d^{-3/2}$ power law which is in agreement with the theory of non-contact friction on a thin metallic film on an insulator

Such model accounts for for the topologically protected electronic structure of the sample.

The phenomenological model for the observed dissipation features is shown in Fig. 21. At a specific voltage and tip-sample distance, the electron tunneling leads to the occupation of the IPS. The IPS charging and subsequent charge relaxation lead to substantial charge fluctuations in the system and thus give rise to an increase of mechanical dissipation. At large tip–sample distances, the AFM dissipation is sensitive to single-electron tunneling process. In AFM mode of operation at larger distances as compared to STM, the tunneling rate is less detectable as compared to STM tunneling current. Similar to the case of single electron charging of a quantum dot [46], the amount of dissipated power is in the range of meV per cycle.

In order to confirm the suppression of Joule dissipation on topologically protected surface Yildiz et al. [48] increase the electron back-scattering by applying magnetic field B which breaks protected character of the surface. External magnetic fields were ranging from B = 0–0.8 T and were oriented perpendicularly to the sample surface. As the magnetic field rises, dissipation maxima become less pronounced, and the overall dissipation background rises as marked by green arrows in Fig. 22. The spectrum obtained for B = 0.8 T resembles the common parabolic shape of Joule dissipation obtained on ordinary metal surfaces. Thus, in the presence of magnetic field the spin-momentum locking appears to be destroyed, back-scattering becomes prominent and restores the expected Joule dissipation process.

The experiment demonstrates that AFM dissipation might address the quantum effects in dissipation spectroscopy performed at tip–sample distances as large as several nanometres. The AFM tip is sensitive to single-electron tunneling into the IPSs and the observed non-contact friction peaks are due to charge fluctuation between tip and IPS in analogy to van der Waals dissipation.

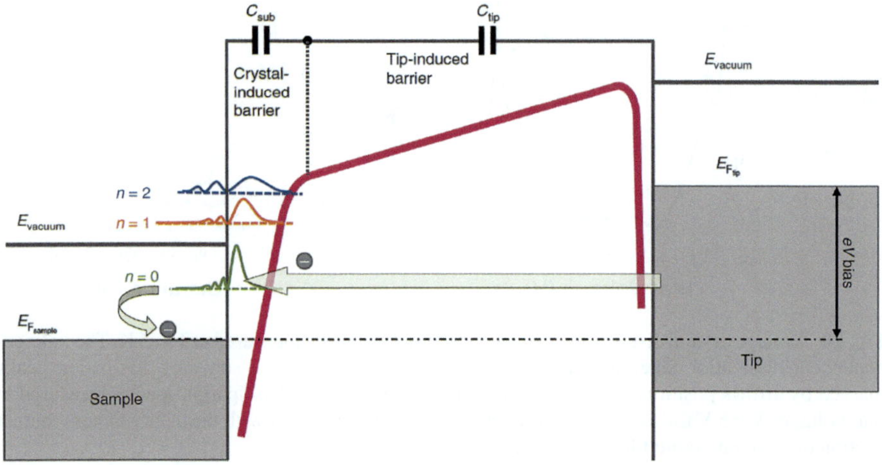

Fig. 21 Single-electron tunneling between IPSs localized above Bi_2Te_3 sample and an oscillating AFM tip. With a positive voltage applied to the tip the single electron tunneling occurs between tip and sample through the IPS and the tunneling process enhances the charge fluctuations in the system leading to the rise of AFM dissipation

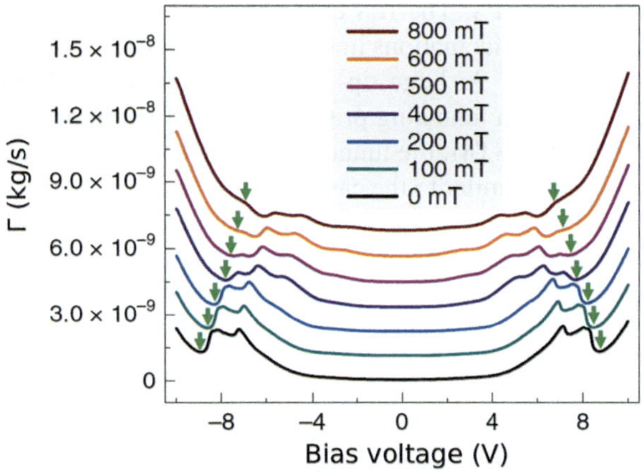

Fig. 22 Non-contact friction as a function of magnetic field B applied perpendicularly to the Bi_2Te_3 surface and ranging from $0\ T < B < 0.8\ T$. The AFM tip oscillates at $d = 5$ nm tip-sample distance. The black curve with well pronounced dissipation peaks from single electron tunneling to IPS is taken for $B = 0$ T. As the magnetic field rises the dissipation peaks are suppressed and overall Joule dissipation parabolic background becomes dominant

4 Summary and Conclusions

Thermal fluctuations and fluctuating electromagnetic fields are the limiting factors for ultrasensitive measurements close to the surface. Annealing the cantilever under UHV-conditions removes adsorbates on the cantilever surface and increases the force sensitivity about an order of magnitude. Cooling to cryogenic temperatures reduces the thermal fluctuations and lowers the internal friction further down. The temperature dependent quality factor can be described by the Zener's theory of thermo-elastic damping. Varying the distance between tip and sample opens new dissipation channels. Up to tip sample separations of 250 nm the long-range Joule dissipation is observed. Joule dissipation is relatively well understood, but the quantitative analysis indicates that both the probing tip and the cantilever contribute to dissipation. By compensating the local electric field Joule dissipation can be minimized and attonewton sensitivity can be achieved at separations down to 30 nm. Below 30 nm and for compensated contact potential the energy losses might be caused by short-range forces such as van der Waals forces. The measurement across the superconducting phase transition allows to distinguish between electronic and phononic contribution to friction. The experiment on Nb surface showed suppression of electronic dissipation in the superconducting state by factor of three. The friction coefficient increases as the critical temperature is approached from below. The rise in friction has a form of longitudinal acoustic phonon attenuation known from BCS theory. The distance and voltage dependence of the friction coefficient showed that friction has a phononic and electronic character respectively in superconducting and normal state. Experiment on $SrTiO_3$ crystal showed that AFM dissipation is sensitive to critical fluctuations of the order parameter (strain fluctuations) during the structural phase transition of the crystal. Those experiments on non-contact friction deal with linear frictional response of the system under study. In that case the energy looses could be accessed only by means of ultra-sensitive probes characterized by very small spring constants. The experiments on $NbSe_2$, oxygen reduced $SrTiO_3$ and Bi_2Te_3 surfaces reports on friction produced by hysteretic effects induced in the sample. The experiment on $NbSe_2$ showed that it is possible to couple phase of charge density wave of $NbSe_2$ to the oscillating tip at few nanometer distance and the observed dissipation peaks are due to series of CDW phase slip events. Experiments on oxygen deficient $SrTiO_3$ surface, with oxygen vacancies behaving like an effective quantum dots, and dissipation through IPS on Bi_2Te_3 crystal, both reported on energy loss due to single electron charging. For nonlinear frictional response the dissipated power is few order of magnitude larger as compared to friction losses measured on γ-irradiated quartz crystal or Nb surface.

Acknowledgements We express our gratitude to E. Gnecco, L. Marot, R. Pawlak, F. Pellegrini, G.E. Santoro, R. Buzio, A. Gerbi, G. Balakrishnan, A. Baratoff O. Brovko and E. Tosatti for collaboration and helpful discussions. This work was supported by the Swiss National Science Foundation and NCCR SPIN (SNSF grant 51NF40-180604). We also thank the European Research Council (ERC) under the European Union's Horizon 2020 research and innovation program (ULTRADISS Grant Agreement No. 834402).

References

1. B. Gottsmann, H. Fuchs, Phys. Rev. Lett. **86**, 2597 (2001)
2. A. Vanossi, N. Manini, M. Urbakh, S. Zapperi, E. Tosatti, Rev. Modern Physics **85**, 529–552 (2013)
3. G. Binnig, C. Quate, C. Gerber, Phys. Rev. Lett. **56**, 930 (1986)
4. S. Morita, R. Wiesendanger, E. Meyer, *Noncontact Atomic Force Microscopy* (Springer, Berlin, Germany, 2002)
5. F.J. Giessibl, Rev. Mod. Phys. **75**, 949 (2003)
6. I. Dorofeyef, H. Fuchs, G. Wenning, B. Gotsmann, Phys. Rev. Lett. **83**, 2402 (1999)
7. W. Denk, D.W. Pohl, Appl. Phys. Lett. **59**, 2173 (1991)
8. B.C. Stipe, H.J. Mamin, T.D. Stowe, T.W. Kenny, D. Rugar, Phys. Rev. Lett. **86**, 2874 (2001)
9. S. Rast, Sensoren mit geringer Dissipation zur Messung kleiner Krafte, Diss. Phil.naturwiss. Fak. Basel (1999)
10. D.W. Lee et al., JMMM (2006)
11. F. Ohnesorge, G. Binnig, Science **260**, 1451 (1993)
12. D. Rugar, R. Budakian, H.J. Mamin, B.W. Chui, Nature **43**, 329 (2004)
13. U. Gysin, S. Rast, M. Kisiel, C. Werle, E. Meyer, Rev. Sci. Instrum. **82**, 023705 (2011)
14. S. Rast, C. Wattinger, U. Gysin, E. Meyer, Nanotechnol. **11**, 169 (2000)
15. M. Roukes et al., J. Appl. Phys. **95**, 2682 (2004)
16. T. Albrecht, T.R. Albrecht, P. Grutter, H.K. Horne, D. Rugar, J. Appl. Phys. **69** (1991)
17. U. Gysin, S. Rast, P. Ruff, D.W. Lee, P. Vettiger, C. Gerber, Phys. Rev. B **69**, 045403 (2004)
18. J. Lübbe, L. Tröger, S. Torbrügge, R. Bechstein, C. Richter, A. Kühnle, M. Reichling, Meas. Sci. Technol. **21**, 125501 (2010)
19. C. Zener, Phys. Rev. **53**, 100 (1938)
20. K.G. Lyon, G.L. Salinger, C.A. Swenson, G.K. White, J. Appl. Phys. **48**, 865–868 (1977)
21. M. Ashegi, Y.K. Leung, S.S. Wong, K.E. Goodson, Appl. Phys. Lett. **71**, 1798 (1997)
22. H. Haucke, X. Liu, J.F. Vignola, B.H. Houston, M.H. Marcus, J.W. Baldwin, Appl. Phys. Lett. **86**, 191903 (2005)
23. J. Yang et al., Appl. Phys. Lett. **77**, 3860 (2000)
24. S. Rast, U. Gysin, P. Ruff, C. Gerber, E. Meyer, D.W. Lee, Nanotech. **17**(7), 189 (2006)
25. W. Denk, D.W. Pohl, Appl. Phys. Lett. **59**, 2171–2173 (1991)
26. A. Volokitin, B. Persson, Phys. Rev. Lett. **94**, 086104 (2005)
27. A.I. Volokitin, B.N.J. Persson, H. PUeba, Phys Rev. B **73**, 165423 (2006)
28. M.J. Naughton et al., Rev. Sci. Instr. **68**, 4061 (1997)
29. B.C. Stipe, H.J. Mamin, T.D. Stowe, T.W. Kenny, D. Rugar, Phys. Rev. Lett. **87**, 096801–096811 (2001)
30. A.A. Chumak, P.W. Milonni, G.P. Berman, Phys Rev. B **70**, 085407 (2004)
31. M. Kisiel, E. Gnecco, U. Gysin, L. Marot, S. Rast, E. Meyer, Nat. Mater. **10**, 119 (2011)
32. J. Bardeen, L.N. Cooper, J.R. Schrieffer, Phys. Rev. **108**, 1175 (1957)
33. B.N.J. Persson, *Sliding Friction* (Springer, Berlin, 2000)
34. B.N.J. Persson, Solid State Commun. **115**, 145 (2000)
35. R. Lifshitz, M. Roukes, Phys. Rev. B **61**, 5600 (2000)
36. M. Kisiel, F. Pellegrini, G.E. Santoro, M. Samadashvili, R. Pawlak, A. Benassi, U. Gysin, R. Buzio, A. Gerbi, E. Meyer, E. Tosatti, Phys. Rev. Lett. **115**, 046101 (2015)
37. T. Riste, E.J. Samuelsen, K. Otnes, J. Feder, Solid State Commun. **9**, 1455 (1971)
38. A. Benassi, A. Vanossi, G.E. Santoro, E. Tosatti, Phys. Rev. Lett. **106**, 256102 (2011)
39. M. Langer, M. Kisiel, R. Pawlak, F. Pellegrini, G.E. Santoro, R. Buzio, A. Gerbi, G. Balakrishnan, A. Baratoff, E. Tosatti, E. Meyer, Nat. Mater. **13**, 173–177 (2014). https://doi.org/10.1038/NMAT3836
40. C. Schlenker, C., *Physics and Chemistry of Low-Dimensional Inorganic Conductors* (North Atlantic Treaty Organization., & NATO Advanced Study Institute on Physics and Chemistry of Low-Dimensional Inorganic Conductors, Plenum Press, New York, 1996)

41. J. Polesel-Maris, A. Piednoir, T. Zambelli, X. Bouju, S. Gauthier, Nanotechnol. **15**, S24–S29 (2004)
42. H. Fukuyama, P.A. Lee, P. A., Dynamics of the charge-density wave. I. Impurity pinning in a single chain. Phys. Rev. B **17**, 535–541 (1978)
43. P.A. Lee, T.M. Rice, Electric field depinning of charge density waves. Phys. Rev. B **19**, 3970–3980 (1979)
44. M. Kisiel, O.O. Brovko, D. Yildiz, R. Pawlak, U. Gysin, E. Tosatti, E. Meyer, Nat. Commun. **9**, 2946 (2018)
45. O.O. Brovko, E. Tosatti, Phys. Rev. Mater. **1**, 044405 (2017)
46. L. Cockins, Y. Miyaharaa, S.D. Bennetta, A.A. Clerka, S. Studenikinb, P. Pooleb, A. Sachrajdab, P. Grutter, Proc. Natl. Acad. Sci. **107**, 9496–9501 (2010)
47. H.-T. He et al., Impurity effect on weak antilocalization in the topological insulator Bi_2Te_3. Phys. Rev. Lett. **106**, 166805 (2011)
48. D. Yildiz, M. Kisiel, U. Gysin, O. Gurlu, E. Meyer, Nat. Mater. **18**, 1201–1206 (2019)
49. K.H. Gundlach, Solid State Elect. **9**, 949–957 (1966)
50. A.I. Volokitin, B.N.J. Persson, in *Fundamentals of Friction and Wear on the Nanoscale*, ed. by E. Gnecco, E. Meyer (Springer, 2007), p. 426

Triboluminescence

Roman Nevshupa, Kenichi Hiratsuka, Alina Tukhbatullin, and Gulnaz Sharipov

Abstract This work seeks to summarize recent advances in experimental studying of triboluminescence and elucidate the basic mechanisms whereby triboluminescence is excited.

1 Introduction and Brief Historical Survey

Triboluminescence (TL) is defined by the Oxford English Dictionary as the emission of light from a substance caused by rubbing, scratching, or similar frictional contact [1] or, more generally, as the quality of emitting light at interacting surfaces in relative motion. It should be noted that the substance referred to in this definition may include both the materials of the contacting bodies and the surrounding gas or liquid, e.g., atmospheric air and lubricating oil.

Triboluminescence can naturally occur when frictional contact is composed of dielectric materials susceptible of intensive frictional electrification. In this case triboluminescence is usually manifested by a shine or lightning around the frictional contact and is related to electrical discharge with gas or liquid. This phenomenon can be observed during precise turning of glassy polymers by a diamond tool in dry air [2], in which process weak glowing occurs around the cutting tool near the cutting edge as well as at the rake surface of the tool (see Fig. 1).

Luminous signals in the atmosphere such as glowing, light flashes and lightning, which result from fracture and slippage of rocks in earthquake-prone regions, are

R. Nevshupa (✉)
Eduardo Torroja Institute of Construction Sciences, CSIC, C/Serrano Galvache 4, 28033 Madrid, Spain
e-mail: r.nevshupa@csic.es

K. Hiratsuka
Faculty of Engineering, Chiba Institute of Technology, 2-17-1 Tsudanuma, Narashino-Shi, Chiba 275-8588, Japan

A. Tukhbatullin · G. Sharipov
Institute of Petrochemistry and Catalysis, Ufa Federal Research Center of the Russian Academy of Sciences, Prospekt Oktyabrya 141, Ufa 450075, Russia

E. Gnecco and E. Meyer (eds.), *Fundamentals of Friction and Wear on the Nanoscale*, NanoScience and Technology, https://doi.org/10.1007/978-3-031-63065-1_16

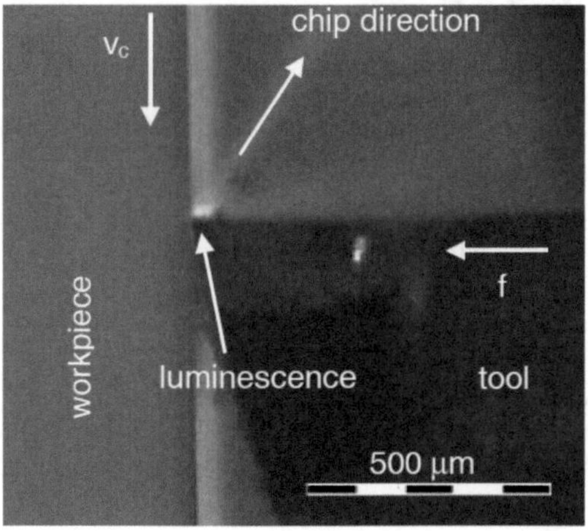

Fig. 1 Side view of the turning process of polycarbonate, showing light emission ($v_c = 1$ m/s, $f = 5$ μm/rev) (reprinted from [2] © 2004, with permission from CIRP)

also attributed by many researchers to triboelectricity and triboluminescence [3–6]. These signals, which occur prior to, concurrently or after severe earthquakes, are a supreme manifestation of triboluminescence which can be observed on the Earth. Even more tremendous are the lightnings and sprites in volcanic plumes and dust devils in planets of our solar system and in the atmospheres of brown dwarfs and gas giant planets outside of our solar system [7]. Triboelectric charging of dust particles is frequently invoked as a mechanism for spectacular lightning discharges in dust clouds of silicate planetary regolith analogues [8].

The art of making fire by rubbing two wooden sticks together is a prehistoric example of how light emission can be produced by friction: first due to incandescent radiation and then due to chemical reaction of wood oxidation [9]. Many other forms of friction-induced "cold" light have been frequently observed by the beginning of the seventeenth century. In 1620, Francis Bacon (lord Verulam) in his discussion about the nature of heat and light in "Novum Organum" [10] described the following examples of triboluminescence: "It is also a well-known fact, and it was almost considered as a miracle, that a few years since a girl's apron sparkled when a little shaken or rubbed, which was, perhaps, occasioned by the alum or salts with which the apron was imbued, and which, after having been stuck together and incrusted rather strongly, were broken by the friction. It is well known that all sugar, whether candied or plain, if it be hard, will sparkle when broken or scraped in the dark. In like manner sea and salt water is sometimes found to shine at night when struck violently by the oar. The foam of the sea when agitated by tempests also sparkles at night..." In 1664 Robert Boyle reported on his notable experiment with a curious diamond: "Being rubb'd upon my Cloaths, as is usual for the exciting of Amber, Wax, and other Electrical Bodies, it did in the Dark manifestly shine like Rotten Wood, or the Scales of Whitings, or other putrified Fish" [11]. He concluded that the light was produced "in a dead and opacous Body, and that not ... by a Violent Destruction of the Texture of the Body, but by so slight a Mechanical Operation upon its Texture...".

The eighteenth century marked a new epoch in electrical research and lightning [12]. Otto von Guericke (1602–1686), renowned for his inventions of vacuum pumps and his experiment demonstrating the force of atmospheric pressure, was also intrigued by electrical phenomena, which he perceived as cosmic immaterial forces or "incorporeal potencies" (*virtutibus mundanis*) [13]. He described these potencies as "not substances born either from or by accidents, but ... the effluents of mundane bodies, which are contained in these bodies and flow from them." The concept of truly immaterial forces (*virtus veró incorporea*) parallels that of "action at a distance" and resembles the modern definition of potential fields: "Truly incorporeal potency, indeed, is that which flows out of the body, and surrounds itself to a finite distance, extends or diffuses, and penetrates all solid and hard bodies". Light and colour were "evidently" attributed as immaterial forces (*virtus lucens atque colorans*), emanating from the Sun. However, the following notion contradicts the definition of light as *virtus incorporea*: "Light, indeed, can be present superficially in all bodies, but not intrinsically, unless this is in a suitable, transparent body".

It's noteworthy to highlight an experiment aimed at demonstrating various immaterial forces achievable through friction in a sulfur globe. The sulfur globe was crafted by casting pre-ground and melted sulfur into a hollow glass sphere. After the sulfur had cooled and solidified, the surrounding glass was broken to unveil the sulfur. Then the globe was perforated, and an iron rod was securely attached in the central hole, serving as an axle. This arrangement facilitated the swift rotation of the globe using a handle, with the entire apparatus anchored on a wooden base (see Fig. 2a).

After describing the prominently manifested immaterial forces such as the conservative (attractive), the impulsive and the expulsive, the author turned his attention to light, sound, and heat, which he found very difficult to comprehend. He describes the light force as follows: "if you bring it [the globe] into a dark room with you, and in the first part of the night put a dry palm on it, you will see how it shines, as if it were pounded with sugar". This is likely the first documented observation of electroluminescence induced by frictional electrification.

In this period, prior to the science was categorized into distinct branches and subfields, the adoption of a holistic approach to studying natural phenomena was fundamental in elucidating the complex relationships between diverse experimental observations. In 1709, Francis Hauksbee produced a considerable amount of light by shaking mercury in a glass vessel, out of which the air was exhausted [15]. He observed strange flushes of pale light, when the mercury was put in motion, and reported triboluminescence upon rubbing of various solids [14]. Hauksbee found that the light produced by rubbing amber against wool was significantly stronger in vacuum than in open air. Similarly, rubbing glass against wool in vacuum yielded a vivid purple light that turned to a pale light. He observed, that a slight friction was sufficient to excite light, and that a greater pressure or a more violent motion did not considerably increase it. However, as noted by Joseph Priestley in his work [15], Hawksebee "did not seem, at that time, to know the reason of the appearance of that light". Notably, to explain the finding that emitted light changed from red to pale after several trials Hauksbee argued that "some sorts of coloured luminous matter

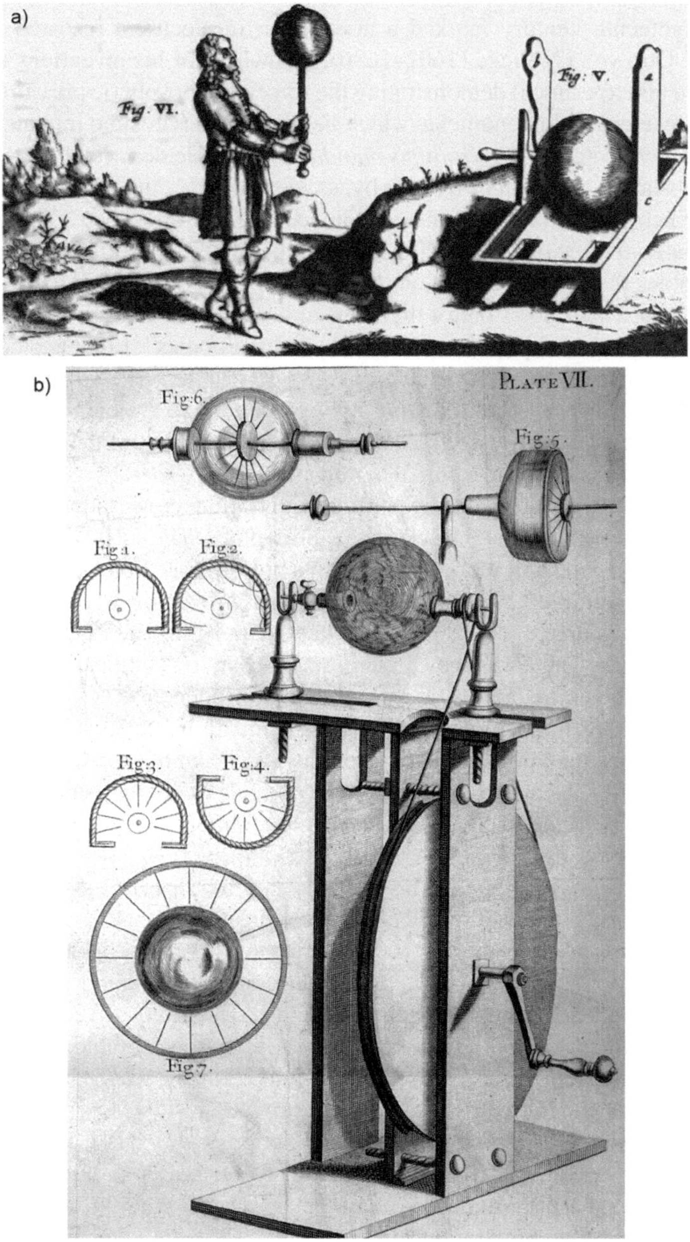

Fig. 2 **a** Otto von Guericke's (1672) experiment on electrification of a sulfur globe (from [13]). **b** Triboelectric generator constructed by F. Hauksbee (1709) to study luminescence (from [14])

are more easily separable from the bodies than others, or that they require different degrees of force and friction in order to their evaporation and discharge".

In 1708, independent of Hauksbee's observations of a noise accompanying triboluminescence, which he termed a snapping, William Wall drew a parallel between the "cracklings and light" produced when amber was subjected to friction and "thunder and lightning" of a storm.

In 1792 H. B. Saussure noted that in certain cases light could be obtained by stroking a substance under such conditions that a pure temperature radiation was altogether unlikely [16]. The same conclusion was reached by Henrich in 1820 [17]. However, along the eighteenth century the attempts to link triboluminescence to the electrical properties of inorganic solids were largely abandoned as they contradicted prevailing theories that explained light and fire as material substances embedded within solids. The following observation of Dessaignes [18] is particularly noteworthy: "the opinion that luminescence in inorganic bodies can be a kind of electric property was treated by scholars as a divorced wife. The famous Bergman[1] first pushed her away from him, to give preference to phlogiston, and all authors at will, witnesses of this refusal, did not dare to throw away a look of kindness on her." It was generally assumed (P. J. Macquer, H. Davy and others) that all forms of luminescence considered as a light-like element could exist in three forms in bodies [19]: (1) chemically bonded light, which is freed on heating, i.e. incandescence, (2) mechanically held invisible light, freed by attrition, friction or collision [18], thereby pressing out the light, i.e. triboluminescence, and (3) mechanically held but visible light, i.e. the phenomena of light reflection and refraction. C. Berthollett proposed that light and heat emitted from the bodies when they are compressed, can be considered as "one and the same substance, which differs only by the state in which it is found" [20].

In 1792 Thomas Wedgwood [21] reported to the Royal Society of London the results of his studies on the intensity and hue of light emitted from a large number of minerals, rocks, glass, ceramics, by "rubbing in the dark two pieces of the same kind against each other" or against a revolving wheel of fine grit. Overall, the approach used by Wedgwood was similar to Hauksbee's, but with a more systematic organization. Although most of his observations can be attributed to incandescence of surface regions heated due to friction, several experiments seem to point at nonthermal triboluminescence. It is remarkable that in most cases luminescence was not altered in different gas atmospheres or in liquids, for example luminescence of hard white sugar rubbed in oil. In parallel with observation of luminescence he also measured, using a fine down of a feather, the electrical charge remained on the solids after attrition.

As the nineteenth century dawned, it became apparent that luminescence phenomena were far more intricate than previously thought and could not be attributed to "the results of one and same action" [20]. In 1809 Dessaignes revived the idea of triboluminescence as an electrical phenomenon, proposing it as the fourth

---

[1] Torbern Olaf Bergman (1735 –1784) was Swedish chemist and mineralogist, who contributed to the study of lightning (D. Müller-Hillebrand, Torbern Bergman as a Lightning Scientist, Daedalus. Tekniska Museets Arsbok, Stokholm, 1963, pp. 35–76.).

plausible explanation for the observed light emission [18]. In 1820 Henrick critically reviewed the experimental findings and suggested the following explanation of triboluminescence: "The light in so far as it is not a burning or glowing is due 1) to the separation of particles caused by friction (to the cracking off of rough places on the crystals); 2) to the electricity resulting from friction; 3) to the chemical decomposition produced by this means" [16].

Throughout the nineteenth century, systematic investigations into triboluminescence were carried out by several scientists, including Homberg, Henrich, Gillet-Laumont, Sheele, Pelletier, Donadei, Burke, Pope, Arnold, and Brugnatelli. These studies involved subjecting various substances to diverse mechanical operations, such as friction, percussion, pressure, traction and fracture. Reminiscent of Bacon's era, the researcher's attention was directed to triboluminscence of minerals and sugar [22] gradually expanding to other crystalline substances such as ortho-benzoic sulfimide (saccharin), candies of different flavours, santonin compounds, and hippuric acid [23]. Although the findings were inconclusive to give a satisfactory explanation of the phenomena, they revealed that triboluminescence is influenced not only by the chemical composition of the material but also by its crystal conformation.

According to Pope [23], the name *triboluminescence* has been applied by E. Wiedemann in 1888 to "those phenomena of light which are not solely conditioned by the rise in temperature". The understanding of triboluminescence significantly advanced with the development of optical spectroscopy. Longchambon [24–26] observed that the triboluminescence emission spectra of sucrose, tartaric acid, cadmium sulfate, uranyl nitrate, and fluorite contained the spectral lines almost identical to the $^3\pi_u$–$^3\pi_g$ bands, or second positive group, of N_2 characteristic of lightning. These findings stressed the connection between the electric phenomena and triboluminescence.

Early experiments indicated that the phenomenon was most commonly observed in noncentrosymmetric materials and was absent in conductors. Longchmbon proposed that charge separation is a necessary first step to initiate a sequence of physical processes laying behind triboluminescence, which can be summarized as following:

(1) separation of charge during mechanical contact of dissimilar materials, strain and fracture;
(2) generation of electric field;
(3) dielectric breakdown of the surrounding medium (gas or liquid);
(4) excitation of gas molecules (e.g. N_2 in the air) in the discharge plasma;
(5) absorption by solids of UV and visible light emitted from the gas discharge;
(6) photoluminescence of solids.

The triboluminescence of wintergreen candy shown in Fig. 3 perfectly falls into the Longchambon's model. It consists of a group of bands in the UV–vis region matching with the N_2 lightning spectra and a broad band centred around 450 nm. The latter corresponds to photoluminescence of the flavour oil, whose mayor component is methyl salicylate. The spectrum of wintergreen oil is shown in Fig. 3 by a dashed green line. Photoluminescence is excited by absorbing UV light from gas discharge.

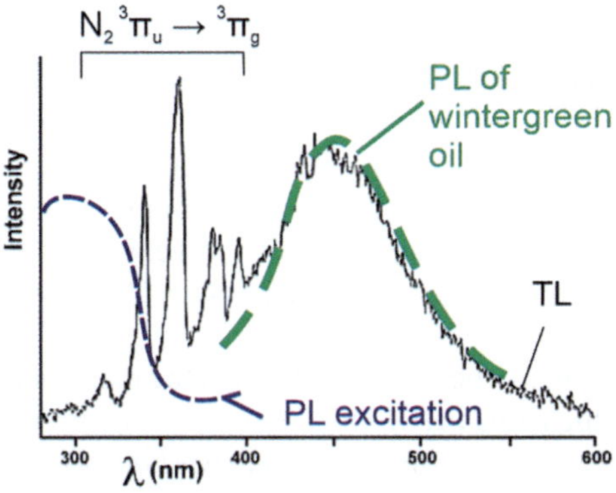

Fig. 3 Triboluminescence spectrum of crashed wintergreen candy. The photoluminescence spectra of wintergreen oil (methyl salicylate) and excitation radiation are also shown. Adapted from the on-line essay "Scientific experiments at home: wintergreen candy and other triboluminescent materials", by of Prof. L. Sweeting, Department of Chemistry, Towson University, Baltimore, MD 21,252, https://tigerweb.towson.edu/ladon/wg/candywww.htm

Later research found that triboluminescence can occur in centrosymmetric crystals, organic and polymer compounds and even in metals. Linda Sweeting [26] summarized the diverse experimental findings as following:

I. only triboluminescent materials that are not photoluminescent emit the N_2 spectrum—lightning with certainty;
II. most photoluminescent materials and minerals that are triboluminescent emit their photoluminescence when fractured.

The knowledge on the triboluminescence developed in the period from the seventeenth century to the third quarter of the twentieth century was reviewed from various perspectives by Walton [27]. Several comprehensive literature surveys on history of luminescence were published for the periods from the earliest times until 1900 [19] and from 1968 to 1998 [28].

In the 1970s it was discovered, that in addition to materials which fall within one of the two groups, many photoluminescent materials exhibit surprising properties such as triboluminescence from non-photoluminescent crystals. Zink [29] noted that this phenomenon is "more than ordinary gas discharge or crystal emissions which are excited in a new way. Instead, the triboluminescence spectra can contain new features which are excited by the mechanical stress but which are absent in the photoluminescence spectra". He also highlighted the possible role of impurities and traps in the triboluminescence excitation process, mainly trapping and detrapping of charge carriers under the influence of mechanical stimuli, and their contributions to the novel triboluminescence spectra. Further details can be found in the review article [30].

Recent studies have revealed novel mechanisms for triboluminescence excitation. Tsuboi et al. [31] demonstrated that laser-induced shock waves can trigger triboluminescence in both crystalline and amorphous sugar, with an intensity 10^5 time higher, than that obtained by conventional grinding. Triboluminescence can also be excited by breaking of adhesive bonds [32, 33]; shaking of liquids [14, 34]; rapid crystallization of certain substances [16]; cavitation [35]; elastic and plastic deformation of solids, scratching, milling, fracture [27, 36] and so on. Overall, a vast array of materials exhibit triboluminescence due to one or more of these mechanisms. Linda Sweeting [26] estimated that around 36% of inorganics, 19% of organics, 37% of aromatic compounds, 70% of alkaloids, and perhaps 50% of all crystalline materials are triboluminescent.

Triboluminescence also depends on the form of the mechanical action. As a consequence, in literature, the name *triboluminescence* is concurrently used as a synonym of *fractoluminescence*, *crystalloluminescence*, *sonoluminescence* and *piezoluminescence,* all of which are different forms of *mechanoluminescence* [37]. Thus, Wiedemann's definition aligns more closely with the contemporary term *fractoluminescence*, while the current meaning of *triboluminescence* is similar to Boyle's observations. This chapter is focused principally on those triboluminescence phenomena which are directly related with tribological interaction between solids.

Although triboluminescence has been studied for centuries, this phenomenon remains somewhat enigmatic at the conceptual and theoretical levels. Triboluminescence is intricately linked to both friction and wear, all of which require an understanding of highly nonequilibrium processes occurring at the molecular level [38]. These processes vary significantly depending on the tribological conditions, environment, and materials. Moreover, it is crucial to comprehend how these elements interact to produce synergistic coupling across different length and time scales.

Triboluminescence is a potential tool for the development of advanced materials, such as those capable of sensing and imaging stress intensity and distribution under applied load [39, 40]. Triboluminescence can also be employed for remote diagnosis of failures in space and vacuum mechanisms [41]. Non-destructive triboluminescent materials can be used for the development of intelligent damage sensors capable of real-time detection of deformation magnitude and location within the material. In the medical field, triboluminescence and triboemission from biological tissues have been proven effective for early cancer and other disease diagnosis [42, 43].

2 Basic Processes and Activation Mechanisms

Generally, all triboluminescence (TL) phenomena can be classified into two broad categories according to the underlying mechanisms: thermal and non-thermal luminescence. The first group encompasses the incandescence of hot surfaces, arising from the direct conversion of mechanical energy into heat or from exothermic chemical reactions triggered by friction and wear, such as chip oxidation, flint sparks, and so on. The second group relies on luminescence driven by secondary processes in the cascade of mechanical energy dissipation, including triboelectrification, gas discharge, excitation of vibrational and electronic states in solids, charge

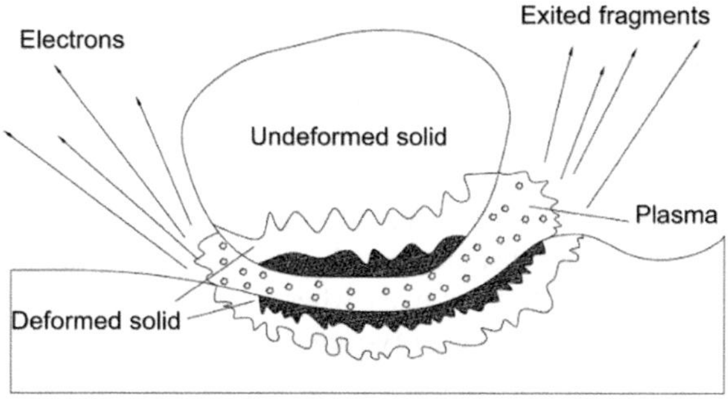

Fig. 4 Magma-plasma concept proposed by Heinicke. Reproduced from [48] Copyright © 2019, The Author(s), under CC BY 4.0 licence

trapping and detrapping, dislocation motion, photoluminescence, ionoluminescence, bremsstrahlung, and many others.

While a substantial body of experimental evidence gathered since the seventeenth century indicated that light emission could occur under conditions where pure thermal radiation was unlikely [16], until the mid-twentieth century the prevailing hypothesis was that even non-thermal triboluminescence is triggered thermally. The culmination of the hypothesis was the "magma-plasma" concept proposed by Heinicke [44] and illustrated in Fig. 4. The core idea behind this concept is that heat is generated in sub-micrometer zones of actual contact, known as Bowden's hot spots, under nearly adiabatic conditions. Although the total generated heat is relatively low, its density is very high, leading to an instantaneous and significant temperature rise, followed by melting, thermionic electron emission, triboplasma, and triboluminescence [45]. Bowden et al. [46] reported that local transient temperatures of the order of 1000 °C were attained at moderate loads and speeds. While these hypotheses have been substantiated for high-speed and relatively heavily loaded contacts, there is still considerable debate regarding the extent of temperature increase in lightly loaded and microscopic tribocontacts, as well as under mild sliding conditions [47].

The emergence of molecular-level simulation methods provided a novel and powerful tool to elucidate how energy is dissipated at hidden mechanical interface. Molecular dynamics simulations demonstrated that the temperature increase on hot spots did not exceed 300 °C even under severe sliding conditions involving contact pressures ranging from 7.6 to 56 GPa and sliding velocity 90.2 ms^{-1} [49, 50]. The temperature flashes are extremely short-time, within picoseconds, and the heated zones are submicrometers in size. Consequently, temperature flashes do not set off thermally activated processes with time constants that are larger than the lifetime of the flashes. For example, modelling gas desorption from amorphous carbon coatings under transient temperature cycles simulating temperature flashes at hot spots yielded a desorption rate that was three to four orders of magnitude lower than the experimentally measured value [51, 52]. The simulation findings were validated through direct measurement of interfacial temperature [53, 54]. Remarkably, the temperature increase even in luminescent triboplasma produced due to triboinduced gas discharge, as measured by Nakayama et al. [55], remained below 1 K. These findings suggest that thermal excitation at hot spots cannot be the sole mechanism behind triboluminescence. In certain cases, the average mechanical energy imparted per atom of a material at hot spots is insufficient to directly excite an atom into emitting visible or UV light [16, 27]. This raises the question about alternative mechanisms of TL excitation.

The alternative concept to the 'magma-plasma' model assumes that mechanical energy is not instantaneously and fully converted into Joule heat. Instead, various relaxation pathways exist where energy is sequentially transformed into different forms [56]. Initial energy-rich excited states arise from stress, strain, bond breaking and stress-induced chemical transformations [36, 57–61]. The relaxation of these excited states triggers a sequence of interconnected chemical reactions, physical processes, and emission events. These include luminescence, emission of electrons and ions, chemical transformations, generation of electric field, and more. Collisions between emitted particles and photons with atoms and molecules can trigger secondary excitation and decay processes, and so on [62]. Energy dissipation at the sliding contact is an intricate process that involves a vast array of elementary physical processes and chemical reactions. These mechanisms encompass incandescent radiation [61, 63], molecular deformation and direct mechanical excitation of electronic levels in mechanically affected zones [36, 61], space charge generation due to cleavage and contact electrification [32, 61, 64, 65], piezoelectric effect [39], recombination of charged particles, holes and radicals [36, 39], electric discharge through surrounding gas or liquid [16, 27, 64], electroluminescence [27, 66], photoluminescence [67], cathode- and ionoluminescence [68], resonance radiation from excited particles [27], phase transformations [61], gas adsorption and chemical reactions [61, 66], unpinning and annihilation of dislocations, defect stripping, motion of charged dislocations [64], along with other mechanisms.

A series of experiments combined with theoretical modelling has significantly advanced our understanding of seemingly disparate phenomena related to the mechanical activation of physical and chemical reactions. These efforts have provided a crucial framework for understanding the mechanisms and behaviour of triboluminescence under principal mechanical actions such as fracture, grinding, and cohesive detachment [16, 26, 27, 36, 39, 61, 65]. It would be remiss if we did not acknowledge the foundational work of Varentsov and Khrustalev [69] who summarized the underlying processes in a comprehensive framework that encompasses the emission phenomena and chemical reactions observed when solids are subjected to external mechanical forces, primarily in the form of deformation or fracture. The schematic in Fig. 5 illustrates the mechanisms and their relationships. The yellow-highlighted boxes represent various luminescence phenomena. This scheme was further refined by incorporating additional non-intuitive processes involved in frictional contact-induced mechanical energy dissipation (blue boxes and lines). One such process is the excitation of metastable states through mere contact without deformation or fracture, such as through contact electrification. These new processes are specifically associated with frictional mechanical interactions and have not been previously considered within the framework of mechanochemistry and mechanoemission.

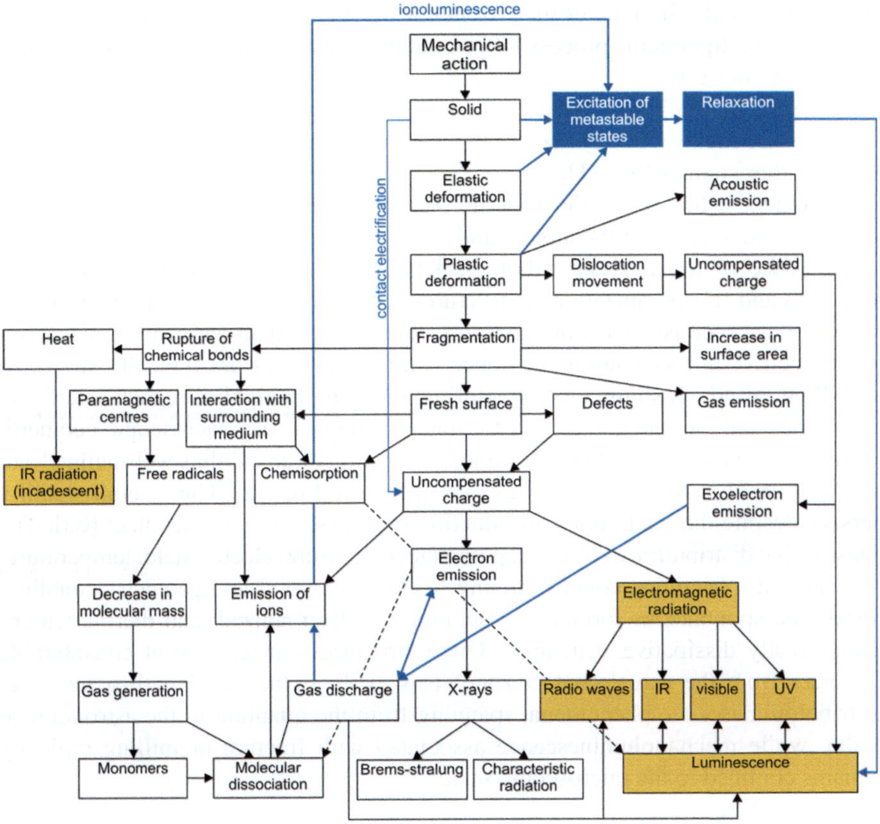

Fig. 5 Schematic illustration of physical processes and chemical reactions involved in mechanical energy dissipation at frictional contact, along with associated triboemission and triboluminescence phenomena. The boxes and lines represent the generalized scheme of emission phenomena and chemical reactions occurring when solids are subjected to external mechanical forces Yellow boxes highlight luminescence phenomena, while blue boxes and lines indicate processes specifically related to frictional mechanical interaction that have been recently investigated. Adapted from [69] © 1995, with permission from the Russian Academy of Sciences and Turpion Ltd.

A variety of the processes shown in Fig. 5 can occur simultaneously. In the case of TL originated from gas discharge (group I in Sweeting's classification), three processes have been observed: excitation of the surrounding gas by electron bombardment, excitation of the material's photoluminescence by electron bombardment, and excitation of the material's photoluminescence by absorption of ultraviolet (UV) emission from the excited gas. These processes can also serve as triggering mechanisms for group II triboluminescence. A detailed discussion of these processes will be provided in Sect. 4. Regarding the second group in Sweeting's classification, researchers have focused their attention on rare-earth complexes that exhibit triboluminescence under deformation without fracture or electric discharge. In these materials, strain energy excites filled electron traps, causing them to release holes into the valence band. These holes then excite rare-earth ions and return to their ground state, emitting light in the process [70]. A detailed discussion of these processes will be provided in Sect. 6.

Materials that fall outside the two main groups, such as doped zinc sulfide, are not discussed in this work. For comprehensive information on these materials, please refer to other sources [30, 71].

It should also be stressed that triboluminescence and mechanoluminescence share a common foundation at the atomic and nanoscale (Fig. 6b). However, as a natural consequence of the intricate tribological behaviour characterized by synergistic processes and self-organization of structures across multiple time and length scales, triboluminescence is a far more complex phenomenon than elementary mechanoluminescence. In mechanoluminescence studies, materials are typically subjected to milling, grinding, straining, or fracturing, resulting in highly irregular interfacial geometries and random contact interactions. In contrast, the macroscopic geometry of tribological contacts, in both machines and nature, often exhibits well-defined and stable or quasi-stable characteristics over time (see Fig. 6a). This overall stability persists despite the continuous introduction of new surfaces into contact [62]. The quasi-stable distribution of physical parameters (pressure, electric field, temperature, etc.) around a frictional contact, arising from its temporal and geometric stability, fosters the spontaneous formation of complex, self-organized, and thermodynamically steady dissipative structures. These structures can emerge at considerable distances from the actual contact zone. This explains the vast length-scale range of triboluminescence phenomena, spanning from the quantum to the astronomical realm, while mechanoluminescence associated with fracture or milling typically remains confined to the microscopic scale.

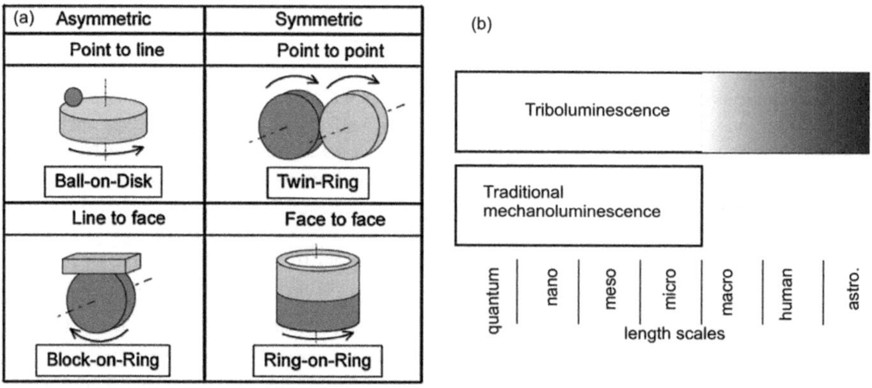

Fig. 6 **a** Classifications of contact configurations of four types of sliding-motion tribometers (reprinted from [72] © 2011, with permission from Elsevier). **b** length scales of traditional mechanoluminescence and triboluminescence processes

3 Experimental Techniques Employed in Studying Triboluminescence

Triboluminescence arises as a consequence of a system's response to the introduction of mechanical energy at a frictional contact. Therefore, an experimental setup for measuring TL must necessarily comprise two main subsystems: (i) a tribological device capable of generating controlled relative motion between two solids under defined velocity and loading conditions, and (ii) an optical system designed to measure the intensity and spectral characteristics of the emitted light across various spectral ranges and time scales.

While several experimental techniques have been developed to measure triboluminescence, only a select few can be considered true tribological systems, where two surfaces slide and/or roll against each other [64, 70, 73, 74]. The majority of other devices are primarily designed for grinding, milling, fracturing, or straining the materials [36, 75, 76]. This distinction stems from the misleading use of the term 'triboluminescence' instead of the more appropriate terms 'deformational luminescence' or 'fractoluminescence' for these latter techniques (Fig. 7).

Simultaneously monitoring both the friction force, which reflects the rate of mechanical energy dissipation, and TL is crucial for identifying the underlying mechanisms that govern these processes. However, this task is not straightforward because in laboratory experiments, TL occurs directly within the buried interface or in close proximity to the contact zone, typically less than 1 mm away. If both solids in frictional contact are opaque in the spectral range of interest for TL studies, emitted photons can only be collected from the sides of the contact zone. If either or both solids are transparent in this spectral range, triboluminescence (TL) can be recorded either from the side or through the transparent solid.

Since the optical system must be focused on a minute luminescent zone embedded within the interface, tribometer configurations with a non-stationary contact zone are not suitable for TL studies. Consequently, the pin-on-disk configuration, with a stationary pin and rotating disk, has been the primary choice for most TL investigations. Figure 4a presents a schematic illustration of a typical experimental setup with a pin-on-disk configuration designed for TL measurement through a transparent disk. Light emitted from the luminescent zone at the frictional contact traverses the quartz disk (Fig. 4b) and a sapphire viewport before being focused by the lenses of an optical microscope. The emitted light can be directed to a high-sensitivity camera for imaging or to a spectrometer for spectral analysis. Quartz and sapphire were selected due to their high transmittance in the ultraviolet (UV) and visible (vis) spectral ranges. The disk and pin are enclosed within a vacuum chamber, enabling precise control over both gas pressure and composition. Introducing specific optical filters into the optical path of the microscope enables the study of TL across different spectral ranges.

Figure 4c illustrates another example of a TL pin-on-disk system where emitted photons are measured from the side. In this setup, a lens, simultaneously serving as

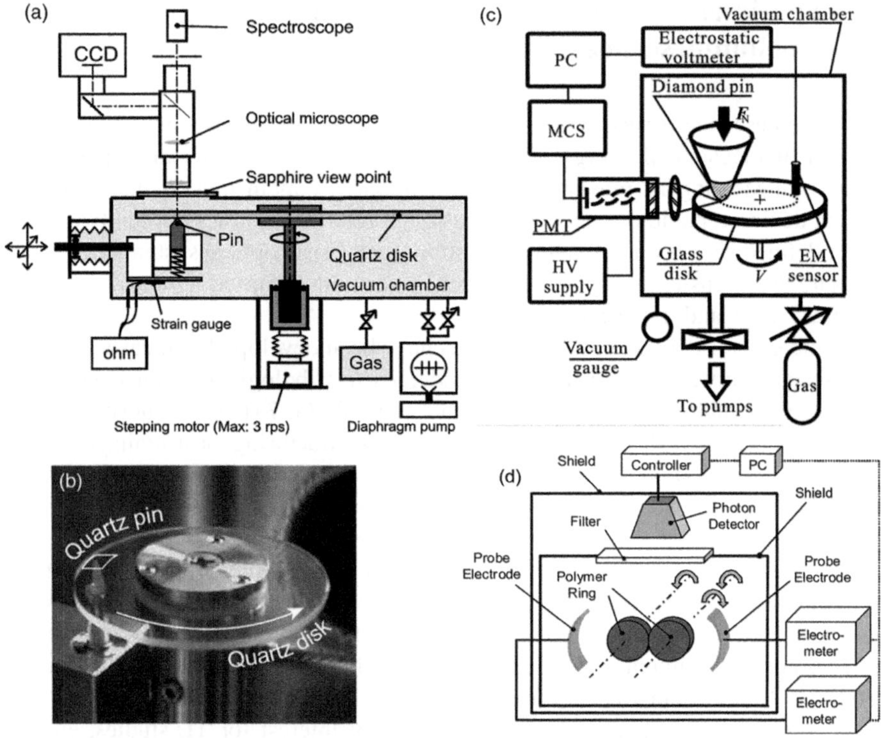

Fig. 7 **a** Schematic drawing of a pin-on-disk set-up in which spatial distributions and spectroscopic characteristics of the TL can be studied. Light passing through the quartz disk is measured from the opposite side of the contact by an optical microscope combined with a charged couple device (CCD) camera and a spectrometer. **b** photograph of the quartz pin and quartz disk. **c** schematic drawing of the experimental vacuum tribometer with a pin-on-disk configuration in which the TL is measured from the side using a photomultiplier tube (PMT). Electric impulses at the anode of a PMT were acquired by a multichannel scaler (MCS). Surface potential was measured by an electrostatic voltmeter using an electromechanical (EM) sensor. **d** the twin-ring tribometer with a chemiluminescence analyzer. Photons passing through a filter are detected by a photomultiplier. Charges induced on the surfaces by rubbing are measured by an electrometer through an air probe electrode which were set at the farthest end of the specimens without contacting them. Figures **a** and **b** reprinted from [73] © 2006, with permission from Elsevier; figure **c** reprinted from [77] © 2003, with permission from American Institute of Physics; figure **d** reprinted from [78] © 2012, with permission from Elsevier

a viewport for the vacuum chamber, condenses the light and guides it to a photomultiplier tube (PMT). The output signal from the PMT is acquired and processed by a multichannel scaler (MCS). The PMT can operate in either current mode or pulse counting mode. In pulse counting mode, the PMT can achieve nanosecond resolution for the TL signal.

The pin-on-disk configuration presents inherent limitations due to the asymmetric sliding conditions between the pin and the disk, leading to differences in wear rates, a significant disparity in the contact zone areas, and varying adsorption conditions [78]. The twin-ring tribometer (Fig. 4d) facilitates pure rolling or rolling/sliding with controlled slip ratio in addition to sliding motion and overcomes the limitations of the pin-on-disk configuration [72]. The symmetrical configuration enables simultaneous measurement of surface charge on both surfaces along with triboluminescence (TL).

4 Characteristics of the Gas-Discharge TL

4.1 Spatial Distribution of TL at a Tribological Contact

K. Nakayama and his research group were likely the first to successfully measure the two-dimensional spatial distribution of triboluminescence (TL) at a sliding contact. They employed a pin-on-disk tribometer operated either in ambient air [64, 79, 80] or in a vacuum chamber [80–82] coupled with an intensified CCD camera equipped with CaF_2 lenses that were transparent across the UV, visible, and IR ranges.

Figure 5a presents the unfiltered pseudo-colour image of the in-plane spatial intensity distribution of triboluminescence (TL) acquired through the disk. The images represent integrated TL intensity over approximately 12 disk revolutions. Triboluminescence emission completely encircled the contact zone, extending up to 200 μm from the contact centre in the direction of relative disk motion, indicated by the arrow. The TL intensity reached its maximum within a zone located approximately 50 μm behind the contact centre. Figure 5b illustrates the intensity distribution of TL emission within the UV range. The overall shapes of the intensity distributions for total TL and UV TL exhibit significant similarity, despite subtle variations in shape attributable to temporal fluctuations in TL intensity [64] as the images in Fig. 5 were acquired sequentially. The total luminescent power was remarkably similar in both cases, suggesting that the TL primarily originated from the excitation of nitrogen molecules in the surrounding air. Spectroscopic measurements generally supported this hypothesis (see Sect. 4.2).

The TL intensity in the IR range was low (Fig. 8c) and exhibited an unusual distribution. Figure 5d presents an enhanced image of the IR TL from Fig. 5c. In addition to the IR radiation at the contact area, which can be attributed to frictional heating, there were two lateral branches extending from the contact zone in the direction of disk rotation, aside from the friction track. The IR TL intensity distribution exhibited an open circular or horseshoe-shaped pattern. Surprisingly, the IR luminescence on the branch zones was significantly more intense than on the friction track behind the contact, despite the fact that these branch zones had not been in direct contact with the pin. These observations suggest that gas discharge at the friction contact plays a significant role in energy dissipation and serves as a source of surface and gas heating within and beyond the friction track.

Miura et al. [79] demonstrated the occurrence of TL in front of the sliding contact. This luminescent zone, designated γ in Fig. 9, is located at a distance from the contact

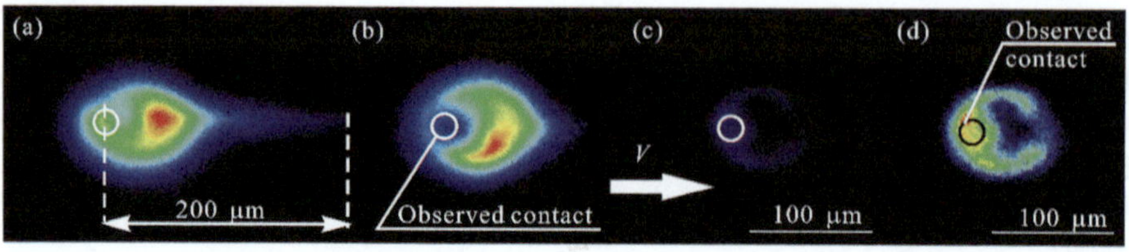

Fig. 8 Two-dimensional spatial distribution of the TL at the sliding contact of a diamond pin (radius of curvature 300 μm) and a sapphire disk in ambient air. **a** unfiltered pseudo-colour image; **b** band-filtered pseudo-colour image of the TL in the range 290–420 nm; **c** band-filtered pseudo-colour image of the TL in the range 720–2800 nm; **d** enhanced image from figure **c**. A circle in each figure denotes the size and position of the nominal contact area determined visually using optical microscopy (reprinted from [80] © 2003, with permission from ASME)

centre nearly twice as large as the luminescent zone β behind the contact. Nitrogen excitation in electrical gas discharge was identified as the primary mechanism for TL in zones β and γ, as evidenced by the predominant emission of light in the UV range. Various authors have reported diverse patterns of triboluminescence intensity distribution, including four distinct emission points linked together and encircling a dark zone immediately behind the contact to form a ring, elongated tails, and irregular strips extending for over a hundred micrometres beyond the contact along the friction track [81, 83, 84]. The characteristics of some of these TL features and the underlying mechanisms responsible for their occurrence will be discussed in the following sections.

The dimensions and shape of the luminescent zone were strongly influenced by the ambient gas pressure. Figure 10 presents a series of pseudo-color images of triboluminescence captured under varying pressures of dry air [81]. Upon decreasing the gas pressure from atmospheric level to 600 Pa, the size of the luminescent zone expanded by more than six-fold. It is noteworthy that when the gas pressure fell below 10 kPa, the zone of maximum triboluminescence intensity shifted from the position behind the contact (β) to lateral positions alongside the contact zone, designated as

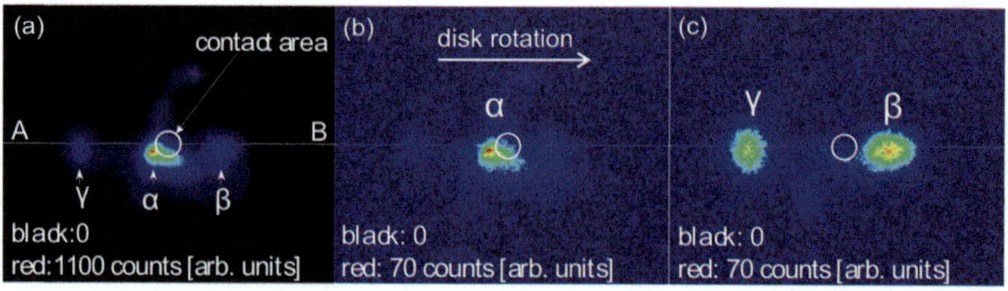

Fig. 9 **a** Unfiltered pseudo-colour image of the TL during sliding in air; **b** band-filtered pseudo-colour image of the TL in the range 610–780 nm; **c** band-filtered pseudo-colour image of the TL in the range 300–420 nm. The size of all images is 250 × 300 μm^2. Red colour corresponds to the maximum intensity indicated in each figure. A circle shows the position and the size of the contact area (reprinted from [79] © 2001, with permission from American Institute of Physics)

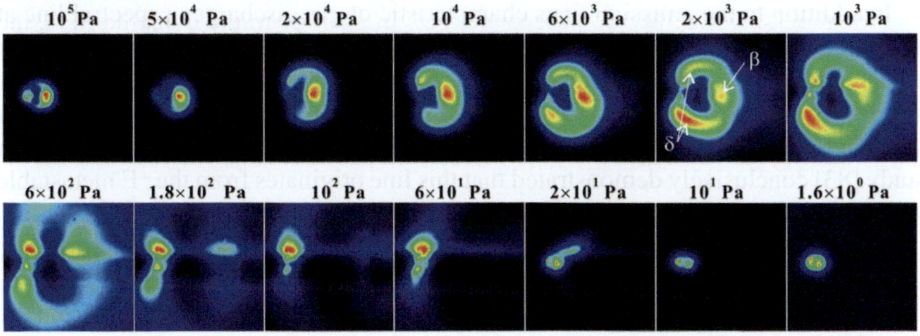

Fig. 10 Unfiltered pseudo-colour images of the TL obtained during sliding of a diamond pin on a sapphire disk under various pressures of dry air (reprinted from [81] © 2004, with permission from Elsevier)

δ from now on. When the gas pressure continued to decrease below 600 Pa, the TL zone quickly shrank. The luminescent zones β, γ, and δ vanished below 20 Pa, while faint light emission could only be observed at the contact zone (zone α). Although the expansion of the luminescent zone with decreasing gas pressure followed by its collapse resembles the typical behaviour of gas discharge, such as glowing discharge in a tube [85] or corona [86], the behaviour of TL is significantly more complex due to the intricate geometry and configuration of the electric field.

4.2 *Effect of the Ambient Gas and the Material of the Counterbodies on Spectral Characteristics and Intensity Distribution of TL*

Brady and Rowell [87] observed that the triboluminescence spectrum measured during attrition or fracturing of rocks was unique to the ambient atmosphere. Spectral lines of N_2 and Ar were found in the TL spectra of alkali halides when they were subjected to mechanical action in the corresponding gas atmospheres [88]. Nakayama and Nevshupa [80] investigated the TL spectra for the diamond pin—sapphire disk configuration under various gas atmospheres and compared the obtained TL spectra to the glowing discharge spectra of the corresponding gases. They observed that the TL emission spectra in air, N_2, O_2, He, and CH_4 closely resembled the spectra of the glowing discharge, although the relative intensities of the spectral peaks exhibited slight variations (compare, for example, Figs. 8a, b and 9a, d). In H_2 atmosphere, the glowing discharge spectrum exhibited a prominent line at $H_\alpha = 656.279$ nm, which was notably absent from the TL emission spectrum. Neither the TL nor the glowing discharge spectra presented any discernible features in ethane and propane atmospheres.

In addition to the emission lines characteristic of gas discharge, a spectral line at approximately 696 nm was observed in all TL spectra except for those obtained in ethane and propane atmospheres. This line was absent from the glowing discharge spectra of the corresponding gases (Figs. 11 and 12a). The 696 nm line was only observed when a sapphire disk was used in the experiment (Fig. 12a–c). A thorough study [83] conclusively demonstrated that this line originates from the ^{2}E metastable state luminescence of chromium impurities in sapphire (694.5 nm). This finding unequivocally demonstrated the secondary luminescence of sapphire. Moreover, the irregular stripes observed alongside the friction track in the TL spatial distribution images were attributed to photoluminescence. These stripes originated from the lateral branches of the TL and extended over 100 μm beyond the contact zone [81, 83]. Whereas the intensity of the gas discharge luminescence stabilized within 2 to 3 min, the intensity of the induced luminescence from sapphire gradually increased over time and reached a steady-state value after approximately 1100 disk revolutions, corresponding to a duration of 7 min [80]. This can be associated with the fact that secondary luminescence requires longer excitation time to set off. Notably, the 696 nm emission was observed in the methane atmosphere but was absent in ethane and propane atmospheres. The spatial mismatch between the photoluminescent zones and the zone of maximum discharge luminescence is surprising. This suggests that the secondary luminescence may not fall under the second Sweeting's group of TL mechanisms and could be triggered by electron [26, 83] or ion [68, 89] bombardment.

The propensity for gas discharge increases with decreasing pressure, but this trend reverses when the density becomes too low, leading to an increased mean free path for electrons that hinders the discharge's sustenance. The suppression of gas discharge

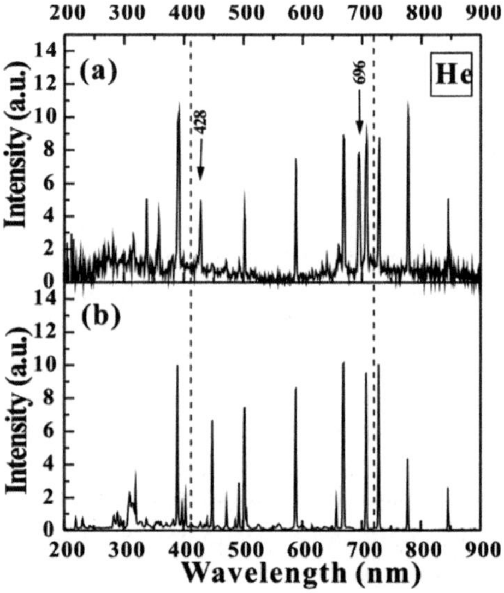

Fig. 11 **a** The TL spectrum for diamond-sapphire sliding in He atmosphere. **b** The luminescence emission spectrum of glowing discharge in He (reprinted from [80] © 2003, with permission from ASME)

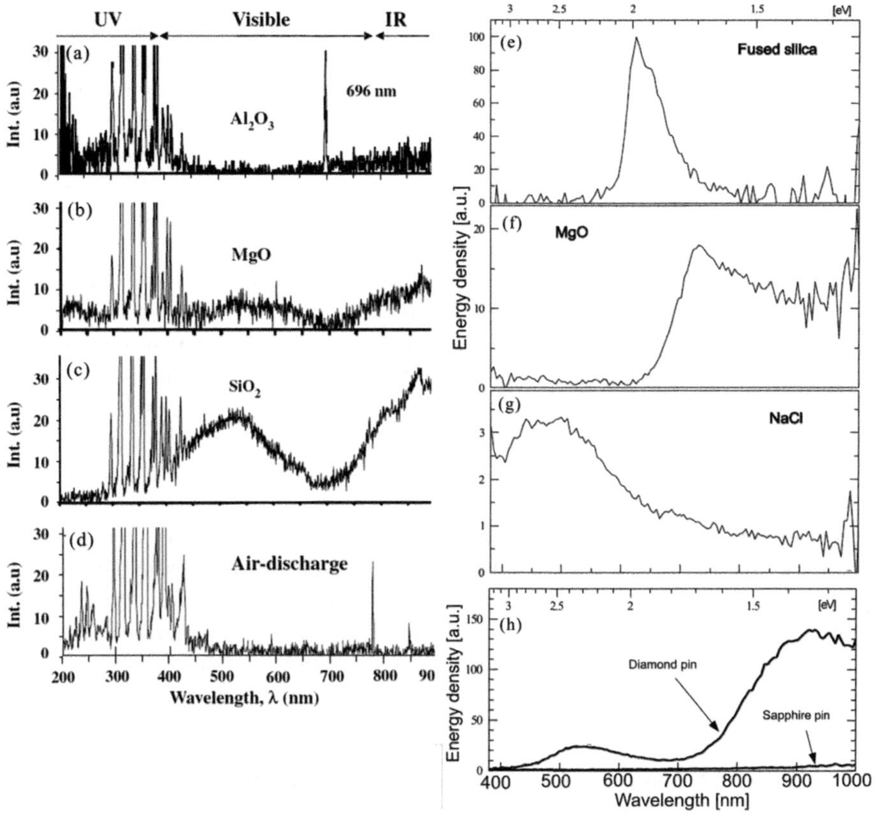

Fig. 12 Effect of material of the disk (**a**–**c**) and the pin (**e**–**h**) on the spectral characteristics of the TL in air. **a** a natural diamond pin and a synthetic sapphire disk; **b** a natural diamond pin and a MgO disk; **c** a natural diamond pin and a crystal SiO_2 disk; **d** spectrum of the luminescence of glowing discharge in air; **e** a synthetic fused silica pin and a synthetic sapphire disk; **f** a MgO single crystal pin and a synthetic sapphire disk; **g** a NaCl pin and a synthetic sapphire disk; **h** a natural diamond pin and a synthetic sapphire disk. **a**–**c** measured in ambient air; **e**–**h** measured in vacuum (10^{-3} Pa). Figures **a**–**d** reprinted from [83] © 2010, with permission from Springer Science and Business Media. Figures **e**–**h** reprinted from [89] © 2009, with permission from Elsevier

under high vacuum conditions, with light emission primarily emanating from the contact zone, satisfactorily explains the experimental findings (Fig. 10).

Miura et al. [89] investigated the spectral characteristics of light emitted from the contact zone in a high vacuum environment for various pin materials. In the diamond pin and quartz disk configuration, they observed three broad bands at wavelengths of 400, 630, and 900 nm. These bands deviated significantly from the theoretical Planck's distribution and remained unaffected by the sliding conditions. These observations led the authors to conclude that thermal radiation was not the primary mechanism responsible for the luminescence excitation. The broad band at 900 nm (Fig. 12h) was attributed to fractoluminescence of diamond. Additionally, a faint band between 350 and 650 nm was observed during diamond abrasion (Fig. 12h). For the ionic crystals investigated in [89] the TL emission spectra closely resembled the corresponding fractoluminescence spectra, suggesting a connection to the electronic excitation of vacancies or impurities during abrasion. For quartz, the band at 630 nm (Fig. 12e) was attributed to fractoluminescence of non-bridging oxygen hole centres. The broad band in the TL emission spectrum of MgO (Fig. 12f) was ascribed to the emission of Mn and Cr^{3+} impurities and resembled the thermoluminescence and photoluminescence spectra of MgO.

Tukhbatullin et al. [90, 91] explored the triboluminescence (TL) of lanthanide, cerium, and terbium salts in hydrocarbon gas atmospheres, including natural gas, methane, ethylene, and acetylene. In methane atmosphere they observed the maximum emission for the CH radical, attributing this to the electronic transitions $C^2\Sigma^+ \rightarrow X^2\Pi$ (314 nm), $B^2\Sigma^- \rightarrow X^2\Pi$ (390 nm), and $A^2\Delta–X^2\Pi$ (431 nm) which align with the methane discharge spectrum. In other hydrocarbon gases, they observed the spectral bands known as Swan bands, which correspond to the $^*C_2$ glow, an excited product of the mechanochemical decomposition of hydrocarbon molecules. The mission from $^*C_2$ dominated over that of $^*CH$ in the TL spectra acquired for lanthanide salts in heavier hydrocarbon gases. This finding agree previous observations indicating a positive correlation between the intensity of the CH emission band and the number of CH_4 molecules generated during the decomposition of higher hydrocarbon gases in an electric gas discharge [92].

4.3 Effect of Friction Type and Humidity on the TL and Triboelectrification of Polymers

Despite extensive research on triboluminescence in ceramics, the field of polymers has received relatively limited attention. Nowadays, polymers are used for numerous machine parts, while the TL phenomena can have undesired effect on their operation life, safety and reliability. Due to the propensity of polymers to accumulate charge during friction, triboluminescence is likely to occur in any situation involving polymers in rubbing contact. The gas discharge associated with polymer triboluminescence releases charged particles and photons of exceptionally high specific energy.

Figure 13 presents time-averaged plasma luminescence intensity images acquired during friction tests on polytetrafluoroethylene (PTFE) and ultrahigh molecular weight polyethylene (UHMWPE) pins against a monocrystalline alumina disk in air [93]. These TL patterns show a general agreement with those observed for single crystal oxides and glasses rubbed with diamond (refer to Figs. 8 and 9). It is essential to emphasize that TL images do not represent instantaneous snapshots of TL emission but rather depict the probability of TL occurrence within a two-dimensional space. The luminescence zone for PTFE exhibits an almost circular shape with a relatively uniform luminosity distribution. In contrast, the TL region for UHMWPE

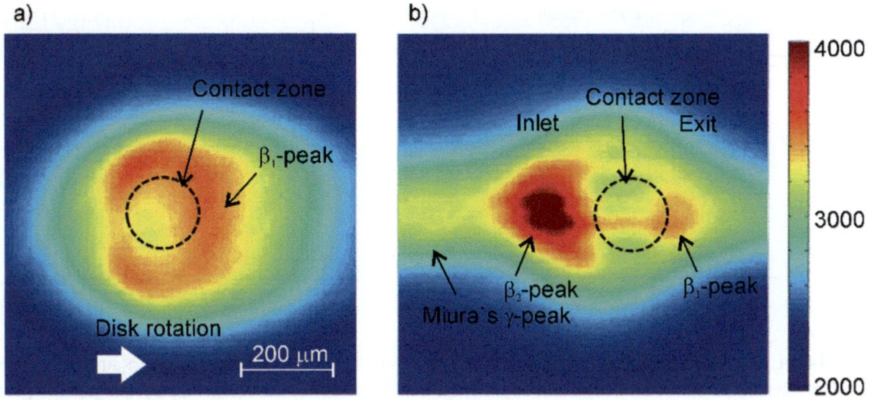

Fig. 13 Time-averaged TL intensities, from **a** PTFE contact, **b** UHMWPE contact. Reproduced from [93] under CC BY 4.0 license

is elongated. Moreover, an additional peak is observed in front of the contact inlet. This feature resembles Miura's γ-peak at the inlet of the contact zone where re-entering charge and gap geometry create a favourable environment for discharge [79]. However, in this case, the peak lies closer to the contact zone. Additionally, both polymers exhibit faint luminescence at the contact areas resembling Miura's α-peak. These discrepancies in comparison to Miura's observations were attributed to the accumulation of adhered wear debris at the outlet of the contact zone. The electric field of this debris could counteract the external electric field generated by the concentrated charge at the pin, thereby hindering the access of electrons to the contact zone, a crucial requirement for sustaining the avalanche process.

Given that triboelectrification (TE) is a fundamental mechanism underlying triboluminescence, surface potential on the disk was measured at 180° position with respect to the stationary pin (refer to Fig. 14a) [93]. At the beginning of sliding, the charge density increases linearly. The initial rates of charge accumulation are 49.6 and 11.0 $nCm^{-1}\ s^{-1}$ for PTFE and UHMWPE, respectively. A deviation from the linear behaviour is observed at approximately 1.5 s (PTFE) and 6.5 s (UHMWPE) which coincides with the discharge ignition. Then, the surface charge density varies slowly with time and tends to stabilize. The linear charge densities at the moment of discharge ignition are 55.6 and 34.0 nCm^{-1} for PTFE and UHMWPE, respectively. This difference can be related to much higher electronegativity (4.0) and electron affinity (328 kJ/mol) of fluorine in comparison with carbon (2.5 and 121.7 kJ/mol, correspondingly) and hydrogen (2.2 and 72.8 kJ/mol, correspondingly). Stabilization of the surface charge density indicates that the rates of charge generation and recombination are close to equilibrium. This suggests that a significant part of the generated charge is dissipated through the gas discharge. If the dynamic equilibrium between charging and gas discharging is achieved, the plasma luminescence signal would remain constant throughout the test, whereas in actuality, it reduces rapidly

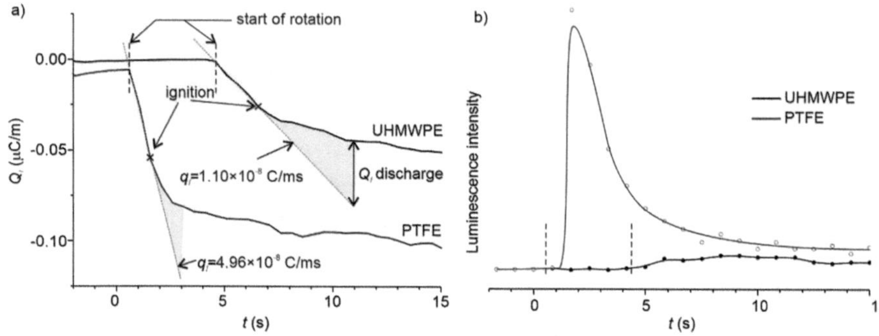

Fig. 14 **a** Variation of surface charge density as a function of time. Straight lines are linear fits. Vertical dashed lines mark the rotation beginning. **b** the maximum luminescence intensity as function of time. Reproduced from [93] under CC BY 4.0 license

(refer to Fig. 14b). Therefore, in the real situation, some other processes through which surface charge is dissipated must be occurring in parallel.

These findings underscore the complexities involved in accurately quantifying the charge generated during triboelectrification (TE). The measured charge on a surface represents the net charge remaining after recombination of charges due to discharge between surfaces and charge leakage through the bulk or on the surface. Despite these challenges, charge measurement becomes feasible under conditions that suppress gas discharge and utilize materials with high electrical resistance. The twin-ring tribometer (Figs. 6a and 7d) is particularly well-suited for this purpose, as it facilitates sequential measurements of both triboelectrification and triboluminescence. Additionally, this test rig has the advantage of realizing the rolling and sliding contacts by just reversing the rotation direction of one specimen.

Figure 15 compares the charge build-up and corresponding photon emission for polyamide (PA66) and PTFE under sliding and rolling contact conditions. Sliding or rolling lasted between 25 and 75 s, as indicated in the data records. The TL intensity observed during sliding was two orders of magnitude higher than that during rolling, despite the comparable saturation charges in both experiments. Notably, gas discharge was absent during the initial revolution of the cylinders in the sliding test, allowing the surface charge to reach its maximum value. Under rolling friction conditions, surface charges exhibited a gradual increase over time. This observation clearly indicates that the rate of charge accumulation during sliding is significantly higher than during rolling. This difference can be attributed to the repeated contact of identical surface areas during sliding. The observed decline in TL intensity during sliding (Fig. 15a) can be attributed to the progressive transfer of PTFE onto PA66, effectively transforming the rubbing pair from PA66/PTFE to PTFE/PTFE. Consequently, the charge generation per contact/separation event between similar materials decreases, leading to the corresponding reduction in TL intensity.

The rapid onset of TL during the sliding contact of polymers suggests that the luminescence is not solely attributable to frictional heating. Instead, photon emission

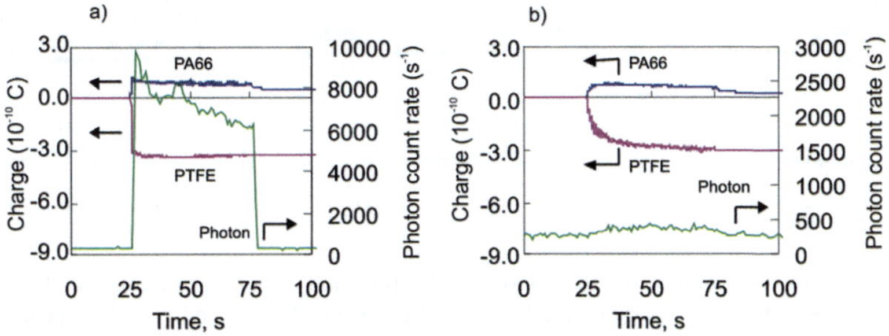

Fig. 15 TE and TL processes during sliding (**a**) and rolling (**b**) between PA66 and PTFE. Reprinted from [78] © 2012, with permission from Elsevier

is likely triggered by electron emission, which could arise from two mechanisms: fracto-emission and field emission. Unlike rolling friction, sliding friction generates continuous fracturing of the material at the contact zone. This fracturing process facilitates electron emission from the newly created surfaces. Frictional electrification further contributes to the accumulation of a strong electric field, which can induce field electron emission.

Environmental humidity is a critical factor influencing TE [94], which in turn affects TL intensity. It is noteworthy that humidity can have dual effects on TE: it can both accelerate surface charge dissipation through conduction through adsorbed water and simultaneously promote more intensive charge generation due to the transfer of surface ions from one contact surface to the other. This interplay between opposing mechanisms highlights the complex relationship between humidity and TE. Figure 16 illustrates this relationship by depicting the effect of humidity on TE and TL during the rubbing of acrylonitrile butadiene styrene (ABS) and PTFE. The increase in relative humidity (RH) from 0 to 58% is accompanied by a corresponding increase in TL intensity. This observation is linked to the enhanced surface charge, which is particularly pronounced between 10 and 58% RH. However, it is noteworthy that the saturation charge on both polymers at 10% RH is higher than that at both 0% and 58% RH. The analysis of the relationship between the charging rate/saturation charge and photon amount has revealed that the TL rate is primarily determined by the charge generated per frictional contact rather than by the overall surface charge. Interestingly, TL was almost undetectable at 0% RH, even though the leakage rate is expected to be the lowest under these conditions. This observation suggests that the limited number of charge carriers at extremely low humidity hinders the subsequent discharge between surfaces, effectively suppressing TL generation. This implies that the charge carriers are ions H_3O^+ and/or OH^- generated from the adsorbed water [94]. However, as relative humidity increases beyond a certain threshold, surface charge diminishes due to discharge and recombination processes.

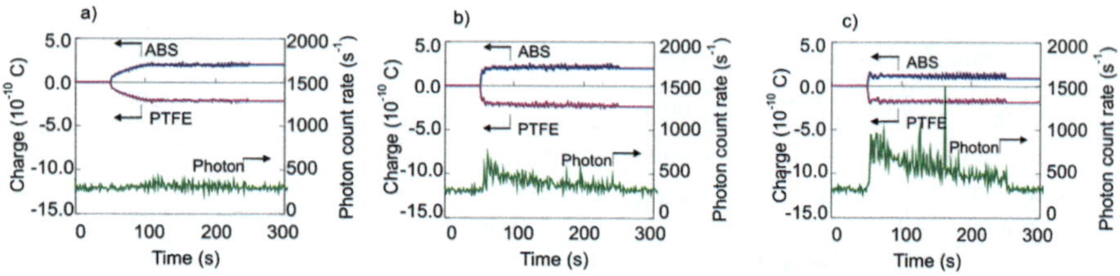

Fig. 16 Effects of humidity on the triboelectrification and triboluminescence in the rubbing between ABS and PTFE at various relative humidity: **a** 0%, **b** 10% and **c** 58% (reprinted from [78] © 2012, with permission from Elsevier)

4.4 Behaviour of the TL on Different Time Scales

Investigations have revealed that gas pressure is a crucial external parameter that influences not only the intensity and spatial distribution of associated with gas discharge, but also its temporal behaviour across various time scales [68, 77]. The experiments involving a TL diamond pin and sapphire disk configuration revealed four distinct patterns of TL time series as a function of argon gas pressure (Fig. 17a) [68]. At pressures below 20 Pa, the TL exhibited low intensity with sporadic and irregular bursts. Between 20 and 300 Pa, the TL exhibited a well-defined pattern characterized by a sequence of groups of nearly evenly spaced bursts (Fig. 17b). The interval between bursts ranged from 20 to 40 ms, while the groups of bursts appeared to coincide with the rotation of the disk. Within the pressure range of 300 Pa to 10 kPa, uniform emission with occasional bursts was observed. At pressures exceeding 10 kPa, only a faint, uniform emission was detected. The occurrence of bursts with a frequency of 33 Hz is intriguing, given that the typical frequencies of gas discharge oscillations are three to four orders of magnitude higher.

Figure 17c depicts the probability density distribution (PDD) functions of the photomultiplier tube (PMT) output time series, representing the number of photon counts per time bin. These PDDs exhibit varying shapes, including exponential-like distributions, bell-shaped distributions, or combinations of both (as shown in Fig. 17d). The exponential-like distribution component was attributed to the burst TL mode, while the bell-shaped component was associated with the uniform mode. Analysis of the probability density distribution function revealed that the intensity of the uniform mode increased as gas pressure decreased from 100 kPa to 300 Pa, disappearing completely below 300 Pa. The burst TL mode also exhibited an increase in intensity as gas pressure decreased, reaching a maximum at 170 Pa. Subsequently, the burst mode intensity gradually declined but did not completely disappear.

These findings corroborate the observation of sparks in TL images captured during polymer-alumina sliding, which coincide with uniform emission in air under atmospheric pressure (refer to Fig. 3 in [93]).

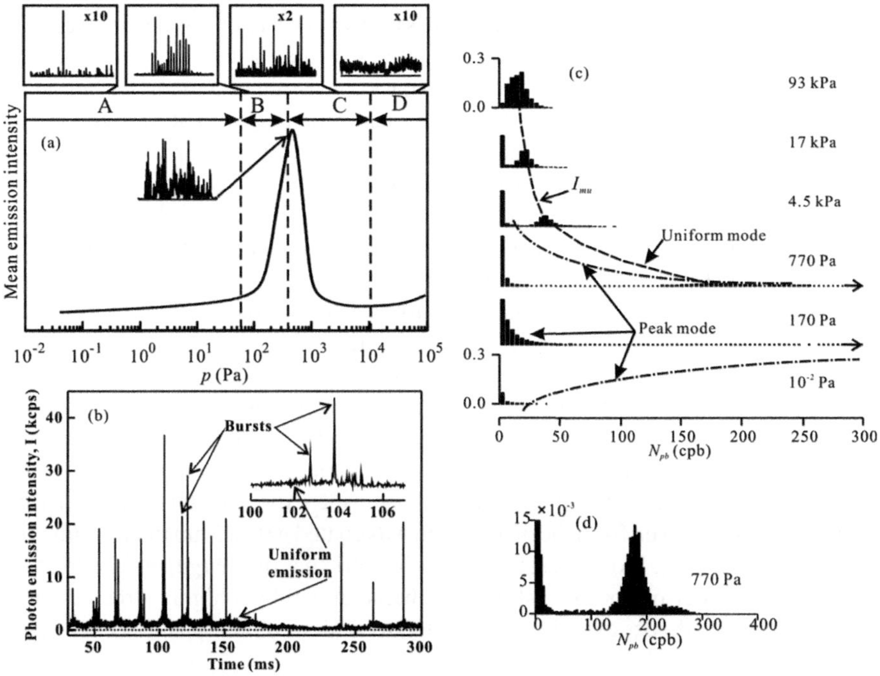

Fig. 17 **a** The effect of the Ar pressure on the total TL intensity and behaviour patterns (reprinted from [68] © 2013, with permission from IOP Publishing Ltd.); **b** fragment of the TL time series measured with a resolution of 20.48 μs (reprinted from [77] © 2003, with permission from American Institute of Physics); **c** histograms of the TL time series distributions on intensity at various pressures of Ar atmosphere; **d** a histogram of the TL time series distributions at gas pressure 770 Pa showing two components: the bell-shape and the exponential-like

5 Mechanisms of Gas Discharge Initiation and Sustenance

Electric gas discharge being an important excitation mechanism for the TL still remains poorly understood. From the point of view of electrostatics, pin-on-disc configuration can be considered as an extreme case of a point-to-plane gap with a point being in contact with a plane. In the case of insulating electrodes the gas discharge, referred to as barrier or silent discharge, is likely to be a very complex process because of the displacement field and very low mobility of charges in the electrodes [95]. Furthermore, in contrast to traditional barrier discharge being driven by alternating feeding potential, the polarities of the pin and the disk are rather stable [68, 77].

When investigating electric gas discharge at a sliding contact, the first question to address concerns the configuration and strength of the electric field. Given the constrained geometry and the potential alteration of the electric field by introducing conductive measuring electrodes, direct measurement of the electric field in the gap around the contact zone is a non-trivial task. The electric field can be roughly estimated if the rate of the charge generation is known. The latter was determined from the surface potential on the disk measured in high vacuum when the gas discharge is hindered (Fig. 18a) [96]. For the sliding contact consisting of a diamond pin with a radius of 1000 μm and a sapphire disk under a normal load of 0.96 N and a sliding velocity of 4.4 cm s^{-1} (1 rps) the charge generation rate was determined to be 3.39 $\times$ 10^{-11} A. Subsequently, considering the geometry of the friction zones of the pin and disk contact and assuming, as a first approximation, that charge leakage through the solid materials and on the surfaces is negligible, the configuration of the electric field was modelled using the finite element method (Fig. 17b).

A notable outcome is that the negative charge on the disk had a negligible influence on the dominant radial component of the electric field and only a minor impact on the axial component. This implies that the overall configuration of the electric field is almost entirely determined by the concentrated positive charge on the pin. These findings contradict the prevailing notion and existing models that assume a bipolar electric field within the gap with the dominating z component [83, 97]. These results provide a plausible explanation for the experimental finding that gas discharge occurred even from the disk, which was coated with a 5 nm thick conductive aluminium layer [98]. From the standpoint of electric gas discharge, the electric field configuration under pin-on-disk sliding is characterized by a highly stressed anode and a low-field cathode, exhibiting characteristics akin to both positive corona and barrier discharges.

Figure 19a depicts the simulated lines of electric field around the contact zone for PTFE pin on Al_2O_3 disk sliding [93]. At a larger distance from the contact zone (several millimeters), the electric field is not axisymmetric because of the displacement field and asymmetric configuration of the setup. These field lines have a bow shape. They begin at the wear track and end on the sides of the contact zone. This configuration of electric field resembles the shape of the burst TL pulses found on time resolved TL images by Puhan et al. [93] (refer to Fig. 19 b). The fact that

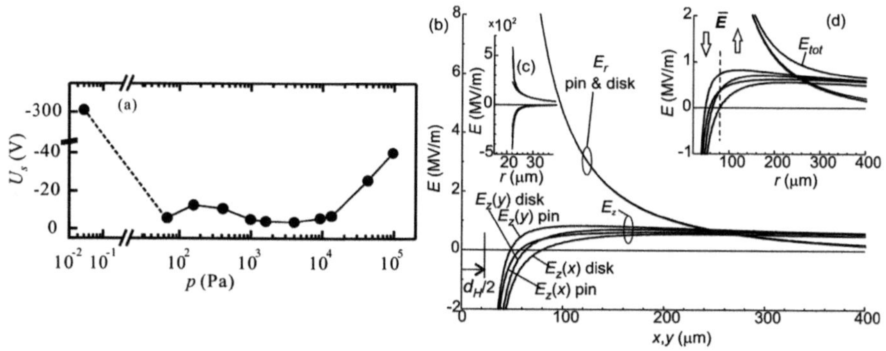

Fig. 18 **a** Mean electrostatic potential measured 2 mm above the friction zone on the disk as function of gas pressure. **b** radial, E_r, and axial, $E_x(x)$ and $E_y(y)$, components of the electric field along the x- and y-axes on the disk and pin surfaces near the contact zone. Insets (**c**, d) show the graphs on different scales (reprinted from [96] © 2013, with permission from IOP Publishing Ltd.)

luminescence can be observed at the inlet of the contact zone can be related with the re-entry of the charged disc surface. This is supported by reports of the half-life of corona charged polyethylene being *ca.* 30 h [99] and that of tribocharged PTFE being *ca.* 60 h [100] (both under 60% RH).

In addition to a sufficiently strong electric field, the initiation and sustenance of electric gas discharge require the fulfilment of conditions for both the initial emission of triggering electrons and secondary processes. The initiation of a positive corona requires an external ionization event within a region of high potential gradient. This ionization can be triggered by a cosmic ray, an electron emitted from the contacting solids due to field emission, exoelectron emission, or photoemission. The initially emitted electrons are attracted toward the positively charged pin, where they are accelerated by the electric field. During this acceleration, they can collide with gas molecules, leading to inelastic collisions and impact ionizations that trigger electron avalanches. Finally, the accelerated electrons reach the positively charged surface zones at and around the pin tip, leaving behind a cloud of positively charged ions. These ions drift in the direction opposite to the pin, driven by the electric field. In the absence of a stressed cathode, secondary electron emission cannot be achieved through cathode bombardment by positive ions [85]. Consequently, in conventional positive corona, secondary emission is primarily driven by photoionization of gas molecules by UV radiation from the plasma concentrated around the tip. In the context of triboplasma, exoelectron emission from mechanically affected surface regions has long been considered a potential mechanism for secondary electron emission [87]. The ingenious experimental setup depicted in Fig. 20a enabled the validation of this hypothesis. The test apparatus simulates a single sliding asperity contact and uses microchannel plates (MCP)—i.e. arrays of electron multipliers—coupled to a

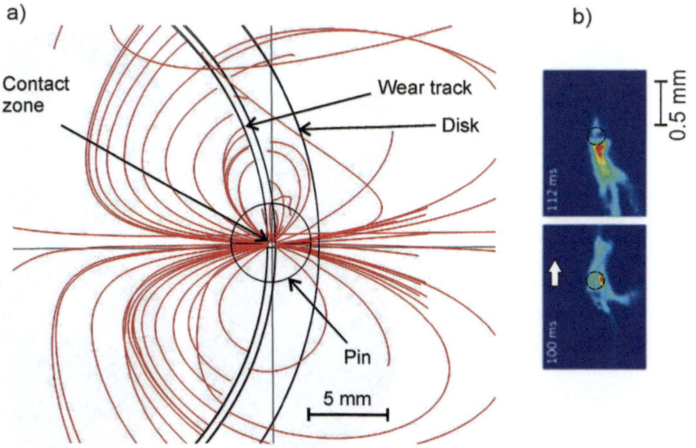

Fig. 19 **a** Simulated lines of the electric field around the contact zone for PTFE/Al_2O_3 sliding just before breakdown. **b** snapshots of TL bursts under UHMWPE (pin)—monocrystalline alumina (disk) sliding in air. Adapted from [93] under CC BY 4.0 license

phosphor screen to capture spatial distributions of the resulting triboemission [101, 102]. The asperity contact is produced by loading a 100 μm radius diamond tip onto the rotating disc specimen. This setup is located within a vacuum chamber maintained at 10^{-5} Torr. Figure 20b presents a series of individual phosphor screen images, each corresponding to a different point in time during a single disk revolution. The top-left image captures an initial burst of emission originating from the tip. Subsequent frames depict the emission at 0.6, 1.06, and 1.6 s following this initial event, thus confirming the persistence of after-emission throughout the entire disk revolution.

The corona model could also reasonably explain the experimentally observed TL behaviour (Sect. 4.4), which encompasses various combinations of two primary modes. The TL emission is uniform when the region of electron impact ionization coincides with the region of secondary processes. At lower pressures, these regions spatially decouple, leading to the instability of the gas discharge and the appearance of bursts [68].

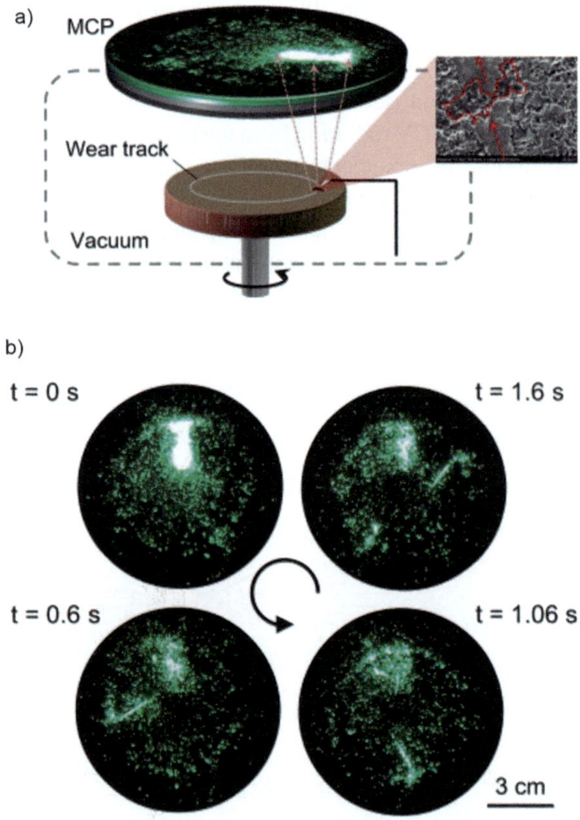

Fig. 20 **a** Schematic drawing of an experimental setup for space- and time-resolved measuring of electron emission from an alumina disk rubbed by a 100 μm radius diamond tip, which simulated a single asperity. **b** snapshots of phosphorus screen displaying electrons emitted from the disk. Adapted from [101] under CC BY 4.0 license

6 Triboluminescence of Lanthanide and Transition Metal Inorganic Compounds

Inorganic TL compounds consist of host materials and doping luminescent centres [103]. A wide range of halides, oxides, sulfates, sulfides, oxysulfides, aluminates, silicates, borates, titanates, niobates, stannates, and oxynitrates have been explored as host materials, while the doping elements primarily include lanthanide ions, magnesium, copper, and titanium. Eu, Ce, and Tb produce bright TL, whereas Pr, Gd, Dy, and Sm exhibit weaker TL.

Since luminescence originates from luminescent centres, TL is highly sensitive to the purity and morphology of the sample. In inorganic compounds, the introduction of "impurities" such as lanthanide ions during the doping process creates traps responsible for light emission. Consequently, the exact composition of the sample must be known to a very detailed level [104]. The sensitivity of TL to crystalline structure is demonstarted by monoclinic α-$SrAl_2O_4$:Eu, which exhibits pronounced elastico-luminescence, that is, luminescence induced by pure elastic deformation. In contrast, other crystal forms of Eu-doped strontium aluminates, including hexagonal, cubic, and orthogonal, do not exhibit this property. [105].

A notable characteristic of TL from lanthanoid-doped crystals is the slow luminescence decay. For TL from α-$SrAl_2O_4$:Eu it can persist for tens of seconds, which has been attributed to the simultaneous presence of Eu^{2+} and Eu^{3+} ions and the influence of surface effects [106]. The emission intensity is usually proportional to the applied load [107].

Lanthanide-doped aluminates typically exhibit both photoluminescence and thermoluminescence. It has been proposed that TL in these crystals arises from the release of trapped charge carriers or from the piezoelectric effect. In certain cases, the spectra of TL exhibit a blue shift relative to the corresponding photoluminescence spectra. This phenomenon has been attributed to the tailoring of the conduction and valence bands by the piezoelectric field [107] or to the influence of surface energy in micro- and nanoparticles [108].

Figure 21 depicts the mechanism of TL in piezoelectric lanthanide-doped calcium niobates. By use of lanthanide ions as substitutional dopants, emission centers and carrier traps can be simultaneously created in the host lattice. This allows to convert nonluminescent piezoelectrics into recoverable TL materials.

TL spectra of lanthanide-doped inorganic compounds may consist solely of photoluminescence bands or a combination of both photoluminescence and lightning bands. Figure 22 illustrates TL produced during the grinding of terbium and europium sulfates [110, 111]. The inset images depict snapshots of the TL captured through the quartz bottom side of the grinder. Similar to photoexcitation, the bands of Tb^{3+} and Eu^{3+} result from transitions from the level $^5D_4$ to sublevels of the main multiplet ^7F_j ($j = 0\ldots6$). It is worth mentioning that Pd^{3+} and Gd^{3+} emit in UV range.

Sharipov et al. [110, 111] demonstrated that isotopic substitution of H_2O by D_2O led to an intensification of the TL of various of lanthanide sulfate hydrates. This effect was proposed to be utilized in the development of highly efficient triboluminophors.

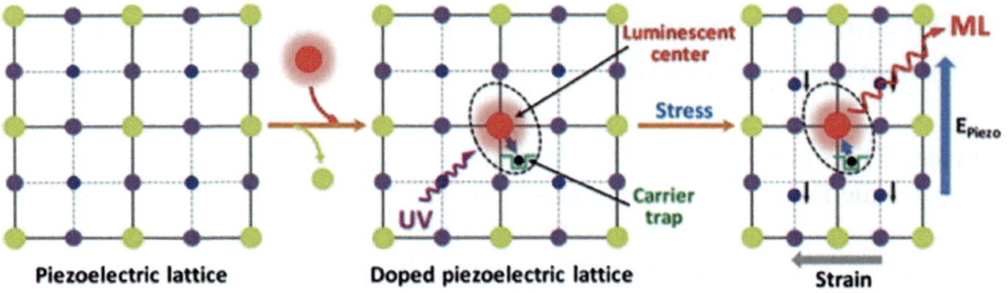

Fig. 21 Schematic illustration showing the design strategy to creating recoverable TL in piezoelectrics and the mechanism of TL under applied stress. Reproduced with permission from [109], Copyright © 2016, American Chemical Society

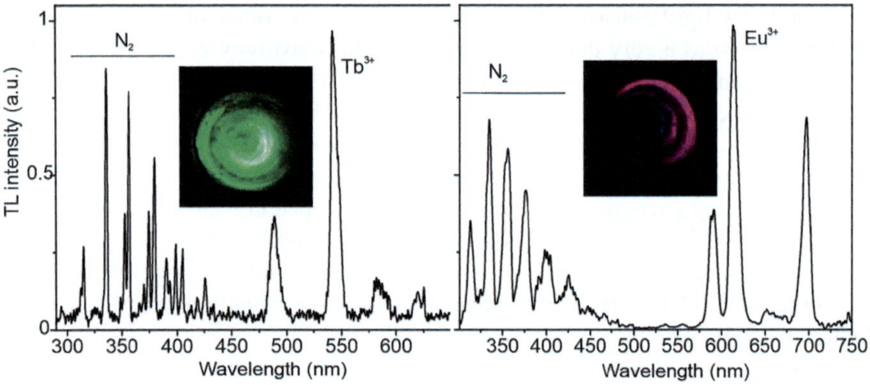

Fig. 22 Triboluminescence spectra during $Tb_2(SO_4)_3{\cdot}8H_2O$ (left) and $Eu_2(SO_4)_3{\cdot}8H_2O$ (right) in air

Recently, there has been a growing interest in transition metal complexes with promising luminescent properties for applications in optical devices [112]. Studies have focused on the tribo-, photo- and electroluminescent properties of metallocene and post-metallocene complexes of Ti, Zr, and Hf, as well as of carborane complexes of Hf [113–115]. TL spectra of these complexes are similar to their photoluminescence spectra. This highlights the similarity of the excited states involved in photoluminescence and triboluminescence (both arising from charge transfer from the ligand to the metal) despite the different excitation mechanisms. As the structural complexity of zirconocenes increases, the maximum of the luminescence band exhibits a bathochromic (red) shift.

A noteworthy finding was the intensification of TL in the *rac*-isomer compared to the *meso*-isomer of the $Me_2C(2\text{-}Me\text{-}4\text{-}Bu^t\text{-}C_5H_2)_2ZrCl_2$ complex [115], likely attributable to the molecular packing arrangement in the crystal. This difference is likely due to the fact that the *meso*-isomer possesses a centrosymmetric crystal structure, while the *rac*-isomer exhibits a non-centrosymmetric structure. During mechanical action, charge separation is expected to occur more efficiently in the non-centrosymmetric *rac*-isomer, leading to increased charge recombination and subsequent excitation of luminescence in the zirconocene. Modifying the ligand structure has proven to be a valuable tool for tailoring the luminescent properties of metal complexes.

More recently, triboluminescence of Pt(II) and Ru(II) complexes, which have prominent photo-, electro-, and chemiluminescent properties was reported [116, 117]. The TL mechanism was attributed to a combination of factors: the bombardment of the crystal surface with charged particles during mechanical stimulation, and the electron–hole recombination of charge carriers within the crystal.

Acknowledgements This work was supported through the grants of the Ministry of Science and Innovation of Spain (grants PID2019-111063RB-I00, PID2020-112770RB-C22, RTI2018-099794-B-I00, and TED2021-129950B-I00) with the participation of European Regional Development Fund (FEDER) and NextGenerationEU.

References

1. Triboluminescence. Oxford Dictionaries (Oxford University Press, 2010)
2. G.P.H. Gubbels, G.J.F.T. van der Beek, A.L. Hoep, F.L.M. Delbressine, H. van Halewijn, Diamond tool wear when cutting amorphous polymers. CIRP Ann. Manuf. Technol. **53**(1), 447–450 (2004)
3. F. Freund, Charge generation and propagation in igneous rocks. J. Geodyn. **33**(4–5), 543–570 (2002)
4. F. St-Laurent, J.S. Derr, F.T. Freund, Earthquake lights and the stress-activation of positive hole charge carriers in rocks. Phys. Chem. Earth, Parts A/B/C **31**(4–9), 305–312 (2006)
5. T.V. Losseva, I.V. Nemchinov, Earthquake lights and rupture processes. Nat. Hazard. **5**, 649–656 (2005)
6. J. Muto, H. Nagahama, T. Miura, I. Arakawa, Frictional discharge at fault asperities: origin of fractal seismo-electromagnetic radiation. Tectonophys. **431**(1–4), 113–122 (2007)
7. C. Helling, M. Jardine, D. Diver, S. Witte, Dust cloud lightning in extraterrestrial atmospheres. Planet. Space Sci. **77**, 152–157 (2013)
8. A.A. Sickafoose, J.E. Colwell, M. Horányi, S. Robertson, Experimental investigations on photoelectric and triboelectric charging of dust. J. Geophys. Res. Space Physics **106**(A5), 8343–8356 (2001)
9. H. Blok, The flash temperature concept. Wear **6**(6), 483–494 (1963)
10. F. Bacon, *Advancement of Learning and Novum Organum* (Colonial Press, 1900)
11. R. Boyle, Experiments and considerations touching colours. First Occasionally Written, Among Some Other Essays, to a Friend; And Now Suffer'd To Come Abroad as The Beginning of an Experimental History of Colours, Anchor on the Lower walk of the New Exchange, London (1664)
12. D. Müller-Hillebrand, Torbern Bergman as a Lightning Scientist, Daedalus. Tekniska Museets Arsbok (Stokholm, 1963), pp. 35–76
13. O. von Guericke, Experimenta nova (ut vocantur) Magdeburgica de vacuo spatio (Würzburg, 1672)
14. F. Hauksbee, Physico-mechanical experiments on various subjects: containing an account of several surprizing phaenomena touching light and electricity, producible on the attrition of bodies (R. Brugis, 1709)
15. J. Priestley, *The History and Present State of Electricity with original Experiments*, 3rd edn (C. Bathurst and T. Lowndes, 1775)

16. H.B. Weiser, Crystalloluminescence II. J. Phys. Chem. **22**(8), 576–595 (1917)
17. P. Heinrich, Die Phosphorescenz der Körper oder die im Dunkeln bemerkbaren Lichtphänomene der anorganischen Natur. 4: Von der durch mechanisches Verfahren oder durch Druck, Bruch, und Reibung bewirkten Phosphorescenz (Schrag, Nürnberg, 1820)
18. J.-P. Dessaignes, Mémoires sur Phosphorescences. Journal de Physique, de Chimie, d'Histoire Naturelle et des Arts **68**, 444–451 (1809)
19. E.N. Harvey, A history of luminescence. From the earliest times until 1900. The American Philosophical Society, Philadelphia (1957)
20. C.-L. Berthollet, An essay on chemical statics; with copious explanatory notes, and an appendix on vegetable and animal substances, Printed for J. Mawman, 22 Poultry, W. Flint, Printer, Old Bailey London, London (1804)
21. T. Wedgwood, J. Banks, III. Experiments and observations on the production of light from different bodies, by heat and by attrition, Phil. Trans. Royal Soc. London **82**, 28–47 (1792)
22. T. Steel, Luminosity of sugar. Nature **59**(1526), 295–296 (1899)
23. W.J. Pope, On triboluminescence. Nature **59**(1539), 618–619 (1899)
24. H. Longchambon, Etude spectro-photographique des phénomènes de triboluminescence et de cristalloluminescence. Bulletin de la Société française de minéralogie et de cristallographie **48**(2/5), 130–214 (1925)
25. H. Longchambon, Étude spectrale de la triboluminescence du saccharose. C. R. Hebd. Seances Acad. Sci. **175**, 1633–1634 (1922)
26. L.M. Sweeting, Triboluminescence with and without air. Chem. Mater. **13**(3), 854–870 (2001)
27. A.J. Walton, Triboluminescence. Adv. Phys. **26**(6), 887–948 (1977)
28. L.J. Kricka, J. Stroebel, P.E. Stanley, Triboluminescence: 1968–1998. Luminescence **14**(4), 215–220 (1999)
29. J.I. Zink, Tribophosphorescence from nonphotophosphorescent crystals. J. Am. Chem. Soc. **96**(21), 6775–6777 (1974)
30. A. Qasem, P. Xiong, Z. Ma, M. Peng, Z. Yang, Recent advances in mechanoluminescence of doped zinc sulfides. Laser Photonics Rev. **15**(12), 2100276 (2021)
31. Y. Tsuboi, T. Seto, N. Kitamura, Laser-induced shock wave can spark triboluminescence of amorphous sugars. J. Phys. Chem. A **112**(29), 6517–6521 (2008)
32. B.V. Derjaguin, N.A. Krotova, Y.P. Toporov, in Emission of high-speed electrons and other phenomena accompanying the process of breaking adhesion bonds, ed. J.M. Georges, *Tribology Series* (Elsevier, 1981), pp. 471–487
33. T. Miura, M. Chini, R. Bennewitz, Forces, charges, and light emission during the rupture of adhesive contacts. J. Appl. Phys. **102**(10), 103509–103516 (2007)
34. Experience faite a l´observatoire sur le barometre simple touchant un nouveau phenomene qu´on ya découvert, Le Journal des sçavans, Académie des inscriptions et belles-lettres (Paris, France, 1676)
35. M.P. Brenner, S. Hilgenfeldt, D. Lohse, Single-bubble sonoluminescence. Rev. Mod. Phys. **74**(2), 425–484 (2002)
36. P.Y. Butyagin, The luminescence accompanying mechanical deformation and rupture of polymers. Vysokomol. soyed **A12**(2), 290–299 (1970)
37. B.P. Chandra, K.K. Shrivastava, Dependence of mechanoluminescence in rochelle-salt crystals on the charge-produced during their fracture. J. Phys. Chem. Solids **39**(9), 939–940 (1978)
38. M. Urbakh, J. Klafter, D. Gourdon, J. Israelachvili, The nonlinear nature of friction. Nature **430**(6999), 525–528 (2004)
39. B.P. Chandra, Mechanoluminescence of Nanoparticles. The Open Nanosci. J. **5**, 45–58 (2011)
40. J.-C. Zhang, C.-N. Xu, S. Kamimura, Y. Terasawa, H. Yamada, X. Wang, An intense elastico-mechanoluminescence material CaZnOS:Mn2+ for sensing and imaging multiple mechanical stresses. Opt. Express **21**(11), 12976–12986 (2013)
41. E.A. Deulin, V.P. Mikhailov, Y.V. Panfilov, R.A. Nevshupa, *Mechanics and physics of precise vacuum mechanisms* (Springer, Dordrecht, 2010)

42. V.E. Orel, I.N. Kadyuk, N.N. Dzyatkovskaya, M.I. Danko, Y.I. Mel'nic, Mechanoluminescence: lymphocyte analysis after exposure to ionizing radiation, Lumin. **15**(1), 29–36 (2000)
43. V.E. Orel, A.V. Romanov, N.N. Dzyatkovskaya, I. Mel'nik, The device and algorithm for estimation of the mechanoemisson chaos in blood of patients with gastric cancer. Med. Eng. Phys. **24**(5), 365–371 (2002)
44. G. Heinicke, *Tribochemistry* (Carl Hanser Verlag, Munchen, 1984)
45. B. Vick, M.J. Furey, An investigation into the influence of frictionally generated surface temperatures on thermionic emission. Wear **254**(11), 1155–1161 (2003)
46. F.P. Bowden, F.H. Thomas, The surface temperature of sliding solids. Proceed. Royal Soc. London. Series A. Math. Phys. Sci. **223**(1152), 29–40 (1954)
47. R. Nevshupa, The role of athermal mechanisms in the activation of tribodesorption and triboluminisence in miniature and lightly loaded friction units. J. Friction Wear **30**(2), 118–126 (2009)
48. I. Tole, K. Habermehl-Cwirzen, A. Cwirzen, Mechanochemical activation of natural clay minerals: an alternative to produce sustainable cementitious binders—review. Mineral. Petrol. **113**(4), 449–462 (2019)
49. J.D. Schall, G. Gao, J.A. Harrison, Effects of adhesion and transfer film formation on the tribology of self-mated DLC contacts. J. Phys. Chem. C **114**(12), 5321–5330 (2010)
50. T.-B. Ma, Y.-Z. Hu, H. Wang, Molecular dynamics simulation of shear-induced graphitization of amorphous carbon films. Carbon **47**(8), 1953–1957 (2009)
51. A. Rusanov, R. Nevshupa, J. Fontaine, J.-M. Martin, T. Le Mogne, V. Elinson, A. Lyamin, E. Roman, Probing the tribochemical degradation of hydrogenated amorphous carbon using mechanically stimulated gas emission spectroscopy. Carbon **81**, 788–799 (2015)
52. A. Rusanov, R. Nevshupa, J.-M. Martin, M.Á. Garrido, E. Roman, Tribochemistry of hydrogenated amorphous carbon through analysis of mechanically stimulated gas emission. Diam. Relat. Mater. **55**, 32–40 (2015)
53. M.L. Rahaman, L. Zhang, Interface temperature during contact sliding of two solids: relationship between predicted flash temperature and the experimentally measured. Proceed. Inst. Mech. Eng., Part J: J. Eng. Tribol. **231**(1), 3–13 (2017)
54. J. Zhang, H. Jiang, C. Jiang, Q. Cheng, G. Kang, In-situ observation of temperature rise during scratch testing of poly (methylmethacrylate) and polycarbonate. Tribol. Int. **95**, 1–4 (2016)
55. K. Nakayama, F. Yagasaki, The temperature of triboplasma. Tribol. Lett. **66**(1), 10 (2017)
56. A.I. Rusanov, Thermal effects in mechanochemistry. Russ. J. Gen. Chem. **72**(3), 327–344 (2002)
57. P.Y. Butyagin, Problems in mechanochemistry and prospects for its development. Russ. Chem. Rev. **63**(12), 965 (1994)
58. P.Y. Butyagin, Structural disorder and mechanochemical reactions in solids. Russ. Chem. Rev. **53**(11), 1025–1038 (1984)
59. T.E. Fischer, Tribochemistry. Ann. Rev. Mat. Sci. **18**, 303–323 (1988)
60. S.L. Craig, Mechanochemistry: a tour of force. Nature **487**(7406), 176–177 (2012)
61. B.P. Chandra, S. Tiwari, M. Ramrakhiani, M.H. Ansari, Mechanoluminescence in centrosymmetric crystals. Cryst. Res. Technol. **26**(6), 767–781 (1991)
62. R.A. Nevshupa, Triboemission: an attempt of generalized classification, in *Tribology: Science and Applications*. ed. by C. Kajdas (PAS, Vienna, 2004), pp.11–25
63. J. Thevenet, M. Siroux, B. Desmet, Measurements of brake disc surface temperature and emissivity by two-color pyrometry. Appl. Therm. Eng. **30**(6–7), 753–759 (2010)
64. K. Nakayama, R.A. Nevshupa, Plasma generation in a gap around a sliding contact. J. Phys. D Appl. Phys. **35**(12), L53–L56 (2002)
65. V.A. Kluev, T.N. Vladikina, Y.P. Toporov, V.J. Anisimova, B.V. Derjaguin, Emission phenomena accompanying the triboelectrification process in vacuum, industry applications. IEEE Trans. IA-14 **6**, 544–546 (1978)
66. G. Heinike, *Tribochemistry* (Carl Hanser Verlag, Munchen, 1984)

67. J.P. Duignan, I.D.H. Oswald, I.C. Sage, L.M. Sweeting, K. Tanaka, T. Ishihara, K. Hirao, G. Bourhill, Do triboluminescence spectra really show a spectral shift relative to photoluminescence spectra? J. Lumin. **97**(2), 115–126 (2002)
68. R.A. Nevshupa, Effect of gas pressure on the triboluminescence and contact electrification under mutual sliding of insulating materials. J. Phys. D Appl. Phys. **46**(18), 185501 (2013)
69. E.A. Varentsov, Y.A. Khrustalev, Mechanoemission and mechanochemistry of molecular organic crystals. Russ. Chem. Rev. **64**(8), 783–797 (1995)
70. C.N. Xu, Coatings, in *Encyclopedia of smart materials*. ed. by M. Schwartz (Willey, New York, 2002), pp.190–201
71. L. Zhang, K. Shi, Y. Wang, L. Su, G. Yang, B. Huang, J. Kong, X. Dong, Z.L. Wang, Unraveling the anomalous mechanoluminescence intensity change and pressure-induced red-shift for manganese-doped zinc sulfide. Nano Energy **85**, 106005 (2021)
72. K. Hiratsuka, T. Yoshida, The twin-ring tribometer—characterizing sliding wear of metals excluding the effect of contact configurations. Wear **270**(11–12), 742–750 (2011)
73. J. Muto, H. Nagahama, T. Miura, I. Arakawa, Frictional discharge plasma from natural semiconductor/insulator junctions: origin of seismo-electromagnetic radiation. Phys. Chem. Earth, Parts A/B/C **31**(4–9), 346–351 (2006)
74. K. Nakayama, H. Hashimoto, Effect of surrounding gas pressure on triboemission of charged particles and photons from wearing ceramic surfaces. Tribol. Trans. **38**(1), 35–42 (1995)
75. G.E. Hardy, J.I. Zink, Triboluminescence and pressure dependence of the photoluminescence of tetrahedral manganese(II) complexes. Inorg. Chem. **15**(12), 3061–3065 (1976)
76. L.M. Sweeting, J.L. Guido, An improved method for determining triboluminescence spectra. J. Lumin. **33**(2), 167–173 (1985)
77. R.A. Nevshupa, K. Nakayama, Triboemission behavior of photons at dielectric/dielectric sliding: Time dependence nature at 10^{-4}–10^{4} s. J. Appl. Phys. **93**(11), 9321–9328 (2003)
78. K.I. Hiratsuka, K. Hosotani, Effects of friction type and humidity on triboelectrification and triboluminescence among eight kinds of polymers. Tribol. Int. **55**, 87–99 (2012)
79. T. Miura, K. Nakayama, Two-dimensional spatial distribution of electric-discharge plasma around a frictional interface between dielectric surfaces. Appl. Phys. Lett. **78**(19), 2979–2981 (2001)
80. K. Nakayama, R.A. Nevshupa, Characteristics and pattern of plasma generated at sliding contact. J Tribol-T ASME **125**(4), 780–787 (2003)
81. K. Nakayama, R.A. Nevshupa, Effect of dry air pressure on characteristics and patterns of tribomicroplasma. Vacuum **74**(1), 11–17 (2004)
82. K. Nakayama, The plasma generated and photons emitted in an oil-lubricated sliding contact. J. Phys. D Appl. Phys. **40**(4), 1103–1107 (2007)
83. K. Nakayama, Triboplasma generation and triboluminescence: influence of stationary sliding partner. Tribol. Lett. **37**(2), 215–228 (2010)
84. K. Nakayama, Mechanism of triboplasma generation in oil. Tribol. Lett. **41**(2), 345–351 (2011)
85. Y.P. Raizer, *Gas Discharge Physics* (Springer, New York, 1991)
86. L.B. Loeb, *Electrical Coronas, Their Basic Physical Mechanisms* (University of California Press, Berkeley, 1965)
87. B.T. Brady, G.A. Rowell, Laboratory investigation of the electrodynamics of rock fracture. Nature **321**(6069), 488–492 (1986)
88. L.M. Belyaev, Y.N. Martyshev, Triboluminescence of some alkali halide crystals. Phys. Status Solid. B **34**(1), 57–62 (1969)
89. T. Miura, E. Hosobuchi, I. Arakawa, Spectroscopic studies of triboluminescence from a sliding contact between diamond, SiO_2, MgO, NaCl, and Al_2O_3 (0001). Vacuum **84**(5), 573–577 (2009)
90. A.A. Tukhbatullin, G.L. Sharipov, A.R. Bagautdinova, New triboluminescence emitters of inorganic lanthanide salts in CH4 and C2H2 atmosphere. J. Lumin. **173**(Complete), 127–129 (2016)

91. A.A. Tukhbatullin, G.L. Sharipov, D.N. Gerasimov, Luminescence of reaction products of mechanochemical decomposition for some gaseous hydrocarbons CxHy during tribodestruction of cerium and terbium salts. J. Lumin. **197**(Complete), 335–338 (2018)
92. U. Fantz, Emission spectroscopy of molecular low pressure plasmas. Contrib. Plasma Phys. **44**(5–6), 508–515 (2004)
93. D. Puhan, R. Nevshupa, J.S.S. Wong, T. Reddyhoff, Transient aspects of plasma luminescence induced by triboelectrification of polymers. Tribol. Int. **130**, 366–377 (2019)
94. L.S. McCarty, G.M. Whitesides, Electrostatic charging due to separation of ions at interfaces: contact electrification of ionic electrets. Angew. Chem. Int. Ed. **47**(12), 2188–2207 (2008)
95. H.E. Wagner, R. Brandenburg, K.V. Kozlov, A. Sonnenfeld, P. Michel, J.F. Behnke, The barrier discharge: basic properties and applications to surface treatment. Vacuum **71**(3), 417–436 (2003)
96. R. Nevshupa, Effect of gas pressure on the triboluminescence and contact electrification under mutual sliding of insulating materials. J. Phys. D: Appl. Phys. (to be published)
97. K. Nakayama, M. Tanaka, Simulation analysis of triboplasma generation using the particle-in-cell/Monte Carlo collision (PIC/MCC) method. J. Phys. D Appl. Phys. **45**(49), 495203 (2012)
98. R.A. Nevshupa, K. Nakayama, Effect of nanometer thin metal film on triboemission of negatively charged particles from dielectric solids. Vacuum **67**(3–4), 485–490 (2002)
99. T.A.L. Burgo, T.R.D. Ducati, K.R. Francisco, K.J. Clinckspoor, F. Galembeck, S.E. Galembeck, Triboelectricity: macroscopic charge patterns formed by self-arraying ions on polymer surfaces. Langmuir **28**(19), 7407–7416 (2012)
100. M.W. Williams, Triboelectric charging of insulating polymers-some new perspectives. AIP Adv. **2**(1), 010701 (2012)
101. A. Ciniero, J. Le Rouzic, I. Baikie, T. Reddyhoff, The origins of triboemission—correlating wear damage with electron emission. Wear **374–375**, 113–119 (2017)
102. J. Le Rouzic, T. Reddyhoff, Spatially resolved triboemission measurements. Tribol. Lett. **55**(2), 245–252 (2014)
103. W. Zhaofeng, W. Fu, in Triboluminescence: materials, properties, and applications, ed. by Sergei, *Luminescence* (IntechOpen, Rijeka, 2018), Ch. 2
104. J.-C.G. Bünzli, K.-L. Wong, Lanthanide mechanoluminescence. J. Rare Earths **36**(1), 1–41 (2018)
105. C.-N. Xu, H. Yamada, X. Wang, X.-G. Zheng, Strong elasticoluminescence from monoclinic-structure $SrAl_2O_4$. Appl. Phys. Lett. **84**(16), 3040–3042 (2004)
106. Y. Liu, C.-N. Xu, Influence of calcining temperature on photoluminescence and triboluminescence of europium-doped strontium aluminate particles prepared by sol−gel process. J. Phys. Chem. B **107**(17), 3991–3995 (2003)
107. L. Li, K.-L. Wong, P. Li, M. Peng, Mechanoluminescence properties of Mn^{2+}-doped BaZnOS phosphor. J. Mat. Chem. C **4**(35), 8166–8170 (2016)
108. Z. Tang, F. Zhang, Z. Zhang, C. Huang, Y. Lin, Luminescent properties of $SrAl_2O_4$: Eu, Dy material prepared by the gel method. J. Eur. Ceram. Soc. **20**(12), 2129–2132 (2000)
109. J.-C. Zhang, Y.-Z. Long, X. Yan, X. Wang, F. Wang, Creating recoverable mechanoluminescence in piezoelectric calcium niobates through Pr^{3+} doping. Chem. Mater. **28**(11), 4052–4057 (2016)
110. G.L. Sharipov, A.A. Tukhbatullin, M.R. Muftakhutdinov, A.M. Abdrakhmanov, Luminescence of OD radical as an evidence for water decomposition under destruction of the deuterated terbium sulfate crystal hydrate. J. Lumin. **148**, 79–81 (2014)
111. G.L. Sharipov, A.A. Tukhbatullin, E.S. Mescheryakova, The H_2O/D_2O isotope effect in crystalline lanthanide sulfates at photo-, radio-, and triboluminescence. Opt. Mater. **52**, 44–48 (2016)
112. Y. Chi, P.-T. Chou, Transition-metal phosphors with cyclometalating ligands: fundamentals and applications. Chem. Soc. Rev. **39**(2), 638–655 (2010)
113. E. Hong, H. Jang, Y. Kim, S.C. Jeoung, Y. Do, Mechano- and electroluminescence of a dissymmetric hafnium carborane complex. Adv. Mater. **13**(14), 1094–1096 (2001)

114. G.L. Sharipov, A.A. Tukhbatullin, P.V. Kovyazin, L.V. Parfenova, P.V. Ivchenko, I.E. Nifant'ev, Triboluminescence of zirconium η5-complexes. Russ. Chem. Bull. **64**(12), 2776–2779 (2015)
115. A.A. Tukhbatullin, P.V. Kovyazin, G.L. Sharipov, L.V. Parfenova, P.V. Ivchenko, I.E. Nifant'ev, Photoluminescence and mechanoluminescence of solid-state zirconocene dichlorides. Lumin. **36**(4), 943–950 (2021)
116. C.-W. Hsu, K.T. Ly, W.-K. Lee, C.-C. Wu, L.-C. Wu, J.-J. Lee, T.-C. Lin, S.-H. Liu, P.-T. Chou, G.-H. Lee, Y. Chi, Triboluminescence and metal phosphor for organic light-emitting diodes: functional Pt(II) complexes with both 2-pyridylimidazol-2-ylidene and bipyrazolate chelates. ACS Appl. Mater. Interf. **8**(49), 33888–33898 (2016)
117. G.L. Sharipov, A.A. Tukhbatullin, Triboluminescence of tris(2,2′-bipyridyl)ruthenium(II) dichloride hexahydrate. J. Lumin. **215**, 116691 (2019)

Sliding Friction in Liquid Environments at the Nanoscale

Ronen Berkovich, Rong An, and Enrico Gnecco

Abstract Friction Force Microscopy (FFM) conducted in liquid environment proves to be a highly effective method for investigating atomic-scale friction on crystalline surfaces. It can probe friction between surface atoms while also providing sublattice resolution, opening doors to new areas of research. The observed similarity in FFM measurements conducted in liquid environments and Ultra-High Vacuum (UHV) is attributed to the lack of capillary bridges in both settings. These bridges usually increase adhesion between the scanning probe and the sample, resulting in surface wear during imaging in ambient conditions. We review various instances of nanotribological phenomena occurring on crystalline surfaces within different liquid environments—specifically water, ethanol, and ionic liquids—as studied by the authors of this chapter. We discuss the influence of the damping state of the sliding contact in the presence of liquids, which is reflected by variations in the slip length of the scanning probe. Finally, we showcase how FFM can be used to investigate sliding friction in ionic liquids. This approach allows us to probe the fascinating interplay

R. Berkovich (✉)
Department of Chemical Engineering, Ben-Gurion University of the Negev, 8410501 Beer Sheva, Israel
e-mail: berkovir@bgu.ac.il

The Ilse Katz Institute for Nanoscale Science and Technology, Ben-Gurion University of the Negev, 8410501 Beer Sheva, Israel

R. Berkovich · E. Gnecco
Marian Smoluchowski Institute of Physics, Jagiellonian University, Lojasiewicza 11, 30-348 Krakow, Poland

R. An
Herbert Gleiter Institute of Nanoscience, School of Materials Science and Engineering, Nanjing University of Science and Technology, Nanjing 210094, China
e-mail: ran@njust.edu.cn

Shandong Laboratory of Advanced Materials and Green Manufacturing, Yantai Zhongke Research Institute of Advanced Materials and Green Chemical Engineering, Yantai 264006, China

E. Gnecco
e-mail: enrico.gnecco@uj.edu.pl

E. Gnecco and E. Meyer (eds.), *Fundamentals of Friction and Wear on the Nanoscale*, NanoScience and Technology, https://doi.org/10.1007/978-3-031-63065-1_17

between friction at the nanoscale and the unique nanostructures formed by confined ionic liquids.

1 Introduction

The term "nanotribology" refers to the study of the atomic-scale and nanoscale interactions between surfaces in relative motion, such as friction, adhesion, lubrication, and wear [1, 2]. In the past three decades, a considerable effort has been invested to provide better understanding on the nanoscale nature of friction and its effect on the macroscale [1, 3, 4]. Within this frame of work, Friction Force Microscopy (FFM) emerged as a technique to measure lateral interactions, and when implemented with Atomic Force Microscopy (AFM) it presents an efficient and useful approach to characterize the tribological response of solid surfaces on the nanometer scale with a single asperity resolution [1, 2]. In such measurement, a sharp tip supported by a cantilever is dragged back and forth at a certain rate and under a specific load across a solid surface. During this scan lateral forces caused by the friction between the tip and the surface are recorded in the form of force-position loops. On a simple crystalline surface, the time evolution of the lateral force usually consists in a saw-tooth signal representing the stick–slip frictional behavior. If scanning is repeated on a series of parallel lines, a friction map resembling the symmetry of the surface lattice is obtained.

AFM based FFM has been widely used to measure friction of sliding contacts, while detecting thermally activated atomic stick–slip and parameters affecting it such as temperature [5–7], sliding velocity [8–10], normal load [11–13], superlubricity [4, 14], environmental conditions [15–18], etc. During the relative motion between a sharp AFM cantilever tip and a surface, forces that result from energy dissipation are measured. This energy is released into the bulk of the substrate as phonons, and electron–hole excitations in metals [1, 19]. Consequently, under high loads higher lateral friction forces are generated, and heat dissipation is increased [11, 13].

In ambient conditions FFM measurements are also affected by the water meniscus formed around the tip due to humidity condensation, whose strength can be much larger than the normal and lateral forces in the solid contact (usually in the nanonewton range) [20, 21]. This makes lattice resolution more difficult to achieve. The problem of capillary condensation can be overcome if the measurements are performed in ultra-high vacuum (UHV) [22]. While measurements in UHV provide a sterile and ideal conditions that pour light on pure unperturbed interactions, in daily life, however, where even extreme conditions may prevail (such as intense mechanical loads, high concentrations of ionic and corrosive molecules, etc.), the interaction between two sliding bodies is different.

While most prominent nanoscale studies were reported under UHV conditions, studies have also been performed in liquid surroundings [15–17, 23–34]. Apart from providing a clean environment, and avoiding capillary condensation, some studies related the effect of the surroundings to the observed frictional signal. For instance, it was shown experimentally that the friction signal on MoS_2 increases with relative humidity, when measurements were performed in ambient conditions [16, 28, 35], and when fully immersed in water [33, 34, 36]. Humidity was also shown to play a role in the friction on graphene [37, 38], graphene oxide [38], and graphite [39]. Considering different liquid environments, FFM measurements on MoS_2 displayed different friction features and trends in organic environments, such as in ethanol [40], oil [41], and dodecane [33]. FFM studies were also conducted on glass and quartz surfaces that were fully immersed in octamethylciclotetrasiloxane and squalene [42], metallic glasses in borate and phosphate buffers [32], and with respect to electrolyte aqueous solutions on mica [29], and on wider scales on calcite [18, 43], all demonstrating differences interactions that are affected by the presence of the surroundings.

This increasing volume of literature the friction in the presence of liquid alludes not only to the possible different channeling mechanisms of the energy dissipation, but also raises the possibility that several individual molecules may be trapped within the sliding contact and affect the overall interaction within it [44, 45]. Additionally, nanoscale friction was also shown to be affected by the presence of possible contamination [46] and adsorption of liquid molecules on the surface [29, 44, 47] and in defects [31], or by local mobility of atoms [48]. The reduction in contact size due to the presence of interfering molecules from the surrounding was also suggested as a possible mechanism that influence friction [49, 50]. These effects are also present in perhaps the most complex environment investigated through FFM, i.e. ionic salts consisting of cations and anions found in a liquid state at room temperature. These ionic liquids (ILs) not only adsorb onto the surfaces of sliding solid contacts, but they also form different nanostructures from the bulk (liquid) phase when confined, which can lead to a significant reduction in friction.

The Chapter is structured as follows: After a short review on lattice resolved FFM imaging in of mineral, synthetic crystalline, and alkali-halide surfaces in water and ethanol, we move on to discuss the damping states that are imposed by the surrounding of the sliding contact on the basis of the Prandtl model. Lastly, we review the layering of ILs on model solid surfaces, as observed by AFM, and ultimately attempt to relate it to the accompanying variation of IL-solid interactions and of the nanoscale friction [51, 52].

2 Lattice Resolved Friction Microscopy in Liquid Environments

FFM has been successfully applied to obtain lattice resolved images of mineral and synthetic crystalline surfaces. A first example is provided in Fig. 1a, b. It shows a friction force map of the (104) cleavage surface of calcite ($CaCO_3$), as recorded while scanning along the $[42\bar{1}]$ or the perpendicular [0 1 0] direction of the surface lattice in water. The zigzag arrangement of the friction peaks suggests that, in both cases, the tip is pinned by the protruding O atoms of the surface. Each rectangular unit cell contains two of such atoms as highlighted in Fig. 1c. Note that the second O atom is not exactly located at the center of the cell. As a consequence, the stick–slip motion and the resulting friction pattern will develop differently depending on the type of atom to which the tip is anchored. Please see [25] for more details. Comparable FFM maps have been reported on kyanite (Al_2SiO_5) [53] and on zabuyelite (Li_2CO_3) films grown on dolomite and kutnohorite from solutions with different saturation [54]. It was also possible to extend these studies to soft materials such as surfaces of organic semiconductors, where not only the friction force, but also the stiffness, was found to depend significantly on the scan direction [55]. Note that, in this case, the surfaces started to be worn off at low loading forces in the nanonewton range, with the threshold value also depending on the scan direction (whereas mineral surfaces could be scanned up to forces of 50 nN and more without visible damage). As a next step, it is also worth mentioning the work by Nita et al., where ultrathin layers of an organic semiconductor (Cu-phthalocyanine) could be resolved simultaneously to the underlying dolomite [$CaMg(CO_3)_2$] substrate using FFM in water [27]. As seen in Fig. 1d the organic molecules were found to stand up and align in parallel rows on the insulating substrate. Considering that the core of the organic molecules consists in a Cu atom, possible applications of the system as an array of conductive nanowires can be foreseen.

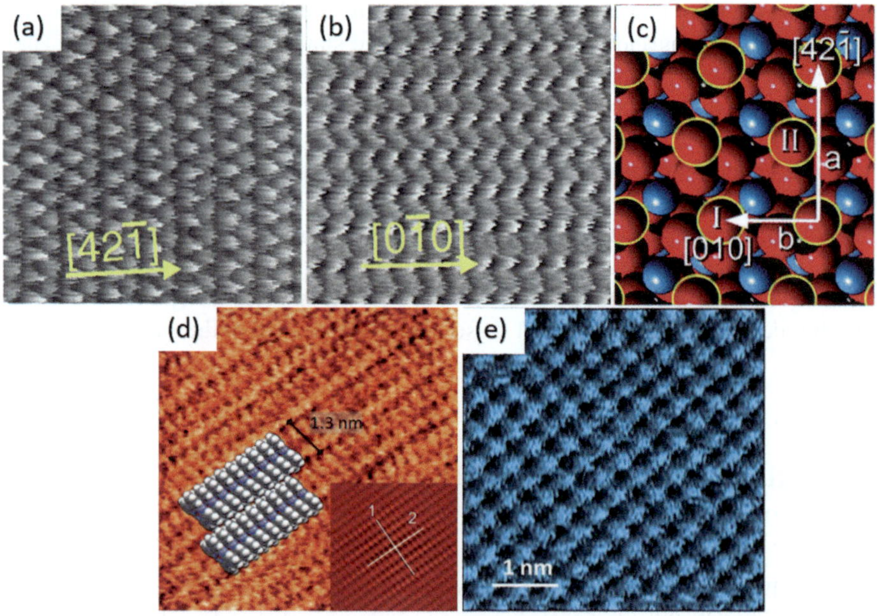

Fig. 1 **a**, **b** Friction maps of a calcite (104) cleavage surface scanned almost parallel to the [421] direction and the [010] direction in water. **c** crystal structure of the same surface. **d**, **e** friction maps of a Cu-phthalocyanine layer on dolomite in water, and of a NaCl (100) surface in ethanol in ethanol. Reprinted (adapted) with permission from [25, 27] and [17]

Altogether, the aforementioned results show that the quality of FFM images recorded in water is not far from that achievable with more demanding experiments in UHV: see e.g. Chap. 4 in this book. A theoretical investigation by Vilhena et al. helped to shed light on this issue [15]. The authors reproduced the friction force acting on a diamond tip sliding on the top of a 3 ML graphene (with the bottom layer fixed) using Molecular Dynamics (MD) simulations. The tip was either immersed in water, or in a vacuum. The stick–slip profiles resulting under otherwise similar loading conditions in the two environments were almost overlapping. The simulated signal was slightly noisier in water, due to the frequent collisions of the water molecules with the tip. Thanks to the versatility offered by MD, it was also possible to investigate the sliding friction on the first hydration layer with the tip pressed by an extremely low force of about 100 pN. The symmetry of the graphene lattice was still visible in this case, although the alternance of maxima and minima in the friction profiles appeared to be inverted. This was possibly due to the fact that water molecules in the layer are preferably located over the hollow sites of graphene [56]. The numerical results were compared with FFM measurements on graphene monolayers on Ni and Cu substrates, and also on HOPG crystal surfaces. The corresponding friction maps were comparable to those reported in UHV conditions by other groups [57]. Interestingly, the graphene lattice could not be resolved on any of the aforementioned systems, when the friction measurements were repeated in ambient conditions, due to the formation of capillary bridges already mentioned in the introduction.

We should also remark that lattice resolved friction maps can be obtained also on water-soluble materials with no need to operate in UHV conditions. To this end, submersing media with respect to which the examined surface has low solubility turned out to be excellent alternative. Figure 1e shows a FFM map of a NaCl surface in ethanol as an example [17]. Also in this case, the similarity with FFM maps acquired in UHV [8] is remarkable. Moreover, the liquid environment prevented the onset of abrasive wear up to normal force values of 60 nN at least, whereas much lower forces in the nN range are known to cause permanent damage of alkali halide surfaces on the atomic scale. The situation may change in the presence of surface defects. In the case of calcite, wear track growth rates in aqueous solutions were accurately measured by scanning the surface across a step edge [58]. For an overview on this topic, the reader is referred to the tribochemistry chapter by Dickinson in the first edition of this book [59]. Here, we will only observe that scanning with sufficiently high loads may also lead to the detachment of adlayers from the surface. From the measurement of the friction force peak accompanying the detachment, divided by the projected area of the adlayer, Pimentel et al. were able to estimate the shear strength τ of calcite on the (104) cleavage surfaces of both dolomite and kutnahorite [$CaMn(CO_3)$] crystals [60]. The measured value of τ turned out to be much lower on dolomite (7 MPa) than on kutnohorite (130 MPa), which can be due to the larger misfit of the unit cells of calcite/dolomite versus calcite/kutnohorite. In this context, the readers are also referred to the book on the Frenkel-Kontorova model by Braun and Kivshar for the role of commensurability on friction and related phenomena [61]. To conclude this Section, we would like to remark that ordered structures have been also recognized on friction force maps of crystal surfaces acquired in ionic liquids

[62]. In this case, the observed features correspond to the formation of ordered layers between the probing tip and the chosen surface. These features, and the accompanying friction force, can be reversibly tuned by using an electrochemical cell and varying the potential through it. We will come back to this topic in Sect. 4.

3 Friction in Liquid Environment: Interaction Energy, Lateral Stiffness, and Damping

As discussed in the previous section, FFM measurements and atomistic MD simulations over graphene with sharp crystalline Si probe under UHV and water [15] demonstrated that the contact between the sliding surfaces remains completely dry, while the presence of the surrounding liquid introduces noise effects in the friction signal. A recent study of the dynamics of nanoscale frictional interaction in the presence of different liquids with respect to the stick–slip behavior and characteristics, showed that the lateral interaction on metallic glasses in corrosive borate and phosphate buffers [32] manifested in different interaction strength (force, contact stiffness), and dynamics (slip length), while exhibiting monotonous trends (increase) with the applied load. Influence of surrounding liquid, ethanol, on the lateral contact stiffness and on slip length dynamics were reported for nanoscale friction on alkali halide, NaCl, with (oxidized) Si probe [17, 63]. The $Si(O_2)$-NaCl frictional behavior in the presence of ethanol displayed different features and dynamics than the ones reported for the same system under UHV conditions [13, 64].

The recorded stick–slip patterns indicate an interaction energy surface, comprising energy barriers determined by the position of, the surface atoms. Each stick event occurs in a local minimum and is followed by an irreversible jump (slip) over a potential barrier defined by the scan direction. This interaction is simply described for ordered-lattice surfaces by the Prandtl-Tomlinson (PT) model [65–67]. This model assumes that the crystal surface has a corrugated periodic potential, where the minimum values correspond to the equilibrium position of a point mass ("tip") pushed against the surface with a given load. The point mass is also dragged along the surface by a lateral force, which is equivalent in its effect to a spring being elongated by a moving support$_V$.

In one dimension (1D) the PT potential, $U(x, t)$, is accordingly described by the interaction periodic potential overlapping with a quadratic elastic potential that is moving with the time t:

$$U(x,t) = -U_0 cos\left(\frac{2\pi x}{a}\right) + \frac{k}{2}(x - Vt)^2 \tag{1}$$

where U_0 is the amplitude of the corrugation interaction energy, a is the periodicity of the lattice, and k is the elastic spring constant at the contact, and V is the scan velocity of the support (different from the local velocity $\dot{x}$ of the tip). This model became highly useful for the interpretation and analysis of FFM of crystalline surfaces since it is formulated in terms of the experimental parameters.

A dimensionless number, known as the PT parameter, quantifies the ratio between the tip-surface interaction and the elastic energy stored in the deformed contact [13, 68, 69]:

$$\eta = \left(\frac{2\pi}{a}\right)^2 \frac{U_0}{k} \tag{2}$$

Under the assumption of the potential in Eq. (1), the characteristic value of the maximal slip force $\langle F \rangle$, and lateral stiffness (slope of the stick phase), $\langle dF/dx \rangle$ [13, 68], allows to estimate η experimentally as $\eta = 2\pi \langle dF/dx \rangle a{-}1)$, and consequently, the contact elasticity $k = \langle dF/dx \rangle \ (1 + 1/\eta)$. The PT parameter can be considered as an order parameter to characterize the frictional dynamics. During the tip motion, friction dynamics ranges from continuous sliding, through stick–slip over a single period (single slip), up to and stick–slip over several lattice periods (multiple/long slip) [13, 64, 70]. For $\eta < 1$, the elastic energy of the system is dominant, and a continuous sliding is likely to occur (there is only a single minimum in the potential), while at $\eta > 1$ the corrugation amplitude of the system is dominant, and stick–slip is expected (two or more minima in $U(x,t)$). At critical value, $\eta = 1$ the corrugation amplitude and the elastic energy are equal and marks the transition between the sliding and the stick–slip regimes.

In addition to the PT parameter, single and multiple/long slips are determined by dissipative phenomena [63, 64, 71, 72]. Using the PT interaction potential, FFM can be described with a basic equation of motion, $m\ddot{x} = -m\,\gamma\,\dot{x} - \nabla U(x, t)$, where m is the effective mass of the cantilever, and γ is a damping coefficient. Under UHV, γ is related to the phonon (and possibly electron) excitation that spreads from contact area into the bulk [1, 2, 73–75]. In ambient/liquid environment viscous and electronic properties of the surroundings introduce additional contributions. Variations in friction regimes have been attributed to the interplay between η that determines the number of possible minima in the interaction potential and γ that characterize the damping state if the system [64, 71, 76].

The damping coefficient is an elusive parameter, which is difficult to assess experimentally [75]. Indirect information about it can be empirically obtained through slip length (or alternatively lateral force variations) distributions [63, 64, 71, 72, 76, 77]. In the harmonic oscillator approximation its critical value is given by:

$$\gamma_c = 2\sqrt{\frac{k}{m}}. \tag{3}$$

This approach generalizes all degrees of freedom of all vibrational modes across the sliding surfaces into a thermal (harmonic) bath. It is used to distinguish between the underdamped and overdamped states that are related to observed slip lengths [63, 64, 71, 72, 76]. In practice, γ in FFM represents an effective overall damping at the contact, since the actual molecular-level modes are considerably faster than those of the measured reaction coordinates, $\boldsymbol{x}(t)$, imposing instantaneous thermal equilibrium in every time step [78, 79]. This means that while the environment is considered in thermal equilibrium, the actual system is not [80]. It is therefore reasonable to assume that the observed effect of liquid surrounding in FFM measurements would be manifested in the value of γ (with respect to its critical value).

In the following, we focus as a case study on the NaCl-Si(O_2) nano-scale contact that was studied under UHV conditions [13, 64] and in liquid (ethanol) environment [17, 63]. In an early work, where FFM measurements were performed in UHV under a low range of normal load ($F_N < 6.5$ nN), the amplitude of the corrugation energy, U_0, monotonously increased with the applied load, while the lateral contact stiffness, k, remained nearly constant [13]. The dependence of the interaction energy is strongly associated with the lateral friction force that depends on the applied load [11, 81–83]. In another FFM study in UHV, the range of the applied load was increased up to 91 nN, also reporting no specific dependence of k with F_N [64]. FFM measurements in liquid surroundings under normal loads ranging from ~ 3 to 60 nN also disclosed an increase of the interaction energy amplitude with the applied load [17]. Here, however, a power-law dependency was observed between k and F_N. Additionally, while the dynamics under UHV displayed increasing of the slip-length with the load [64], in liquid surroundings the dynamics remained around a single slip [63].

These differences in the ratios between U_0 and k, as mentioned above, are reflected in the dependence of the PT parameter on the applied load. The evolution of η with F_N for NaCl-Si(O_2) contact is plotted in Fig. 2a under UHV and liquid surroundings. The linear increase of η in UHV is reasoned by the monotonic increase of U_0 and almost no dependence of k with the applied load. The non-monotonic behavior of η in liquid is related to the different rates of increase of U_0 and k with the load [30], where up to ~ 20 nN U_0 grows faster than k, and then reverse trends. The different trends of η with the load highlight a specific role of the damping state of the system on the slip length [64, 71].

For dry nano-scale friction, damping at the contact is expected to increase with the applied load [73], however the environmental conditions can affect it. Theoretically, with the increase of η, more states in the energy landscape of the interaction potential become available, thus introducing the possibility for long distance transitions (i.e., long slip lengths) [70]. The slip dynamics, however, will be determined by the damping at a given value of η [63, 64, 70, 71, 84, 85]. When the system's vibrations are underdamped ($\gamma < \gamma_c$) long transitions can occur instantaneously over two lattice constants and up to the possible maximal states enabled by the value of η. When the system is overdamped ($\gamma > \gamma_c$), even at high values of η, mostly single transitions across local energy barrier will be observed (single slips).

To reproduce damping effects, the PT model has to be investigated with the Langevin equation (LE). The LE is a stochastic differential equation describing momentum balance with a fluctuating thermal force. It reflects the momentum change as a sum of its counter forces that induce motion: a dissipative friction force, the force resulting from the potential of the system (the PT potential), and a stochastic term describing a randomly fluctuating forces acting on it (due to the thermal energy of the system). In a two-dimensional (2D) problem it reads:

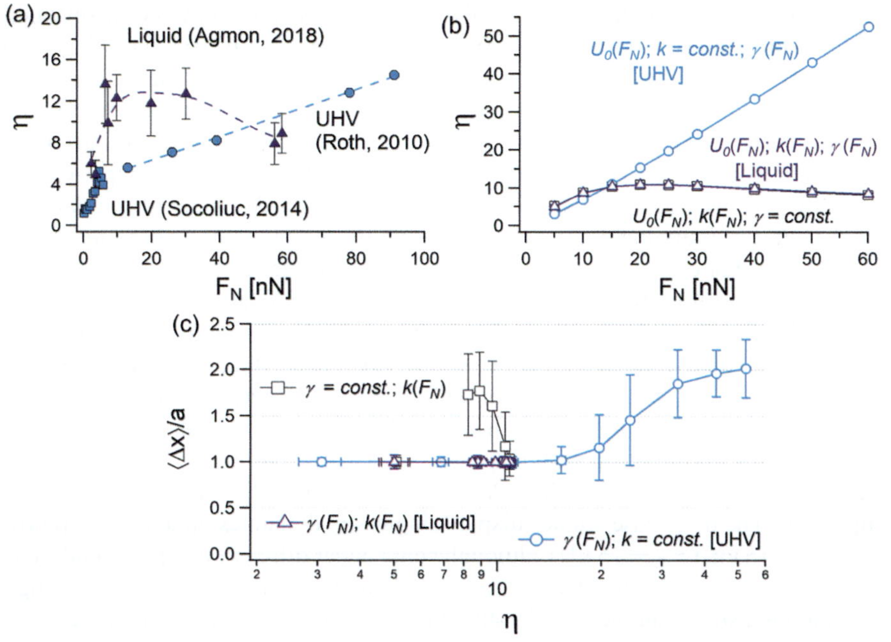

Fig. 2 Nanoscale friction dynamics in UHV and in liquid. **a** variation of the PT parameter, η, with the applied load for sliding NaCl-Si(O_2) contact in UHV (light blue squares [13], and circles [64]), and in liquid (ethanol) surroundings (purple triangles [17]). The dashed lines are linear fits (UHV), and spline interpolation (Liquid) to guide the eye. **b** Variation of η obtained from numerical solutions of 1D LE for case studies representing UHV (empty light blue circles), and liquid (empty purple triangles) conditions, with a complementary situation, in which the damping was held constant (empty black squares) (adapted from [63]). **c** mean normalized slip lengths for the systems shown in **b** as a function of η (adapted from [63])

$$m\ddot{\boldsymbol{r}}(t) = -m\gamma\dot{\boldsymbol{r}}(t) - \nabla U(\boldsymbol{r}, t) + \Gamma(t) \tag{4}$$

where $\boldsymbol{r} = (x, y)$ is the time dependent position vector of the tip (in 1D: $\boldsymbol{r} = x$). The thermal activation force, $\Gamma(t)$, represents the stochastic contributions emerging from the fluctuation–dissipation theorem $\langle\Gamma(t)\rangle = 0$, and $\left\langle\Gamma(t)\Gamma\left(t'\right)\right\rangle = 2m\gamma k_B\delta\left(t - t'\right)$, with k_B being the Boltzmann constant, T is the absolute temperature, $\delta(t)$ is the Dirac delta function, and $\langle\ldots\rangle$ denotes ensemble averaging. The averaging of the thermal fluctuations of the fast modes (that are not considered in Eq. (4)) are manifested via environmental conditions ("thermal bath") that couple the reaction coordinate $\boldsymbol{r}(t)$ with γ in the noise term. This suggests that the system of the sliding contact is close to be critically damped, where the damping can be associated with the vibrational frequencies in the system (Eq. 3) [2, 85].

The numerical solutions of Eq. (4), with the potential U defined in Eq. (1) also extended to 2D, result in time dependent trajectories across the lattice, from which the frictional forces can be calculated as $F_L(t) = -k[\boldsymbol{r}(t) - Vt]$: see below. In this way the LE provides a simplistic, yet highly useful platform to study the involvement of dissipation during sliding in the single asperity level [2, 63, 64, 71, 72, 76, 80, 86, 87]. Back to the different behavior of η for the NaCl-Si(O_2) contact under UHV conditions and in liquid surroundings [30, 63], the assumption that $U_0 = f(F_N)$, $k = const.$ [13] in UHV, and $U_0 = f(F_N)$, $k = f(F_N)$ in liquid [17, 63] was introduced in numeric solutions of the (1D) PT model based on LE [63]. The dissipative state for both systems was calculated using Eq. (3) to define the value of the damping coefficient, i.e. assuming that $\gamma = \gamma_c(F_N)$, with the definition of k characteristic for each system. For comparison γ was also taken as a constant.

From the resulting friction (maximal) slip forces and local stiffnesses, the calculated η displayed trends similar to those observed experimentally (Fig. 2b). Interestingly, the liquid-counter system, in which $\gamma = const.$ (black squares) displayed similar behavior as the liquid system with $\gamma = \gamma(F_N)$ (purple triangles). Furthermore, Fig. 2c plots the load dependency of the means and standard deviations of the slip-lengths normalized with respect to the lattice constant, a. The slip lengths were estimated using Hooke's law, $\Delta x = \Delta F/k$, where ΔF is the force drop in the slip phase. The three case studies display single-slip dynamics up to η ~ 10 (corresponding to a load $F_N = 15$ nN), although being under different damping conditions. Above this value of η, the "UHV" system, where k was maintained constant (light blue circles), shows an increase in slip length with the load, and γ increased from being slightly underdamped (below $\eta = 11$) and becomes overdamped as η grows [64]. For the "Liquid" system (purple triangles), k increased with the load, and η persisted below 11. Consequently, the system remained at critical damping, and multiple slips were suppressed. For the situation in which $\gamma = const.$, while $k(F_N)$ (black squares), below $\eta = 11$ the system went from being overdamped (below $F_N = 15$, at the transition point of η) and became underdamped with the increase of the load (although η decreased). Hence, even at values of $\eta < 11$, multiple jumps began to appear.

From this simplistic modeling, under the assumption that the contact remains dry [15], it is reasonable to assume that a substantial fraction of the dissipated heat is released into the liquid surroundings. As a result, it is possible that the liquid surrounding serves a large heat bath (imposing quasi-isothermal conditions) that absorbs and removes the released heat by convection, and the contact gets stiffer with the increased load. Under UHV conditions, the generated heat has no other place to dissipate but into the contacting surfaces, thus varying its effective stiffness very little. The damping, however, appears to vary from being slightly underdamped into overdamped [64, 71]. Although the 1D PT model can provide insights, it may overlook aspects that add to the complexity of the description of the system [88], and also take into account that the role of the damping along a single reaction coordinate in 1D [89] and ignores possible contributions resulting from the curvature of the potential [90].

For this reason the effect of the damping coefficient of the system was further investigated on an hexagonal lattice [72], where not only the dimensionality but also the scanning angle takes a role in the frictional dynamics [30, 83, 87, 91–98]. The attention here will be reduced to the (relatively) simple ideal (yet insightful) case of 2D nano-scale friction at absolute zero temperature under UHV conditions ($k = const.$). The dynamics for this system were explored on a specific scanning angle of 12°, at $\eta = 10$ for $\gamma = \gamma_c$, and $\gamma = 0.1\gamma_c$, i.e., at critical and underdamped oscillation values. On a periodic 1D potential, the solutions of Eq. (1) at the quasistatic limit, the solutions to $\nabla U = 0$ define the number of possible states in the energy landscape of the system, and consequently the values of η at which the slip length vary (number of lattice constants) when the system is overdamped [70, 71, 99]. In 2D this description defines lattice constant positions confined into triangle and hexagonal landing patches on the x–y plane at $\eta > 1$, that get rounder as η grows [100]. These positions correspond to the "landing points" that mark the end of the slip phase. Figure 3a shows the possible landing points for $\eta = 10$ at the angle of 12°.

Trajectories over hexagonal lattice for the conditions specified above at critical and underdamped oscillations are shown in Fig. 3b, where the numbers of landing positions are labeled (corresponding to the attractor patterns in Fig. 3a). At high underdamping the number of accessible landing positions dramatically increases from three (4, 5, and 7) that are close to the "shot point" to twelve. For both situations

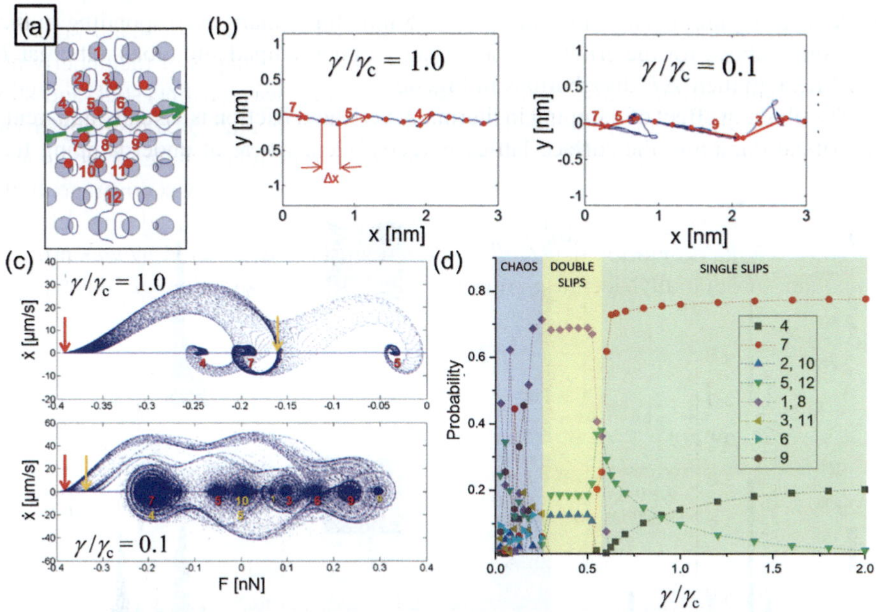

Fig. 3 Critical and strongly underdamped oscillation effects on nano-scale friction at $T = 0$ K at a scanning angle of 12° and PT parameter $\eta = 10$. **a** theoretically accessible landing sites (red dots) from an initial position (green dot) across scanning direction (green dashed arrow) over the stability region patches (light purple). **b** position trajectories on the x–y plane at critical damping (left), and underdamped oscillations (right). Landing points are marked by red points. **c** pathways in the x–F subspace at critical damping (up), and underdamped oscillations (bottom). **d** damping coefficients probabilities of the possible landing sites. Adapted from [72]

single slips are mostly observed (type 7), however for 0.1 γ_c also double (type 8) and triple (type 9) with the same orientations can be recognized. As in 1D [71], the stick slip dynamics in 2D at strong underdamping becomes chaotic. This behavior can be represented by a phase plot of the local velocity, $\dot{x}$ and lateral force, $F = F_x$ (=$k(x\text{–}Vt)$), plane (Fig. 3c). Upon reaching the shot point the $\dot{x}$-F plane is explored by the tip in a spiral trajectory [101] until the tip ends in the new pinning site. Here, at critical damping the three landing points are distinct, while at 0.1 γ_c they become more difficult to recognize.

The damping states of the system were indirectly estimated (Fig. 3d) by comparison of the peak distribution of the force drops, ΔF, along the slip events with those who were numerically obtained by solving Eq. (4) for 2D hexagonal potential at varying values of γ. At $T = 0$ K, for a given scan line angle (here 12°) some peaks in the force drop distributions are more prominent than others and can be used to identify similar peaks in the distributions obtained when γ is modified.

The damping coefficient changes the probabilities of accessing the possible landing sites, which is reflected in the location and intensity of the peaks in the force drop probabilities (Fig. 4a). The introduction of thermal vibrations reduces static friction when the process described above occurs at higher temperatures, and slip becomes thermally activated [6]. Consequently, the extent of long slip events is reduced. Figure 4b shows simulated 5 × 5 nm^2 force maps corresponding to the case study above for the critically and strongly underdamped situations, albeit at $T = 70$ K, with their ΔF distributions in Fig. 4c.

The blurring effect of damping in thermally activated friction is visually apparent. At critical damping, the surface lattice is recognized, where at underdamping the

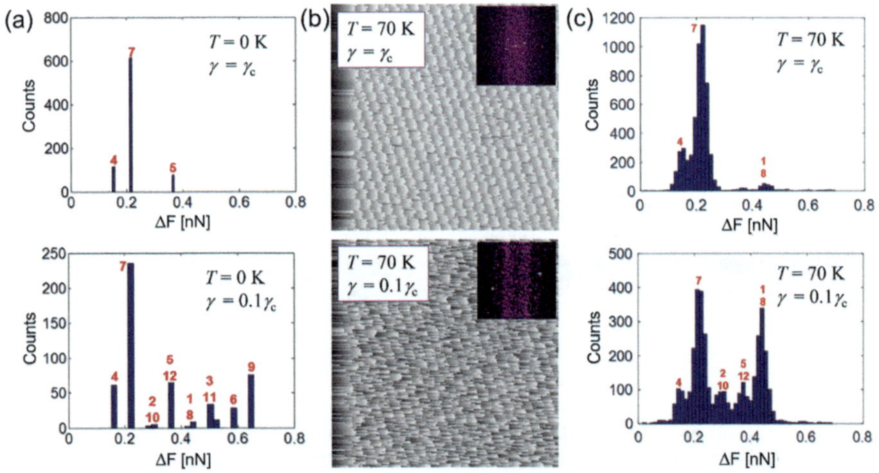

Fig. 4 The effects of thermal activation at critical damping (all upper panels), and at underdamped oscillations (all lower panels) on nano-scale friction. **a** Slip phase force drop distributions, ΔF, occurring in the slip phases at zero temperature. **b** friction maps simulated at T = 70 K. **c** slip phase force drop distributions, ΔF, occurring in the slip phases at $T = 70$ K. Adapted from [72]

friction image gets hazy. At zero temperature the increase in the accessible number of neighboring locations when the system becomes underdamped is evident in Fig. 4a. At 70 K the effect of the damping is still apparent, however the peak distribution is broadened, the peak intensities vary, and the borders of the dominant peaks mix with those of less probable slip events. The inclusion of thermal oscillations thus makes the landing sites less pronounced and consequently more difficult to explicitly detect.

4 Sliding Friction and Interfacial Nanostructures in Ionic Liquid Environments

Composed primarily of organic cations and inorganic anions, ionic liquids (ILs) represent a class of molten salts with melting points typically below 100 °C. Their potential as high-performance lubricants was first demonstrated in 2001 [102, 103]. ILs based lubricants [102, 104] offer several advantages over conventional oils stemming from their unique properties. These lubricants are typically non-flammable, exhibit excellent thermal stability, possess negligible vapor pressure, and boast intrinsic ionic conductivity. Additionally, their molecular structures can be tailored for specific applications, and they function within a wide electrochemical window [105–109]. In their application as lubricants, ILs necessarily interact directly with solid surfaces [51, 109]. This intimate contact considerably alters the nanoscopic organization of the ILs at the interface [110], resulting in properties that can deviate substantially from those observed in the bulk liquid [51, 52, 109, 111–114].

ILs demonstrate superior friction reduction capabilities compared to conventional molecular lubricants. This advantage is attributed to their ability to form robust boundary films and highly ordered interfacial structures [115–121]. These well-organized structures exhibit enhanced resistance to squeeze-out under high load conditions. At the surface, ILs self-assemble into ordered ion layers, exhibiting strong attractive forces with the surface [122]. These forces encompass long-range electrostatic (Coulombic) interactions, short-range dispersion forces (van der Waals), hydrogen bonding, dipole–dipole and various solvophobic interactions [62]. The formation of ordered ion layers at the interface is a critical factor in enhancing the lubrication performance of IL–solid interfaces [51, 123]. For instance, studies using AFM have demonstrated exceptional super-lubrication with ultra-low friction coefficient (μ) of 0.001 at pressures up to 78 MPa for imidazolium ILs on graphite, even without the application of an electric field regulation (Fig. 5a) [124]. This finding highlights the potential of ILs to achieve superlubricity. "Atomic-scale locking" between carbon atoms on the alkyl chain of ILs and graphite was supposed to be the underlying mechanism responsible for the ordered striped structure that was observed.

The self-assembled monolayer structure formed by imidazolium, 1-Octadecyl-3-methylimidazolium bis((trifluoromethyl)sulfonyl)imide ([C18mim][NTf2]), on the

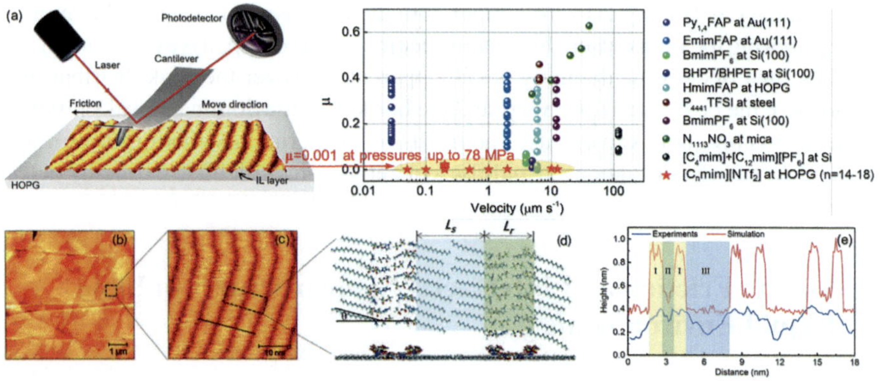

Fig. 5 **a** Friction coefficients, μ, of ILs on various solid surfaces as estimated from AFM experiments. **b–e** characterization of [C_{18}mim][NTf_2] on HOPG based on AFM and numeric simulations: **b** AFM topography image of a [C_{18}mim][NTf_2] lubricant layer and **c** high-resolution AFM image revealing the fine structure of the layer. **d** top and side representations obtained from simulations, depicting the arrangement of for cations and anions within the lubricant layer. L_s and L_r represent the respective lengths of the alkyl chain and anion region. The underlying graphite surface is also shown here. **e** comparison of the cross-sectional profiles obtained from AFM experiment (black line in **c**) and simulation results. From [124]

graphene surface is shown in Fig. 5b. The film exhibits a smooth morphology with a uniform thickness of approximately 0.5 nm, consistent with a monolayer IL with the imidazole rings oriented parallel to the surface. A high-resolution image (Fig. 5c) further unveils a well-ordered striped structure with a periodicity of about 6 nm. Here, the bright striped regions correspond to the densely packed ILs assemblies on the surface. Accompanying simulations suggest the presence of alternating ordered and disordered ionic regions within the IL film (Fig. 5d). Notably, the simulations reveal a head-to-head structural motif, where two distinct imidazole ring configurations alternate within the same vertical column. The cross-sectional profiles obtained from both AFM experiments and simulations (Fig. 5e) reveal a three-region periodicity within a single structural period. Regions I, II, and III correspond to imidazole rings, anions, and alkyl chains, respectively. The observed discrepancy between experimental and simulated results can be primarily attributed to the out-of-plane flexibility of the warped imidazole rings. During AFM tip scanning, these rings likely experience deformation are pushed closer to the underlying graphite surface.

Although significant advancements in understanding IL layering on surfaces have been made, the precise influence of the layered structure of ILs on friction continues to be a topic of discussion. Layering structures of 1-hexyl-3-methylimidazolium tris(pentafluoroethyl) trifluorophosphate ([HMIM][FAP]) and trihexyl-(tetradecyl)phosphonium bis(trifluoromethane) sulfonamide ([$P_{6,6,6,14}$][TFSI]), confined between a sharp AFM silicon probe and a stainless-steel surface, were compared using AFM force curves and friction force measurements (Fig. 6a) [125]. The force curves in Fig. 6a (left) reveal more ion layers for [HMIM][FAP] compared to $[P_{6,6,6,14}]$[TFSI] near the surface. This difference arises because the bulky $[P_{6,6,6,14}]^+$ cation in $[P_{6,6,6,14}]$[TFSI] weakens the electrostatic attraction to the anion, leading to less stable layered structures. Consequently, the sliding AFM probe can disrupt the weak interfacial structure of the $[P_{6,6,6,14}]$[TFSI] ionic liquid with minimal energy dissipation and low friction (Fig. 6a, right). Fig. 6b shows the presence of a "quantized" friction behavior for the ionic liquid 1-butyl-1-methylpyrrolidinium bis[(trifluoromethane)sulfonyl]imide $[C_4C_1Pyrr][NT_f2]$ confined between mica sheets in a surface force apparatus (SFA). Within each quantized friction regime, the friction force shows a roughly linear dependence on the applied normal load. Interestingly, the slope of this dependence, which defines the friction coefficient μ, increases as the number of the confined ion layers decreased. This trend is attributed to a stronger interlocking of the more ordered layers closer to the surface. These more ordered and densely packed ion layers result in a 'stronger inter-layer attraction for the same value of contact area and a greater barrier for unlocking the surfaces to allow shear' [126]. AFM studies of imidazole-based ILs on mica and carbon surfaces have also reported a correlation between a stronger layered structure of IL ions at the surface and a higher friction force [127–129]. This can be again attributed to the increased energy dissipation required to overcome the stronger interlayer bonds within the most well-organized structures.

However, Atkin's group reported contrasting findings [130, 131]. Their research suggested that increased ion ordering can lead to a smoother sliding plane at the interface, resulting in reduced friction. This assumption was supported by simulation results [132]. Atomistic nonequilibrium molecular dynamics (NEMD) simulations revealed indeed a correlation between the structure ofthe ion layering of imidazolium ILs on surfaces and the corresponding friction forces (Fig. 6c) [132, 133]. The left panel of Fig. 6c depicts the number density profiles of 1-ethyl-3-methylimidazolium bis-(trifluoromethylsulfonyl)imide ($[C_2MIM][NTf_2]$) confied between two positively charged slabs under a 400 MPa load. The profiles reveal the presence of five distinct ionic layers. The characteristic alternating charge pattern of the IL under confinement is evidentFurthermore, the time-averaged 2D structure factor, $S_m(k_x, k_y)$ [132], reveals crystalline ordering of ions within the first three layers closest to the slab surface, as shown in the right panel of Fig. 6c. Layer 1 (not shown in Fig. 6c) corresponds to the anion-rich layer closest to the fixed wall. The adjacent cation-rich layer 2 is followed by layer 3, the anion-rich layer in the middle of the IL film. The well-defined bright spots in layer 3 of $[C_2MIM][NTf_2]$ (Fig. 6c, right panel) indicate crystallization throughout the confined IL film caused by the positively charged slab surface. In contrast, layer 2 exhibits a clear disordered, liquid-like structure. The observed crystalline ordering of the IL ions has a close connection to the friction forces. When the ordered ion structures are weakly commensurate or incommensurate with the slab structure, the reduction in friction is facilitated. Thus, the relationship between IL ordering and friction remains a topic of debate. Further investigation is needed to fully understand the complex interplay between IL structure and tribological behavior.

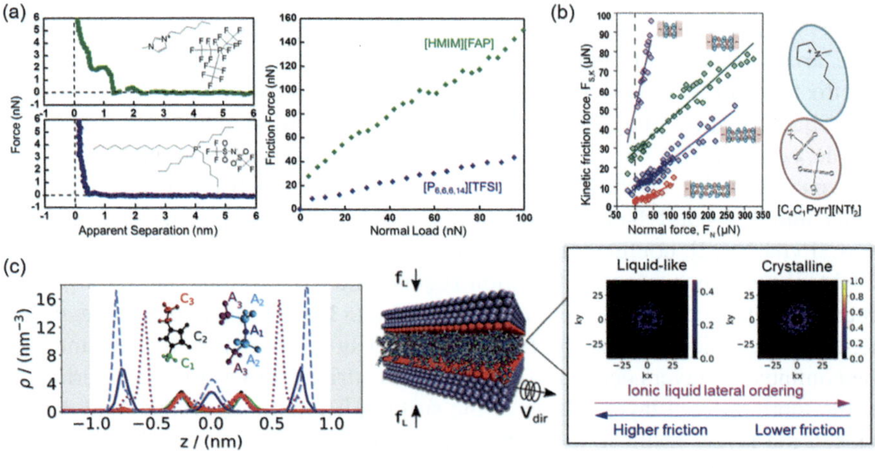

Fig. 6 **a** AFM data for [HMIM][FAP] and [$P_{6,6,6,14}$][TFSI] on a stainless-steel surface: force curves (left) and friction force vs normal load (right). From [125]. **b** Plots of kinetic friction force, $F_{S,k}$, versus normal load, F_N, for various confined ion layer thicknesses of the IL [C_4C_1Pyrr][NTf_2] measured using a SFA with mica surfaces. From [126]. **c** left: number density profiles obtained from nonequilibrium MD simulations for the IL [C_2MIM][NTf_2] confined between two positively charged slabs (+32 μC/cm^2) under a 400 MPa load. The line colors correspond to the different ionic groups within the IL (see legend). Middle: This panel depicts the slabs confining the IL. A normal load, f_L, is applied, and the top slab is moving at a constant velocity, V_{dir}, in the y direction, while the bottom slab remains stationary. Right: 2D structure factor $S_m(k_x, k_y)$ for layers 2 and 3 of the first three ion layers of the confined [C_2MIM][NTf_2] closest to the slab surface. This plot reveals crystalline ordering of the ions. k_x and k_y represent the wave-vectors in the *x* and *y* directions, respectively. Bright spots in the plot indicate the formation of 2D crystalline structures. Layer 1 (not shown here) would be the anion-rich layer closest to the fixed wall. Layer 2 (depicted in the figure) is the adjacent cation-rich layer. Layer 3 (also shown here) is the anion-rich layer in the middle of the confined IL film. From [132]

5 Conclusions and Outlook

In summary, AFM and, more specifically, FFM in water is a very suitable technique for measuring atomic-scale friction on mineral and synthetic crystalline surfaces and, at the same time, for obtaining sublattice resolution on them in a scarcely explored way. The quality of the FFM images achieved on carbonate, graphene, and organic semiconductor surfaces is indeed comparable to those obtained in UHV. This is ultimately due to the absence of capillary bridges in both environments, which tend to increase the adhesion between scanning probe and sample and cause surface wear when imaging in ambient conditions. Friction maps also allow one to monitor the growth of heteroepitaxial overgrowths in aqueous solutions, and to identify the structure and orientation of ultrathin films of organic molecules previously assembled on a crystal surface. Other liquids than water have been also addressed successfully. An example is the use of ethanol for imaging water-soluble NaCl. In this case FFM not only allowed to resolve the square lattice of the alkali halide surface, but also to do that with normal forces values well above those causing surface wear in dry conditions. A key parameter, when dealing with atomic-scale friction in liquids, is the damping coefficient γ limiting the slip length of the scanning probe. The issue is presented in the chapter by means of measurements, and of numeric solutions of the Prandtl-Tomlinson model at low temperature, which may serve as a basis for interpretations of experimental results still to come. Another promising topic is the application of FAFM to sliding friction in ILs. In this context, the combination of FFM and force curves acquired by AFM has allowed to investigate the interplay between nanoscale friction and nanostructures formed by the confined ILs. Some contradictory results are nevertheless present, which stresses the importance of continuing the development of experimental investigations in close connection with complementary theoretical methods such as MD simulations.

References

1. J.Y. Park, M. Salmeron, Fundamental aspects of energy dissipation in friction. Chem. Rev. **114**(1), 677–711 (2014)
2. S.Y. Krylov, J.W.M. Frenken, The physics of atomic-scale friction: Basic considerations and open questions. Phys. Status Solid. B-Basic Solid State Phys. **251**(4), 711–736 (2014)
3. A. Vanossi et al., Recent highlights in nanoscale and mesoscale friction. Beilstein J. Nanotechnol. **9**, 1995–2014 (2018)
4. O. Hod et al., Structural superlubricity and ultralow friction across the length scales. Nature **563**(7732), 485–492 (2018)
5. Y. Sang, M. Dube, M. Grant, Thermal effects on atomic friction. Phys. Rev. Lett. **87**(17) (2001)
6. L. Jansen, et al., Temperature dependence of atomic-scale stick-slip friction. Phys. Rev. Lett. **104**(25) (2010)
7. I. Barel et al., Temperature dependence of friction at the nanoscale: when the unexpected turns normal. Tribol. Lett. **39**(3), 311–319 (2010)
8. E. Gnecco et al., Velocity dependence of atomic friction. Phys. Rev. Lett. **84**(6), 1172–1175 (2000)
9. R. Bennewitz et al., Atomic-scale stick-slip processes on Cu(111). Phys. Rev. B **60**(16), R11301–R11304 (1999)
10. E. Granato, S.C. Ying, Non-monotonic velocity dependence of atomic friction. Tribol. Lett. **39**(3), 229–233 (2010)
11. S. Fujisawa et al., Load dependence of two-dimensional atomic-scale friction. Phys. Rev. B **52**(7), 5302–5305 (1995)
12. R.W. Carpick et al., Variation of the interfacial shear strength and adhesion of a nanometer-sized contact. Langmuir **12**(13), 3334–3340 (1996)
13. A. Socoliuc, et al., Transition from stick-slip to continuous sliding in atomic friction: Entering a new regime of ultralow friction. Phys. Rev. Lett. **92**(13) (2004)
14. M. Dienwiebel, et al., Superlubricity of graphite. Phys. Rev. Lett. **92**(12) (2004)
15. J.G. Vilhena et al., Atomic-scale sliding friction on graphene in water. ACS Nano. **10**(4), 4288–4293 (2016)
16. X.Y. Zhao, S.S. Perry, The role of water in modifying friction within MoS_2 sliding interfaces. ACS Appl. Mater. Interf. **2**(5), 1444–1448 (2010)
17. L. Agmon, et al., Estimation of interaction energy and contact stiffness in atomic-scale sliding on a model sodium chloride surface in ethanol. Scientif. Rep. **8** (2018)
18. Y.J. Diao, R.M. Espinosa-Marzal, The role of water in fault lubrication. Nat. Comm. **9** (2018)

19. U. Landman, W.D. Luedtke, A. Nitzan, Dynamics of tip-substrate interactions in atomic force microscopy. Surf. Sci. **210**(3), L177–L184 (1989)
20. R.D. Piner, C.A. Mirkin, Effect of water on lateral force microscopy in air. Langmuir **13**(26), 6864–6868 (1997)
21. L. Sirghi, Effect of capillary-condensed water on the dynamic friction force at nanoasperity contacts. Appl. Phys. Lett. **82**(21), 3755–3757 (2003)
22. T. Palamarciuc et al., Spin crossover materials evaporated under clean high vacuum and ultra-high vacuum conditions: from thin films to single molecules. J. Mater. Chem. **22**(19), 9690–9695 (2012)
23. F. Ohnesorge, G. Binnig, True atomic resolution by atomic force microscopy through repulsive and attractive forces. Science **260**(5113), 1451–1456 (1993)
24. A. Labuda, et al., High-resolution friction force microscopy under electrochemical control. Rev. Scient. Inst. **81**(8) (2010)
25. C.M. Pina, R. Miranda, E. Gnecco, Anisotropic surface coupling while sliding on dolomite and calcite crystals. Phys. Rev. B **85**(7) (2012)
26. B.J. Robinson, N.D. Kay, O.V. Kolosov, Nanoscale interfacial interactions of graphene with polar and nonpolar liquids. Langmuir **29**(25), 7735–7742 (2013)
27. P. Nita et al., Molecular resolution friction microscopy of Cu phthalocyanine thin films on dolomite (104) in water. Nanoscale **6**(14), 8334–8339 (2014)
28. E. Serpini et al., Nanoscale frictional properties of ordered and disordered MoS_2. Tribol. Int. **136**, 67–74 (2019)
29. C. Cafolla, K. Voïtchovsky, Lubricating properties of single metal ions at interfaces. Nanoscale **10**(25), 11831–11840 (2018)
30. S. Skuratovsky, L. Agmon, R. Berkovich, Comparative study of dimensionality and symmetry breaking on nanoscale friction in the Prandtl-Tomlinson model with varying effective stiffness. Tribol. Lett. **68**(4) (2020)
31. C. Cafolla, W. Foster, K. Voitchovsky, Lubricated friction around nanodefects. Sci. Adv. **6**(14) (2020)
32. H.R. Ma, R. Bennewitz, Atomic-scale stick-slip friction on a metallic glass in corrosive solutions. Tribol. Int. **171** (2022)
33. D.L. Zhang et al., Liquid-phase friction of two-dimensional molybdenum disulfide at the atomic scale. ACS Appl. Mater. Interfaces. **15**(17), 21595–21601 (2023)
34. C.L. Yi et al., Frictional properties of MoS_2 on a multi-level rough wall under starved lubrication. Phys. Chem. Chem. Phys. **25**(20), 14348–14358 (2023)
35. K.B. Jinesh, J.W.M. Frenken, Experimental evidence for ice formation at room temperature. Phys. Rev. Lett. **101**(3) (2008)
36. T. Arif, et al., Understanding the independent and interdependent role of water and oxidation on the tribology of ultrathin molybdenum disulfide (MoS_2). Adv. Mat. Interf. **6**(22) (2019)
37. Z.J. Ye et al., Load-dependent friction hysteresis on graphene. ACS Nano. **10**(5), 5161–5168 (2016)
38. T. Arif, G. Colas, T. Filleter, Effect of humidity and water intercalation on the tribological behavior of graphene and graphene oxide. ACS Appl. Mater. Interf. **10**(26), 22537–22544 (2018)
39. A. Khajeh et al., Effect of ambient chemistry on friction at the basal plane of graphite. ACS Appl. Mater. Interf. **11**(43), 40800–40807 (2019)
40. L. Agmon et al., Nanoscale contact mechanics of the interactions at monolayer MoS_2 interfaces with Au and Si. Tribol. Int. **174**, 107734 (2022)
41. J.F. Li et al., Boundary slip of oil molecules at MoS_2 homojunctions governing superlubricity. ACS Appl. Mater. Interf. **14**(6), 8644–8653 (2022)
42. R. Buzio, C. Boragno, U. Valbusa, Friction laws for lubricated nanocontacts. J. Chem. Phys. **125**(9) (2006)
43. B.X. Fu, Y.J. Diao, R.M. Espinosa-Marzal, Nanoscale insight into the relation between pressure solution of calcite and interfacial friction. J. Colloid Interface Sci. **601**, 254–264 (2021)

44. G. He, M.H. Müser, M.O. Robbins, Adsorbed layers and the origin of static friction. Science **284**(5420), 1650–1652 (1999)
45. M.H. Müser, in *Atomistic Simulations of Solid Friction, in Bridging Time Scales: Molecular Simulations for the Next Decade*, ed. by P. Nielaba, M. Mareschal, G. Ciccotti (Springer, Berlin, Heidelberg, 2002), pp. 289–317
46. W.G. Ouyang, A.S. de Wijn, M. Urbakh, Atomic-scale sliding friction on a contaminated surface. Nanoscale **10**(14), 6375–6381 (2018)
47. J. Krim, Friction and energy dissipation mechanisms in adsorbed molecules and molecularly thin films. Adv. Phys. **61**(3), 155–323 (2012)
48. M.H. Müser, L. Wenning, M.O. Robbins, Simple microscopic theory of Amontons's laws for static friction. Phys. Rev. Lett. **86**(7), 1295–1298 (2001)
49. J.F. Li et al., Fluorination to enhance superlubricity performance between self-assembled monolayer and graphite in water. J. Colloid Interface Sci. **596**, 44–53 (2021)
50. X.Y. Ren et al., Superlubricity under ultrahigh contact pressure enabled by partially oxidized black phosphorus nanosheets. NPJ 2d Mat. Appl. **5**(1) (2021)
51. R. Hayes, G.G. Warr, R. Atkin, Structure and nanostructure in ionic liquids. Chem. Rev. **115**(13), 6357–6426 (2015)
52. R. An et al., Atomic force microscopy probing interactions and microstructures of ionic liquids at solid surfaces. Nanoscale **14**(31), 11098–11128 (2022)
53. C. Pimentel, E. Gnecco, C.M. Pina, High-resolution imaging of (100) kyanite surfaces using friction force microscopy in water. Surf. Sci. **635**, 123–127 (2015)
54. C. Pimentel, et al., Formation of zabuyelite, Li2CO3, on dolomite and kutnohorite {10.4} surfaces from supersaturated and undersaturated solutions: Growth behavior and anomalous friction at the nanoscale. J. Crystal Growth **550** (2020)
55. R. Alvarez-Asencio, et al., Molecular-scale shear response of the organic semiconductor β-DBDCS (100) surface. Phys. Rev. B **96**(11) (2017)
56. G. Tocci, L. Joly, A. Michaelides, Friction of water on graphene and hexagonal boron nitride from Ab initio methods: very different slippage despite very similar interface structures. Nano Lett. **14**(12), 6872–6877 (2014)
57. C. Lee et al., Frictional characteristics of atomically thin sheets. Science **328**(5974), 76–80 (2010)
58. N.S. Park et al., Tribological enhancement of $CaCO_3$ dissolution during scanning force microscopy. Langmuir **12**(19), 4599–4604 (1996)
59. J.T. Dickinson, in *Single Asperity Nanometer-Scale Studies of Tribochemstry*, in *Fundamentals of Friction and Wear*, ed. E. Gnecco, E. Meyer (Springer, Berlin, Heidelberg, 2007), pp. 481–520
60. C. Pimentel, C.M. Pina, E. Gnecco, Epitaxial growth of calcite crystals on dolomite and kutnahorite (104) surfaces. Cryst. Growth Des. **13**(6), 2557–2563 (2013)
61. O.M.B.Y.S. Kivshar, in *The Frenkel-Kontorova Model, Concepts, Methods, and Applications* (Theoretical and Mathematical Physics, Springer Berlin, Heidelberg, 2004), p. 472
62. J. Sweeney, et al., Control of nanoscale friction on gold in an ionic liquid by a potential-dependent ionic lubricant layer. Phys. Rev. Lett. **109**(15) (2012)
63. S. Skuratovsky et al., Surroundings affect slip length dynamics in nanoscale friction through contact stiffness and damping. Friction **11**(2), 216–227 (2023)
64. R. Roth et al., Multiple slips in atomic-scale friction: an indicator for the lateral contact damping. Tribol. Lett. **39**(1), 63–69 (2010)
65. L. Prandtl, Ein Gedankenmodell zur kinetischen Theorie der festen Körper. ZAMM—J. Appl. Math. Mech./Zeitschrift für Angewandte Mathematik und Mechanik **8**(2), 85–106 (1928)
66. G.A. Tomlinson, CVI. a molecular theory of friction. The London, Edinburgh, and Dublin Philos. Magazine J. Sci. **7**(46), 905–939 (1929)
67. V.L. Popov, J.A.T. Gray, Prandtl-Tomlinson model: history and applications in friction, plasticity, and nanotechnologies. Zamm-Zeitschrift Fur Angewandte Mathematik Und Mechanik **92**(9), 683–708 (2012)

68. E. Gnecco et al., Friction experiments on the nanometre scale. J. Physics-Cond. Matter **13**(31), R619–R642 (2001)
69. J.S. Helman, W. Baltensperger, J.A. Holyst, Simple model for dry friction. Phys. Rev. B **49**(6), 3831–3838 (1994)
70. S.N. Medyanik, et al., Predictions and observations of multiple slip modes in atomic-scale friction. Phys. Rev. Lett. **97**(13) (2006)
71. E. Gnecco, R. Roth, A. Baratoff, Analytical expressions for the kinetic friction in the Prandtl-Tomlinson model. Phys. Rev. B. **86**(3) (2012)
72. E. Gnecco, L. Agmon, R. Berkovich, Friction and chaos: Influence of the damping coefficient on atomic-scale stick-slip on hexagonal crystal lattices. Phys. Rev. B **105**(23) (2022)
73. B.N.J. Persson, in *Sliding Friction: Physical Principles and Applications*. (NanoScience and Technology, Springer Berlin, Heidelberg, 2000), p. 516
74. G. Witte et al., Damping of molecular motion on a solid substrate: Evidence for electron-hole pair creation. Phys. Rev. Lett. **80**(1), 121–124 (1998)
75. S.Y. Krylov, J.W.M. Frenken, Atomistic mechanisms for frictional energy dissipation during continuous sliding. Scient. Rep. **11**(1) (2021)
76. Y. Dong, A. Vadakkepatt, A. Martini, Analytical models for atomic friction. Tribol. Lett. **44**(3), 367–386 (2011)
77. W.G. Conley, A. Raman, C.M. Krousgrill, Nonlinear dynamics in Tomlinson's model for atomic-scale friction and friction force microscopy. J. Appl. Phys. **98**(5) (2005)
78. P. Hänggi, P. Talkner, M. Borkovec, Reaction-rate theory: fifty years after Kramers. Rev. Mod. Phys. **62**(2), 251–341 (1990)
79. P. Reimann, Brownian motors: noisy transport far from equilibrium. Phys. Rep.-Rev. Sect. Phys. Lett. **361**(2–4), 57–265 (2002)
80. P. Reimann, M. Evstigneev, Description of atomic friction as forced Brownian motion. New J. Phys. **7** (2005)
81. M.G. Rozman, M. Urbakh, J. Klafter, Controlling chaotic frictional forces. Phys. Rev. E **57**(6), 7340–7343 (1998)
82. C. Fusco, A. Fasolino, Power-law load dependence of atomic friction. Appl. Phys. Lett. **84**(5), 699–701 (2004)
83. P. Steiner, et al., Two-dimensional simulation of superlubricity on NaCl and highly oriented pyrolytic graphite. Phys. Rev. B. **79**(4) (2009)
84. J. Nakamura, S. Wakunami, A. Natori, Double-slip mechanism in atomic-scale friction: Tomlinson model at finite temperatures. Phys. Rev. B. **72**(23) (2005)
85. D.W. van Baarle, et al., On the non-trivial origin of atomic-scale patterns in friction force microscopy. Tribol. Lett. **67**(1) (2019)
86. A. Vanossi et al., Colloquium: modeling friction: from nanoscale to mesoscale. Rev. Mod. Phys. **85**(2), 529–552 (2013)
87. C. Fusco, A. Fasolino, Velocity dependence of atomic-scale friction: A comparative study of the one- and two-dimensional Tomlinson model. Phys. Rev. B. **71**(4) (2005)
88. S. Maier, et al., Fluctuations and jump dynamics in atomic friction experiments. Phys. Rev. B. **72**(24) (2005)
89. Y.L. Dong et al., The roles of statics and dynamics in determining transitions between atomic friction regimes. Tribol. Lett. **42**(1), 99–107 (2011)
90. O.Y. Fajardo, J.J. Mazo, Effects of surface disorder and temperature on atomic friction. Phys. Rev. B. **82**(3) (2010)
91. S. Fujisawa et al., Atomic-scale friction observed with a two-dimensional frictional-force microscope. Phys. Rev. B **51**(12), 7849–7857 (1995)
92. W. Karino, H. Shindo, Frictional force microscopic detection of anisotropy at NaCl (100), (110) and (111) surfaces. Tribol. Int. **40**(10–12), 1568–1573 (2007)
93. O.Y. Fajardo, E. Gnecco, J.J. Mazo, Anisotropy effects and friction maps in the framework of the 2d PT model. Phys. B-Cond. Matt. **455**, 44–48 (2014)
94. C.S. Takoutsing, G.D. Kenmoé, T.C. Kofané, Effects of anisotropy and substrate shape on atomic friction force in two-dimensional model. Tribol. Lett. **65**(3) (2017)

95. O.E. Dagdeviren, Exploring load, velocity, and surface disorder dependence of friction with one-dimensional and two-dimensional models. Nanotech. **29**(31) (2018)
96. M.R. Vazirisereshk et al., Friction anisotropy of MoS_2: effect of tip-sample contact quality. J. Phys. Chem. Lett. **11**(16), 6900–6906 (2020)
97. Q.Z. Yao, et al., Atomic stick-slip friction as a two-dimensional thermally activated process. Phys. Rev. B. **105**(16) (2022)
98. J. Konrad, et al., Anisotropy of nanoscale friction: influence of lattice structure, temperature, and wear. Phys. Rev. B. **107**(19) (2023)
99. K.L. Johnson, J. Woodhouse, Stick-slip motion in the atomic force microscope. Tribol. Lett. **5**(2–3), 155–160 (1998)
100. E. V, *Quasi-isotropy of static friction on hexagonal surface lattices.* EPL **91**(6) (2010)
101. G.D. Kenmoé, A.K. Jiotsa, T.C. Kofané, Nonlinear spring model for frictional stick-slip motion. Europ. Phys. J. B **70**(3), 353–361 (2009)
102. F. Zhou, Y.M. Liang, W.M. Liu, Ionic liquid lubricants: designed chemistry for engineering applications. Chem. Soc. Rev. **38**(9), 2590–2599 (2009)
103. C.F. Ye et al., Room-temperature ionic liquids: a novel versatile lubricant. Chem. Commun. **21**, 2244–2245 (2001)
104. M.R. Cai et al., Ionic liquid lubricants: when chemistry meets tribology. Chem. Soc. Rev. **49**(21), 7753–7818 (2020)
105. J.K. Kim et al., An imidazolium based ionic liquid electrolyte for lithium batteries. J. Power Sources **195**(22), 7639–7643 (2010)
106. N. Karimi, et al., Nonfluorinated ionic liquid electrolytes for lithium metal batteries: ionic conduction, electrochemistry, and interphase formation. Adv. Energy Mat. **11**(4) (2021)
107. J.Z. Feng et al., Ion regulation of ionic liquid electrolytes for supercapacitors. Energy Environ. Sci. **14**(5), 2859–2882 (2021)
108. M. Kar et al., Ionic liquid electrolytes for reversible magnesium electrochemistry. Chem. Commun. **52**(21), 4033–4036 (2016)
109. S.G. Zhang et al., Nanoconfined ionic liquids. Chem. Rev. **117**(10), 6755–6833 (2017)
110. M.A. Gebbie et al., Long range electrostatic forces in ionic liquids. Chem. Commun. **53**(7), 1214–1224 (2017)
111. Y.L. Wang, Z.Y. Lu, A. Laaksonen, Heterogeneous dynamics of ionic liquids in confined films with varied film thickness. Phys. Chem. Chem. Phys. **16**(38), 20731–20740 (2014)
112. Y.L. Wang, A. Laaksonen, Z.Y. Lu, Influence of ionic liquid film thickness on ion pair distributions and orientations at graphene and vacuum interfaces. Phys. Chem. Chem. Phys. **15**(32), 13559–13569 (2013)
113. M.Y. Dong, et al., Stable two-dimensional nanoconfined ionic liquids with highly efficient ionic conductivity. Small. **18**(14) (2022)
114. X.H. Qiu et al., Probing the nanofriction of non-halogenated phosphonium-based ionic liquid additives in glycol ether oil on titanium surface. Friction **10**(2), 268–281 (2022)
115. R. Atkin et al., An in situ STM/AFM and impedance spectroscopy study of the extremely pure 1-butyl-1-methylpyrrolidinium tris(pentafluoroethyl) trifluorophosphate/Au(111) interface: potential dependent solvation layers and the herringbone reconstruction. Phys. Chem. Chem. Phys. **13**(15), 6849–6857 (2011)
116. K. Shimizu et al., 2D or not 2D: Structural and charge ordering at the solid-liquid interface of the 1-(2-hydroxyethyl)-3-methylimidazolium tetrafluoroborate ionic liquid. Faraday Discuss. **154**, 155–169 (2012)
117. R. Hayes et al., Double layer structure of ionic liquids at the Au(111) electrode interface: an atomic force microscopy investigation. J. Phys. Chem. C **115**(14), 6855–6863 (2011)
118. R. Hayes, G.G. Warr, R. Atkin, At the interface: solvation and designing ionic liquids. Phys. Chem. Chem. Phys. **12**(8), 1709–1723 (2010)
119. M. Mezger et al., Molecular layering of fluorinated ionic liquids at a charged sapphire (0001) surface. Science **322**(5900), 424–428 (2008)
120. R. Atkin, G.G. Warr, Structure in confined room-temperature ionic liquids. J. Phys. Chem. C **111**(13), 5162–5168 (2007)

121. R.G. Horn, D.F. Evans, B.W. Ninham, Double-layer and solvation forces measured in a molten salt and its mixtures with water. J. Phys. Chem. **92**(12), 3531–3537 (1988)
122. S. Perkin, T. Albrecht, J. Klein, Layering and shear properties of an ionic liquid, 1-ethyl-3-methylimidazolium ethylsulfate, confined to nano-films between mica surfaces. Phys. Chem. Chem. Phys. **12**(6), 1243–1247 (2010)
123. S. Perkin, Ionic liquids in confined geometries. Phys. Chem. Chem. Phys. **14**(15), 5052–5062 (2012)
124. Y.M. Lu, et al., Ultralow friction and high robustness of monolayer ionic liquids. ACS Nano. (2022)
125. P.K. Cooper et al., Ionic liquid lubrication of stainless steel: friction is inversely correlated with interfacial liquid nanostructure. ACS Sustain. Chem. Eng. **5**(12), 11737–11743 (2017)
126. A.M. Smith et al., Quantized friction across ionic liquid thin films. Phys. Chem. Chem. Phys. **15**(37), 15317–15320 (2013)
127. R. An et al., Controlling the nanoscale friction by layered ionic liquid films. Phys. Chem. Chem. Phys. **22**(26), 14941–14952 (2020)
128. R. An et al., Ionic liquids on uncharged and charged surfaces: In situ microstructures and nanofriction. Friction **10**(11), 1893–1912 (2022)
129. B.C. Wang et al., Nanometer-thick fluorinated ionic liquid films as lubricants in data-storage devices. ACS Appl. Nano Mat. **2**(8), 5260–5265 (2019)
130. H. Li, M.W. Rutland, R. Atkin, Ionic liquid lubrication: influence of ion structure, surface potential and sliding velocity. Phys. Chem. Chem. Phys. **15**(35), 14616–14623 (2013)
131. H. Li et al., Boundary layer friction of solvate ionic liquids as a function of potential. Faraday Discuss. **199**, 311–322 (2017)
132. S. Di Lecce et al., Lateral ordering in nanoscale ionic liquid films between charged surfaces enhances lubricity. ACS Nano **14**(10), 13256–13267 (2020)
133. F. Bresme et al., Electrotunable friction with ionic liquid lubricants. Nat. Mater. **21**(8), 848–858 (2022)

Liquid-Solid Friction on Crystalline Surfaces

Mathieu Lizée and Alessandro Siria

Abstract Liquids flowing against solid surfaces experience friction. While solid friction is familiar to anyone with a sense of touch, liquid friction is much more exotic. At macroscopic scales indeed, the assumption of inifinite friction, i.e. that interfacial liquid molecules stick to solid surfaces, is hard to disprove. Still, it has been known for a few decades that some materials exhibit very strong liquid slippage, leading to a dramatic increase in the permeability of nanoscale tubes to liquid flow. Harnessing liquid friction holds the promise of high-efficiency membrane separation processes, heat recovery systems or blue energy harvesting, making it a highly strategic field for reducing carbon emissions and addressing the climate emergency. In this chapter we review the history of liquid-solid friction measurements, mainly driven by the advent of new techniques and materials. We highlight the most established results and point out some directions that seem to us to be particularly dynamic and promising for the field.

1 Introduction

Liquid-solid friction occupies a special place in the field of tribology. While solid friction is part of everyone's experience and has been ubiquitous in engineering for millennia, liquid-solid friction is a much more exotic phenomenon that has only been the subject of quantitative study for a few decades. In fact, the common assumption that liquids have zero velocity near solid surfaces holds surprisingly well for a wide range of materials, liquids and flow geometries, making infinite friction the standard and most common assumption. Deviations from this behaviour, in particular what is known as 'liquid slippage', have only been demonstrated in recent decades. Based on the development of very specific and diverse experimental platforms, the study of liquid-solid interface friction (or simply 'liquid friction') is still a young field of research, which is beginning to unveil an impressive wealth of effects, following in this way the broad phenomenology of solid-solid friction.

M. Lizée (✉) · A. Siria
Laboratoire de Physique de l'Ecole Normale Supérieure, Paris, France
e-mail: mathieu.lizee@ens.fr

E. Gnecco and E. Meyer (eds.), *Fundamentals of Friction and Wear on the Nanoscale*, NanoScience and Technology, https://doi.org/10.1007/978-3-031-63065-1_18

We will begin by reviewing the main techniques for measuring liquid friction and discussing their respective contributions to the field. We will highlight important results such as the link between friction and wettability, the liquid superlubricity of graphene or the dependence of friction on temperature. We will then give some theoretical insights using liquid state theory and rationalise the link between slippage and contact angle. Finally, we will highlight several topics that seem to be particularly dynamic in the field and that the authors believe should be the focus of further work. In particular, we will discuss friction in ultra-confined environments, slippage on charged surfaces and the fate of liquid friction when the liquid is supercooled towards the glassy state.

Although the technological implications of liquid friction are beyond the scope of this chapter, they are crucial in the field of membrane processes. Indeed, slippery membrane materials could dramatically reduce energy costs and change the paradigm of separation processes, which are often based on highly energy-intensive distillation. For a review of membrane processes, see for example Ref. [1].

Slip length To account for the possible mobility of liquid molecules near the surface and go beyond the infinite friction picture, Navier introduced the liquid-solid friction coefficient :

$$\lambda = \frac{\sigma_{interface}}{v_x} \tag{1}$$

Equation 1 is a Newtonian hypothesis for the liquid-solid interface in analogy to the bulk Newtonian assumption relating tangential stress to velocity gradients $\sigma = \eta \partial_z v_x$ where η is independent of the tangential stress itself.

Writing the continuity of tangential stress with the bulk liquid yields

$$\lambda = \frac{\sigma_{interface}}{v_x} = \frac{\eta \partial_z v_x}{v_x} \tag{2}$$

From Eq. 2, we define the slip length as the distance behind the wall at which the velocity profile extrapolates to zero : $b = \eta/\lambda$ as sketched on Fig. 1a. This length scale is key in estimating the impact of surface friction on fluid transport. Indeed, solving the Poiseuille problem for a cylinder of radius R with this boundary condition leads to an increase of flux of order b/R. We will see in the following that slip lengths are nanometric for water on flat surfaces, meaning that liquid-solid friction controls transport in nanoscale channels but is largely negligible in wider geometry. This last observation explains why the study of liquid friction is tightly bound to liquid flow measurement in nanometric confinement, firstly in biological trans-membrane proteins and more recently in artificial channels.

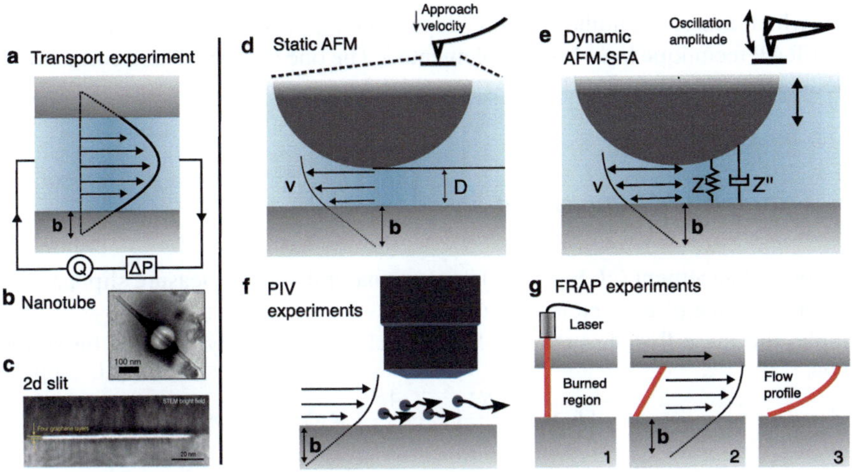

Fig. 1 Overview of the main experimental techniques used to measure liquid-solid friction On panel **a**, we define the slip length b, in the context of pressure driven water transport, as the distance inside the solid at which the linear extrapolation of tangential velocity vanishes. One should note that whereas positive slip length denotes a non zero tangential velocity at the solid surface, negative slip length makes sense as well as it occurs if a layer of liquid rests immobile at the surface. On panels **b, c**, we show electron microscope images of solid state crystalline nanochannels in which liquid friction can be extracted from transport measurements, respectively a transmembrane nanotube on **b** [2] and a 2d nanometric slit [3] on **c**. On panels **d, e**, we describe liquid friction measurements with static and dynamic atomic force microscopes (AFM). In the first case, the sphere approaches the surface at constant velocity. In the dynamic mode however, a nanometric oscillation allows to extract the complex mechanical impedance in confinement at the oscillation frequency. In each case, an approach curve is used to infer slippage. On panel **f**, we sketch the PIV experiments: tracer particles allow to image the liquid flow close to the wall. The FRAP technique is similar but requires no tracers (**g**), a laser marks the surface and thus materializes the velocity profile. One should be aware that there are various different FRAP schemes and that we sketch here the one which has produced most of recent results

2 Experimental Techniques

In this section, we will briefly sketch the history of liquid friction measurements, and introduce the dedicated techniques and instruments along with their respective strengths and weaknesses. On Fig. 1a–c, we show the principle of slippage measurement from transport experiment and electronic microscope images of BN-nanotubes and 2d nanochannels from Refs. [2, 4]. On Fig. 1d, we sketch the static CP-AFM technique in which a sphere glued at the end of a cantilever approaches a surface at fixed velocity while the lever's deflection is recorded. The alternative dynamic AFM/SFA scheme, sketched on Fig. 1e, takes advantage of a vertical oscillation to extract the dissipative impedance caused by viscous dissipation in an axissymmetric drainage flow. On panels Fig. 1f, g, we show the concepts of particle imaging velocimetry (PIV) and Fluorescence Recovery After Photobleaching (FRAP) techniques which image the liquid flow profile to infer slippage. Note that many variants of the FRAP technique have been implemented. The one described here corresponds to the setup of Hénot et al. [5].

2.1 Transport Measurement

Transport enhancement (*cf.* Fig. 1a–c) is a very natural way to measure slippage and was therefore one of the first techniques successfully used for this purpose [6]. For example, the mass flux through a cylindrical channel strongly depends on slip as follows :

$$Q_c = \frac{\rho\pi R^4}{8\eta L}\left(1 + \frac{4b}{R}\right) \times \Delta P \tag{3}$$

where η is the liquid viscosity, R and L the channel's radius and length. To ensure a measurable slip-induced flow enhancement $4b/R$, one needs $R \sim b$ which is typically nanometric for water. To realize such nanometric confinement one can fabricate a nanoporous material and measure its permeability averaged over thousands of channels. An estimation of porosity then allows to infer liquid-solid friction. In the following, we will use the seminal example of carbon nanotubes (CNT). Permeability measurements on CNT membranes have brought the first hints of ultra-low water friction [7, 8] and attracted a considerable interest in the field. Experiments in such membranes however raise legitimate questions: How to evaluate the active surface area or avoid leaks? Is there any cross-talk between these closely packed tubes or are there independent? Moreover, contamination by a linking agent or the catalyst used for nanotube growth has been reported to hinder measurements. The considerable scatter of CNT membrane permeability has required single-channel experiments which are the object of the next discussion.

Carbon nanotube friction Following the reports of ultra low friction in carbon nanotube membranes [7–9], huge efforts were devoted to the study of water confined in carbon nanotubes and especially to the friction of water on the inner tube's wall. We summarize the most important results on Fig. 2d, focusing in particular on the better controlled experiments : those on single nanotubes. For the sake of comparison, the results of Holt et al. [8] on CNT membranes ($b \sim 1$ μm) is shown as a grey dot on panel Fig. 2d.

Tube filling dynamics The single-tube permeability was first measured for very thin ($R < 1$ nm) nanotubes deposited on a silicon wafer by Qin et al. [11] (*cf.* Fig. 2b). The tube can be opened by a high current between the left electrodes (FET 1). The water-front is then detected by conductivity measurement at several positions (FET 2 and 3), allowing to estimate the filling velocity. Unfortunately, as the driving force was not precisely known, the flow profile remains somehow uncertain. Nevertheless, the obtained slip lengths ranging from 100 nm to 1 μm are shown as black squares on our summary panel Fig. 2d.

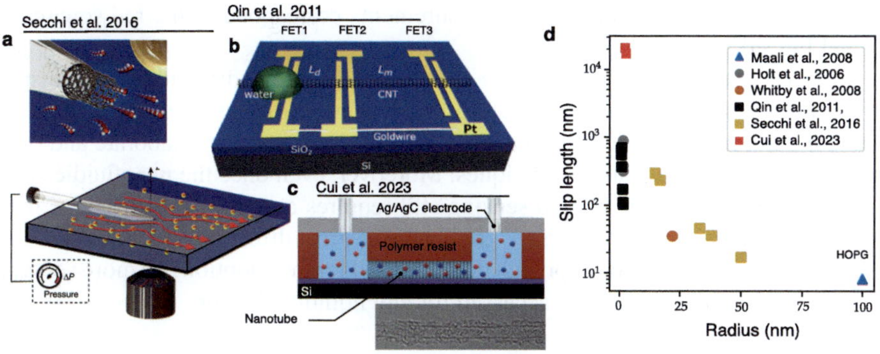

Fig. 2 Slippage in carbon nanotubes On panel **a**, we show the experimental setup used by Secchi et al. [10]. The tube is inserted in a glass capillary pipette. As a pressure is applied in the pipette, water flow is imaged by PIV by imaging micrometric colloids under a microscope. On panel **b**, reproduced from Ref. [11], we sketch the experimental system used by Qin *et al.* A droplet is placed on an end of a nanotube crossing several gold contacts. After cracking the tube open with FET1, wetting velocity in the tube is probed between FET2 and FET3 and the interfacial friction coefficient is subsequently inferred. On panel **c**, we sketch the nanotube system made in the group of Cui et al. [12]. Voltage and concentration drops are applied across the nanotube while the ionic current is recorded between the electrodes. A TEM image of the double wall tube is also shown. Finally, we summarize on panel **d** results from the best controlled experiments. A drastic increase of slippage upon decreasing diameter is clearly observed up to extremely high values of 20 μm from Ref. [12]. Note that in most studies, the electronic properties of the tubes (semi-conducting or metallic character) are unknown

True flow measurement with PIV detection A few years later, Secchi et al. inserted tubes in a quartz nanopipette. This geometry allowed to track the pressure-driven mass flow with microscopic colloids close to the tube's opening [10] (*cf.* Fig. 2a). The authors reported a very low and diameter-independent slippage on boron-nitride nanotubes. Interestingly, the results were radically different for carbon : not only is slippage extremely high but it is also strongly diameter dependent (yellow squares on panel Fig. 2d). These results are consistent with the massive permeability of 2 nm carbon nanotube membranes [7, 8]. The strong discrepancy between carbon and boron-nitride despite identical crystallographic properties is believed to be caused by a difference in the mobility of adsorbed OH^- ions [13]. The dramatic increase of slippage in smaller tubes is no less puzzling and it was recently proposed that differences in the electronic structure of walls of different radii could be the cause of this discrepancy [14] (see Sect. 3.2 for details).

Enhanced control on the tubes properties In a very recent work, the authors managed to build a microfluidic system to address a single double-walled nanotube [12] (*cf.* panel Fig. 2c) with a 5 nm inner diameter and well established electronic properties. Although the geometry does not allow to apply large pressure drops across the channel, one can apply concentration and voltage drops across the tube and measure the induced ionic current. Upon several assumptions, the authors estimated the slip length to be as high as 20 μm in semi-conducting tubes, corresponding to a dramatic drop in friction. These measurements correspond to the red squares on panel Fig. 2d.

Intrinsic weakness of transport measurements The main difficulty with transport measurements is their non-local nature that makes them sensitive to any impurity, contamination, or surface heterogeneity. They also require very elaborate and often material-specific fabrication techniques. Moreover, even once the nanofluidic channel has been fabricated, friction sensing still requires an exquisitely sensitive mass flow measurement under a pressure gradient. The great difficulty and intrinsic weaknesses of this experimental approach therefore call for the adoption of a more versatile scheme, where friction can be accessed for any liquid/solid pair.

2.2 Hydrodynamic Flow Imaging: FRAP and PIV

Among the early attempts to a more versatile liquid friction measurement scheme, the Fluorescence Recovery After Photobleaching (FRAP) was developed in the group of Pit et al. [15]. This technique, which can be viewed as a tracer-less Particle Imaging Velocimetry (PIV) scheme, offers a local measurement and provides the full flow profile in a planar geometry. A few years later, promising PIV-based slippage measurements have yielded a slip length of water on silanized glass of $b \sim 16$ nm [16]. The FRAP scheme has further been used in recent years with a new configuration, detailed on Fig. 1g, to measure the friction of polymer melts as a function of temperature and chain length [5]. A laser is used to burn a line in a fluorescent polymer (functionalized PDMS). A shear motion is then applied, allowing to image the flow and infer slippage. In this study, Henot et al. tried to disentangle the two contributions to the slip length $b = \eta/\lambda$: viscosity and liquid-solid friction. By varying temperature, the authors identified two independent activation energies for viscosity (E_a^η) and liquid-friction coefficient (E_a^λ). They report the independency of E_a^λ on chain-length N and conclude that liquid friction is controlled by the monomer's structure only, in drastic contrast with viscosity which depends on chain length as $\eta \sim N^3$. An interesting consequence, already pointed out by de Gennes [17], is that slip length of polymer melts is proportional to viscosity while it is expected to be independent of η in simple liquids [18]. Overall, the PIV and FRAP techniques have several drawbacks : they require a functionalization step or the addition of tracers to liquids and their resolution is much lower than dynamic SFA and colloidal probe-AFM. However, the simplicity of the FRAP process is a major great advantage that has allowed Restagno et al. to collect extensive data on polymer friction.

2.3 Surface Force Apparatus

The investigation of the liquid-solid boundary condition has taken a new turn in the 1970s with the introduction of the Surface Force Apparatus (SFA). The SFA has allowed to routinely confine various liquids between atomically smooth surfaces while measuring shear or normal forces [19]. Beyond the measurements of surface forces, the SFA was also used to probe confined hydrodynamics and in particular interfacial slippage. Early experiments [20, 21] have set a landmark in confined liquids physics by showing that continuous hydrodynamics hold down to the molecular scale (1 nm) which is in itself a most surprising result. In these works, the authors report an immobile layer of molecules adsorbed at the surface resulting in a negative slip length.

The dynamic SFA To enable systematic slippage measurements with the SFA, a stiffer system was needed. Built around a glass sphere oscillating at a frequency $f \sim 10 - 300$ Hz, the dynamic SFA (d-SFA) built in the group of Elisabeth Charlaix has allowed to measure liquid friction for various kinds of surfaces. The dual detection mechanism (optical and capacitive) ensures a sub-nanometric resolution in both static and dynamical modes. The d-SFA is described elsewhere [22] and we only briefly sketch its design on Fig. 3a and the measurement concept on Fig. 1e. The millimetric sphere is brought in close proximity to a planar surface and probes a region of size $\sqrt{RD} \sim 10\ \mu$m. The sphere is subjected to a vertical oscillation at a chosen frequency ω and nanometric amplitude a while its distance to the plane is measured by interferometry. The viscous dissipation in the drainage flow between the sphere and planar surface is given by the Reynolds formula :

$$\mathrm{Im}(Z) = \frac{6\pi\eta R^2\omega}{D+b} \tag{4}$$

where we assumed a no-slip boundary condition on the sphere and a slip length b on the planar surface. The simplicity of this geometry allows to take into account a wide variety of effects in the force curves. For instance, the impact of slippage on the flow (correction to Eq. 4) [23], bulk visco-elasticity [24] and elastic surface deformation [25] have been accounted for in an extended analytical theory, brilliantly confirmed by experiments.

Slippage measurements on silanized glass The d-SFA can be credited of some of the most systematic slippage measurements, on various surfaces and different liquids, backed up by the thorough understanding of dissipation curves which depart from the Reynolds formula at short distance (*cf.* Fig. 3b). A major result is the quantitative confirmation that hydrophobicity promotes slippage, first observed by Churaev [6], but measured unambiguously with the d-SFA by Cottin et al. [26]. In this experiment, the glass sphere faces either a bare hydrophilic glass plane or its silanized, strongly hydrophobic, counterpart. As shown on Fig. 3b, the slip length for water is measured at zero for the native surface and 16 nm for the silanized one. Here, not only do the authors establish hydrophobic enhanced slippage, they also confirm with excellent resolution the no-slip boundary condition on an hydrophilic substrate of sub-nanometric roughness. Despite the excellent sensitivity of the d-SFA, it remains very hard to operate for two main reasons : (1) the long time stability required by the ultra-slow approach velocity and of (2) the large contact area making it very vulner-

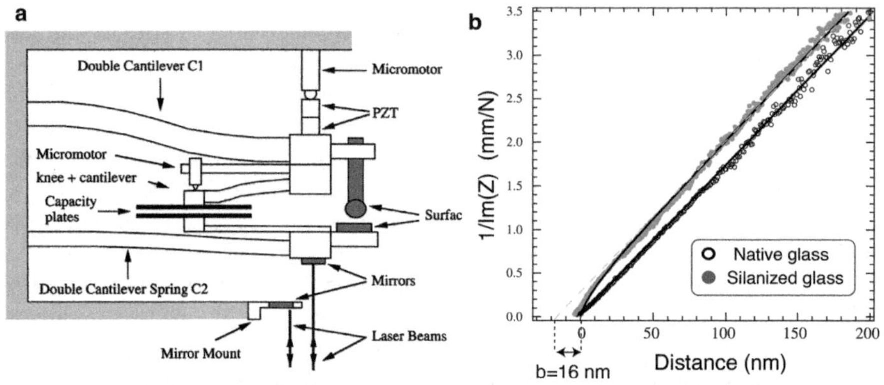

Fig. 3 **Dynamic Surface Apparatus (dSFA)** used to probe nano-confined liquids rheology and interfacial friction. Panel **a** decribes the system and is extracted from Ref. [22]. On panel **b**, we reproduce the curves of the inverse dissipation *versus* distance on native and on silanized glass, extracted from Ref. [26]. The curvature shown on silanized glass denotes a reduced friction at long distance and an increased slope of dissipation close to the contact for slipping liquids. This behavior is in excellent agreement with the theory by Vinogradova [23] and is a key element to discard the presence of possible artefacts—like impurities—wich could lead to errors in the slippage estimate

able to contamination. Another experimental technique for slippage measurement shows none of these shortcomings and is now established as the central technique for liquid-solid friction : the colloidal-probe atomic force microscope (CP-AFM).

2.4 Colloidal Probe AFM

One of the most prolific instrument in the measurement of liquid slippage is the atomic force microscope (AFM) and its micrometric sphere version, the colloidal probe AFM. A micrometric sphere is glued on an AFM cantilever and immersed in liquid close to the chosen surface. Whereas the SFA requires millimeter scale samples, the CP-AFM allows to probe friction on any material, including micrometric single crystals. The small probe size further ensures a good robustness against contamination and one can easily move on the sample surface to check reproducibility at various positions. Once the sphere is glued, the experimentalist approaches the flat surface at constant velocity $v \sim 10 - 100\mu$m/s. The friction force is given by the Reynolds equation $F \sim \eta R^2 v/(D + b)$ and allows to extract the slip length b in complete analogy with the SFA. For these reasons, the colloidal probe AFM has been extensively used to probe the dependency of slippage on surface properties like roughness, wettability or surface charges. In analogy to the dynamic SFA, the CP-AFM has also been used in dynamical mode [27] and confirmed that $b \sim 0$ for mica. The authors obtained $b = 43 \pm 10$ nm for an alcane self-assembled layer and report ultra high slippage on super-hydrophobic composite surfaces which are out of the scope of this chapter. In the following we focus on recent efforts to measure water friction on graphene with the CP-AFM.

From graphene to graphite friction Among the important measurements which are hard to conceive without the CP-AFM is that of the 8 nm slip length of water on graphite by Maali et al. Investigations of water-carbon friction have recently been pushed forward to few-layer graphene samples on SiO_2 in a genuine experimental *tour de force* [28]. In this study, the authors provide a measurement of water slippage as a function of graphene thickness—down to the single layer resolution—with hydrophilic SiO_2 and hydrophobic silanized surfaces. Their results point towards a transluency of graphene for liquid-solid friction and are reproduced on Fig. 4d. For thinner flakes, the substrate dictates liquid friction and one obtains high slippage for silanized substrate while the no-slip boundary condition holds for SiO_2. As the flake thickens, graphene increasingly screens the substrate and ultimately, graphite friction coefficient is recovered above a few nanometers in rough agreement with the 8 nm value of Maali et al.

Link with hydrated solid friction In recent years, attempts to use the AFM to unify hydrated-solid friction and slippage measurements have brought extremely interesting results to the field. The authors of Ref. [29] apply Eyring's theory of transition states to liquid-solid friction [30]. In this framework, interfacial diffusion is controlled by an activation energy E_a and assisted by tangential stress. The activation energy can be measured using velocity-dependent hydrated friction : the solid friction between a sharp AFM tip and the surface is measured in the presence of a nanometric water lubrication layer as sketched on Fig. 4a. Here, the tip is shearing the surface, exactly like in solid-solid friction AFM measurements. By combining these results with standard CP-AFM drainage experiment yielding slip length b for various liquids on graphene (*cf.* Fig. 4b), the authors correlate liquid-solid friction with the activation energy E_a (see panel Fig. 4c). Here, the authors investigate an aqueous sucrose solution and an ionic liquid (EMIM-FAP) on CVD graphene deposited on three different substrates : copper, gold or silicon. They report a drastic drop of slip length from 16 nm down to negative values as E_a increased from 0.05 to 0.15 $k_B T$. In another work, Diao et al. have systematically investigated E_a and b for salted water solutions on graphene [31]. This original approach to liquid-solid friction seems very promising to us as it disentangles the activated equilibrium contribution $E_a/k_B T$ from the stress-assisted one which is often negligible in conventional drainage setups.

3 From Numerical Techniques to Theoretical Insights

From a theoretical perspective, the liquid-solid interface is a notoriously difficult problem, where statistical mechanics of liquids and solid-state physics mix together at surfaces.

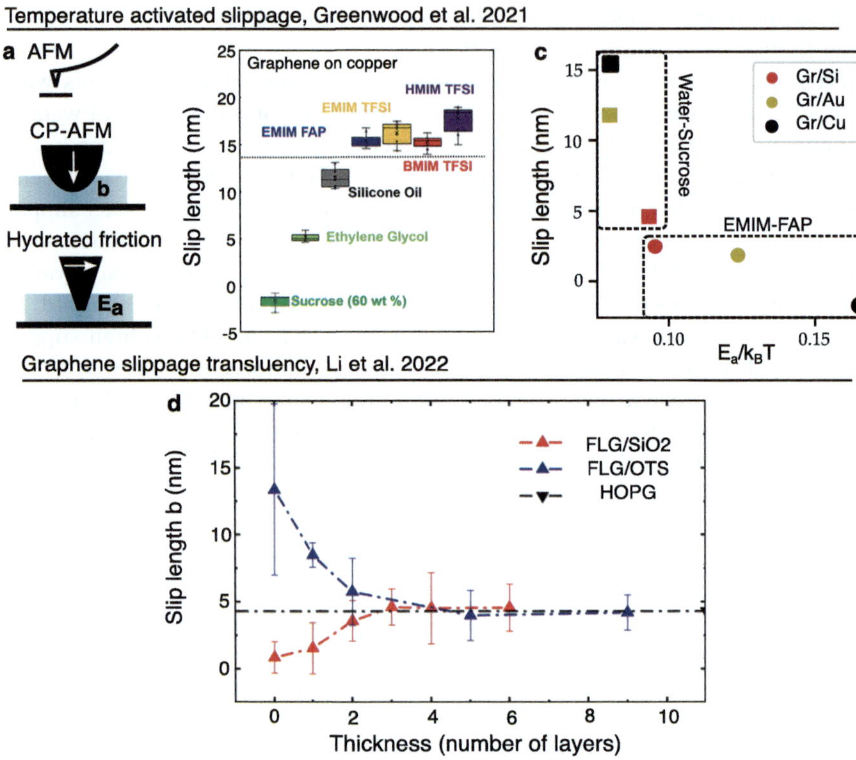

Fig. 4 **Recent experimental results obtained with the CP-AFM** On panel **a–c**, we plot results from the group of Greenwood [29]. In this work, the authors measure the slip lengh of various liquids on three different graphene surfaces. The originality of their work lies in the combination of liquid friction (CP-AFM) with hydrated solid friction.measurements under well defined tangential stress. These two experiments allow to plot on panel (**c**), the negative correlation of slip length and local activation energy in the sense of Eyring. On panel (**d**), we display the experimental results from the group of Li et al. [28] on graphene crystals of various thickness. Here, graphene is deposited either on a hydrophillic (SiO_2) or a hydrophobic substrate (OTS). The results show that graphene plays the role of a screening layer protecting the liquid flow from the underlying substrate while the latter still controls most of the liquid friction behavior

3.1 Static Wall Friction

A first attempt to estimate liquid friction on a corrugated potential can be made within the linear response theory. Using the Green-Kubo formula, one relates the interfacial momentum transport coefficient to the auto-correlation function of the force (*ie:* momentum flux) at equilibrium [32] on a surface $\mathscr{A}$.

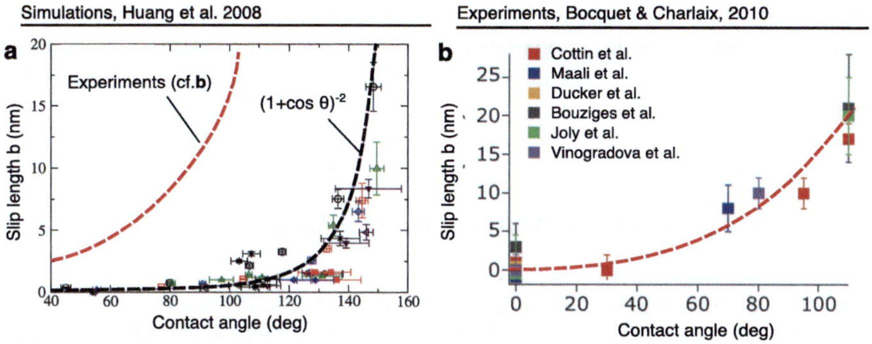

Fig. 5 **Liquid slippage *versus* contact angle** On panel **a** reproduced from Ref. [35], we show the results of molecular dynamic simulations on various surfaces and liquid models. The dashed line denotes the following formula : $(1 + \cos\theta)^{-2}$. On panel **b**, we reproduce a figure from Ref. [36] showing that current experimental results are in semi-quantitative agreement with the theoretical prediction

$$\lambda = \frac{1}{\mathcal{A} k_B T} \int_0^\infty dt \langle f_x(t) f_x(0) \rangle \tag{5}$$

The interfacial friction coefficient can eventually be rewritten in terms of the fluid-solid interaction potential, the liquid's structure factor at the interface $S_{q_\parallel}$, its density ρ and its typical lateral diffusion coefficient at the scale of corrugations $D_{q_\parallel}$, which can all be obtained from equilibrium molecular dynamics simulations.

$$\lambda = \frac{S_{q_\parallel}}{D_{q_\parallel} k_B T} \int_0^\infty dz \rho(z) V_{FS}(z)^2 \tag{6}$$

The obtained slippage can then be compared with the velocity profile from out of equilibrium simulations [33, 34].

Contact angle dependency Across the long story of liquid slippage investigations, the role of hydrophobicity as a slippage enhancer has been often discussed and remains one of the most established findings in the field. On Fig. 5a—reproduced from Ref. [35]—we show the slip length obtained from molecular dynamics simulations on various substrates as a function of the contact angle θ. Considering Eq. 6, one uses the Young equation to relate the liquid-solid interaction potential to the contact angle and finds $V_{FS} \sim 1 + cos(\theta)$. Indeed, on panel Fig. 5a, the dashed line scales like $(1 + cos(\theta))^{-2}$ and denotes a 'quasi-universal relationship' between contact angle and slip length, in good agreement with the theory. On panel Fig. 5b, reproduced from Ref. [18] we plot selected experimental data clearly showing the strong increase of slippage with hydrophobicity in semi-quantitative agreement with the theoretical scaling. Interestingly, we observe on panel Fig. 5a a systematically higher slippage in experiments than in molecular dynamics simulations. Despite the success of this theory, it is apparently defied by radius-dependent slippage in carbon nanotubes (*cf.* Sect. 2.1) for which an alternative mechanism has recently been proposed.

3.2 Wall Fluctuations: Van der Walls Friction

Multiwall carbon nanotube slippage The strongly radius-dependent slippage of carbon nanotubes measured by Secchi et al. [10] was immediatly identified as an extremely puzzling observation: indeed the crystalline structure of the tubes is independent of their diameter to the exception of their curvature, which was shown to have negligible effect on slippage for radii larger than 10 nm [37]. Recently, an original mechanism involving the *electronic* structure of such multiwall carbon nanotubes was proposed by Kavokine et al. [14].

On Fig. 6a, we sketch water slippage on the inner surface of a multiwall cabon nanotube, a curved graphite-like interface. Interestingly, it is known that in these tubes, the interlayer electronic couplings strongly depends on curvature. As sketched on panel Fig. 6b, we expect that small radius tubes should behave like a single graphene sheet whereas large radius tubes in which layers are transversally coupled should exhibit the properties of graphite.

Theoretical result The idea, sketched on Fig. 6a–d, is that momentum transfer from liquid flow to the solid wall can be mediated by the Coulomb interaction in addition to the standard molecular collisions. In that scenario, charge fluctuations in the liquid induce a dielectric response in the solid, which in turn acts on the liquid. Under a flow, this reactive force is retarded and tends to oppose movement, effectively dissipating the mechanical energy into the solid's electronic degrees of freedom. For this process to efficiently exchange momentum, a matching in energy and wavevector of dielectric fluctuation modes between the liquid and solid is needed. This mechanism is perfectly analogous to the electronic friction between two metallic surfaces which was shown to vanish at the superconducting transition [38]. The Van der Walls friction formula, derived by Kavokine et al. [14] writes as an integral over the (q, ω) plane of the product of the dielectric surface response functions of respectively the liquid (g_{w}) and the metallic electrons (g_{e}) multiplied by a q^3 contribution:

$$\lambda_{\mathrm{vdW}} = \frac{1}{8\pi^2} \int_0^{+\infty} dq \int_0^{+\infty} \frac{\hbar d\omega}{k_{\mathrm{B}}T} \frac{\hbar q^3}{\sinh^2(\frac{\hbar\omega}{2k_{\mathrm{B}}T})} \times \frac{g_{\mathrm{e}}(q,\omega) g_{\mathrm{w}}(q,\omega)}{|1 - g_{\mathrm{e}}(q,\omega) g_{\mathrm{w}}(q,\omega)|^2} \tag{7}$$

We mention here that the surface response functions g_{w} and g_{e} are anolog to electromagnetic reflection coefficients and are related to bulk dielectric response functions ϵ_{w} and ϵ_{e} by simple analytical formulae.

Van der Walls contribution to water-graphite friction Following the aforementioned argument, we now compare the excitation spectra of graphene and graphite in

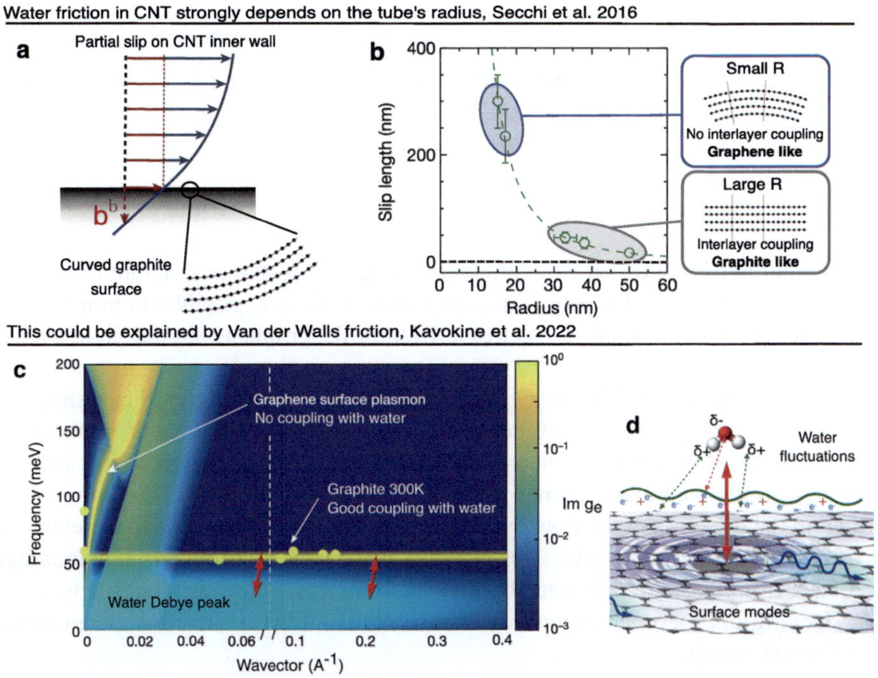

Fig. 6 Van der Walls friction We sketch on panel **a** liquid slippage on the inner wall of multiwall carbon nanotubes. **b** The experimental measurements of Secchi et al. yield a strong decrease in friction (increase in slippage) for the thinner tubes. This observation is explained by the quantum friction theory of Kavokine et al. [14], using that wider tubes are electronically similar to graphite while narrow tubes, lacking interlayer couplings, are closer to graphene from the electronic point of view. **c** Electronic excitations for graphite and graphene in the (q, ω) space. Water's Debye peak is added in light blue. The red arrows denote momentum exchange following Eq. 7. **d** Sketch of the Van der Walls friction which couples water charge fluctuation modes to their solid counterpart through Coulomb interaction. NB : the arrow shown here denotes a surface plasmon but the one of graphite is rather an interlayer, transverse excitation

the frequency-wavector plane (q, ω) (*cf.* panel Fig. 6c). We also add—in light blue—water's Debye peak, described in the previous section, and mention that efficient momentum transfer gives a strongly dominant role to high wavevector modes. This matching in high q regions, shown on panel Fig. 6c between water's and graphite's excitation spectra is absent in the case of graphene. We therefore expect a much higher Van-der-Walls friction on graphite than on graphene and explain the large slippage observed in small radius carbon nanotubes. The Van der Walls friction is an additional contribution to the classical roughness-induced friction and shall thus be measurable only on molecularly smooth surfaces. More precisely, the total liquid-friction coefficient writes as the sum of liquid friction on a static wall, free of any fluctuations, and of the Van der Walls term:

$$\lambda = \lambda_{\text{stat}} + \lambda_{\text{vdW}} \tag{8}$$

A detailed derivation of the static wall contribution to friction (λ_{stat}) was estimated from Green-Kubo formula at Eq. 6. Considering the overlap in modes of water and graphite, we compute the Van der Walls formula and find

$$\lambda_{\text{vdW}} \sim 1.3 \times 10^5\,\text{Pa.s} \qquad b_{\text{vdW}} = \eta/\lambda_{\text{vdW}} = 10\text{nm} \tag{9}$$

in excellent agreement with experimental findings for water on graphite [39]. This calculation suggests that classical friction (static wall) is negligible on graphite. In consequence, the vanishing of electronic friction for graphene should yield a very high slip length.

Negative Van der Walls friction? Recent theoretical works have predicted that Van der Walls friction could become negative, if one assumes an efficient momentum transfer from the liquid to phonons and from phonons to electrons. In this picture where electrons eventally acquire a mean velocity superior to that of the liquid, they drag it in the direction of the flow by the Van-der Wals friction mechanism, resulting in an enhanced slip length [40]. This negative Van der Walls friction is a momentum feedback from the solid, mediated by the Coulomb interaction. Although it is, as of now, lacking an experimental confirmation, it could motivate more efforts to engineer low friction metals.

4 Perspectives in Liquid-Solid Friction

In this last section, we dwell on a handful of selected topics which seem to us particularly interesting, dynamic and worthy of investigations in the field of liquid friction.

4.1 Super-Lubricity or the Puzzle of Graphene Friction

The measurement of ultra-fast water flow in carbon nanotubes has sparked a revolution in nanofluidics and given the water-carbon interface a place of its own. The massive slippage suggested by these early measurements is in apparent contradiction with the moderate hydrophilicity of graphite. As detailed above, friction on the inner wall of carbon nanotubes is extremely small and strongly radius dependent. For water on graphene, although some results obtained with an original micro-channel system by Duan et al., suggest a possibly huge slip length of hundreds of nanometers [41], it has still never been measured consistently. Li et al. [28] (*cf.* Fig. 4d) have demonstrated a slippage transluency of graphene, analog to the previously reported wetting transluency [42]. Thus, graphene on a substrate is radically different from its suspended counterpart, from both the contact angle and the slippage perspective.

It is thus unsurprising that the inner-wall of carbon nanotubes behaves so strikingly differently to graphene crystals deposited on flat surfaces. To this day, liquid superlubricity remains an exclusive property of carbon nanotubes inner walls and it is a huge challenge to extend it to other surfaces. For technological purposes, it appears necessary to leverage these results in order to fabricate affordable highly permeable CNT-based membranes.

4.2 Charged Interfaces and Electrolytes

It is crucial for any practical application to consider the effect of surface charge on liquid friction. Real surfaces are often charged in solution and an ionic double layer of the scale of the Debye length is usually present. The nature and mobility of surface charges is often mysterious and concentration dependent.

Experiments On the experimental side, very few experiments have linked surface charge and slippage. Streaming measurement (ionic current under pressure drop) measurements in angstrom scale two-dimensional channels have allowed to extract surface charge and infer slippage indirectly [43]. In 2-nm carbon nanotubes, a negative surface charge along with a large slip length result in a subtle interplay between pressure driven flows and ionic conductance [44]. Measurements on single carbon nanotubes by Seechi et al. have allowed to estimate surface charge as functions of salt concentration [45] and demonstrated surface charge regulation effects. Unfortunately, the slippage measurements were not extended to salty solutions. The consensus is that graphite is much more slippery than boron-nitride while both have a negative surface charge in agreement with the OH^- adsorption obtained in DFT calculations [46]. These results are also consistent with the prediction of a much higher mobility of the surface charge on graphite compared to BN.

Theory In a recent numerical study [47], the authors tried to explain the slippage difference between graphene and BN with a simple model. They compared two types of surfaces with either an homogenous (delocalized) or heterogenous surface charge corresponding respectively to metallic and insulating surfaces. The simulations show that the heterogeneity of surface charge is very detrimental to slippage as it allows ion adsorption and thus slows down the liquid flow close to the surface. Finally, the coupling of ion motion with free electrons in solids through the Van der Walls friction mechanism remains unexplored to this day.

Perspective Despite encouraging results, the problem of liquid friction on charged surfaces remains unsolved. Firstly, the nature of the surface charge—species and type of bound—is a difficult problem in itself. Then comes its unknown dependency on salt concentration and pH through complex regulation mechanisms which exhibit intriguingly long timescales [48]. Finally, one has to estimate the friction between the liquid and the surface charge which largely depends on the nature of the material (*cf.* carbon *vs* BN [46]). A last problem which will hinder the experimental studies is the possible streaming currents induced by liquid flows on charged surfaces which may lead to complicated flow structures including swirls [49]. Concentration gradients and osmotic currents can also couple to mass flow and generate unexpected counter-flow structures which were recently measured with the d-SFA [50]. These multiple couplings between electrostatics, quantum chemistry, and hydrodynamics make it a considerable challenge to study liquid friction on charged surfaces. We believe that experimental efforts in this field are greatly needed. Hydrated friction measurements *versus* salt concentration by Diao et al. [31] are already a promising attempt at systematic investigations.

4.3 Liquid Slippage and Confinement

In nanometric channels, hydrodynamic and molecular length scales meet. As deviations from standard hydrodynamics start to develop, it is natural to expect that the friction coefficient could be modified compared to the standard half-plane geometry of liquid slippage.

Experiments Probing liquid friction at fixed confinement discards all scanning probe techniques which involve fitting a z-spectroscopy curve with a hydrodynamic model. This leaves us with transport experiments only to probe slippage in confined systems. Direct mass flow transport for liquids in nanometric (below 10 nm) confinement and under well-controlled pressure difference has, to our knowledge, never been achieved. Several techniques have been developed to circumvent this constraint in carbon nanotubes as discussed on Fig. 2.

Theory On the theoretical side, the mode coupling approach allows to tackle the problem of confined friction. In a recent study, the authors of Ref. [51] have computed the friction coefficient for two-dimensional slits as a function of the slit's thickness. Confinement induces a coupling of plasmonic modes in the face to face metallic surfaces yielding a different dispersion to the single surface mode. The effect of this cross-plasmon on friction is relatively weak but not the only consequence of confinement. Indeed, the restricted geometry of a 7 angstrom slit also modifies the dynamical structure factor of water with respect to the bulk. This results in a modified Van der Walls friction force which can result in a 2-fold enhancement of friction coefficient for 7 angstrom slits compared to thicker ones. Note that this mode coupling theory computes only the 'Van der Walls friction' contribution caused by the coupling of water dipoles to free electrons in metallic walls.

Perspective In parallel to this direction, it is now possible to estimate slippage using an analysis of ion transport measurements under concentration and voltage drops (or pressure and voltage) [43, 52]. Eliminating the need for exquisite mass flow measurements, this technique trades experimental difficulty for a challenging rationalization and several hypotheses. This approach is particularly promising for studies of slippage in ultra-confined systems and has been applied to that of small-diameter carbon nanotubes [12], angstrom-scale two-dimensional channels [43], and activated carbon nano-conduits [52]. Despite these interesting proposals, true friction measurements for confinements below 10 nm are still lacking. We point out that, more generally, high resolution mass flow measurement techniques in ultimate confinement are extremely rare but would make a huge contribution not only to the confined-friction problem but to nanofluidics in general.

4.4 Liquid Dynamics and Non-newtonian Interfaces

As evidenced in Green-Kubo equation Eq. 6, liquid friction intrinsically depends on molecular timescales needed for force correlations to relax. Supercooled liquids offer a unique platform to tune molecular timescales on demand. Indeed, a mere temperature drop allows to drastically shift the liquid's dielectric relaxation peak or in other words its molecular relaxation time which is close to the Maxwell time $\tau \sim \eta/G_\infty$ (G_∞ is the high-frequency elastic modulus). These measurements would be especially interesting in the Van der Walls friction framework, in which an overlap of liquid and solid excitations in the (q, ω) space enhances friction. Moreover, like many biological fluids, ionic liquids are often already supercooled at room temperature and are routinely confined in battery electrodes, super-capacitors [53] and membranes used to harvest temperature gradients [54].

Interfacial visco-elasticity? Finally, in slow liquids under shear, the Newtonian hypothesis may no longer hold. Whereas the complex and frequency dependent rheology of the bulk is often well known (complex frequency dependent viscosity $\tilde{\eta}(\omega)$), it is a fully open question to know how the interface behaves at high frequency. First experiments in poly-electrolyte solutions have shown that liquid-solid friction remained real [24] whereas the bulk deviated strongly from newtonian behavior but we could assume that the interfacial friction coefficient becomes complex and frequency dependent $\tilde{\lambda}(\omega)$ at high shear rate.

First results On Fig. 7, we show the results of two studies of supercooled liquid friction on solid surfaces. On panel Fig. 7a, we plot the temperature-dependent friction coefficient extracted from molecular dynamics simulations of water and methanol on either graphene or a Lennard-Jones wall [55]. The solid lines are fits with an Arrhenius law. On panel Fig. 7b, we reproduce experimental results from Ref. [5] on polymer melts. In this work, the authors report an Arrhenius dependency for both viscosity and friction. Having in mind that viscosity η describes liquid-liquid friction while λ denotes the interfacial friction, these two different activation energies account for two separate mechanisms for friction in these two situations. It is interesting to note that a variations of the slip length $b = \eta/\lambda$ implies a bulk-surface decoupling upon cooling. In this latter study, the authors compared for each surface the polymer melt's liquid friction $k_{melt} = \eta/b$ with the solid friction $k_{elastomer} = \sigma_\parallel/V_\parallel$ of the corresponding reticulated polymer. When both are plotted together as a function of temperature on Fig. 7b, it is striking that they have roughly the same temperature dependency and order of magnitude. This observation strengthens the claim that liquid friction depends on the monomers only and that its dynamical behavior is radically decoupled from the bulk's for which viscosity scales as the chain's length to a power close to 3.5 [56]. Finally, we mention a recent investigation, by the authors, of the friction of supercooled glycerol on mica. Using a dynamic CP-AFM, we reported

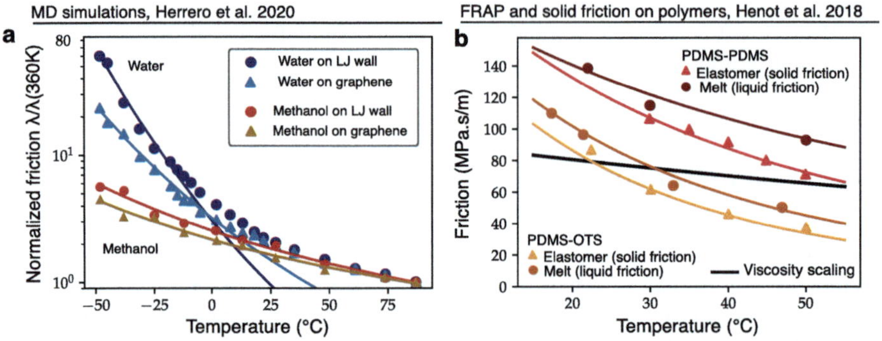

Fig. 7 **Supercooled liquid-solid friction** On panel **a**, we replot the data from Ref. [55] where the authors use molecular dynamics simulations to compute the friction coefficient and slip length of water and methanol on graphene and a Lennard-Jones wall *versus* temperature. They observe a strong increase of friction upon cooling, the solid lines are fits with an Arrhenius law. Note that the friction coefficient is normalized by its value at 360K. On panel **b**, we display the experimental results from Ref. [5]. In this work, the authors use a FRAP setup to measure the dependency of friction coefficient on temperature for polymer melts of various molar masses

drastic deviations to Arrhenius behavior and proposed that the strong shift of the liquid's collective modes upon cooling is tuning liquid-solid friction in a similar way to the Van der Walls mechanism described above [57].

5 Conclusion

Liquid-solid friction is now established as a mature field of research with enormous technological relevance and several key experimental techniques. Due to its extreme sensitivity to surface roughness, slippage is rare for water. Nevertheless, a handful of materials exhibit drastically reduced friction, which suggests that highly permeable—yet selective—membranes may be available in the near future. The emerging Van der Walls friction for liquids promises to control slippage by tuning electronic properties in the solid state. We argue that the respective influence of surface charge, nanometric confinement and liquid supercooling are open questions that deserve special attention and for which the experimental techniques are mature.

Finally, we emphasize that, like the combined AFM measurements of liquid and hydrated solid friction, which provide the activation energy E_a and the slip length b, the study of polymer friction for both the melt (liquid) and the reticulated polymer (solid) has provided groundbreaking physical insights. We argue that the coupling of liquid and solid friction—which now appear to be complementary—will be key to unravelling the dynamical processes at the liquid-solid interface.

References

1. E. Drioli, L. Giorno, *Encyclopedia of Membranes* (Springer, Berlin Heidelberg Berlin, 2016)
2. A. Siria et al., Giant osmotic energy conversion measured in a single transmembrane boron nitride nanotube. Nature **494**, 455–458 (2013)
3. B. Radha et al., Molecular transport through capillaries made with atomic-scale precision. Nature **538**, 222–225 (2016)
4. Q. Yang et al., Capillary condensation under atomic-scale confinement. Nature **588**, 250–253 (2020)
5. M. Hénot et al., Temperature-controlled slip of polymer melts on ideal substrates. Phys. Rev. Lett. **121**, 177802 (2018). https://link.aps.org/doi/10.1103/PhysRevLett.121.177802
6. N. Churaev, V. Sobolev, A. Somov, Slippage of liquids over lyophobic solid surfaces. J. Colloid Interface Sci. **97**, 574–581 (1984)
7. M. Majumder, N. Chopra, R. Andrews, B.J. Hinds, Enhanced flow in carbon nanotubes. Nature **438**, 44–44 (2005)
8. J.K. Holt et al., Fast mass transport through sub-2-nanometer carbon nanotubes. Science **312**, 1034–1037 (2006)
9. M. Majumder, N. Chopra, B.J. Hinds, Mass transport through carbon nanotube membranes in three different regimes: ionic diffusion and gas and liquid flow. ACS Nano **5**, 3867–3877 (2011). https://pubs.acs.org/doi/10.1021/nn200222g
10. E. Secchi et al., Massive radius-dependent flow slippage in carbon nanotubes. Nature **537**, 210–213 (2016). http://www.nature.com/articles/nature19315
11. X. Qin, Q. Yuan, Y. Zhao, S. Xie, Z. Liu, Measurement of the rate of water translocation through carbon nanotubes. Nano Lett. **11**, 2173–2177 (2011)
12. G. Cui et al., Enhanced osmotic transport in individual double-walled carbon nanotube. Nat. Commun. **14**, 2295 (2023)
13. E. Mangaud, M.-L. Bocquet, L. Bocquet, B. Rotenberg, Chemisorbed vs physisorbed surface charge and its impact on electrokinetic transport: carbon vs boron nitride surface. J. Chem. Phys. **156** (2022)
14. N. Kavokine, M.-L. Bocquet, L. Bocquet, Fluctuation-induced quantum friction in nanoscale water flows. Nature **602**, 84–90 (2022). https://www.nature.com/articles/s41586-021-04284-7
15. R. Pit, H. Hervet, L. Leger, Direct experimental evidence of slip in hexadecane: solid interfaces. Phys. Rev. Lett. **85**, 980 (2000)
16. L. Joly, C. Ybert, E. Trizac, L. Bocquet, Hydrodynamics within the electric double layer on slipping surfaces. Phys. Rev. Lett. **93**, 257805 (2004)
17. P.-G. de Gennes, Ecoulements viscométriques de polymères enchevêtrés, in *Simple Views On Condensed Matter* (World Scientific, 2003), pp. 203–205
18. L. Bocquet, E. Charlaix, Nanofluidics, from bulk to interfaces. Chem. Soc. Rev. **39**, 1073–1095 (2010)
19. J. Israelachvili et al., Recent advances in the surface forces apparatus (SFA) technique. Rep. Prog. Phys. **73**, 036601 (2010)
20. D.Y. Chan, R. Horn, The drainage of thin liquid films between solid surfaces. J. Chem. Phys. **83**, 5311–5324 (1985)
21. J. Georges, S. Millot, J. Loubet, A. Tonck, Drainage of thin liquid films between relatively smooth surfaces. J. Chem. Phys. **98**, 7345–7360 (1993)
22. F. Restagno, J. Crassous, E. Charlaix, C. Cottin-Bizonne, M. Monchanin, A new surface forces apparatus for nanorheology. Rev. Sci. Instr. **73**, 2292–2297 (2002)
23. O.I. Vinogradova, Drainage of a thin liquid film confined between hydrophobic surfaces. Langmuir **11**, 2213–2220 (1995). https://pubs.acs.org/doi/abs/10.1021/la00006a059
24. B. Cross, et al., Wall slip of complex fluids: interfacial friction versus slip length. Phys. Rev. Fluids **3**, 062001 (2018). https://link.aps.org/doi/10.1103/PhysRevFluids.3.062001

25. S. Leroy, E. Charlaix, Hydrodynamic interactions for the measurement of thin film elastic properties. J. Fluid Mech. **674**, 389–407 (2011). https://www.cambridge.org/core/product/identifier/S0022112010006555/type/journal_article
26. C. Cottin-Bizonne, B. Cross, A. Steinberger, E. Charlaix, Boundary slip on smooth hydrophobic surfaces: intrinsic effects and possible artifacts. Phys. Rev. Lett. **94**, 056102 (2005). https://link.aps.org/doi/10.1103/PhysRevLett.94.056102
27. B. Bhushan, Y. Wang, A. Maali, Boundary slip study on hydrophilic, hydrophobic, and superhydrophobic surfaces with dynamic atomic force microscopy. Langmuir **25**, 8117–8121 (2009)
28. H. Li, Z. Xu, C. Ma, M. Ma, Translucency and negative temperature-dependence for the slip length of water on graphene. Nanoscale **14**, 14636–14644 (2022)
29. G. Greenwood et al., Effects of layering and supporting substrate on liquid slip at the single-layer graphene interface. ACS Nano **15**, 10095–10106 (2021). https://pubs.acs.org/doi/10.1021/acsnano.1c01884
30. H. Eyring, Viscosity, plasticity, and diffusion as examples of absolute reaction rates. J. Chem. Phys. **4**, 283–291 (1936)
31. Y. Diao, G. Greenwood, M.C. Wang, S. Nam, R.M. Espinosa-Marzal, Slippery and sticky graphene in water. ACS Nano **13**, 2072–2082 (2019)
32. J.-L. Barrat et al., Influence of wetting properties on hydrodynamic boundary conditions at a fluid/solid interface. Faraday Discuss. **112**, 119–128 (1999)
33. P.A. Thompson, M.O. Robbins, Shear flow near solids: epitaxial order and flow boundary conditions. Phys. Rev. A **41**, 6830 (1990)
34. J.-L. Barrat, L. Bocquet, Large slip effect at a nonwetting fluid-solid interface. Phys. Rev. Lett. **82**, 4671 (1999)
35. D.M. Huang, C. Sendner, D. Horinek, R.R. Netz, L. Bocquet, Water slippage versus contact angle: a quasiuniversal relationship. Phys. Rev. Lett. **101**, 226101 (2008). https://link.aps.org/doi/10.1103/PhysRevLett.101.226101
36. L. Bocquet, E. Charlaix, Nanofluidics, from bulk to interfaces. Chem. Soc. Rev. **39**, 1073–1095 (2010). http://xlink.rsc.org/?DOI=B909366B
37. K. Falk, F. Sedlmeier, L. Joly, R.R. Netz, L. Bocquet, Molecular origin of fast water transport in carbon nanotube membranes: superlubricity versus curvature dependent friction. Nano Lett. **10**, 4067–4073 (2010)
38. M. Kisiel et al., Suppression of electronic friction on nb films in the superconducting state. Nat. Mater. **10**, 119–122 (2011)
39. A. Maali, T. Cohen-Bouhacina, H. Kellay, Measurement of the slip length of water flow on graphite surface. Appl. Phys. Lett. **92**, 053101 (2008). http://aip.scitation.org/doi/10.1063/1.2840717
40. B. Coquinot, L. Bocquet, N. Kavokine, Quantum feedback at the solid-liquid interface: Flow-induced electronic current and its negative contribution to friction. Phys. Rev. X **13**, 011019 (2023). https://link.aps.org/doi/10.1103/PhysRevX.13.011019
41. Q. Xie et al., Fast water transport in graphene nanofluidic channels. Nat. Nanotechnol. **13**, 238–245 (2018)
42. T. Ondarçuhu et al., Wettability of partially suspended graphene. Sci. Rep. **6**, 24237 (2016)
43. T. Mouterde et al., Molecular streaming and its voltage control in ångström-scale channels. Nature **567**, 87–90 (2019)
44. A. Marcotte, T. Mouterde, A. Niguès, A. Siria, L. Bocquet, Mechanically activated ionic transport across single-digit carbon nanotubes. Nat. Mater. **19**, 1057–1061 (2020)
45. E. Secchi, A. Niguès, L. Jubin, A. Siria, L. Bocquet, Scaling behavior for ionic transport and its fluctuations in individual carbon nanotubes. Phys. Rev. Lett. **116**, 154501 (2016)
46. B. Grosjean, M.-L. Bocquet, R. Vuilleumier, Versatile electrification of two-dimensional nanomaterials in water. Nat. Commun. **10**, 1656 (2019)
47. Y. Xie, L. Fu, T. Niehaus, L. Joly, Liquid-solid slip on charged walls: the dramatic impact of charge distribution. Phys. Rev. Lett. **125**, 014501 (2020)
48. P. Robin et al., Long-term memory and synapse-like dynamics in two-dimensional nanofluidic channels. Science **379**, 161–167 (2023)

49. G. Hu, M. Mao, S. Ghosal, Ion transport through a graphene nanopore. Nanotechnology **23**, 395501 (2012)
50. L. Garcia, Ph.D. thesis, Université Grenoble Alpes (ComUE), 2016
51. B. Coquinot, M. Becker, R. Netz, L. Bocquet, N. Kavokine, Collective modes and quantum effects in two-dimensional nanofluidic channels. Faraday Discuss. (2023)
52. T. Emmerich et al., Enhanced nanofluidic transport in activated carbon nanoconduits. Nat. Mater. **21**, 696–702 (2022). https://www.nature.com/articles/s41563-022-01229-x
53. J. Chmiola et al., Anomalous increase in carbon capacitance at pore sizes less than 1 nanometer. Science **313**, 1760–1763 (2006)
54. M. Pascual, et al., Waste heat recovery using thermally responsive ionic liquids through tio 2 nanopore and macroscopic membranes. Energy Environ. Sci. (2023)
55. C. Herrero, G. Tocci, S. Merabia, L. Joly, Fast increase of nanofluidic slip in supercooled water: the key role of dynamics. Nanoscale **12**, 20396–20403 (2020). http://xlink.rsc.org/?DOI=D0NR06399A
56. R.H. Colby, L.J. Fetters, W.W. Graessley, The melt viscosity-molecular weight relationship for linear polymers. Macromolecules **20**, 2226–2237 (1987)
57. M. Lizée, B. Coquinot, G. Mariette, A. Siria, L. Bocquet, Anomalous friction of supercooled glycerol on mica (2024). *arXiv preprint*arXiv:2403.18693

Micro- and Nanotribology at the Insect-Plant Interface

Elena V. Gorb and Stanislav N. Gorb

Abstract As a result of evolutionary arm race between insects and plants, numerous plant surfaces that reduce insect attachment have been evolved. These surfaces provide an effective repelling effect against herbivores, sap-sucking insects and nectar robbers due to the reduction of adhesive and frictional forces in contact between the plant surface and insect attachment devices. This review summarizes literature data and own results on tribological aspects of insect-plant interactions. First, we provide a short introduction to attachment systems of insects. Second, tribological effects of three-dimensional micro- and nanoscopical epicuticular waxes of plants are demonstrated. The contact force reduction mechanisms of plant wax structures (roughness effect, contamination effect, fluid-adsorption effect, and wax-dissolving hypothesis) and their potential implications for biology, agriculture and engineering are discussed.

1 Introduction

The widely spread association between insect species and plants is based on the insect climbing ability supported by specialized attachment devices of their legs [1]. In the contact between insect legs and plant surface, the materials in contact may differ in their surface geometry and physico-chemical properties. Additionally, insect leg pads produce a fluid secretion in the contact zone. Also plants possess strong diversity of surface features that make the contact between two biological surfaces even more complex. The plant cuticle is an interface between the organism and its surroundings and its properties mirror various influences from the specific environment to which the plant is adapted. These environmental factors impact on both the structure and chemical composition of the cuticle. In the context of the co-evolution between insects and flowering plants, the latter have adapted their surface chemistry and structure to facilitate the proper attachment and movement of pollinating insects. And

E. V. Gorb · S. N. Gorb (✉)
Department of Functional Morphology and Biomechanics, Kiel University, Am Botanischen Garten 9, D-24118 Kiel, Germany
e-mail: sgorb@zoologie.unikiel.de

E. Gnecco and E. Meyer (eds.), *Fundamentals of Friction and Wear on the Nanoscale*, NanoScience and Technology, https://doi.org/10.1007/978-3-031-63065-1_19

conversely, in cases of antagonistic co-evolution, plants have developed specialized surfaces aimed at reducing insect attachment. This latter type of interaction has led to the emergence of surface-related defense mechanisms against herbivorous insects and nectar thieves. In some remarkable cases, plants even prevent insects from escaping from highly specialized insect traps known from some carnivorous plants.

In this chapter, we present a comprehensive overview of literature data and our own prior experimental findings about surface adaptations in plants linked to interactions with insects. Our primary objective is to illustrate the profound impact that plant surfaces can exert on the tribological capabilities of insect attachment pads. Specifically, we emphasize the pivotal role played by three-dimensional epicuticular waxes of plants in reducing insect forces in contact. Furthermore, we delve into the structural features and physical mechanisms behind the tribological characteristics exhibited by plant cuticles covered with epicuticular waxes.

2 Insect Attachment Devices

The mechanisms governing insect attachment are intricately tied to the specific surface characteristics of the plant substrate. Typically, insects employ their sharply pointed claws to grasp onto mesoscale surface irregularities (Fig. 1a, d). The successful interlocking of the claws with surface roughness relies on the claw tip diameter [2]. Insects have evolved two distinct types of specialized structures capable of establishing secure contact and, consequently, achieving robust adhesion and friction, even on smooth surfaces. These structures are commonly referred to as "smooth pads" (Fig. 1a–c) and "hairy pads" (Fig. 1d–f). Both types have independently emerged multiple times throughout the evolutionary history of insects. The noteworthy softness of the smooth pad material and the combination of specific geometry with exceptional flexibility of fine insect surface structures (setae or acanthae), often termed "tenent setae," within the hairy pads contribute to an increased contact area on various substrate textures [3]. These attachment devices can be associated with various leg structures, including claws, segments of the pretarsus, ventral surfaces of tarsomeres, or tibio-tarsal joints. Phylogenetic analyses of insects have unveiled the independent origins of these structures in different insect lineages [4].

Insect attachment devices generate secretory fluids within the contact zone. This fluid consists of non-volatile compounds having lipid-like characteristics, and its presence has been detected in footprints using various histological techniques and analytical chemistry methods [5–9]. High-resolution microscopy techniques have revealed that some insects produce fluids resembling micro-emulsions composed of two immiscible groups of substances: water-soluble and lipid-soluble components [3, 8, 10].

Several fundamental physical forces contribute to friction and adhesion in contact. In the experiment involving the bug *Rhodnius prolixus*, Edwards and Tarkanian [11] observed a significant reduction in adhesion, when the insect's tarsi were treated with organic solvents. Stork's experiments with leaf beetles [12] led him to conclude

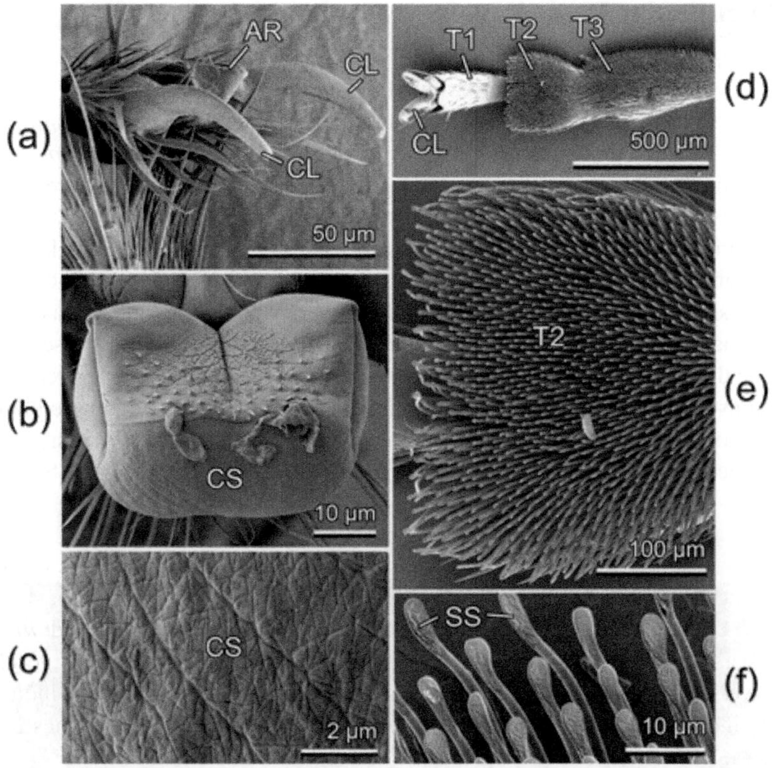

Fig. 1 Pretarsus of the ant *Lasius niger* (**a–c**) having smooth adhesive pad (arolium) and hairy attachment devices of the beetle *Coccinella septempunctata* (female) **d–f** (SEM). Lateral view of the ant claws and the folded (**a**), frontal view of the spread arolium (**b**), arolium surface frozen in contact with a substrate (**c**), ventral views of the foreleg tarsus (**d**), its second proximal tarsomere (**e**), and tenent setae of the second proximal tarsomere (**f**). AR, arolium; CL, claws; CS, arolium surface, which contacted a substrate; SS, spatulate setae (with flattened and widened tips); T1–T3, tarsomeres. From [36] (**a–c**) and [90] (**d–f**)

that cohesive forces, surface tension, and molecular adhesion may contribute to the mechanism of insect attachment mediated by the pad fluid. Langer et al. [13] used atomic force microscopy to investigate forces at the tips of individual tenent setae in the fly *Calliphora vicina.* Their findings provided clear evidence that adhesion is heavily dependent on the presence of the pad fluid surrounded by air. This gives strong support for the idea that insects employ attractive capillary interactions facilitated by the pad secretion between the tips of individual hairs and the plant substrate.

3 Epicuticular Plant Waxes

Cuticular waxes constitute a hydrophobic component of the cuticle in terrestrial plants. According to Jeffree [14], cuticular waxes can be either impregnated within the cuticle (*intracuticular waxes*) or deposited on its surface (*epicuticular waxes*). Epicuticular waxes are typically composed of a complex mixture of cyclic compounds, such as triterpenoids, and long-chain aliphatic substances, including primary and secondary alcohols, primary aldehydes, fatty acids, and alkanes [15, 16]. In higher plants, these waxes cover all exposed primary surfaces and often form relatively smooth two-dimensional films. However, in some specialized cases, three-dimensional micro- or nanoscale projections (Fig. 2) can emerge from these 2D films [17]. The thickness of these films can vary significantly, ranging from extremely thin films, consisting of just a few molecular layers on the surfaces of aquatic plants, to thicker wax crusts of up to 0.5 mm in some terrestrial plants. The dimensions of 3D wax projections can also vary widely, ranging from 0.5 to 100 μm [15, 18].

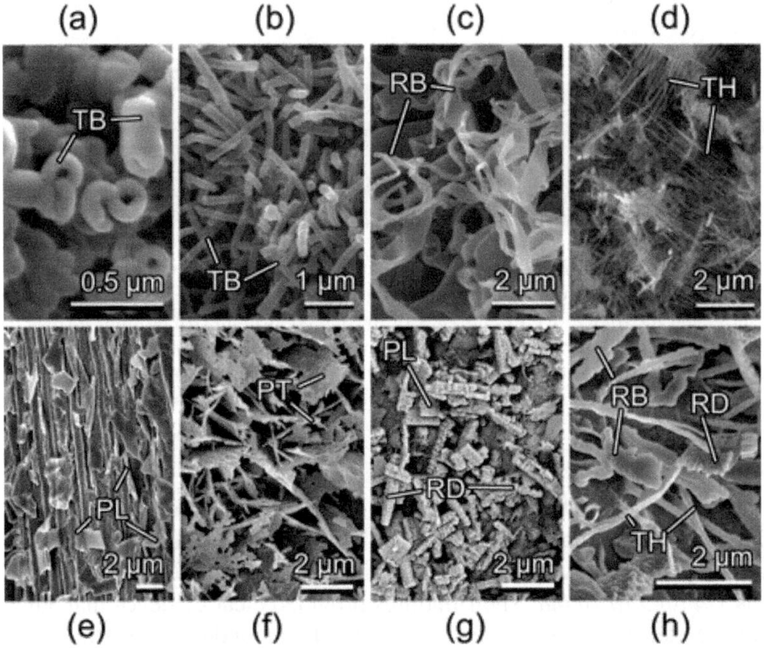

Fig. 2 Epicuticular wax projections on the surfaces of the *Prunus domestica* fruit (**a**), adaxial (upper) leaf side of *Chelidonium majus* (**b**), *Acer negundo* young stem (**c**), abaxial (lower) leaf side of *Brassica oleracea* (**d**), flower stems of *Dahlia pinnata* (**e**), *Tagetes patula* (**f**), and *Anethum graveolens* (**g**), and abaxial leaf side of *Pisum sativum* (**h**) (SEM). RB, ribbons; RD, rodlets; TB, tubules; TH, threads; PL, plates; PT, platelets. From [67] (**a–d**), [36] (**e–g**) and [69] (**h**)

Both 2D and 3D waxes exhibit a crystalline structure, which has been recently revealed through electron and X-ray diffraction analyses [19, 20]. Wax projections (called also "crystals") take on a variety of shapes, including rodlets, threads, platelets, tubules, etc. (Fig. 2) [15]. This diversity in shapes arises from the self-assembly of molecules on the cuticle surface [20–24]. The morphology of these crystals is strongly correlated with the presence of specific dominant chemical compounds within the wax [15, 25]. Epicuticular waxes at the interface between the plant cuticle and the environment serve multiple functions. They can enhance the mechanical stability of the surface, influence transpiration and gas exchange, and provide thermal isolation to control surface temperature. Additionally, these wax projections are known to reduce the wettability of the cuticle surface and prevent surface contamination by dirt particles or pathogenic spores [15, 17, 25].

It has also been repeatedly observed and experimentally supported in various plant species (e.g., *Eucalyptus, Pisum, Brassica*) that 3D wax coverage serves a protective function against herbivorous insects [26–30]. Recent observations on young culms of several *Bambusa* species, which are often heavily infested by sap-sucking insects, particularly aphids, suggest that the wax coverage on the culm surface, often combined with trichomes, impedes the access of ants to aphid colonies, indirectly protecting the plants from aphids (Fig. 3). This phenomenon is referred to as '*stem guard syndrome'* [31].

Plant surfaces with 3D wax coverages can also deter the attachment, locomotion, and foraging behavior of insect predators and parasitoids [32]. Flowering stems of plants with wax coverings may be part of the '*greasy pole syndrome' (GPS)* preventing nectar robbery by ants and safeguarding other plant resources through the slippery properties of the stems. This syndrome is prevalent in numerous plant genera, such as *Salix, Hypenia,* and *Eriope* [33–36]. Some myrmecophytic (obligatory associated with ants) plant species also bear wax crystals on their stems. In the genus *Macaranga*, such wax coverages act as selective barriers for ants. In this case, they protect the plant against non-specialized ant species while allowing access to ants with mutualistic associations with the plant [37–40].

Certain plants, like *Aristolochia* and *Arisaema*, have slippery wax-covered surfaces on their kettle trap flowers that can temporarily capture tiny insect pollinators [41–43]. Some plants have highly specialized slippery surfaces for capturing insects for further digestion. Well-known examples include the inner pitcher surfaces of carnivorous plants from genera *Nepenthes, Sarracenia,* and *Darlingtonia*, as well as certain insect-trapping Bromeliaceae [43–54]. These surfaces are typically anisotropic due to the presence of lunate cells or tiny pointed trichomes, but may also be additionally covered by 3D epicuticular waxes.

As mentioned earlier, wax crystals have the ability to reduce insect attachment to the plant cuticle surface [55, 56]. Numerous studies, employing various experimental approaches, consistently demonstrate that insects can generate strong attachment forces on smooth plant surfaces either without wax or with the wax removed. However, a significant reduction in attachment force is almost always observed on surfaces covered with 3D waxes [27, 38, 39, 46, 47, 49, 54, 57–73].

Fig. 3 Surface in the basal part of a culm sheath and in distal part of an internode in young culms of *Bambusa chungii* (**a**), *B. emeiensis* (**b**) and *B. multiplex* (**c**) showing whitish epicuticular wax coverage in **a** and **b** and aphid infestation on the young culm (**d**, **f**). From [31]

Insect attachment is influenced not only by the presence of 3D wax, but also by the shape and size of the wax projections as well as the density of their distribution. For instance, experiments with the ladybird beetle *Cryptolaemus montrouzieri* walking on *Pisum sativum* plants with wild-type waxes and with reduced waxes revealed significant correlations between attachment forces and wax crystal length (negative correlation) and density (positive correlation) [69]. Experiments conducted with the other ladybird beetle *Coccinella septempunctata* on bioinspired wax surfaces created using four alkanes with varying chain lengths also demonstrated stronger insect attachment on surfaces with a higher density of wax coverage consisting of smaller plate-like crystals [74].

A comparative study of the attachment ability of the chrysomelid beetle *Gastrophysa viridula* on 16 different plant species clearly demonstrated that 3D waxes only temporarily reduce attachment [67]. In many cases, no lasting impact on subsequent insect attachment ability was observed, or the insects rapidly recovered their attachment ability. Additionally, force tests conducted with the aphid *Acyrthosiphon pisum* revealed that it could fully restore its attachment ability shortly after walking on wax-covered leaf surfaces of *Trifolium pratense, Medicago sativa,* and *Pisum sativum* [73].

4 Mechanisms of Insect Adhesion Reduction on Waxy Plant Surfaces

The phenomenon of reduced insect adhesion on plant cuticles covered with 3D waxes is not easily explained and is likely influenced by multiple mechanisms. Four following hypotheses have previously been proposed to explain these contact phenomena (Fig. 4).

Roughness hypothesis: This hypothesis suggests that the physical roughness of the wax crystals on the plant surface reduces the real contact area between the insect and the surface, thus diminishing adhesion and friction.

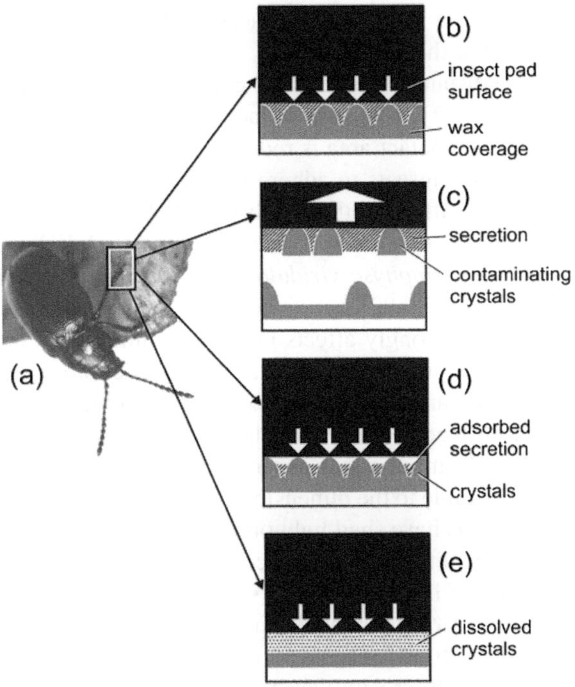

Fig. 4 Schematic presentation of hypotheses explaining anti-adhesive properties of plant cuticle covered with 3D wax projections. The chrysomelid beetle *Gastrophysa viridula* attached to the surface of its host plant *Rumex obtusifolius* (**a**), roughness hypothesis (**b**), contamination hypothesis (**c**), fluid-adsorption hypothesis (**d**), and wax-dissolving hypothesis (**e**). Adapted from [1]

Contamination hypothesis: According to this hypothesis, the wax crystals on the plant surface act as a barrier, preventing insect's adhesive organs from making direct contact with the substrate, leading to reduced attachment.

Fluid-adsorption hypothesis: This hypothesis posits that the wax crystals absorb fluids from the adhesive organs of the insect, reducing the effectiveness of the adhesive organs.

Wax-dissolving hypothesis: The wax-dissolving hypothesis proposes that the wax crystals on the plant surface are susceptible to dissolution by the adhesive fluids of the insect, leading to an increase of the fluid thickness on the plant surface and consequently, to reduced adhesion.

Each of these hypotheses provides a potential explanation for the observed reduction in insect adhesion on wax-covered plant surfaces, and the relative importance of these mechanisms may vary depending on the specific plant–insect interaction and environmental conditions.

Roughness hypothesis

The micro- and nanoscale surface roughness generated by wax crystals on plant cuticles can significantly affect contact forces and therefore insect attachment. In tribological literature, it is well-documented that surface roughness leads to a reduction in the real contact area between the plant surface and the insect's attachment pad. Contact forces, such as adhesion and friction, rely on the actual contact area between the interacting surfaces, and this area is influenced by the topographical features of the surfaces.

The distances between the contacting surfaces are in the order of atomic dimensions [75–77]. When the surface is micro-rough due to the presence of wax crystals, the actual contact area is reduced. This reduction in contact area can make it challenging for insects to adhere to the micro-rough surfaces created by wax crystals, ultimately affecting their ability to stick to these surfaces. Indeed, the experimental evidence from studies on various insects, including the fly *Musca domestica* and beetles *Gastrophysa viridula, Leptinotarsa decemlineata,* and *Coccinella septempunctata* [3, 78–82], consistently supports the idea that micro- and nanoscale surface roughness strongly affects insect attachment. In experiments involving centrifugal and traction tests, these insects demonstrated stronger attachment forces, when tested on smooth artificial substrates compared to their performance on nano- and micro-rough surfaces (Fig. 5). The discovery that the lowest attachment forces occurred on fine rough surfaces with asperity sizes ranging from 0.3 to 1.0 μm [3, 78–80], which is comparable to the dimensions of plant wax crystals, is quite intriguing. Numerical simulations have shed light on this phenomenon by proposing an explanation based on the reduction of the real contact area between the rough surface profile and the setal tips of individual insect setae responsible for intimate contact with the surface. The model demonstrated that there is a maximum distance and, consequently, a minimum attractive force between the setal tip of the insect pad and the substrate for surfaces with asperity sizes falling within the 0.3–1.0 μm range. In contrast, surfaces with larger asperity sizes (3–12 μm) did not lead to such a significant reduction in attachment forces [3, 78, 79, 81].

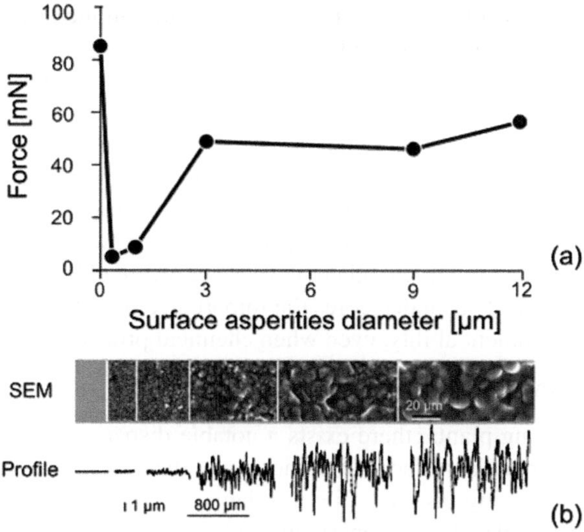

Fig. 5 Attachment forces of the Colorado potato beetle *Leptinotarsa decemlineata* obtained in centrifugal friction experiments on artificial substrates with different surface micro-roughness (**a**) generated by asperities with nominal sizes ranging from 0.3 to 12.0 µm (**b**). Adapted from [79]

This phenomenon can be attributed to the unique features of insect tenent setae. These setae possess dimensions in the range of 2–10 µm and have a spatula-like geometry at their tips. This specific morphology of tenent setae allows them to generate sufficient contact area even on substrates with relatively large dimensions of surface asperities. Essentially, the setal tips conform to the contours of the substrate, making intimate contact with relatively smooth areas on the tops of the asperities [78, 79].

The critical roughness, which is the substrate roughness, where attachment force is the lowest, has been observed in various studies, including experiments with the stick insect *Carausius morosus* and the carnivorous plant *Nepenthes alata*. In the case of *C. morosus,* the lowest attachment ability was found on the finest roughness with 3 µm asperity size, when compared to smoother and other micro-rough substrates with asperity sizes of 10, 25, and 46 µm [51]. This critical roughness displayed similar geometrical parameters to the wax-covered slippery zone found in *N. alata* pitchers.

The results from these experiments and theoretical models suggest that surface roughness within a certain range, referred to as the critical roughness, can prevent insect adhesion. In the case of *Nepenthes alata*, the presence of wax crystals on the plant surface leads to a reduction in insect attachment due to micro- and nanoscale surface roughness. Similar effects were observed in traction force experiments with the coccinellid beetle *Adalia bipunctata* walking on the lower wax layer of *N. alata* [49]. Pull-off adhesion tests conducted on this wax coverage showed intermediate adhesion force values on the artificial 3 µm micro-rough surface compared to the lower and upper wax layers of the plant [83, 84]. These experiments utilized deformable tacky polymer beads with an elasticity modulus similar to that of the smooth adhesive pads in insects, and they applied a normal load force that simulated the pressure generated by an insect weight.

Traction experiments conducted with the beetle *Coccinella septempunctata* on eight different types of artificial surfaces, each characterized by distinct chemical properties and topography, provided clear evidence for the dominant influence of surface roughness on the beetles' attachment ability [82]. Among the four micro-rough samples tested, all exhibited the lowest attachment force, with no statistically significant differences observed between them. Surprisingly, the chemical surface properties did not have a significant effect on the beetles' attachment performance. This underscores the importance of surface roughness in determining insect attachment ability, even when chemical properties vary.

Contamination hypothesis

Among plants, there exists a notable disparity in the adherence of wax crystals to the plant's surface. In some cases, these wax crystals are firmly anchored to the underlying layers, while in others, they exhibit a remarkable propensity to detach easily. The latter scenario presents a noteworthy challenge for insects that come into contact with these detachable wax structures. These wax crystals can readily stick to the attachment organs of insects, thereby causing contamination (Fig. 6c). This effect is particularly pronounced, when insects possess adhesive pads coated with fluid secretions. The contamination of these adhesive pads by wax can significantly impede their normal functioning. As a consequence, insects may experience a substantial reduction in their ability to adhere to surfaces.

Experimental evidences support this hypothesis because insects demonstrate grooming behavior to remove wax contamination from their attachment organs after moving on waxy plant surfaces. For instance, the beetle *Paropsis charybdis* has been observed grooming itself after traversing young leaves of *Eucalyptus nitens* [27]. Similarly, the fly *Drosophila melanogaster* and the ant *Iridomyrmex humilis* have exhibited grooming behavior following their attempts to climb surfaces with wax, such as the waxy zone of *Nepenthes alata* pitchers [46]. Earlier studies have also provided direct evidence that loosely connected 3D wax crystals can contaminate insect attachment pads. Using scanning electron microscopy (SEM), researchers have demonstrated the ability of fragile wax crystals from *Brassica* spp. to contaminate the attachment organs of the herbivorous beetle *Phaedon cochleariae* and three species of predatory insects [58, 59, 62]. These findings underscore the dynamic interplay between plants and insects, where insects have developed mechanisms to mitigate the challenges posed by wax contamination during locomotion on waxy plant surfaces. The contaminating effect of the upper wax crystals found on *Nepenthes plants* on the attachment devices of the fly *Musca domestica* has been confirmed through the use of a transmission electron microscope (TEM). This research conducted by Juniper and Burras [44] and further elaborated upon by Juniper et al. [45] provided insights into how the wax crystals from *Nepenthes* plants can affect the attachment ability of flies. In SEM studies, it was observed that wax platelets from carnivorous plants such as *N. ventrata, Brocchinia reducta,* and *Catopsis berteroniana* can become attached to the attachment pads of flies, like *Calliphora vomitoria.* This attachment of wax

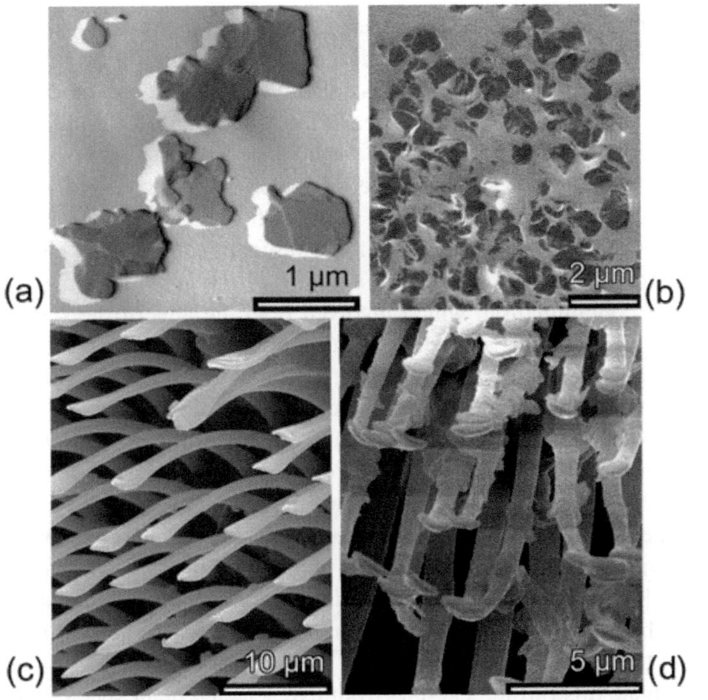

Fig. 6 Evidences for the contamination hypothesis. Isolated wax crystals from the upper wax layer of the pitcher of the carnivorous plant *Nepenthes alata* (TEM) (**a**), the tacky polymer probe made of polydimethylsiloxane after adhesion test with the upper wax layer (cryo-SEM) (**b**), and tarsal tenent setae of the coccinellid beetle *Adalia bipunctata* after walking on the wax-free (treated with warm chloroform) (**c**) and wax-bearing slippery zone of the *N. alata* pitcher (**d**) (both SEM). Adapted from [49] (**a**, **c**, **d**)

platelets further highlights the potential challenges caused by wax contamination on insect attachment organs.

The contamination hypothesis has received experimental support in various insect species with hairy-type adhesive pads and a range of plant species. One notable example comes from the slippery zone of *Nepenthes alata* pitchers, where the outermost wax layer exhibits a remarkable ability for contaminating adhesive pads with wax crystals [49]. This outer wax layer is comprised of fragile, stalked crystals with a distinctive leaf-like shape, making them susceptible to detachment or fragmentation (Fig. 6a). Traction force experiments involving the coccinellid beetle *Adalia bipunctata* revealed a significant reduction in attachment force when interacting with this wax-covered surface. Subsequent SEM examinations of the beetle's pads following these experiments provided clear evidence that the force reduction was indeed attributed to the contamination of the adhesive pads with the upper wax crystals (Fig. 6d). Pull-off force tests, utilizing tacky polymer probes composed of polydimethylsiloxane [84] further supported the concept of the upper wax layer's strong anti-adhesive properties. Moreover, these experiments demonstrated the crystals' ability to easily exfoliate under minimal preload conditions and contaminate sticky surfaces (Fig. 6b). These findings collectively underscore the effectiveness of the upper wax layer in preventing insect adhesion through contamination of their attachment organs [49, 83].

Attachment pads of chrysomelid beetles, specifically *Chrysolina fastuosa*, were examined after walking on 12 different plant surfaces covered with 3D waxes, and the results revealed varying degrees of contamination by wax crystals [85]. The contamination was characterized by the nature of the contaminating material and the degree of contamination, which involved both the area of the setal tip surface covered with wax particles and the proportion of contaminated setae (Fig. 7).

These variations in contamination were attributed to the differing shapes and dimensions of the wax crystals present in each plant species. Notably, crystals with the largest aspect ratios, such as rod-like and particularly thin platelet-like crystals, showed the strongest contamination effect on the attachment pads. In contrast, tube-like crystals had a significantly lower impact on contamination.

Based on experimental findings, a theoretical mechanics approach was applied to explain the varying degrees of insect pad contamination by plant wax of different plant species [86]. The hypothesis proposed that the extent of contamination primarily depends on the fracture behavior of wax crystals, which is in turn influenced by the geometry of these crystals.

To investigate this further, crystals from five selected plant species (*Aristolochia fimbriata, Aquilegia vulgaris, Berberis vulgaris, Chelidonium majus,* and *Prunus domestica*) were carefully measured using SEM images. The analysis revealed that during the contact between insect adhesive pads and the plant surface, the mechanisms of wax crystal fracture could vary significantly, depending on the slenderness ratio of the crystals. High slenderness ratio crystals could undergo elastic or elastic–plastic buckling.

For most of the crystals examined, the calculated values for critical buckling (26.5 for elastic buckling and 18.7 for elastic–plastic buckling) indicated that these crystals were unlikely to buckle under the applied loads. However, a closer examination revealed that insects were capable of fracturing wax crystals of certain plant species through elastic–plastic bending. This approach is generally applicable to crystals of different shapes, even though the models were initially developed for tubular crystals [86].

The contamination of insect adhesive pads by wax crystals can have a temporary impact on the insects' further attachment ability. For instance, in an experimental study with the chrysomelid beetle *Chrysolina fastuosa*, it was observed that while most of the 11 waxy plant surfaces examined either did not affect or only temporarily reduced the subsequent attachment ability of the insects, the wax-covered stems of *Acer negundo*, which exhibited the highest contaminating capability, disabled beetle attachment for a certain period [67].

In traction experiments involving the aphid *Acyrthosiphon pisum* and waxy leaf surfaces of *Trifolium pratense, Medicago sativa*, and *Pisum sativum*, aphids were tested on a clean reference glass surface immediately after walking on the waxy

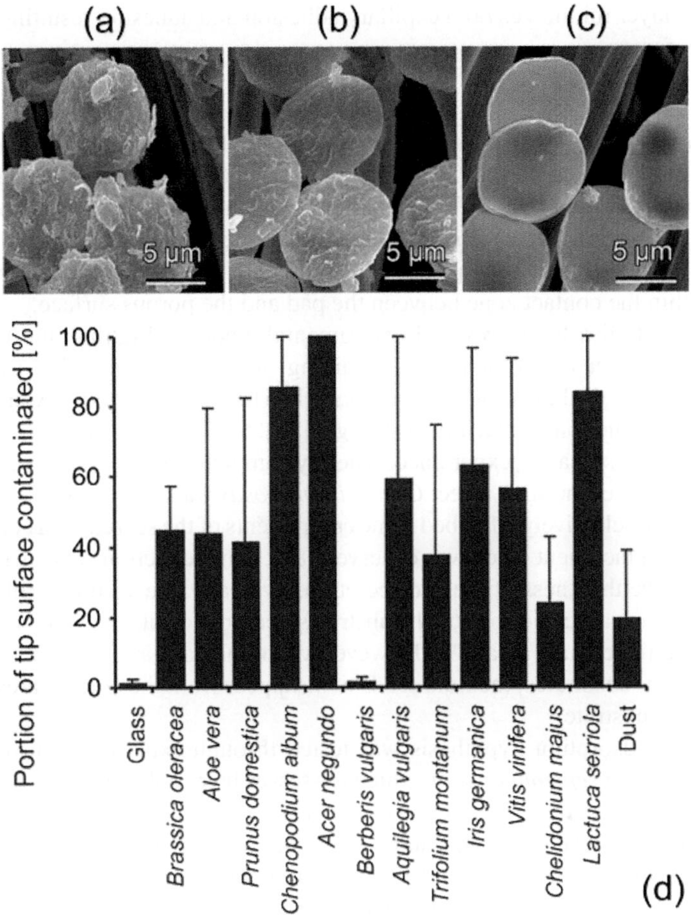

Fig. 7 Degree of contamination of adhesive pads of the chrysomelid beetle *Chrysolina fastuosa* by wax crystals of various plant species. Beetle adhesive setae after walking on a young stem of *Acer negundo* (**a**), adaxial (upper) leaf side of *Chenopodium album* (**b**) and *Berberis vulgaris* (**c**) (all SEM), and the portion of setal tip areas contaminated with wax material after beetles having walked for the same period of time on different plant surfaces (**d**). From [85]

plant surfaces [73]. The first three consecutive pulling events showed very low force values. However, by the third trial, the aphids had regained their attachment ability and pulled with the same force as they did before walking on the waxy plants. Interestingly, adhesion tests on the upper wax layer in *Nepenthes alata*, where strong contaminating properties of the upper wax crystals have been confirmed, did not showed a subsequent reduction in adhesion due to contamination [83]. Based on these experimental findings, it was suggested that in such cases, broken-off wax crystals serve as a separation layer between an insect pad and a plant surface. This separation layer minimizes both capillary adhesion and adhesion resulting from van der Waals forces.

Fluid-adsorption hypothesis

This hypothesis proposes that the three-dimensional wax coverage can adsorb the adhesive fluid from the surface of the insect pad. This occurs due to the high porosity of the wax coverage and its affinity to the lipid-rich pad secretion. The adsorption of the fluid causes a decrease in the thickness of the fluid layer between the insect pads and the substrate. Consequently, this reduction leads to a decrease in capillary forces within the contact zone between the pad and the porous surface.

Previous studies have provided experimental evidence highlighting the importance of insect pad secretion in maintaining adhesion to smooth surfaces. For example, pads of the bug *Rhodnius prolixus*, when treated with lipid solvents, showed a significant reduction in attachment strength [11]. Similarly, the aphid *Aphids fabae*, after walking on silica gel, experienced a nearly complete loss of its attachment ability [87]. In the case of the stick insect *Carausius morosus* walking on smooth polyimide substrates that selectively adsorbed some components of the secretion, a considerable reduction in attachment force was observed [88]. This reduction was attributed to a decrease in the thickness of the pad secretion. Typically, the secretion fills the gaps between surface asperities on rough substrates, leading to adhesion enhancement by maximizing the contact area [89]. However, when the gaps are continuous as in the case of porous substrates, the fluid tends to disappear from the contact area between the pad and substrate.

The fluid-adsorption hypothesis was tested through experiments involving the beetle *Coccinella septempunctata,* which possesses hairy adhesive pads. The beetles walked on nano-porous substrates with the same pore diameter (220–250 nm) but varying porosity levels (28, 42, and 51%), and smooth solid surfaces were used as control samples [90]. Traction force measurements revealed significantly higher attachment forces on the solid surfaces compared to all porous substrates. Among the porous substrates, only those with the highest and lowest porosity demonstrated significant differences in attachment forces. Experimental evaluations of the contact angle using both water and oil showed that nano-porous surfaces had a strong adsorption capacity for both polar and non-polar fluids (Fig. 8). These experiments collectively concluded that the high porosity of nano-porous membranes led to fluid adsorption from beetle adhesive pads, thereby reducing capillary interactions and insect attachment ability.

More recently, the adsorption ability of 3D epicuticular plant waxes was experimentally confirmed [91]. Using a cryo SEM approach and high-speed video recordings of water and oil drop behavior, followed by numerical analysis of the data, it was demonstrated that the wax coverage of *Nepenthes alata* is capable of adsorbing oil. Both intact (two-layered) and treated (one-layered, bearing only the lower wax layer) wax samples exhibited changes in the base, height, and volume of oil drops over time (Fig. 9). The thickness of the wax layers on plant samples had no significant effect on oil drop behavior. Based on these findings, it was concluded that 3D

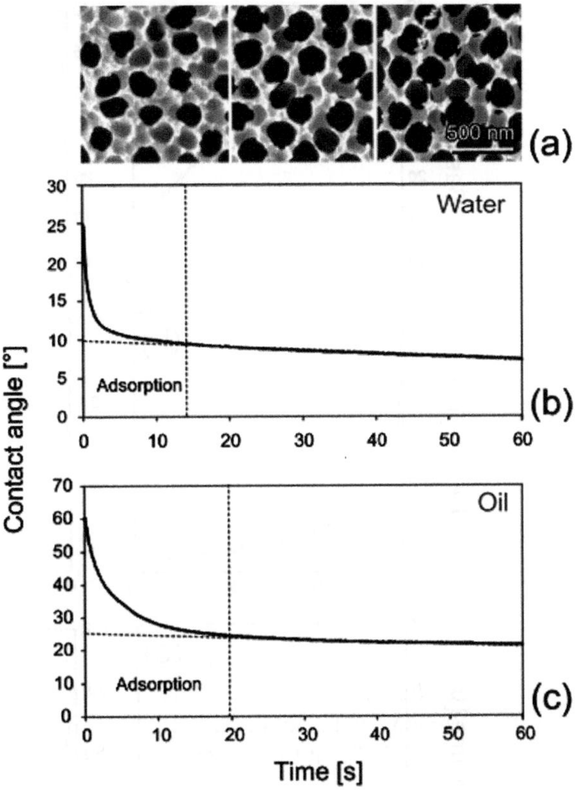

Fig. 8 Evidences for the fluid-adsorption hypothesis. SEM images of nano-porous substrates with different porosities used in force experiments with *Coccinella septempunctata* beetles (SEM) (**a**) and changes in contact angles of water (**b**) and oil (**c**) on the nano-porous sample with a porosity of 42%. Dotted lines demarcate parts of curves showing the effect of fluid adsorption. Adapted from [90]

wax coverages on plant surfaces, due to their adsorption capability, generally exhibit anti-adhesive properties for insects that rely on wet adhesion.

Wax-dissolving hypothesis

It is hypothesized that certain components of insect pad secretion, which contain a significant proportion of lipids and lipophilic substances, may have the ability to dissolve plant wax crystals. This dissolution would result in an increased thickness of the fluid layer on the plant surface. As a consequence, the contact forces (friction and adhesion) between the insect pads and the surface would decrease, making the substrate slippery for insects due to the newly-formed thick fluid layer in the contact area. While there is currently only indirect evidence for pad secretion's ability to dissolve wax, just some observations support this hypothesis.

For example, when the fly *Calliphora vomitoria* walked on wax-bearing surfaces of carnivorous plants like *Nepenthes ventrata* and *Brocchinia reducta*, their tenent setae were found to be glued together with an amorphous substance [47]. This substance could potentially be a product of waxes losing their crystalline structure due to chemical interactions with the pad secretion. Similar effects were observed in a study on the contamination of the hairy adhesive pads of the beetle *Chrysolina*

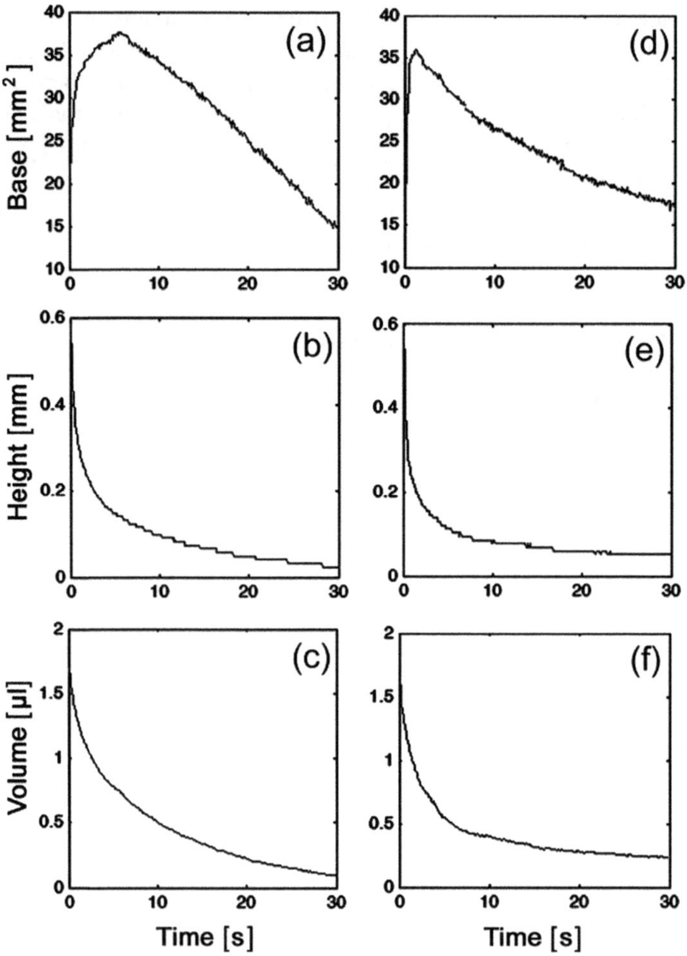

Fig. 9 Oil drops behaviour on wax-covered surfaces of the *Nepenthes alata* pitcher. Time-dependant values of the base (**a**, **d**), height (**b**, **e**), and volume of the drop (**c**, **f**) on the one-layered (**a**–**c**) and two-layered wax samples (**d**–**f**). Adapted from [91]

fastuosa by the wax coverage of 12 different plant surfaces [85]. In some cases, such as *Aquilegia vulgaris, Trifolium montanum, Aloe vera,* and *Berberis vulgaris*, the wax contamination exhibited a relatively homogeneous texture where wax crystals were not recognizable. This uniform, almost unstructured contamination pattern was also attributed to the possible dissolving of wax crystals by the insect secretory fluid.

5 Combination of Plant Cuticle Sculpture and Wax Coverage

Many plant species exhibit a combination of cuticle microsculpture and crystalline wax on their surfaces, creating a hierarchical structure with variations in shape, size, and orientation [92]. This hierarchical structure can be observed across multiple levels, each with its own distinctive features. *Macro-scale level:* At the macro-scale level, structures can have dimensions of several millimeters, leading to specific waviness or contours of a leaf surface. *Micro-scale level*: Moving down the hierarchy, the micro-scale level includes structures like multi-cellular trichomes and curved outlines of individual cells. *Nano-scale level:* The third hierarchical level comprises cuticular folds and wax projections, with dimensions ranging from hundreds of nanometers to a few micrometers [92, 93]. This level can introduce additional complexity because cuticular folds are often coated with wax crystals. Wax crystals themselves may form hierarchical sub-levels, creating sub-layers with crystals of different sizes, shapes, and orientations [1, 49, 71, 94, 95]. This hierarchical organization contributes to the diverse surface properties observed in plants, including their adhesion and friction characteristics, as well as their interaction with insects and other organisms.

The functional importance of additional structural complexity on plant surfaces has only recently become the subject of systematic study, despite repeated observations of hierarchically-structured plant surfaces, particularly in plants capable of trapping insects (e.g., Oelschlägel et al. [42]). It is logical to assume that such hierarchical surface architecture further reduces insect attachment. A comparative SEM study of insect pitfall traps in 53 plant species unveiled a wide diversity of combinations of epidermal cell curvatures, trichomes, cuticular folds, and epicuticular wax crystals [43]. This diversity was categorized into 12 types of anti-adhesive surfaces based on the composition of their structural elements and the observed hierarchy. Among the examined plant species, 28 had only a single level of hierarchy, characterized by features such as convex cell shapes or tabular cells with superimposed waxes. Interestingly, the largest group of plants comprising 24 species exhibited two-level or three-level hierarchies. These surfaces typically had papillate epidermal cells covered with waxes or cuticular folds (two-level) or papillate cells with cuticle folds and superimposed waxes (three-level). It was suggested that such hierarchical surfaces represent a combination of two or more strategies to induce surface slipperiness, potentially enhancing the effectiveness of insect trapping. Based on existing literature on insect attachment, the authors proposed an intriguing hypothesis that the slipperiness of a plant surface is correlated with its hierarchical complexity, specifically with an increased number of strategies implemented in a given surface.

Previous research on carnivorous plant *Nepenthes plants* has provided experimental evidence regarding the impact of hierarchical plant surface structures on insect attachment. In the case of intact *N. alata* surfaces with both lunate cells and the 3D wax coverage, attachment of ants *Iridomyrmex humilis* was entirely disabled [46]. Conversely, chemically de-waxed pitchers containing only lunate cells were much less effective and allowed ants to escape. Beetles *Coccinella septempunctata* exhibited significantly lower traction forces on chemically de-waxed pitchers and their replicas compared to smooth glass surfaces [54]. This effect was especially pronounced when insects moved in a direction out of the pitcher. In such cases, their claws struggled to properly grip the downward-pointed asymmetric lunate cells (Fig. 10c). The intact waxy zone of the pitcher, which featured a three-level hierarchical structure combining lunate cells and two levels of wax crystals (Fig. 10a), drastically reduced beetle forces in both pulling directions when compared to forces measured on glass and wax-free pitcher surfaces (Fig. 10b–d). The effectiveness of the waxy pitcher surface in reducing beetle adhesion can be attributed to contamination of adhesive pads with wax crystals and further reduced contact between contaminated insect attachment pads and lunate cells due to additional anisotropic surface micro-roughness [46, 49, 50, 53].

In some *Nepenthes* species, the wax coverage is composed of two layers of wax crystals (Fig. 11a) [49, 95]. These individual wax crystals from the two layers exhibit

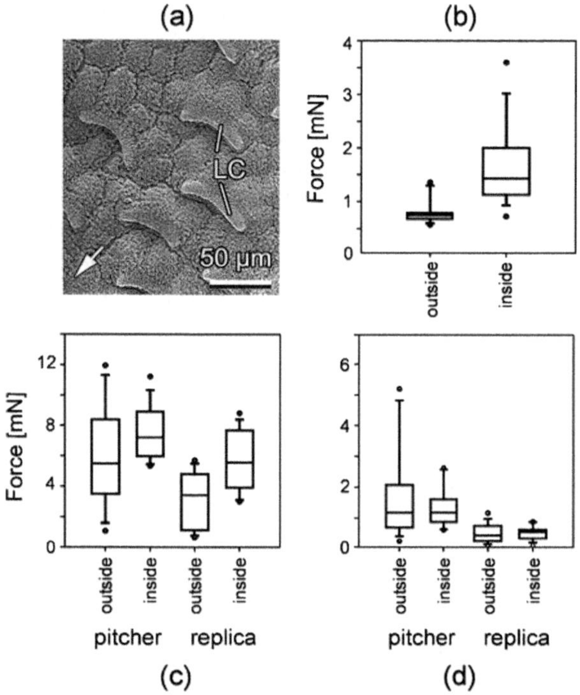

Fig. 10 The waxy zone of the *Nepenthes alata* pitchers and its effect on the attachment ability of the beetle *Coccinella septempunctata*. Three-levelled intact waxy zone with lunate cells and epicuticulat wax coverage in SEM (**a**) and the results of force measurements with intact insects and intact outside and inside pitcher surfaces (**b**). The intact (**c**) and clawless insects (**d**) were also tested on de-waxed (treated with warm chloroform) pitchers and their polymer replicas. Inside, movement in the direction outside the pitcher; LC, lunate cells; outside, movement in the direction toward the pitcher bottom. Arrowhead in (A) indicates the direction toward the pitcher bottom. Adapted from [54]

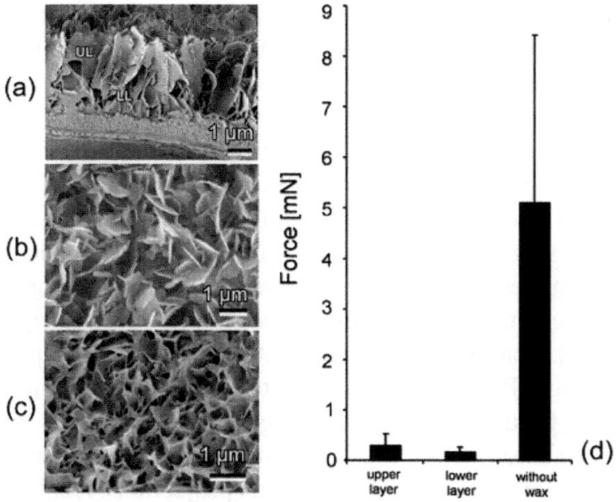

Fig. 11 The waxy zone of the *Nepenthes alata* pitcher and its effect on attachment of the coccinellid beetle *Adalia bipunctata*. Side view of the wax coverage (cryo-SEM, fractured pitcher sample) (**a**), the intact upper wax layer (above aspect) (**b**) (SEM) and lower wax layer after treatment of an intact pitcher with the fluid silicone elastomer and removal of the upper wax layer by peeling off (**c**) (SEM). Traction forces produced by beetles on the different exposed wax layers and on the completely de-waxed pitcher surface treated with warm chloroform (**d**). LL, lower wax layer; UL, upper wax layer. Adapted from [49]

notable differences in morphology, chemical composition, as well as mechanical material properties [49]. Interestingly, coccinellid beetles *Adalia bipunctata* with hairy pads displayed a similar degree of adhesion reduction on both the upper and lower wax layers (Fig. 11d). This finding is quite surprising and can only be explained by the assumption that the two wax layers achieve a similar anti-adhesive effect through two different mechanisms. Wax crystals in the upper wax layer (Fig. 11b) are easily eroded, leading to contamination of the beetle's adhesive pads. In contrast, the lower wax layer (Fig. 11c) is mechanically more stable and reduces adhesion mainly by decreasing the actual contact area between the insect's feet and the plant surface.

6 Case Study with *Hippeastrum*

As mentioned above, 3D wax projections of plants have a remarkable ability to render plant surfaces slipperiness, acting as a protective shield against intruders from various insect lineages. Experimental evidence supporting the protective function of three-dimensional epicuticular waxes has predominantly been observed on plant leaves across various plant families (f.e., [67, 96, 97]). Interestingly, when it comes to the flower stems, their protective function primarily targets ants that attempt to steal flower nectar and do not contribute to cross-pollination—a phenomenon termed "greasy pole syndrome" (GPS) [33–36, 98]. Through field observations and experiments, it has been previously demonstrated that ants have troubles to get a grip on waxy stems due to the dislodgment of wax crystals, causing them to lose their footing and fall to the ground or to strongly decelerate their locomotion speed.

The impact of wax coverage on stems can be attributed to ants' inability to adhere effectively to a wax bloom, as described in previous research [37]. Generalist ants tend to avoid climbing stems covered in wax, not necessarily because they are entirely incapable of traversing slippery vertical surfaces, but due to the additional effort required to walk on these slippery stems. Even when ants manage to traverse such surfaces, their movement remains cautious and slowed down, likely incurring higher energetic costs [36]. Consequently, in their natural habitat, ants typically prefer plant stems devoid of the GPS [98]. Given that stems with the GPS are predominantly oriented vertically, frictional forces play a significant role in ants' grip and locomotion. Frictional and shear forces, arising from the specific geometry of wax projections, can lead to the breakage of individual wax projections, resulting in the slippery nature of the GPS surfaces [86]. Indeed, our previous studies have demonstrated that 3D wax projections on stems exhibit distinct morphological features that may differ from the microstructure of other plant organs, even within the same plant species [36, 98].

In the case study reported below [99], we measured the frictional properties of three-dimensional wax projections specific to flower stems in the plant *Hippeastrum reginae* (Amaryllidaceae) (Fig. 12a) having relatively large flower stems. Using a microtribometer, we conducted experiments to assess the frictional coefficients of intact stems covered by 3D wax projections and those, from which the wax layers had been mechanically removed. Furthermore, we employed a cryo-scanning electron microscopy approach, to capture images of frozen plant surfaces with both intact and smeared wax layers.

Microstructure of the plant surface wax

While both studied plant surfaces, namely the flower stem and the upper leaf side, bear epicuticular wax coverage consisting of microscopic 3D wax projections, the micromorphology of the wax structures differs between these surfaces. On the leaf, which has a relatively glossy appearance, wax projections take the form of flat, irregularly-shaped, thin platelet-like structures (Fig. 12g, h). They densely cover the surface, resembling closely packed scales. In contrast, the flower stem has a pruinose appearance due to the numerous wax projections that are distinctly different. These projections are regularly distributed across the surface. They appear as separate, transversely ridged rodlets that protrude perpendicularly from the surface (Fig. 12d). These rodlets vary in length and exhibit different, but relatively regular shapes and dimensions in cross-section. Along their longitudinal axis, the rodlets display characteristic alternating furrows and ridges, which clearly indicate a layered nanostructure (Fig. 12e). Delamination between these layers results in characteristic damage and exfoliation of the rodlets.

Fig. 12 **a** *Hippeastrum reginae* flowers with the leaf and flower stems. **b**, **c** Two individual flower stems with 3D wax coverage (matt appearing sites, white arrows) and with some damages in the 3D wax coverage (glossy appearing sites, black arrows). **d**, **e** 3D wax projections of intact coverage of the flower stem and an individual wax projection with clearly seen layered structure and characteristic damage failure (arrowheads) due to the delamination between layers **e**. **f** 3D wax projections peeled off by ScotchTM tape. **g**, **h** H. 3D wax projections of the upper (adaxial) surface of the leaf. LA, layers constituting wax projections; WP, wax projections. From [99]

Frictional properties

On the intact flower stems covered with 3D wax projections, despite noticeable force fluctuations within a single sliding cycle, frictional forces remained relatively low, when compared to samples from which the wax had been removed (Fig. 13a). The significant fluctuations in force during individual sliding cycles can be attributed to the presence of some coarse roughness resulting from the shapes of the underlying epidermal cells. Additionally, in the absence of wax, there was a distinct stick-and-slip behavior. This behavior arose due to the relatively large contact area resulting from the smooth surface of the epidermal cells and the soft cellular nature of the underlying fresh plant material. However, on the intact stems with 3D wax coverage, both frictional force and stick-and-slip behavior were significantly reduced (Fig. 13b).

These results provide evidence that flower stem surfaces covered by wax projections can substantially decrease friction. Cryo-SEM examination of intact plant

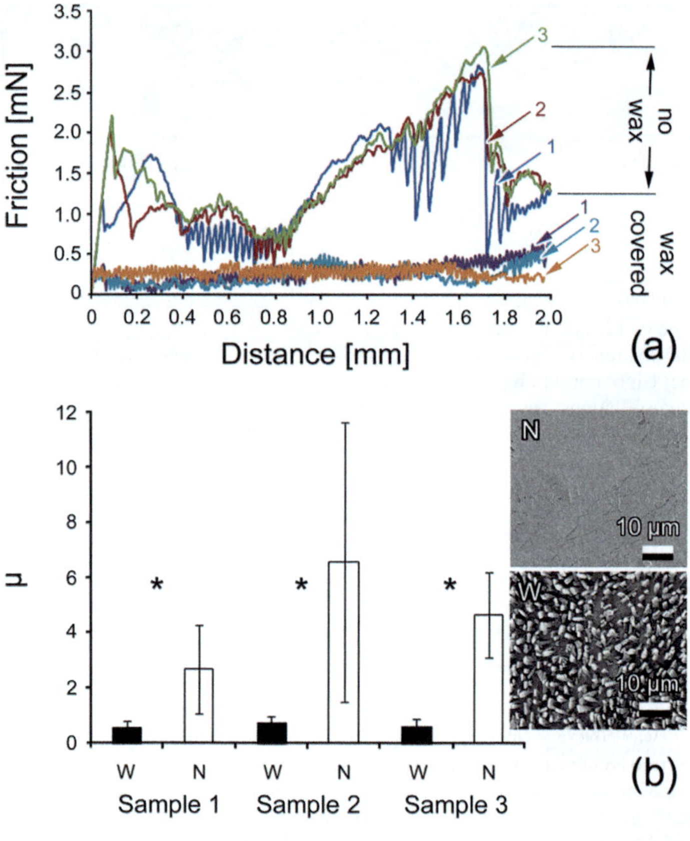

Fig. 13 Frictional properties of flower stems in *Hippeastrum reginae*. **a** force-distance diagrams of three first subsequent sliding cycles (same direction) on the intact stem covered by 3D wax projections (wax covered) and of those on the stem with wiped off wax projections (no wax). **b** frictional coefficients (μ) obtained for 10 sliding cycles on three individual intact flower stems (W) and on the same stems with removed wax coverage (N). Asterisks indicate statistically significant differences between two different conditions of each sample. Insets show intact sample (W) and sample with removed wax coverage (N). From [99]

surfaces after frictional experiments (Fig. 14) reveals that the slipperiness mechanism is based on the ease of delamination of the layered structure of individual wax projections (Fig. 14c). Ultimately, this leads to the creation of smeared surfaces with presumably lower intrinsic friction between individual layers (Fig. 14).

In *Hippeastrum*, there are mechanisms that actively limit self-pollination. One such mechanism is self-incompatibility, where seeds can only be produced through pollination from other plants. Additionally, the plant typically releases its pollen approximately two days before its stigma becomes receptive, thereby increasing the likelihood of cross-pollination [100]. The pollinators for *Hippeastrum* include hummingbirds and moths. These aspects collectively suggest that ants, with their nectar-robbing behavior and limited ability to facilitate cross-pollination, are not particularly beneficial to this plant. Furthermore, our cryo-SEM results, which reveal

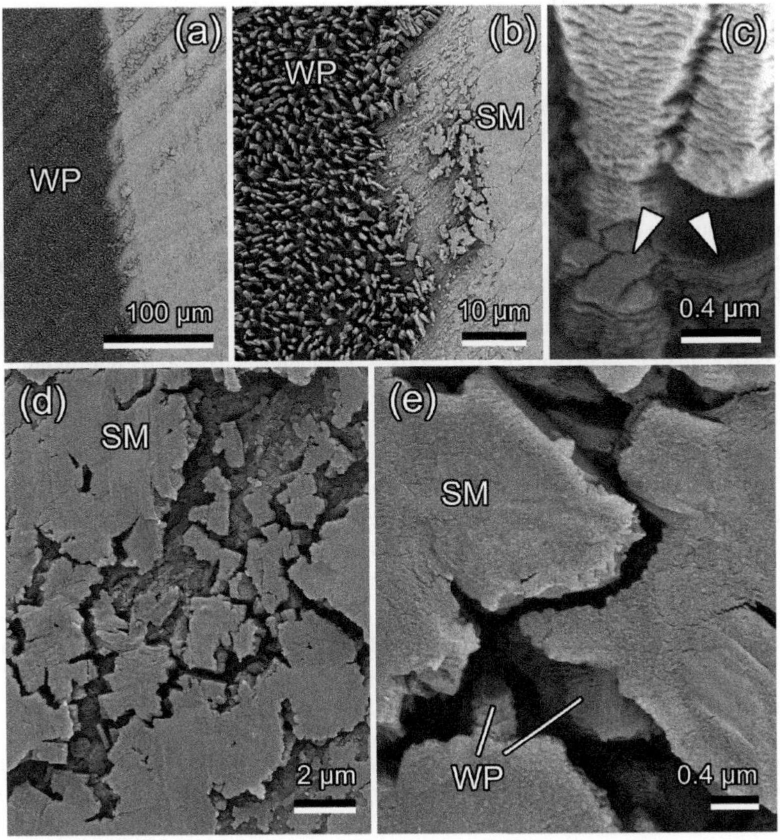

Fig. 14 *Hippeastrum reginae,* damages of 3D wax coverage on flower stems **a, b** A smeared track in the 3D wax coverage (right side of the image). **c** Fractured single wax projection due to the delamination between layers (arrowheads). **d, e** Smeared track in the 3D wax coverage at high magnifications. SM, wax projections smeared into a wax layer; WP, individual wax projections. From [99]

substantial differences in the structure of 3D wax coverings between leaves and flower stems, indicate a pronounced specialization of the wax projections found on the stems.

Previous studies have observed that ants readily walked on stems from plants like *Anethum graveolens* (Apiaceae), *Dahlia pinnata* (Asteraceae), and *Tagetes patula* (Asteraceae) after wiping the stems and removing the epicuticular wax coverage [36]. However, they did not venture onto intact flower stems. Flower stems of *Tulipa gesneriana*, which had prominent 3D wax coverage, were also avoided by ants during their foraging behavior [98]. Previous experimental studies with various insect and plant species, as mentioned above and in [97], have demonstrated that 3D epicuticular wax coverages typically reduce insect attachment through different mechanisms, including (1) reducing the real contact area, (2) contaminating insect adhesive organs, (3) adsorbing fluid from insect pads, (4) facilitating hydroplaning, and (5) forming a separation layer [67, 83].

The microtribological experiments show that surfaces covered by flake-like wax nanoparticles on flower stems can significantly reduce friction and act as a form of dry lubricant. This slipperiness mechanism is based on the easy delamination of layers and the lower intrinsic friction between individual layers. Dry or solid lubricants in engineering are typically powder-like materials that reduce friction between surfaces without the need for a liquid medium. Two well-known dry lubricants are graphite and molybdenum disulfide, both of which have low-friction characteristics attributed to their layered structure at the nanostructural or molecular level, with weak bonding between layers. This allows the layers to slide relative to each other with minimal applied force, resulting in their low frictional properties [101].

Recent study [102] investigated the effect of surfaces covered by kaolin nanoparticles on reducing the attachment ability of two economically important polyphagous insect pests, each characterized by different types of attachment devices. Kaolin particle films consist of flat, flake-like nanoparticles that can easily erode and contaminate insect tarsal attachment devices, thereby interfering with their adhesion to surfaces. The traction of these insects has been significantly affected by the presence of a kaolin particle film. Subsequent SEM analysis of the insects' tarsal attachment devices after walking on kaolin-covered surfaces revealed significant contamination by the kaolin nanoflakes. This mechanism of action of kaolin particles in reducing insect adhesion is akin to that proposed by the contamination hypothesis explaining the effect of epicuticular plant waxes in reducing insect attachment [67]. The strongly flattened structure of kaolin particles might also enable them to function as dry lubricants in addition to their contaminating effect.

7 Some Implications and Perspectives

An important avenue for future research is the exploration of animal adaptations that allow them to overcome the slipperiness of specialized plant surfaces. Some highly specialized insect species have evolved counter specializations to thrive on plant surfaces covered by wax projections. Typically, their adaptations are behavioral, but it is not out of the question that these specialist animals possess remarkable contact elements with specific shapes or dimensions on their attachment pads to increase real contact area with challenging plant surfaces. Also, they can potentially generate highly-specialized fluid compositions on their attachment devices compared to generalist insects.

Numerous structural aspects of the self-assembly of plant wax projections and their dependence on environmental conditions await experimental verification for the majority of wax types. Information on the mechanical properties of individual crystals and complex wax coverages is available for only a handful of plant species. Since insects apply shear forces while gripping plant surfaces, direct measurements of the frictional properties of various wax coatings, similar to the experiments reported in our case study on *Hippeastrum* could shed light on the differences in how different waxes interact with insect feet. Currently, most insights are derived from theoretical predictions based on the geometry of individual crystals. Recent advancements in force probe microscopy offer a promising tool for addressing these questions.

The knowledge about friction and adhesion reducing mechanisms of plant waxes has the potential to contribute to the development of future physical control barriers based on dry lubricants against pest insects. This is particularly relevant due to the need to reduce the negative impacts of pesticides on the environment and human health.

Lastly, an unconventional avenue of exploration of the current knowledge involves the use of plant waxes for creating templates in materials science and nanotechnology. This could lead to the development of surface coatings for technical applications with a wide range of beneficial properties.

References

1. E.V. Gorb, S.N. Gorb, in *Proceedings of the IUTAM Symposium on Scaling in Solid Mechanics (25–29 June 2007, Cardiff, UK)*, ed. by F.M. Borodich (Springer, Berlin, 2009)
2. Z. Dai, S.N. Gorb, U. Schwarz, J. Exp. Biol. **205**, 2479–2488 (2002)
3. S.N. Gorb, *Attachment Devices of Insect Cuticle* (Kluwer Academic Publishers, Dordrecht, 2001)
4. R. Beutel, S.N. Gorb, J. Zoolog. Syst. Evol. Res. **39**, 177–207 (2001)
5. S. Ishii, Appl. Entomol. Zool. **22**, 222–228 (1987)
6. A. Kosaki, R. Yamaoka, Jpn. J. Appl. Entomol. Zool. **40**, 47–53 (1996)
7. T. Eisner, D.J. Aneshansley, Proc. Natl. Acad. Sci. U.S.A. **97**, 6568–6573 (2000)
8. W. Vötsch, G. Nicholson, R. Müller, Y.-D. Stierhof, S.N. Gorb, U. Schwarz, Insect Biochem. Mol. Biol. **32**, 1605–1613 (2002)
9. S.F. Geiselhardt, S. Geiselhardt, K. Peschke, Chemoecology **19**, 185–193 (2009)
10. W. Federle, M. Riehle, A.S.G. Curtis, R.J. Full, Integr. Comp. Biol. **42**, 1100–1106 (2002)
11. J.S. Edwards, M. Tarkanian, Proc. R. Entomol. Soc. Lond. A **45**, 1–5 (1970)
12. N.E. Stork, J. Exp. Biol. **88**, 91–107 (1980)
13. M.G. Langer, J.P. Ruppersberg, S. Gorb, Proc. R. Soc. Lond. B **271**, 2209–2215 (2004)
14. C.E. Jeffree, in *Biology of the Plant Cuticle*, ed. by M. Riederer, C. Müller (Oxford, Blackwell, Oxford, 2006)

15. W. Barthlott, C. Neinhuis, D. Cutler, F. Ditsch, I. Meusel, I. Theisen, H. Bot, J. Linn. Soc. **126**, 237–260 (1998)
16. R. Jetter, L. Kunst, A.L. Samuels, in *Biology of the Plant Cuticle*, ed. by M. Riederer, C. Müller (Oxford, Blackwell, Oxford, 2006)
17. C.F. Jeffree, in *Insects and the Plant Surface*, ed. by B. Juniper, R. Southwood (Edward Arnold Publishers, London, 1986)
18. K. Koch, B. Bhushan, W. Barthlott, in *Handbook of Nanotechnology*, ed. by B. Bhushan (Springer, Heidelberg, 2010)
19. I. Meusel, C. Neinhuis, C. Markstadter, W. Barthlott, Plant Biol. **2**, 462–470 (2000)
20. H.J. Ensikat, M. Boese, W. Mader, W. Barthlott, K. Koch, Chem. Phys. Lipids **144**, 45–59 (2006)
21. C.E. Jeffree, E.A. Baker, P.J. Holloway, New Phytol. **75**, 539–449 (1975)
22. R. Jetter, M. Riederer, Planta **195**, 257–270 (1994)
23. R. Jetter, M. Riederer, Bot. Acta **108**, 111–120 (1995)
24. K. Koch, H.J. Ensikat, Micron **39**, 759–772 (2008)
25. H. Bargel, K. Koch, Z. Cerman, C. Neinhuis, Funct. Plant Biol. **33**, 893–910 (2006)
26. T.H. Anstey, J.F. Moore, J. Hered. **45**, 39–41 (1954)
27. P.B. Edwards, Austral Ecol. **7**, 347–352 (1982)
28. R.P. Bodnaryk, Can. J. Plant Sci. **72**, 1295–1303 (1992)
29. E.B. Brennan, S.A. Weinbaum, Entomol. Exp. Appl. **100**, 179–185 (2001)
30. G.C. Chang, J. Neufeld, S.D. Eigenbrode, Entomol. Exp. Appl. **119**, 197–205 (2006)
31. E.V. Gorb, Z. Dai, S.N. Gorb, Flora **230**, 14–25 (2017)
32. S. Eigenbrode, Arthr. Struct. Dev. **33**, 91–102 (2004)
33. R.M. Harley, in *Proceedings of a Workshop on Neotropical Distributions* (Academia Brasileira de Ciencias, Rio de Janeiro, 1988)
34. R. Harley, in *Ant-Plant Interactions*, ed. by C.R. Huxley, D.E. Cutler (Oxford University Press, Oxford, 1991)
35. B.E. Juniper, in *Waxes: Chemistry, Molecular Biology and Functions*, ed. by R.J. (Oily, West Ferry, Dundee, 1995)
36. E. Gorb, S. Gorb, Arthropod-Plant Interact. **5**, 297–306 (2011)
37. W. Federle, U. Maschwitz, B. Fiala, M. Riederer, B. Hölldobler, Oecologia **112**, 217–224 (1997)
38. W. Federle, B. Fiala, U. Maschwitz, Trop. Zool. **11**, 83–84 (1998)
39. W. Federle, K. Rohrseitz, B. Hölldobler, J. Exp. Biol. **203**, 505–512 (2000)
40. C. Markstädter, W. Federle, R. Jetter, M. Riederer, B. Hölldobler, Chemoecology **10**, 33–40 (2000)
41. S. Vogel, J. Martens, Bot. J. Linn. Soc. **133**, 61–100 (2000)
42. B. Oelschlägel, S.N. Gorb, S. Wanke, C. Neinhuis, New Phytol. **184**, 988–1002 (2009)
43. S. Poppinga, K. Koch, H.F. Bohn, W. Barthlott, Func. Plant Biol. **37**, 952–961 (2010)
44. B.E. Juniper, J.K. Burras, New Sci. **269**, 75–77 (1962)
45. B.E. Juniper, R.J. Robins, D.M. Joel, *The Carnivorous Plants* (Academic Press, London, 1989)
46. L. Gaume, S. Gorb, N. Rowe, New Phytol. **156**, 476–489 (2002)
47. L. Gaume, P. Perret, E. Gorb, S. Gorb, J.-J. Labat, N. Rowe, Arthr. Struct. Dev. **33**, 103–111 (2004)
48. B. Di Giusto, M. Gueroult, N. Rowe, L. Gaume, in *Functional Surfaces in Biology* (II), ed. by S.N. Gorb (Springer, Dordrecht, 2009)
49. E. Gorb, K. Haas, A. Henrich, S. Enders, N. Barbakadze, S. Gorb, J. Exp. Biol. **208**, 4651–4662 (2005)
50. U. Bauer, C. Willmes, W. Federle, Ann. Bot. **103**, 1219–1226 (2009)
51. U. Bauer, B. Di Giusto, J. Skepper, T.U. Grafe, W. Federle, PLoS ONE **7**, e38951 (2012)
52. I. Scholz, M. Bückins, L. Dolge, T. Erlinghagen, A. Weth, F. Hischen, J. Mayer, S. Hoffmann, M. Riederer, M. Riedel, W. Baumgartner, J. Exp. Biol. **213**, 115–1125 (2010)
53. L. Wang, Q. Zhou, Adv. Nat. Sci. **3**, 152–160 (2010)

54. E.V. Gorb, S.N. Gorb, Beilstein J. Nanotechnol. **2**, 302–310 (2011)
55. S.D.Eigenbrode, in *Plant Cuticles—An Integrated Functional Approach*, ed. by G. Kerstiens (BIOS Scientific Publishers, Oxford, 1996)
56. C. Müller, in *Biology of the plant cuticle,* ed. M. Riederer, C. Müller (Blackwell, Oxford, 2006)
57. M.J. Way, G. Murdie, Ann. Appl. Biol. **56**, 326–328 (1965)
58. N.E. Stork, Entomol. Exp. Appl. **26**, 100–107 (1980)
59. N.E. Stork, in *Insects and the Plant Surface*, ed. by B.E. Juniper, T.R.E. Southwood (Edward Arnold, London, 1986)
60. S.D. Eigenbrode, N.N. Kabalo, Entomol. Exp. Appl. **91**, 125–130 (1999)
61. S.D. Eigenbrode, N.N. Kabalo, K.A. Stoner, Entomol. Exp. Appl. **90**, 225–235 (1999)
62. S.D. Eigenbrode, T. Castognola, M.-B. Roux, L. Steljes, Entomol. Exp. Appl. **81**, 335–343 (1999)
63. S.D. Eigenbrode, L. Rayor, J. Chow, P. Latty, Entomol. Exp. Appl. **97**, 161–166 (2000)
64. S.D. Eigenbrode, W.E. Snyder, G. Clevenger, H. Ding, S.N Gorb, in *Functional Surfaces in Biology* (II), ed. by S.N. Gorb (Springer, Dordrecht, 2009)
65. E.B. Brennan, S.A. Weinbaum, Aust. J. Entomol. **40**, 270–277 (2001)
66. S.D. Eigenbrode, R. Jetter, Integr. Comp. Biol. **42**, 1091–1099 (2002)
67. E.V. Gorb, S.N. Gorb, Entomol. Exp. Appl. **105**, 13–28 (2002)
68. C.E. Rutledge, S.D. Eigenbrode, Can. Entomol. **135**, 93–101 (2003)
69. E. Gorb, D. Voigt, S.D. Eigenbrode, S. Gorb, Arthropod-Plant Interact. **2**, 247–259 (2008)
70. D. Voigt, E. Gorb, S. Gorb, Arthropod-Plant Interact. **1**, 221–243 (2007)
71. B. Prüm, R. Seidel, H.F. Bohn, T. Speck, J.R. Soc, Interface **9**, 127–135 (2012)
72. B. Prüm, R. Seidel, H.F. Bohn, T. Speck, Beilstein J. Nanotechnol. **3**, 57–65 (2012)
73. K. Friedemann, G. Kunert, E. Gorb, S.N. Gorb, R.G. Beutel, Ecol. Entomol. **40**, 732–740 (2015)
74. E. Gorb, S. Böhm, N. Jacky, L.-P. Maier, K. Dening, S. Pechook, B. Pokroy, S. Gorb, Beilstein J. Nanotechnol. **5**, 1031–1041 (2014)
75. F.P. Bowden, D. Tabor, *The Friction and Lubrication of Solids* (Oxford University Press, New York, 2001)
76. K. Kendall, *Molecular Adhesion and its Applications—The Sticky Universe* (Kluwer Academic Publishers, New York, 2001)
77. M. Varenberg, B. Murarash, Y. Kligerman, S.N. Gorb, Appl. Phys. A **103**, 933–938 (2011)
78. A. Peressadko, S. Gorb, in *Proceedings of the 1st International Industrial Conference Bionic (22–23 April 2004, Hannover, Germany)*, ed. by I. Boblan, R. Bannasch (VDI Verlag, Düsseldorf, 2004)
79. D. Voigt, J.M. Schuppert, S. Dattinger, S.N. Gorb, J. Insect Physiol. **54**, 765–776 (2008)
80. J.M. Bullock, W. Federle, J. Exp. Biol. **212**, 1876–1888 (2009)
81. E. Gorb, S. Gorb, Entomol. Exp. Appl. **130**, 222–228 (2009)
82. M.W. England, T. Sato, M. Yagihashi, A. Hozumi, S.N. Gorb, E.V. Gorb, Beilstein J. Nanotechnol. **7**, 1471–1479 (2016)
83. E.V. Gorb, J. Purtov, S.N. Gorb, Sci. Rep. **4**, 5154 (2014)
84. J. Purtov, E.V. Gorb, M. Steinhart, S.N. Gorb, Appl. Phys. A **111**, 183–189 (2013)
85. E.V. Gorb, S.N. Gorb, in *Ecology and Biomechanics – a Mechanical Approach to the Ecology of Animals and Plants*, ed. by A. Herrel, T. Speck, N.P. Rowe (CRC Press, Boca Raton, 2006)
86. F.M. Borodich, E.V. Gorb, S.N. Gorb, Appl. Phys. A **100**, 63–71 (2010)
87. A.F.G. Dixon, P.C. Croghan, R.P. Gowing, J. Exp. Biol. **152**, 243–253 (1990)
88. J.-H. Dirks, C.J. Clemente, W. Federle, J.R. Soc, Interface **7**, 587–593 (2010)
89. P. Drechsler, W. Federle, J. Comp. Physiol. A **192**, 1213–1222 (2006)
90. E.V. Gorb, N. Hosoda, C. Miksch, S.N. Gorb, J.R. Soc, Interface **7**, 1571–1579 (2010)
91. E.V. Gorb, P. Hofmann, A.E. Filippov, S.N. Gorb, Sci. Rep. **7**, 45483 (2017)
92. K. Koch, in *Proceedings of the Beilstein Symposium on Functional Nanoscience (17–21 May 2010, Bozen, Italy)*, (Beilstein-Institut, Frankfurt am Main, 2010)

93. W. Barthlott, N. Ehler, in *Tropische und Subtropische Pflanzenwelt 19*, ed. by W. Rauh (Franz Steiner Verlag, Wiesbaden, 1977)
94. E. Gorb, S. Gorb, in *Functional Surfaces in Biology (II)*, ed. by S.N. Gorb (Springer, Dordrecht, 2009)
95. M.J. Benz, E.V. Gorb, S.N. Gorb, Arthropod-Plant Interact. **6**, 147–158 (2012)
96. E.V. Gorb, S.N. Gorb, in *Materials Design Inspired by Nature: Function Through Inner Architecture*, ed. by P. Fratzl, J.W.C. Dunlop, R. Weinkamer (RSC Publishing, Cambridge, 2013)
97. E.V. Gorb, S.N. Gorb, J. Exp. Bot. **68**, 5323–5337 (2017)
98. S.N. Gorb, E.V. Gorb, Arthropod-Plant Interact. **13**, 311–320 (2019)
99. E.V. Gorb, S.N. Gorb, Appl. Phys. A **126**, 549 (2020)
100. M. Williams, Amaryllis Bullet. **1**, 20 (1980)
101. W. Bollmann, J. Spreadborough, Nature **186**, 29–30 (1960)
102. G. Salerno, M. Rebora, A. Kovalev, E.V. Gorb, S.N. Gorb, J. Pest. Sci. **93**, 315–327 (2020)

Batch number: 10755214

Printed by Printforce, the Netherlands